STUDENT SOLUTIONS MANUAL

Edward A. Robinson

ATOMS, MOLECULES, AND REACTIONS

An Introduction to Chemistry

STUDENT SOLUTIONS MANUAL
Edward A. Robinson

ATOMS, MOLECULES, AND REACTIONS

An Introduction to Chemistry

Ronald J. Gillespie
McMaster University

Donald R. Eaton
McMaster University

David A. Humphreys
McMaster University

Edward A. Robinson
University of Toronto

Prentice Hall
Englewood Cliffs, New Jersey 07632

Editorial/production supervision: ***Amy Jolin***
Editor in Chief: ***Tim Bozik***
Acquisitions editor: ***Paul Banks***
Supplements acquisitions editor: ***Mary Hornby***
Production coordinator: ***Trudy Pisciotti***

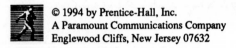

© 1994 by Prentice-Hall, Inc.
A Paramount Communications Company
Englewood Cliffs, New Jersey 07632

All rights reserved. No part of this book may be reproduced, in any form or by any means, without permission in writing from the publisher.

Printed in the United States of America

10 9 8 7 6 5 4 3 2 1

ISBN 0-13-107087-8

Prentice-Hall International (UK) Limited, London
Prentice-Hall of Australia Pty. Limited, Sydney
Prentice-Hall Canada Inc. Toronto
Prentice-Hall Hispanoamericana, S.A., Mexico
Prentice-Hall of India Private Limited, New Delhi
Prentice-Hall of Japan, Inc., Tokyo
Simon & Schuster Asia Pte. Ltd., Singapore
Editora Prentice-Hall do Brasil, Ltda., Rio de Janeiro

PREFACE

The ***Student Solutions Manual*** contains detailed answers to all of the Problems marked with *blue* numbers at the end of each Chapter of "**Atoms, Molecules, and Reactions**" by Gillespie, Eaton, Humphreys, and Robinson, (Prentice-Hall, Inc., 1994), short answers to which are to be found at the end of the text. It also contains complete answers to the **Exercises** in each Chapter.

Care has been taken to avoid errors in the solutions to the numerical problems. Should any still persist, I should be grateful to have them brought to my attention. Suggestions for alternative strategies in solving particular problems would also be welcomed. You may write to me at the address below. In the answers to the problems based upon descriptive or conceptual material, you are not expected to reproduce my answer word for word; read the background material in the text and express your answer in your own style, but make sure that you have a logical development of ideas, and write in proper grammatical sentences.

Many have contributed to the problem sets over many years. My particular thanks to my colleague at Erindale College, University of Toronto, Mrs. J. C. Poë, to Professors R. J. Gillespie and D. R. Eaton, and to many others to whom I am indebted for suggestions. I am grateful to Mrs. Elizabeth Kobluk of Erindale College for advice on Word Processing and for the printing of the manuscript.

Professor E. A. (Peter) Robinson
Department of Chemistry
Erindale College
University of Toronto
Mississauga, Ontario
CANADA L5L 1C6

February, 1994.

TO FIND A PARTICULAR PROBLEM OR EXERCISE

Use the codes at the top corner of each page.

For example, **08.37** at the top of Page 97 tells you that PROBLEM 37 of CHAPTER 8 is the first problem whose complete solution starts on that page.

CONTENTS

ANSWERS TO SELECTED PROBLEMS

CHAPTER 1.

Elements, Compounds and Formulas

1. Give the symbol for each of the following elements: (a) sodium (b) potassium (c) mercury (d) gold (e) tin (f) lead (g) iron

---

(a) Na (b) K (c) Hg (d) Au (e) Sn (f) Pb (g) Fe

---

3. Give the names of the elements with each of the following symbols: (a) H (b) He (c) Ne (d) F (e) Mg (f) Al (g) P (h) S (i) K (j) Na

---

(a) hydrogen (b) helium (c) neon (d) fluorine (e) magnesium (f) aluminum (g) phosphorus (h) sulfur (i) potassium (j) sodium

---

4. What formulas represent each of the following? (a) a molecule of solid sulfur containing 8 S atoms (b) a molecule of liquid sulfuric acid containing 2 H, 1 S, and 4 O atoms (c) liquid sulfuric acid (d) a solution of sulfuric acid in water (e) solid silicon carbide (carborundum) containing an infinite array of Si and C atoms in the ratio 1:1

(a) S_8 (b) H_2SO_4 (c) $H_2SO_4(\ell)$ (d) $H_2SO_4(aq)$ (e) $SiC(s)$

---

5. Write molecular and empirical formulas, where possible, and the formula for a sample of the bulk substance for each of the following:
(a) solid white phosphorus with molecules containing 4 P atoms
(b) solid red phosphorus containing infinite sheets of P atoms
(c) a gaseous hydrocarbon with molecules containing 3 C atoms and 6 H atoms
(d) a liquid oxide of sulfur with molecules containing 3 S atoms and 6 O atoms

---

The molecular formula gives the total numbers of each type of atom in a molecule; the empirical formula is the simplest formula that gives the correct ratios of the different atoms in one molecule as whole numbers, the bulk formula also gives the physical state under specified conditions, most usually ordinary temperature and pressure.

(a) molecular formula P_4; empirical formula P; bulk formula $P_4(s)$
(b) molecular formula P_n; empirical formula P; bulk formula P(s)
(c) molecular formula C_3H_6; empirical formula CH_2; bulk formula $C_3H_6(g)$
(d) molecular formula S_3O_9; empirical formula SO_3; bulk formula $S_3O_9(\ell)$

---

7. Give the formula of a bulk sample of each of the following substances: (a) hydrogen (b) oxygen (c) nitrogen (d) water (e) hydrogen peroxide (f) methane (g) carbon dioxide (h) magnesium oxide (i) silicon dioxide

---

(a) $H_2(g)$ (b) $O_2(g)$ (c) $N_2(g)$ (d) $H_2O(\ell)$ (e) $H_2O_2(\ell)$ (f) $CH_4(g)$ (g) $CO_2(g)$ (h) $MgO(s)$ (i) $SiO_2(s)$

---

10. (a) Explain the terms "mixture" and "impure"; (b) Although it is commonly used, why is "pure" strictly speaking a redundant term? (c) What is meant when a substance is described as 99.98% pure?

(a) A mixture is a blend of different substances; pure describes a single substance, and impure describes a substance contaminated with other substances.

(b) "Pure" is redundant because naturally occurring substances, substances made in the laboratory, or industrially, are always mixtures that are never 100% pure. Even substances that have been carefully purified still contain small amounts of other substances. Purity is a relative term.
(c) 99.98% pure means that the sample contains 0.02 g of impurities in 100 g of the sample.

---

11. Classify each of the following as an element, a compound, or a mixture: (a) water (b) iron (c) beer (d) sugar (e) wine (f) silicon dioxide (g) sulfur (h) concrete (i) air (j) magnesium oxide.

Elements: (b) iron, (g) sulfur; Compounds: (a) water, (d) table sugar, (f) silica, (j) magnesium oxide; Mixtures: (c) beer, (e) wine, (h) concrete, (i) air.

---

12. Classify each of the following as a pure substance or as a mixture. Divide the pure substances into elements and compounds and the mixtures into homogeneous or heterogeneous mixtures:
(a) nitrogen (b) gasoline (c) sterling silver (d) sodium chloride (e) carbon dioxide (f) black coffee (g) diamond (h) distilled water (i) filtered sea-water (j) carbon (k) vegetable soup (ℓ) concrete

---

(a) nitrogen, $N_2(g)$, is an **element**; (b) gasoline, is a **homogeneous mixture** (solution) of liquid hydrocarbons; (c) sterling silver is a **homogeneous mixture** (an alloy of silver and copper); (d) sodium chloride (common salt), $NaCl(s)$, is a **compound**; (e) carbon dioxide, $CO_2(g)$, is a **compound**; (f) black coffee is a **homogeneous mixture**; (g) diamond is a pure form of the **element** carbon; (h) distilled water is pure water, $H_2O(\ell)$, a **compound**; (i) filtered sea-water is a **homogeneous mixture** (solution) of salts (mainly $NaCl(s)$) in water;
(j) carbon is an **element**; (k) vegetable soup is a **heterogeneous mixture** of liquids and solids, and (ℓ) concrete is a **heterogeneous mixture**, an aggregate of solids.

The Separation of Mixtures

14. Briefly explain each of the following: (a) filtration (b) distillation (c) homogeneous mixture (d) solution (e) heterogeneous mixture

---

(a) Filtration is a process whereby a solid is separated from a liquid by passing the mixture through a porous filter through which the liquid (filtrate) can pass but solid particles are retained.

(b) Distillation is a method of separating liquids by evaporation and condensation, utilizing their differences in boiling point (ease of vaporization). On heating a mixture of liquids, the vapor produced is richer in the lower boiling components than the original mixture, so that when it condenses to a liquid the ensuing mixture (condensate) is also enriched in lower boiling components. Repeated distillation, or **fractional distillation** using a distillation column, increases the proportion of the lowest boiling point component in the condensate and eventually very efficient separation is achieved.

(c) Homogeneous mixture describes a mixture of substances with uniform properties throughout. None of the individual components are discernible; it consists of one phase (solid, liquid, or gas) and has a uniform composition in all parts.

(d) Solution is the common term used to describe a homogeneous mixture, most usually a mixture in the liquid state such as a solution in water (an aqueous solution).

(e) Heterogeneous mixture describes a mixture of substances with a non-uniform composition, usually discernible by eye or under a microscope. Different parts have different properties and different compositions.

---

16. Suggest possible ways to separate the components of each of the following mixtures: (a) sugar and water (b) water and gasoline (c) iron filings and wood sawdust (d) sugar and powdered glass (e) food coloring and water.

(a) Sugar and water could be separated by distilling off the water, leaving a residue of sugar. For a sample of pure sugar, this would not be very satisfactory because sugar decomposes if heated strongly. Slow evaporation of the water would give a pure sample of sugar, but some would remain in solution.

(b) Water and gasoline are immiscible (do not dissolve in each other) and form two distinct layers with the gasoline layer floating on top of the water. The gasoline could be decanted or, in the laboratory, a separatory funnel with a tap is used. When very pure water is required, the separated water layer could then be purified by distillation, and the gasoline layer (a mixture itself) could be dried, using a suitable drying agent, such as calcium chloride, $CaCl_2(s)$. It could then be fractionally distilled to separate it into its constituents.

(c) Iron filings and sawdust could be separated using a special property of iron, namely that it is strongly attracted to a magnet.

(d) Sugar and powdered glass could be separated using the property that sugar is soluble in water while glass is insoluble. When water is added and the mixture warmed, a solution of sugar containing solid powdered glass results, which can be separated by filtration, and the glass dried. The sugar could be recovered from the filtrate as described in (a) above.

(e) Food coloring and water could be separated by one of a number of chromatographic methods, depending on the amount of food coloring present. Paper chromatography would effectively separate small amounts. After separation into different bands, the paper could be carefully dried, cut into strips, and a food color removed from each strip by dissolving in a suitable organic solvent, such as ether. For larger amounts, the solution could be passed down a column containing powdered aluminum oxide. Very small amounts are conveniently removed using activated charcoal, which has the property of absorbing "impurities" from solutions.

Chemical Reactions and Chemical Equations

17. (a) What are meant by the *reactants* and the *products* of a chemical reaction? (b) What information is needed before a balanced chemical equation can be written to describe a reaction between carbon and oxygen?

(a) The reactants are the substances that react together in a chemical reaction; the products are the new substances that are formed.

(b) Before the chemical equation for a reaction can be balanced, all of the reactants and all of the products, and their formulas, must be known. You should remember, for example, that carbon and oxygen can combine to form either carbon monoxide, $CO(g)$, or carbon dioxide, $CO_2(g)$. Thus, for the reaction between $C(s)$ and $O_2(g)$, the product could be $CO(g)$ or $CO_2(g)$, (or even a mixture of them), so exact information about the actual product is also required before the equation can be balanced. (Note that some reactions are more complex than implied by the balanced equation commonly given, which represents only the dominant reaction that takes place under the stated conditions.)

---

19. Balance those among the following equations that are not already balanced: (a) $2SO_2 + H_2O + O_2 \rightarrow 2H_2SO_4$; (b) $CH_3OH + 2O_2 \rightarrow CO_2 + 2H_2O$; (c) $H_2O_2 \rightarrow H_2O + O_2$; (d) $H_2SO_4 + KOH \rightarrow KHSO_4 + H_2O$

(a) Unbalanced as written, $2SO_2 + H_2O + O_2 \rightarrow 2H_2SO_4$

Only the S atoms are balanced, however if we balance the H atoms by adding one H_2O molecule to the left-hand side, both sides are then balanced:

$$2SO_2 + 2H_2O + O_2 \rightarrow 2H_2SO_4 \quad \text{balanced}$$

(b) Unbalanced as written, $CH_3OH + 2O_2 \rightarrow CO_2 + 2H_2O$

The C and H atoms are balanced, and the O atoms can be balanced by reducing the number on the LHS to 4:

$$CH_3OH + 1\tfrac{1}{2}O_2 \rightarrow CO_2 + 2H_2O$$

Since we do not normally write fractional numbers of molecules, this is now multiplied throughout by 2, to give

$$2CH_3OH + 3O_2 \rightarrow 2CO_2 + 4H_2O \quad \underline{\text{balanced}}$$

(c) Unbalanced as written, $H_2O_2 \rightarrow H_2O + O_2$

The H atoms are balanced but not the O atoms, which can be balanced by writing

$$H_2O_2 \rightarrow H_2O + \tfrac{1}{2}O_2$$

Now multiply throughout by 2 to obtain integral coefficients

$$2H_2O_2 \rightarrow 2H_2O + O_2 \qquad \underline{\text{balanced}}$$

(d) This equation is already balanced.

---

21. Balance each of the following equations: (a) $S + O_2 \rightarrow SO_3$; (b) $C_2H_2 + O_2 \rightarrow CO_2 + H_2O$; (c) $Cu_2S + O_2 \rightarrow Cu_2O + SO_2$; (d) $Na_2SO_4 + H_2 \rightarrow Na_2S + H_2O$; (e) $Na_2CO_3 + Ca(OH)_2 \rightarrow NaOH + CaCO_3$.

(a) $S + O_2 \rightarrow SO_3$, unbalanced

S atoms are balanced; balance O atoms and then multiply throughout by 2, to give

$$S + 1\tfrac{1}{2}O_2 \rightarrow SO_3 \quad \underline{\text{unbalanced}}$$

$$2S + 3O_2 \rightarrow 2SO_3 \quad \underline{\text{balanced}}$$

(b) $C_2H_2 + O_2 \rightarrow CO_2 + H_2O$, unbalanced

Balance C atoms and then O atoms, and then multiply throughout by 2, to give

$$C_2H_2 + 2\tfrac{1}{2}O_2 \rightarrow 2CO_2 + H_2O \quad \underline{\text{unbalanced}}$$

$$2C_2H_2 + 5O_2 \rightarrow 4CO_2 + 2H_2O \quad \underline{\text{balanced}}$$

(c) $Cu_2S + O_2 \rightarrow Cu_2O + SO_2$, unbalanced

Only the O atoms are unbalanced; balance them by writing

$$Cu_2S + 1\tfrac{1}{2}O_2 \rightarrow Cu_2O + SO_2 \quad \underline{\text{unbalanced}}$$

and then multiply throughout by 2, to give

$$2Cu_2S + 3O_2 \rightarrow 2Cu_2O + 2SO_2 \quad \underline{\text{balanced}}$$

(d) $Na_2SO_4 + H_2 \rightarrow Na_2S + H_2O$, unbalanced

Na and S atoms are balanced. The four O atoms on the LHS must appear as H_2O on the RHS; add $3H_2O$ to the RHS, and then balance the H atoms on the LHS, to give

$$Na_2SO_4 + H_2 \rightarrow Na_2S + 4H_2O \quad \underline{\text{unbalanced}}$$

$$Na_2SO_4 + 4H_2 \rightarrow Na_2S + 4H_2O \quad \underline{\text{balanced}}$$

(e) $Na_2CO_3 + Ca(OH)_2 \rightarrow NaOH + CaCO_3$, unbalanced

Balance Na atoms, to give

$$Na_2CO_3 + Ca(OH)_2 \rightarrow 2NaOH + CaCO_3 \quad \underline{\text{balanced}}$$

The Structure of Atoms

23. What are the chemical symbols and the atomic numbers of atoms of each of the following? (a) hydrogen (b) oxygen (c) fluorine (d) neon (e) magnesium (f) phosphorus (g) chlorine (h) calcium (i) zinc

---

The atomic number, Z, of an element is obtained from its position in the periodic table, where the elements are arranged from left to right in any row in order of increasing Z. Z is the charge on the nucleus; it is equal to the number of protons in the nucleus, and to the number of electrons surrounding the nucleus in a neutral atom:

	Element	Symbol	Z		Element	Symbol	Z
(a)	hydrogen	H	1	(f)	phosphorus	P	15
(b)	oxygen	O	8	(g)	chlorine	Cl	17
(c)	fluorine	F	9	(h)	calcium	Ca	20
(d)	neon	Ne	10	(i)	zinc	Zn	30
(e)	magnesium	Mg	12				

---

25. How many protons are there in the nucleus of one atom of each of the following? (a) boron (b) nitrogen (c) hydrogen (d) neon (e) chlorine (f) oxygen (g) sulfur (h) potassium (i) iron

---

In the periodic table, the elements are arranged in order of increasing atomic number Z, where Z is the number of nuclear protons, (which is balanced in a neutral atom by an equal number of electrons surrounding the nucleus).

	Element	Symbol	Z		Element	Symbol	Z
(a)	boron	B	5	(f)	oxygen	O	8
(b)	nitrogen	N	7	(g)	sulfur	S	16
(c)	hydrogen	H	1	(h)	potassium	K	19
(d)	neon	Ne	10	(i)	iron	Fe	26
(e)	chlorine	Cl	17				

---

27. Give the symbols of the isotopes with each of the following atomic numbers (Z) and mass numbers (A): (a) Z = 19, A = 40; (b) Z = 14, A = 30; (c) Z = 18, A = 40; (d) Z = 7, A = 15; (e) Z = 16, A = 32; (f) Z = 11, A = 23; (g) Z = 13, A = 27.

The atomic number, Z, identifies the position of the element in the periodic table, and is placed as a <u>subscript</u> before the symbol of the element. The mass (nucleon) number is placed as a <u>superscript</u> before the symbol of the element:

(a) ${}^{40}_{19}K$ (b) ${}^{30}_{14}Si$ (c) ${}^{40}_{18}Ar$ (d) ${}^{15}_{7}N$ (e) ${}^{32}_{16}S$ (f) ${}^{23}_{11}Na$ (g) ${}^{27}_{13}Al$.

---

29. Give the numbers of protons, neutrons, and electrons in each of the following atoms: (a) ${}^{2}H$ (b) ${}^{19}F$ (c) ${}^{40}Ca$ (d) ${}^{112}Cd$ (e) ${}^{117}Sn$ (f) ${}^{235}U$

For the method, see Problem 28.

(a) ${}^{2}_{1}H$	A = 2	1 proton, 1 neutron, 1 electron
(b) ${}^{19}_{9}F$	A = 19	9 protons, 10 neutrons, 9 electrons
(c) ${}^{40}_{20}Ca$	A = 40	20 protons, 20 neutrons, 20 electrons
(d) ${}^{112}_{48}Cd$	A = 112	48 protons, 64 neutrons, 48 electrons

(e) $^{117}_{50}Sn$	A = 117	50 protons, 67 neutrons, 50 electrons
(f) $^{235}_{92}U$	A = 235	92 protons, 143 neutrons, 92 electrons

---

30. Complete the following table:

Atomic symbol	^{9}Be	^{15}N	^{18}O	-	-
Mass number	9	-	-	-	23
Atomic number	4	-	-	-	11
Number of protons	4	-	-	6	-
Number of electrons	4	-	-	6	-
Number of neutrons	5	-	-	6	-

Atomic symbol	$^{9}_{4}Be$	$^{15}_{7}N$	$^{18}_{8}O$	$^{12}_{6}C$	$^{23}_{11}Na$
Mass number	9	**15**	**18**	**12**	23
Atomic number	4	**7**	**8**	**6**	11
Number of protons	4	**7**	**8**	6	**11**
Number of electrons	4	**7**	**8**	6	**11**
Number of neutrons	5	**8**	**10**	6	**12**

---

32. Write the symbols for each of the isotopically different molecules of water, given that the isotopes of hydrogen and oxygen are ^{1}H, ^{2}H, ^{16}O, ^{17}O, and ^{18}O.

The molecular formula of water is H_2O, and the two H atoms could be ^{1}H and ^{1}H, ^{2}H and ^{2}H, or ^{1}H and ^{2}H, and the single O atom could be ^{16}O, ^{17}O, or ^{18}O, giving the possibilities:

$^{1}H^{1}H^{16}O$ $^{2}H^{2}H^{16}O$ $^{1}H^{2}H^{16}O$ $^{1}H^{1}H^{17}O$ $^{2}H^{2}H^{17}O$ $^{1}H^{2}H^{17}O$ $^{1}H^{1}H^{18}O$ $^{2}H^{2}H^{18}O$ $^{1}H^{2}H^{18}O$

For a total of nine isotopically distinct H_2O molecules.

33.(a) What information is given by the symbol ^{24}Mg? (b) From the data in the accompanying table, calculate the average atomic mass of magnesium:

Mass Number	Abundance (%)	Atomic mass (u)
24	78.99	23.985
25	10.00	24.986
26	11.01	25.983

(a) From its position in the periodic table, Mg has Z = 12, so the complete symbol is:

$$^{24}_{12}\mathbf{Mg}$$

The symbol tells us that this is the Mg-24 isotope, with a mass number, A, of 24, and an atomic number, Z, of 12, so the nucleus contains 12 protons and 12 neutrons, and there are 12 electrons.

(b) We first express the percent abundances as fractions of 1, by dividing the % abundances by 100. Then for all the isotopic species:

Average mass = Σ (isotope fraction x isotope mass)

isotope	mass	fractional abundance	abundance x mass
^{24}Mg	23.985 u	0.7899	18.9458 u
^{25}Mg	24.986 u	0.1000	2.4986 u
^{26}Mg	25.983 u	0.1101	2.8607 u

Average mass = Σ (abundance x mass) = 24.31 u

---

35. Uranium has an average atomic mass of 238.03 and consists of ^{235}U, mass 235.044 u, and ^{238}U, mass 238.051 u. The ^{235}U isotope is the fuel used in nuclear power reactors. What is the percentage abundance (by mass) of ^{235}U in natural uranium?

If the fractional abundance of ^{235}U is x, than that of ^{236}U is 1-x, and we can write:

$$x(235.044\ u) + (1+x)(238.051\ u) = 238.03\ u;\ x = 0.0070$$

Thus, the fractional abundance of ^{235}U in natural uranium is 0.0070, and the % abundance is 0.70 %.

---

37. Bromine atoms and chlorine atoms combine to give BrCl molecules. BrCl(g) consists of molecules with approximate masses 114, 116, and 118 u, and chlorine has just two isotopes, with mass numbers 35 and 37. (a) Deduce the possible isotopes of bromine, and write their symbols. (b) Give the formulas of each of the isotopically different BrCl molecules.

The mass of a molecule is the sum of the masses of the constituent atoms, given approximately by the sum of their mass numbers, so that

mass of BrCl = mass of Br atom + mass of Cl atom

Since, there are BrCl molecules with masses 114, 116, and 118 u, and two isotopes of Cl, ^{35}Cl with mass 35 u and ^{37}Cl with mass 37 u, there are apparently six possibilities for the mass of the Br atom, each corresponding to subtracting one of the possible masses of the Cl atom (35 u or 37 u) from each of the three observed BrCl masses (114, 116, or 118 u):

(1) (114-35) = 79 u (2) (114-37) = 77 u (3) (116-35) = 81 u

(4) (116-37) = 79 u (5) (118-35) = 83 u (6) (118-37) = 81 u

However of these four possibilities for the mass of the Br atom, (77, 79, 81, and 83 u), only 79 u and 81 u are admissable, since any one Br isotope can form molecules containing either ^{35}Cl or ^{37}Cl, and thus only the Br masses that appear above twice are valid.

Thus, (a) The isotopes of Br are ^{79}Br and ^{81}Br.

(b) The isotopically different BrCl molecules are:

^{79}Br^{35}Cl (mass 114 u); ^{79}Br^{37}Cl (mass 116 u)

^{81}Br^{35}Cl (mass 116 u); ^{81}Br^{37}Cl (mass 118 u)

The Mole: Counting by Weighing

39. If scientists had selected 1 lb (453.6 g) as the basic unit for measuring mass, and defined the mole on this basis, what difference would this have had on (a) the definition of the mole, and (b) the value of Avogadro's number?

(a)

Avogadro's number, N_A, results from taking the atomic mass unit, u, as 1/12 of the mass of a ^{12}C atom, because it is the number of ^{12}C atoms that has a mass of exactly 12 g. If 1 lb (453.6 g) had been selected as the basic mass unit, rather than 1 g, the mole would have been defined as **the number of ^{12}C atoms in 12 lb of ^{12}C atoms**, rather than 12 g, giving the value:

$$\left(\frac{6.022 \times 10^{23} \text{ C atoms}}{12 \text{ g C atoms}}\right)\left(\frac{453.6 \text{ g}}{1 \text{ lb}}\right)(12 \text{ lb}) = \underline{2.732 \times 10^{26} \text{ entities mol}^{-1}}$$

---

41. An atom of element X has a mass of 3.155×10^{-23} g. What are (a) the atomic mass of X compared to 12 u for one ^{12}C atom, (b) the molar mass of X, and (c) the symbol for X?

(a) The atomic mass unit, u, is one-twelfth of the mass of a ^{12}C atom, and 1 mol of ^{12}C atoms has a mass of exactly 12 g, Thus, 1 mol of atomic mass units is exactly 1 g, so that in grams:

$$1 \text{ u} = \left(\frac{1 \text{ g}}{1 \text{ mol u}}\right)\left(\frac{1 \text{ mol u}}{6.022 \times 10^{23} \text{ u}}\right) = \underline{1.6606 \times 10^{-24} \text{ g u}^{-1}}$$

Thus, for element X

$$\text{mass of X} = (3.155 \times 10^{-23} \text{ g atom}^{-1})\left(\frac{1 \text{ mol u}}{1.660\,6 \times 10^{-24} \text{ g}}\right) = \underline{19.00 \text{ u atom}^{-1}}$$

The atomic mass of X is **19.00 u**.

(b) The molar mass is 19.00 g mol^{-1} and reference to the periodic table gives X as <u>fluorine</u>, which has only one natural isotope, $^{19}_{9}F$.

---

42. Using atomic masses, calculate the molar mass of each of the compounds: (a) $H_2O(\ell)$ (b) $H_2O_2(\ell)$ (c) NaCl(s) (d) $MgBr_2(s)$ (e) CO(g) (f) $CO_2(g)$ (g) $CH_4(g)$ (h) $C_2H_6(g)$ (i) $NH_3(g)$ (j) HCl(g)

For a compound, the molar mass is simply the sum of the molar masses of all of the constituent atoms, taken from the periodic table and given to <u>4 significant figures</u>:

(a) H_2O, 18.02 g mol^{-1}; (b) H_2O_2, 34.02 g mol^{-1}; (c) NaCl, 58.44 g mol^{-1}; (d) $MgBr_2$, 184.1 g mol^{-1}; (e) CO, 28.01 g mol^{-1}; (f) CO_2, 44.01 g mol^{-1}; (g) CH_4, 16.04 g mol^{-1}; (h) C_2H_6, 30.08 g mol^{-1}; (i) NH_3, 17.03 g mol^{-1}; (j) HCl, 35.46 g mol^{-1}.

---

45. The artificial sweetener "Nutrasweet" is the compound aspartamine with the molecular formula $C_{14}H_{18}N_2O_5$: (a) What is the mass of 1.000 mol of aspartamine? (b) How many moles of aspartamine are in 6.22 g of the substance? (c) What is the mass, in grams, of 0.245 mol of aspartamine? (d) How many molecules are in 4.28 mg of aspartamine?

(a) Mass of 1.000 mol of aspartamine, $C_{14}H_{18}N_2O_5$,(its molar mass) = [14(12.01)+18(1.008)+2(14.01)+5(16.00)]

$$= \underline{294.3 \text{ g mol}^{-1}}.$$

(b) Mol aspartamine = (6.22 g aspartamine)$\left(\frac{1 \text{ mol aspartamine}}{294.3 \text{ g aspartamine}}\right)$ = $\underline{2.11 \times 10^{-2} \text{ mol aspartamine}}$

(c) Mass aspartamine = (0.245 mol aspartamine)$\left(\frac{294.3 \text{ g asp}}{1 \text{ mol aspartamine}}\right)$ = $\underline{72.1 \text{ g aspartamine}}$

(d) Molecules of aspartamine = (4.28 mg aspartamine)$\left(\frac{10^{-3} \text{ g}}{1 \text{ mg}}\right)\left(\frac{1 \text{ mol aspartamine}}{294.3 \text{ g aspartamine}}\right)\left(\frac{6.022 \times 10^{23} \text{ molecules}}{1 \text{ mol asp}}\right)$

= $\underline{8.76 \times 10^{18} \text{ molecules}}$

---

47. (a) How many moles of sulfur dioxide are in 0.028 g $SO_2(g)$? (b) What mass contains 3 mol of $SO_2(g)$?

Molar mass of SO_2 = [32.07 + 2(16.00)] = 64.07 g mol^{-1}

(a) 0.028 g SO_2 = (0.028 g SO_2)$\left(\frac{1 \text{ mol } SO_2}{64.07 \text{ g } SO_2}\right)$ = $\underline{4.4 \times 10^{-4} \text{ mol}}$

(b) 3 mol SO_2 = (3 mol SO_2)$\left(\frac{64.07 \text{ g } SO_2}{1 \text{ mol } SO_2}\right)$ = $\underline{192.2 \text{ g}}$

---

49. The pain reliever aspirin is the compound acetylsalicylic acid, $C_9H_8O_4(s)$. (a) What is the molecular mass of aspirin? (b) How many moles, and how many molecules of $C_9H_8O_4$ are there in a 500-mg aspirin tablet?

(a) Molecular mass aspirin = $\underline{180.2 \text{ u}}$

(b) moles aspirin = (500 mg)$\left(\frac{1 \text{ g}}{10^3 \text{ mg}}\right)\left(\frac{1 \text{ mol aspirin}}{180.2 \text{ g aspirin}}\right)$ = $\underline{2.77 \times 10^{-3} \text{ mol}}$

molecules aspirin = (2.77x10^{-3} mol aspirin)$\left(\frac{6.022 \times 10^{23} \text{ molecules}}{1 \text{ mol aspirin}}\right)$ = $\underline{1.67 \times 10^{21}}$

---

51. (a) Assuming that the human body contains 6x10^{13} cells and that the earth's population is 4x10^9 persons, approximately how many moles of human body cells are there on the earth? (b) Assuming that the human body is 80% water by mass, how many water molecules are there in the body of a person with a mass of 65 kg?

(a) This question graphically illustrates the enormous magnitude of the number of entities in 1 mol. You may find the answer surprising!

Cells = (4x10^9 bodies)$\left(\frac{6 \times 10^{13} \text{ cell}}{1 \text{ body}}\right)\left(\frac{1 \text{ mol cells}}{6.02 \times 10^{23} \text{ cells}}\right)$ = $\underline{0.4 \text{ mol}}$

Remarkably the total number of human body cells in the entire world is only 0.4 mol; 0.4 mol of water (only ~7 mL of water) contains the same number of water molecules.

(b) First convert kg to grams, and multiply by 80% to give the mass of water, then convert to moles, and finally to molecules, which can be done in one step:

$$\left(\frac{80}{100}\right)(65\ \text{kg})\left(\frac{10^3\ \text{g}}{1\ \text{kg}}\right)\left(\frac{1\ \text{mol H}_2\text{O}}{18.02\ \text{g H}_2\text{O}}\right)\left(\frac{6.022 \times 10^{23}\ \text{H}_2\text{O molecules}}{1\ \text{mol H}_2\text{O}}\right) = \underline{1.7 \times 10^{27}\ \text{H}_2\text{O molecules}}$$

---

52. Carbon monoxide, CO(g), taken into the lungs reduces the ability of the blood to transport oxygen. An amount of CO as high as $2.38 \times 10^{-4}\ g{\cdot}L^{-1}$ is fatal. Calculate the number of CO molecules that must be emitted from an automobile exhaust to produce a fatal concentration of CO in a garage of volume 150 m^3. (1 L = 1 dm^3.)

First calculate the volume of the garage in liters (using 10 dm = 1 m and 1 L = 1 dm^3):

$$\text{Volume} = (150\text{m}^3)\left(\frac{10\ \text{dm}}{1\ \text{m}}\right)^3\left(\frac{1\ \text{L}}{1\ \text{dm}^3}\right) = 1.50\text{x}10^5\ \text{L}$$

Next we calculate the fatal dose of CO(g) in this volume in grams, convert grams of CO to moles of CO, and then to the number of molecules, using the unit factors: $2.38\text{x}10^{-4}$ g CO = 1 L and 1 mol CO = 28.01 g CO. Thus, the number of CO molecules in the garage is given by:

$$(1.50\text{x}10^5\ \text{L})\left(\frac{2.38 \times 10^{-4}\ \text{g CO}}{(1\ \text{L})}\right)\left(\frac{1\ \text{mol CO}}{28.01\ \text{g CO}}\right)\left(\frac{6.022 \times 10^{23}\ \text{CO molecules}}{1\ \text{mol CO}}\right) = \underline{7.68 \times 10^{23}\ \text{CO molecules}}$$

Unstable Atoms: Radioactivity; Conservation of Mass and Energy

53. Explain (a) why some nuclei are unstable, and (b) why the mass of any atom is not exactly the same as the mass of its constituent protons, neutrons, and electrons.

(a) A nucleus is stable only if the internuclear forces holding the protons and neutrons together in the nucleus are very strong. Thus, the binding energy has to be sufficiently strong to overcome the large repulsions between the closely packed positively charged protons, which is not true for the heavier nuclei.

(b) When electrons, protons, and neutrons combine to form an atom, the combination of the protons and neutrons to form the nucleus results in the release of a very large amount of energy, because of the very strong internuclear forces. This energy has a significant mass equivalent, so that the mass of an atom is always less than that of the sum of its constituent protons, neutrons, and electrons.

55.* Explain why the product of the decay of a radioactive atom by (a) emission of an α particle gives an atom in which the mass number has decreased by 4 and the atomic number has decreased by 2, and (b) emission of a β particle gives an atom in which the mass number is unchanged but the atomic number has increased by 1.

(a) An α particle is a helium nucleus, ${}^4_2He$, so its loss from the nucleus of a radioactive atom decreases the mass number of the resulting atom by 4. Since this also results in the removal of two protons from the nucleus, the atomic number Z of the resulting atom also decreases by 2.

(b) A β particle is an electron, ${}^0_{-1}e$, with negligible mass, and in its loss from the nucleus of a radioactive atom a neutron is converted into a nuclear proton, ${}^1_1H$, so the mass number of the resulting atom is unchanged but its atomic number Z is increased by 1.

General Problems

57. Ammonium sulfate, $(NH_4)_2SO_4(s)$ ("sulfate of ammonia"), is an important commerical fertilizer produced by the reaction of hydrogen, $H_2(g)$, with nitrogen, $N_2(g)$, to give ammonia, $NH_3(g)$. The ammonia is passed into aqueous sulfuric acid, $H_2SO_4(aq)$, until reaction is complete, and the ammonium sulfate formed is then crystallized from the solution. Write balanced equations for these reactions.

---

1. Ammonia, $NH_3(g)$, is produced by the reaction $N_2(g) + 3H_2(g) \rightarrow 2NH_3(g)$ (Haber Process)
2. $2NH_3(g) + H_2SO_4(aq) \rightarrow (NH_4)_2SO_4(aq)$

(Note that crystallization from solution to give $(NH_4)_2SO_4(s)$ is a physical process.)

---

59. (a) Write the symbol for the isotope that contains as many protons as neutrons and has a mass number 28. (b) The element with the isotope in (a) also has isotopes of mass numbers 29 and 30. Write their symbols. (c) The isotopes of mass numbers 28, 29, and 30 have atomic masses of 27.976 93 u, 28.976 50 u, and 29.973 77 u, with relative abundances of 92.18%, 4.710%, and 3.120% by mass, respectively. What is the average atomic mass of the element?

(a) The mass number A is 28, so for equal number of protons and neutrons Z = 14 and A-Z = 14. The element with Z = 14 is Si, and the symbol of this isotope is $^{28}_{14}Si$. (b) $^{29}_{14}Si$ and $^{30}_{14}Si$.

(b) The fractional abundances are, respectively, 0.9218, 0.04710, and 0.03120. Thus, we have:

isotope	contribution to average atomic mass	
^{28}Si	(27.976 93 u)(0.921 8)	= 25.7891 u
^{29}Si	(28.976 50 u)(0.047 10)	= 1.3648 u
^{30}Si	(29.937 77 u)(0.031 20)	= 0.9341 u
	Average atomic mass	= 28.09 u

---

60. Heavy water, $^2H_2O(\ell)$, or $D_2O(\ell)$, occurs to the extent of 0.003% by mass in natural water. Pure heavy water, which is used to reduce the energies of fast neutrons in the CANDU nuclear reactor, is a colorless liquid which forms hexagonal crystals upon freezing and has a density of 1.105 $g\cdot cm^{-3}$, a melting point of 3.8°C, and a boiling point of 101.4°C. It is separated from ordinary water by fractional distillation or by electrolysis, and reacts with $SO_3(g)$ to give deuterosulfuric acid, $D_2SO_4(\ell)$. From this description list (a) the physical properties and (b) the chemical properties of heavy water that are mentioned.

(a) Physical properties: colorless, liquid, forms hexagonal crystals, density 1.105 $g\cdot cm^{-3}$, mp 3.8°C, bp 101.4°C.

(b) Chemical properties: Electrolysis (gives $D_2(g)$ and $O_2(g)$); reaction with $SO_3(g)$ gives $D_2SO_4(\ell)$ according to the equation $D_2O(\ell) + SO_3(g) \rightarrow D_2SO_4(\ell)$

(Note that fractional distillation is a physical process whereby D_2O is separated from HDO and H_2O.)

---

61. (a) Explain why the molecular mass of water in grams is the same as the mass of one water molecule in atomic units. (b) How many molecules are there in 1.000 g of water. (c) Given that the density of water is 0.997 $g\cdot cm^{-3}$ at 25°C, what is the volume of one water molecule in liquid water (in units of pm^3) at this temperature?

(a) When 1 atomic unit (1 u) is multiplied by Avogadro's number N_A, the answer is 1 g. Since a molar mass is the mass of one mole (N_A molecules) of a substance in grams, the mass of 1 molecule in u is the same numerically as the molar mass in grams.

(b) The molar mass of water, H_2O, is [2(1.008) + 16.00] = 18.02 $g\cdot mol^{-1}$. Thus, the number of water molecules in 1 g of water is given by:

$$(1.000 \text{ g } H_2O)\left(\frac{1 \text{ mol } H_2O}{18.02 \text{ g } H_2O}\right)\left(\frac{6.022 \times 10^{23} \text{ molecules}}{1 \text{ mole}}\right) = \underline{3.342 \times 10^{22} \text{ molecules}}$$

(c) 1 g of water at 25°C has a volume of 0.997 cm^3, so the volume of 1 molecule at 25°C is given by:

$$V = \left(\frac{0.997 \text{ cm}^3}{3.342 \times 10^{22} \text{ molecules}}\right)\left(\frac{1 \text{ m}}{10^2 \text{ cm}}\right)^3\left(\frac{10^{12} \text{ pm}}{1 \text{ m}}\right)^3 = \underline{2.98 \times 10^7 \text{ pm}^3}$$

63.* Gunpowder is a mixture of potassium nitrate, $KNO_3(s)$, sulfur, and charcoal (carbon). (a) How could a sample of gunpowder be separated into its components? (b) Write the balanced equation for the reaction between the components of gunpowder, to give potassium nitrite, $KNO_2(s)$, sulfur dioxide, $SO_2(g)$, and carbon dioxide, $CO_2(g)$, when heated.

(a) If you look up the properties of the substances in gunpowder in your text, you will discover that potassium nitrate, $KNO_3(s)$, is soluble in water (as are all metal nitrates), while sulfur and charcoal are insoluble. Thus, KNO_3 can be separated from sulfur and charcoal by heating gunpowder with water to dissolve the potassium nitrate. The insoluble sulfur and charcoal can then be separated by filtration. You will also discover that sulfur is soluble in a solvent such as carbon disulfide, $CS_2(\ell)$, while charcoal (and $KNO_3(s)$) are insoluble. Thus, after the mixture of sulfur and charcoal left after filtration is dried, warming with $CS_2(\ell)$ dissolves only the sulfur, and charcoal can then be separated from the solution, again by filtration. $KNO_3(s)$ and S(s) could then be recovered by careful evaporation of the respective solutions.

(b) The unbalanced equation is

$$KNO_3(s) + S(s) + C(s) \rightarrow KNO_2(s) + SO_2(g) + CO_2(g)$$

where K, N, S, and C are balanced, leaving only O to be balanced. KNO_3 and KNO_2 differ by 1 O atom and formation of 1 molecule each of SO_2 and CO_2 from 1 S and 1 C requires 4 O atoms, so the equation can be balanced with $4KNO_3$ on the LHS, giving:

$$4KNO_3(s) + S(s) + C(s) \rightarrow 4KNO_2(s) + SO_2(g) + CO_2(g)$$

CHAPTER 2

The Atmosphere; Oxygen; Nitrogen; Hydrogen

2. (a) What are the three most abundant gases in the atmosphere? (b) Name three gases in air that are chemically reactive. (c) Explain why carbon dioxide is an essential constituent of the atmosphere.

(a) The three most abundant gases in the atmosphere are, in order, nitrogen, oxygen, and the noble gas argon.

(b) Of the most abundant elements, only oxygen is chemically highly reactive; in the atmosphere nitrogen reacts with oxygen only at very high temperatures, such as in lightning discharges, to form NO(g), and Ar forms no compounds. Among the minor component gases, carbon dioxide, CO_2, methane, CH_4, and hydrogen, H_2, are all very reactive, as are water vapor, H_2O(g), nitrogen monoxide, NO(g), and ozone, O_3(g), among the variable components.

(c) Carbon dioxide is an essential component in the photosynthesis-respiration **carbon cycle** on which life is dependent. CO_2 produced by respiration in animals is converted by green plants back to O_2, a waste product of photosynthesis, the reaction by which they synthesize organic compounds from CO_2 and water using sunlight. Since nearly all other forms of life directly or indirectly depend on plants for food and oxygen for energy, photosynthesis is essential for life on earth; virtually all atmospheric oxygen has come from photosynthesis and it is the way in which a constant replenishment of O_2(g) in the atmosphere is ensured.

---

5. Write balanced equations to decribe the reactions that occur when: (a) sulfur burns in air, (b) magnesium burns in oxygen, (c) methane burns in air, and (d) hydrogen burns in oxygen.

In combustion reactions with oxygen in the air, the products are oxides of the elements:

(a) sulfur reacts only with the oxygen in air to give SO_2(g):

$$S(s) + O_2(g) \rightarrow SO_2(g)$$

(b) magnesium reacts with both the O_2(g) and the N_2(g) in air, but in oxygen alone

$$2Mg(s) + O_2(g) \rightarrow 2MgO(s)$$ magnesium oxide

(c) methane reacts completely with the O_2(g) in air (burns) to give carbon dioxide and water:

$$CH_4(g) + 2O_2(g) \rightarrow CO_2(g) + 2H_2O(g)$$

(d) hydrogen burns in oxygen to give water:

$$2H_2(g) + O_2(g) \rightarrow 2H_2O(g)$$

---

7. Write a balanced equation for the reaction of steam at high temperature with each of the following: (a) iron, (b) magnesium, (c) methane

---

(a) $2Fe(s) + 3H_2O(g) \rightarrow Fe_2O_3(s) + 3H_2(g)$; (b) $Mg(s) + H_2O(g) \rightarrow MgO(s) + H_2(g)$;

(c) $CH_4(g) + H_2O(g) \rightarrow CO(g) + 3H_2(g)$

---

8. Write balanced equations to describe the reaction of nitrogen at high temperature with the following, and name the products: (a) hydrogen, (b) oxygen.

Nitrogen, N_2(g), is unreactive at ordinary temperatures but takes part in some nitrogen fixation reactions at high temperatures:

(a) Hydrogen reacts with nitrogen at high temperature to give ammonia, NH_3(g):

$$N_2(g) + 3H_2(g) \rightarrow 2NH_3(g)$$

(b) Nitrogen reacts with O_2(g) at high temperature to give nitrogen monoxide, NO(g):

$$N_2(g) + O_2(g) \rightarrow 2NO(g)$$

---

10. Write the balanced equations for two reactions by which hydrogen may be formed from water.

Hydrogen is obtained from water in any reaction that reduces H_2O to H_2. For example:

(i) Using a hydrocarbon such as methane, CH_4, as the reducing agent gives synthesis gas:

$$H_2O(g) + CH_4(g) \rightarrow 3H_2(g) + CO(g)$$

(ii) Hydrogen is produced from water by electrolysis:

$$2H_2O(\ell) \rightarrow 2H_2(g) + O_2(g)$$

---

12. Write a balanced equation for each of the following: (a) the reaction of carbon when heated in air to give carbon monoxide, (b) the reaction of calcium with oxygen to give calcium oxide, CaO(s), (c) the combustion of propane, $C_3H_8(g)$, in excess oxygen, (d) the combustion of ethanol, $C_2H_6O(l)$, in excess oxygen.

(a) $$2C(s) + O_2(g) \rightarrow 2CO(g)$$

Note: This reaction gives CO(g) as the main product when the amount of $O_2(g)$ is limited; in excess $O_2(g)$ the product is carbon dioxide, $CO_2(g)$.

(b) $$2Ca(s) + O_2(g) \rightarrow 2CaO(s)$$

(This reaction is analogous to that of Mg(s) with $O_2(g)$)

(c) propane, $C_3H_8(g)$, like methane, $CH_4(g)$, is a hydrocarbon and burns in excess oxygen to give carbon dioxide and water:

$$C_3H_8(g) + 5O_2(g) \rightarrow 3CO_2(g) + 4H_2O(g)$$

(d) Ethanol is an alcohol and like all alcohols burns in excess oxygen to give carbon dioxide and water:

$$C_2H_6OH(\ell) + 3O_2(g) \rightarrow 2CO_2(g) + 3H_2O(g)$$

Solids, Liquids, and Gases

14. Give the physical state (gas, liquid, or solid) at room temperature of each of the following elements: (a) magnesium (b) nitrogen (c) oxygen (d) sulfur (e) copper (f) hydrogen (g) bromine

---

(a) magnesium, **solid** (b) nitrogen, **gas** (c) oxygen, **gas** (d) sulfur, **solid** (e) Copper, **solid** (f) hydrogen, **gas** (g) bromine, **liquid.**

The Gas Laws

NOTE ON GAS LAW PROBLEMS

In all problems, such as those below, involving the Gas Laws, you will find it convenient to start with one equation, namely the Ideal Gas Law, PV = nRT:

1. When n and T are constant, PV = Constant, **Boyle's Law;**
2. When n and P are constant, V = Constant x T, **Charles' Law;**
3. When P and T are constant, V = Constant x n, **Avogadro's Law;**
4. For a constant number of moles of a gas, n, (a fixed amount):

$$\frac{P_1V_1}{T_1} = \frac{P_2V_2}{T_2}$$

---

17. An automobile tire of volume 28 L is filled with air at a pressure of 2.3 atm. What volume does this air occupy at atmospheric pressure (1.00 atm) at the same temperature?

PV = nRT, and the amount of gas in the tire, n, and T, remain constant:

so $P_1V_1 = P_2V_2$ (Boyle's law)

i.e., (2.3 atm)(28 L) = (1.0 atm)V_2; **V_2 = 64 L.**

The air in the tire would occupy a volume of **64 L** at atmospheric pressure.

---

19. A cylinder in a gasoline engine initially has a volume of 0.50 L and contains gases that exert a pressure of 1.00 atm. What is the pressure of the gases when the volume of the cylinder is reduced to 0.20 L, assuming that there is no change in the temperature?

PV = nRT, and the amount of gases in the cylinder, n, and the temperature, T, remain constant:

so $P_1V_1 = P_2V_2$ (Boyle's law)

i.e., (1.00 atm)(0.50 L) = P_2(0.20 L); **P_2 = 2.5 atm**

The pressure in the cylinder would increase to **2.5 atm.**

---

21. A sample of hydrogen gas has a pressure of 0.98 atm when it is confined in a bulb of volume 2.00 L. A tap connects the bulb to another bulb that has a volume of 5.00 L and has been completely evacuated. What is the pressure in the two bulbs after the tap is opened?

PV = nRT, and the amount of gas in the system, n, and T remain constant:

so $P_1V_1 = P_2V_2$ (Boyle's law), and V_1 is 2.00 L, and V_2 is (2.00 + 5.00) = 7.00 L.

i.e., (0.98 atm)(2.00 L) = P_2(7.00 L); **P_2 = 0.28 atm**

The final pressure of gas in each bulb is the **same** at **0.28 atm.**

---

23. To what temperature must 30 L of helium at 25 °C be cooled at 1.00 atm pressure for its volume to be reduced to 1.00 L at the same pressure?

PV = nRT, and the amount of gas in the system, n, and the pressure, P, remain constant:

$$\frac{V_1}{T_1} = \frac{V_2}{T_2}; \quad V_1 T_2 = V_2T_1 \quad \text{Charles' law}$$

but we must remember to express temperatures in kelvins, that is, T_1 = (25+273) = 298 K.

Thus: (30.0 L)T_2 = (1.00 L)(298.1 K); **T_2 = 10 K**, or **-263 °C.**

Molecules and Volume: Avogadro's Law

25. (a) Which sample contains more molecules: 1.00 L of O_2(g) at STP or 1.00 L of H_2(g) at STP. (c) Which sample has the greater mass?

(a) No calculation is needed; both gases have the same volume and are at the same temperature and pressure (STP = 0°C and 1 atm), thus they must contain the same number of moles (and number of molecules).

(b) However, the molecular masses are different (H_2, 2.016 u, and O_2, 32.00 u). Thus, since the numbers of molecules are the same, **1.00 L of O_2(g) at STP has a greater mass** than 1.00 L of H_2(g) at STP.

---

27. The anesthetic cyclopropane is a gas containing only carbon and hydrogen. Suppose 0.550 L of cyclopropane at 120 °C and 0.900 atm react with oxygen to give 1.65 L of $CO_2(g)$ and 1.65 L of $H_2O(g)$ at the same temperature and pressure. What are (a) the molecular formula of cyclopropane and (b) the balanced equation for its reaction with oxygen?

From the data given, in terms of volumes of gases:

0.550 L cyclopropane(g) → 1.65 L $CO_2(g)$ + 1.65 L $H_2O(g)$

1 vol 3 vol 3 vol

and since equal volumes of gases at constant T and P contain equal numbers of moles (molecules), we can write:

1 cyclopropane(g) → $3CO_2(g)$ + $3H_2O(g)$

1 mol → 3 mol 3 mol

or 1 molecule → 3 molecules 3 molecules

and since atoms are conserved in the reaction, and cyclopropane contains only C atoms and H atoms, there must be 3 C and 6 H atoms in one cyclopropane molecule, so the molecular formula must be C_3H_6. Thus, the balanced equation for the reaction is:

$$2C_3H_6(g) + 9O_2(g) \rightarrow 6CO_2(g) + 6H_2O(g)$$

The Kinetic Theory of Gases

30. Explain why the pressure of a gas (a) doubles when the volume is halved, and (b) increases with increasing temperature.

(a) Halving the volume doubles the concentration of molecules and twice as many molecules collide with the walls of the container in a given time; thus, the force exerted by the molecules on unit area of the container wall doubles; i.e., the pressure, doubles.

(b) When the temperature is raised, the average kinetic energy of the molecules, $\frac{1}{2}mv^2$, also increases. In other words, the average velocity of the molecules, v, increases, which leads to an increasing number of collisions with the walls of the container in a given time and with a greater force. Thus, the pressure increases with increased temperature.

Using the Ideal Gas Law

Note: When you use PV = nRT, remember that the value for the gas constant that is normally used is

R = 0.0821 atm L mol^{-1} K^{-1}

Thus, pressures, P, have to be in **atm**, volumes, V, in **liters**, L , amounts of gases in **moles, n,** and the temperature, T, in **kelvins**, K.

---

31. How many moles of methane, $CH_4(g)$, occupy a volume of 4.00 L at 1.00 atm pressure and 25°C?

This is a straightforward application of the ideal gas law:

$$PV = nRT;\ n = \frac{PV}{RT} = \frac{(1.00\ \text{atm})(4.00\ \text{L})}{(0.0821\ \text{atm L mol}^{-1}\ \text{K}^{-1})(298\ \text{K})} = \underline{0.163\ \text{mol}}$$

---

33. The temperature of a closed vessel containing air at a pressure of 1.02 atm is raised from 20 to 200°C. What is the pressure inside the vessel if its volume increases by 10% as the temperature is increased from 20 to 200°C?

Here the number of moles of gas is constant, the volume is increased from V_1 to $V_2 = 1.10\ V_1$, and the temperature from $T_1 = (20+273) = 293$ K, to $T_2 = (200+273) = 473$ K. P_1 is 1.02 atm, and we have to calculate P_2:

$$PV = nRT \ ; \ \frac{P_1 V_1}{T_1} = \frac{P_2 V_2}{T_2} \ ; \ \frac{(1.02 \text{ atm})V_1}{293 \text{ K}} = \frac{P_2(1.10\ V_1)}{473 \text{ K}} \ ; \ \underline{P_2 = 1.50 \text{ atm}}$$

---

35. An automobile tire of fixed volume is inflated to a pressure of 2.50 atm at 20°C. What is the pressure inside the tire at (a) 30°C, and (b) -10°C?

The volume is constant and so is the number of moles (of air).

$$PV = nRT \ ; \ \frac{P_1}{T_1} = \frac{P_2}{T_2}$$

(a) $\frac{2.50 \text{ atm}}{293 \text{ K}} = \frac{P_2}{303 \text{ K}} \ ; \ \underline{P_2 = 2.59 \text{ atm}}$; (b) $\frac{2.50 \text{ atm}}{293 \text{ K}} = \frac{P_2}{263 \text{ K}} \ ; \ \underline{P_2 = 2.24 \text{ atm}}$

---

37. What mass of hydrogen chloride, HCl(g), is needed to provide a pressure of 0.240 atm in a container of volume 250 mL at 37°C?

We use PV = nRT to first calculate n, the number of moles of HCl(g) required to fill the 250 mL flask at 37°C (310 K), remembering to convert mL to L. Then we convert mol of HCl to grams of HCl (molar mass $(1.008+35.45) = 36.46$ g mol^{-1}):

$$PV = nRT; \ n = \frac{PV}{RT} = \frac{(0.240 \text{ atm})(250 \text{ mL})(\frac{1.00 \text{ L}}{1000 \text{ mL}})}{(0.0821 \text{ atm L mol}^{-1} \text{ K}^{-1})(310 \text{ K})} = \underline{2.357 \times 10^{-3} \text{ mol}}$$

$$\text{Mass of HCl} = (2.357 \times 10^{-3} \text{ mol})(\frac{36.46 \text{ g HCl}}{1 \text{ mol HCl}}) = \underline{0.0859 \text{ g HCl}}$$

HCl(g) of mass **0.0859** g would fill the flask to a pressure of 0.240 atm at 37°C.

---

39. As a publicity stunt, a water-bed retailer filled a water bed with helium and floated it above his store. Calculate the mass of helium required to fill a water-bed bag at a pressure of 1.03 atm at 23°C, if its dimensions are 2.00 m x 1.50 m x 0.20 m.

We first calculate the volume of the water bed in liters, using the unit conversion factor ($1 \text{ L} = 1 \text{ dm}^3$), then use the ideal gas law, PV = nRT, to find the moles of He to fill this volume, remembering that the temperature is $(23+273) = 296$ K, and finally, convert moles of He to grams of He:

$$V = (2.00 \text{ m})(1.50 \text{ m})(0.20 \text{ m})(\frac{10 \text{ dm}}{1 \text{ m}})^3(\frac{1 \text{ L}}{1 \text{ dm}^3}) = \underline{600 \text{ L}}$$

$$n_{He} = \frac{PV}{RT} = \frac{(1.03 \text{ atm})(600 \text{ L})}{(0.0821 \text{ atm L mol}^{-1} \text{ K}^{-1})(296 \text{ K})} = \underline{25.43 \text{ mol}} = (25.43 \text{ mol He})(\frac{4.003 \text{ g He}}{1 \text{ mol He}}) = \underline{102 \text{ g He}}$$

---

41. What is the density of the chlorofluorocarbon (CFC) of molecular formula CF_2Cl_2 at 1.00 atm and 20°C?

The units of gas density are g L^{-1}, so we need to calculate the mass 1 L of $CF_2Cl_2(g)$ at 1.00 atm and 20°C (293 K), which we do by first calculating the moles of CF_2Cl_2 in 1 L under these conditions.

$$PV = nRT \; ; \quad n = \frac{PV}{RT} = \frac{(1.00 \text{ atm})(1.00\text{L})}{(0.0821 \text{ atm L mol}^{-1}\text{ K}^{-1})(293 \text{ K})} = \underline{4.157 \times 10^{-2} \text{ mol}}$$

$$\text{Mass of CFC} = (4.157 \times 10^{-2} \text{ mol CFC})(\frac{120.9 \text{ g CFC}}{1 \text{ mol CFC}}) = \underline{5.03 \text{ g}} \; ; \text{ i.e., Density} = \underline{5.03 \text{ g L}^{-1}}$$

Molar Mass

42. A sample of a noble gas has a mass of 0.20 g and exerts a pressure of 0.48 atm in a container of volume 0.26 L at 27 °C. Is the gas helium, neon, argon, krypton, or xenon?

We first calculate the number of moles of gas using PV = nRT:

$$n = \frac{PV}{RT} = \frac{(0.48 \text{ atm})(0.26 \text{ L})}{(0.0821 \text{ atm L mol}^{-1}\text{ K}^{-1})(300 \text{ K})} = \underline{5.07 \times 10^{-3} \text{ mol}} \; ; \quad \text{Molar mass} = (\frac{0.20 \text{ g}}{5.07 \times 10^{-3} \text{ mol}}) = \underline{39 \text{ g mol}^{-1}}$$

The molar mass of 39 g mol^{-1} identifies the gas as **Argon**, (molar mass 39.95 g mol^{-1}).

---

44. A gas at a pressure of 740 mm Hg at 20°C occupies a volume of 1.00 L and has a mass of 1.134 g. (a) What is its molar mass? (b) If the empirical formula of the gas is CH_2, what is its molecular formula?

Use PV = nRT first to calculate the moles of gas in 1 L, and then the mass (1.134 g) to calculate the molar mass:

$$n = \frac{PV}{RT} = \frac{(740 \text{ mm Hg})(\frac{1 \text{ atm}}{760 \text{ mm Hg}})(1.00 \text{ L})}{(0.0821 \text{ atm L mol}^{-1}\text{ K}^{-1})(293 \text{ K})} = \underline{0.0405 \text{ mol}} \; ; \quad \text{Molar mass} = \frac{1.134 \text{ g}}{0.0405 \text{ mol}} = \underline{28.0 \text{ g mol}^{-1}}$$

The empirical formula mass of CH_2 = [12.01+2(1.008)] = 14.03 u, and the molecular mass is 28.0 u, which is close to twice the empirical formula mass. Thus, the molecular formula is $2(CH_2)$.

i.e., the gas is **C_2H_4** (molar mass 28.06 g mol^{-1}).

Dalton's Law of Partial Pressures

46. Air is composed of 78% nitrogen, 21% oxygen, and 1% argon by volume. What are the partial pressures of each gas when the atmospheric pressure is 758 mm Hg?

Rephrasing the question, the fraction of the total number of moles (molecules) in air of each type is 0.78 N_2, 0.21 O_2 and 0.01 Ar. Since the partial pressure of each gas is proportional to the number of moles:

partial pressure of N_2 = 0.78(758 mm Hg) = 5.9×10^2 mm Hg

partial pressure of O_2 = 0.21(758 mm Hg) = 1.6×10^2 mm Hg

partial pressure of Ar = 0.01(758 mm Hg) = 8 mm Hg

---

48. A mixture of 0.200 g of helium and 0.200 g of hydrogen is confined in a vessel of volume 225 mL at 27°C. (a) What is the partial pressure of each gas? (b) What is the total pressure exerted by the gaseous mixture?

In any mixture of gases, the partial pressure of each gas is proportional to the number of moles of that gas, and the total pressure is proportional to the total moles of gas. Thus, we first calculate the number of moles of each gas:

$$\text{moles of He} = \left(\frac{0.200 \text{ g He}}{4.003 \text{ g mol}^{-1}}\right) = \underline{0.0500 \text{ mol}} \;; \quad \text{moles of } H_2 = \left(\frac{0.200 \text{ g } H_2}{2.016 \text{ g mol}^{-1}}\right) = \underline{0.0992 \text{ mol } H_2}$$

(a) Now we can calculate the partial pressure of each gas from PV = nRT:

$$p_{He} = \frac{n_{He}RT}{V} = \frac{(0.0500 \text{ mol})(0.0821 \text{ atm L mol}^{-1} \text{ K}^{-1})(300 \text{ K})}{(225 \text{ mL})(\frac{1 \text{ L}}{10^3 \text{ mL}})} = \underline{5.47 \text{ atm}}$$

$$p_{H_2} = \frac{n_{H_2}RT}{V} = \frac{(0.0992 \text{ mol})(0.0821 \text{ atm L mol}^{-1} \text{ K}^{-1})(300 \text{ K})}{(225 \text{ mL})(\frac{1 \text{ L}}{10^3 \text{ mL}})} = \underline{10.9 \text{ atm}}$$

(b) Total pressure = the sum of the partial pressures (Dalton's law):

$$P_{total} = p_{He} + p_{H_2} = (5.47 + 10.9) \text{ atm} = \underline{16.4 \text{ atm}}$$

---

50. The partial pressure of oxygen in air at 37°C is 159 mm Hg. When a human exhales a breath, the partial pressure of oxygen is only 115 mm Hg. How many oxygen molecules per liter of inhaled air are used by the lungs? (Take body temperature to be 37°C).

The partial pressure of the $O_2(g)$ used by the lungs is (159-115) = 44 mm Hg, at 37 °C (310 K). We can use this data and the volume of 1 L to calculate the moles of $O_2(g)$ used on inhalation, and then convert the number of moles to the number of molecules:

$$\text{mol } O_2 = \frac{PV}{RT} = \frac{(44 \text{ mm Hg})(\frac{1 \text{ atm}}{760 \text{ mm Hg}})(1 \text{ L})}{(0.0821 \text{ atm L mol}^{-1} \text{ K}^{-1})(310 \text{ K})} = \underline{2.28 \times 10^{-3} \text{ mol}}$$

$$\text{molecules of } O_2 = (2.28 \times 10^{-3} \text{ mol})\left(\frac{6.022 \times 10^{23} \text{ molecules}}{1 \text{ mol}}\right) = \underline{1.4 \times 10^{21}}$$

---

52. The mass percentage composition of the atmosphere of Mars is 95% CO_2, 3% N_2, and 2% other gases, principally Ar. What are the partial pressures of each gas at the surface, where the total pressure is 5.0 torr?

The total pressure of 5.0 torr (mm Hg) is proportional to the total number of moles of all the gases present and the partial pressure of each gas is determined by the number of moles of that gas (Dalton's law). Thus, conveniently, we can consider, any specific atmospheric mass, say 100 g, and then first convert the mass of each gas (CO_2, 95 g; N_2, 3 g, and Ar, 2 g) to moles:

$$\text{mol } CO_2 = (95 \text{ g } CO_2)\left(\frac{1 \text{ mol } CO_2}{44.01 \text{ g } CO_2}\right) = \underline{2.16 \text{ mol}} \;; \quad \text{mol } N_2 = (3 \text{ g } N_2)\left(\frac{1 \text{ mol } N_2}{28.02 \text{ g } N_2}\right) = \underline{0.11 \text{ mol}}$$

$$\text{mol Ar} = (2 \text{ g Ar})\left(\frac{1 \text{ mol Ar}}{39.95 \text{ g Ar}}\right) = \underline{0.050 \text{ mol}}$$

Thus, the total is (2.16+0.11+0.050) = $\underline{2.32 \text{ mol}}$ of gases and the total pressure of 5.0 torr is made up of the following partial pressures:

$$p_{CO_2} = (\frac{2.16\ mol}{2.32\ mol})(5.0\ torr) = \underline{4.7\ torr}\ ;\quad p_{N_2} = (\frac{0.11\ mol}{2.32\ mol})(5.0\ torr) = \underline{0.2\ torr}\ ;\quad p_{Ar} = (\frac{0.050\ mol}{2.32\ mol})(5.0\ torr) = \underline{0.1\ torr}$$

for a total atmospheric pressure of (4.7 + 0.2 + 0.1) = 5.0 torr, as observed.

Diffusion and Effusion

53. The average speed of oxygen molecules at 25°C is 450 m s^{-1}. What are the average speeds of each of the following gaseous molecules at 25°C? (a) hydrogen, (b) chlorine, Cl_2, (c) carbon monoxide, (d) water, and (e) carbon dioxide.

The average speed of any particular gas depends on its molar (molecular) mass. Their average kinetic energies depend only on the absolute temperature. Thus, for two gases we can write:

$$\frac{1}{2}m_1v_1^2 = \frac{1}{2}m_2v_2^2\ ;\quad \frac{m_1}{m_2} = \frac{v_2^2}{v_1^2}\ ;\quad \frac{v_2}{v_1} = (\frac{m_1}{m_2})^{\frac{1}{2}}$$

and since the molecular mass of a molecule is proportional to its molar mass, M g mol^{-1}, it follows that:

$$\frac{v_2}{v_1} = (\frac{M_1}{M_2})^{\frac{1}{2}}\ ;\quad v_2 = v_1(\frac{M_1}{M_2})^{\frac{1}{2}}$$

Now we use the last formula to calculate the average speeds of particular molecules from the average speed of $O_2(g)$ molecules at 25 °C:

(a) $v_{H_2} = (450\ m\ s^{-1})(\frac{32.00\ g\ mol^{-1}}{2.016\ g\ mol^{-1}})^{\frac{1}{2}} = \underline{1.79\times10^3\ m\ s^{-1}}$

(b) $v_{Cl_2} = (450\ m\ s^{-1})(\frac{32.00\ g\ mol^{-1}}{70.90\ g\ mol^{-1}})^{\frac{1}{2}} = \underline{302\ m\ s^{-1}}$

(c) $v_{CO} = (450\ m\ s^{-1})(\frac{32.00\ g\ mol^{-1}}{28.01\ g\ mol^{-1}})^{\frac{1}{2}} = \underline{481\ m\ s^{-1}}$

(d) $v_{H_2O} = (450\ m\ s^{-1})(\frac{32.00\ g\ mol^{-1}}{18.02\ g\ mol^{-1}})^{\frac{1}{2}} = \underline{600\ m\ s^{-1}}$

(e) $v_{CO_2} = (450\ m\ s^{-1})(\frac{32.00\ g\ mol^{-1}}{44.01\ g\ mol^{-1}})^{\frac{1}{2}} = \underline{384\ m\ s^{-1}}$

---

55. $H_2(g)$ and $D_2(g)$ are separated by utilizing their different rates of diffusion. What is the expected ratio of these rates?

For the method, see Problem 51. (The molar masses are $D_2(g)$ 4.028 g mol^{-1} and $H_2(g)$ 2.016 g mol^{-1}).

The ratio of rates is given by: $\frac{r_{H_2}}{r_{D_2}} = (\frac{4.028 \text{ mol}^{-1}}{2.016 \text{ mol g}^{-1}})^{\frac{1}{2}} = \underline{1.414}$

---

57. The average speed of helium atoms is 0.707 miles s^{-1} at 25°C. What is the average speed at 25°C of: (a) hydrogen, (b) nitrogen, and (c) oxygen molecules?

This problem is similar to Problem 50.

$$\frac{v_{H_2}}{v_{He}} = \frac{v_{H_2}}{0.707 \text{ mile s}^{-1}} = (\frac{4.003 \text{ g mol}^{-1}}{2.016 \text{ g mol}^{-1}})^{\frac{1}{2}} = \underline{1.409}; \quad v_{H_2} = \underline{0.996 \text{ mile s}^{-1}}$$

$$\frac{v_{N_2}}{v_{He}} = \frac{v_{N_2}}{0.707 \text{ mile s}^{-1}} = (\frac{4.003 \text{ g mol}^{-1}}{28.02 \text{ g mol}^{-1}})^{\frac{1}{2}} = \underline{0.3780}; \quad v_{N_2} = \underline{0.267 \text{ mile s}^{-1}}$$

$$\frac{v_{O_2}}{v_{He}} = \frac{v_{O_2}}{0.707 \text{ mile s}^{-1}} = (\frac{4.003 \text{ g mol}^{-1}}{32.00 \text{ g mol}^{-1}})^{\frac{1}{2}} = \underline{0.3537}; \quad v_{O_2} = \underline{0.250 \text{ mile s}^{-1}}$$

---

58. A gaseous element that exists as a diatomic gas at STP effused through a small hole at 0.324 times the rate of effusion of helium gas under the same conditions. Identify the element.

The effusion rate of the unknown gas, X_2, is 0.324 r_{He}:

$$\frac{r_1}{r_2} = (\frac{M_2}{M_1})^{\frac{1}{2}}; \quad \frac{r_{X_2}}{r_{He}} = (\frac{4.003 \text{ g mol}^{-1}}{M_{X_2} \text{ g mol}^{-1}})^{\frac{1}{2}}; \quad M_{X_2} = \underline{38.1 \text{ g mol}^{-1}}$$

Since $X_2(g)$ is diatomic, the atomic mass of X must be close to ½(38.1) = $\underline{19.0 \text{ g mol}^{-1}}$ which identifies X as **fluorine** (atomic mass 19.00 g mol^{-1}). The **unknown gas** is **F_2(g)**.

General Problems

60. (a) Calculate the average volume effectively occupied by each molecule in an ideal gas at 27°C and 1 atm pressure. (b) Assume the molecule to be spherical and to have a radius of 100 pm and calculate its actual volume. (c) Compare the actual volume with the volume effectively occupied by a molecule in an ideal gas.

Conveniently, we can consider a volume of 1 L of ideal gas at 1 atm and 27°C, calculate the number of moles of gas in this volume, then convert moles of gas to the number of molecules in 1 L, and calculate the volume effectively occupied by 1 molecule, in units of pm^3 (1 pm = 10^{-12} m):

$$n = \frac{PV}{RT} = \frac{(1 \text{ atm})(1 \text{ L})}{(0.0821 \text{ atm L mol}^{-1} \text{ K}^{-1})(300 \text{ K})} = \underline{0.0406 \text{ mol}}$$

and the number of molecules in 1 L is given by:

$$(0.0406 \text{ mol})(6.022 \times 10^{23} \text{ mol}^{-1}) = \underline{2.44 \times 10^{22} \text{ molecules}}$$

so that the volume, V_1, occupied by 1 molecule is:

$$V_1 = \frac{(1\ L)(\frac{1\ dm^3}{1\ L})(\frac{1\ m}{10\ dm})^3(\frac{10^{12}\ pm}{1\ m})^3}{2.44\ x\ 10^{22}\ molecules} = \underline{4.10\ x\ 10^{10}\ pm^3}$$

(b) For an actual molecule, radius 100 pm, the volume V_2 is given by:

$$V_2 = \frac{4}{3}\pi r^3 = \frac{4}{3}\pi(100\ pm)^3 = \underline{4.19\ x\ 10^6\ pm}$$

(c) Taking the ratio of V_1 to the actual volume V_2 we have:

$$\frac{V_1}{V_2} = \frac{4.10\ x\ 10^{10}\ pm^3}{4.19\ x\ 10^6\ pm^3} = \underline{9.78\ x\ 10^3}$$

The available volume per molecule is about 10 000 times the actual molecular volume.

---

61. By what factor does water expand when converted from liquid at 100°C to vapor at 100 °C at exactly 1 atm pressure, given that the density of water at 100°C is 0.96 g cm^{-3}?

Conveniently we can compare the volume of 1 mole of gaseous water at 100°C (373 K) and 1 atm to that of 1 mol of liquid water at the same temperature:

$$\underline{H_2O\ gas}:\ V_1 = \frac{nRT}{P} = \frac{(1\ mol)(0.0821\ atm\ L\ mol^{-1}\ K^{-1})(373\ K)}{1\ atm} = \underline{30.62\ L}$$

$$\underline{H_2O\ liquid}:\ V_2 = (1\ mol\ H_2O)(\frac{18.02\ g\ H_2O}{1\ mol\ H_2O})(\frac{1\ cm^3}{0.96\ g}) = \underline{18.8\ cm^3}$$

$$\underline{Expansion\ Factor} = (\frac{30.62\ L}{18.8\ cm^3})(\frac{1\ cm^3}{1\ mL})(\frac{10^3\ mL}{1\ L}) = \underline{1.63\ x\ 10^3}$$

---

63. Gases collected over water contain water vapor, $H_2O(g)$, the partial pressure (vapor pressure) of which depends only on the temperature. The pressure in a 250 mL collecting bulb filled with wet $Cl_2(g)$ is 754.3 mm Hg at 25.0°C and the pressure falls to 644.7 mm of Hg when the bulb is cooled to to -10.0°C, at which temperature the water vapor has all condensed to ice on the walls of the bulb. What is the vapor pressure of water at 25.0°C?

The pressure in the bulb at 25.0°C is the sum of the partial pressures of $Cl_2(g)$ and $H_2O(g)$. Since at -10°C the only gas in the bulb is $Cl_2(g)$, and its amount is unchanged, the pressure is now entirely due to $Cl_2(g)$ and we can calculate its partial pressure at 25°C from the pressure of 644.7 mm Hg at -10°C:

$$PV = nRT;\quad \frac{P_1}{T_1} = \frac{P_2}{T_2};\quad \frac{644.7\ mm\ Hg}{263.1\ K} = \frac{P_2}{298.1\ K};\quad P_2 = \underline{730.5\ mm\ Hg}$$

Thus, the partial pressure (vapor pressure) of water at 25°C is (754.3 - 730.5) = **23.5 mm Hg.**

---

64. Igniting ammonium dichromate, $(NH_4)_2Cr_2O_7(s)$, results in a spectacular "chemical volcano" due to its decomposition to give $Cr_2O_3(s)$, $N_2(g)$, and water. (a) Write the balanced equation for the reaction. If 2.000 g (7.933 x 10^{-3} mol) of ammonium dichromate is ignited in an evacuated 1.00-L flask, when the temperature of the flask is 150°C, what will be (b) the total pressure in the flask after reaction; (c) the partial pressures of $N_2(g)$ and $H_2O(g)$?

(a) $(NH_4)_2Cr_2O_7(s) \rightarrow Cr_2O_3(s) + N_2(g) + 4H_2O(g)$ <u>balanced</u>

(b) $(NH_4)_2Cr_2O_7(s) \rightarrow Cr_2O_3(s) + N_2(g) + 4H_2O(g)$

	$(NH_4)_2Cr_2O_7(s)$	$Cr_2O_3(s)$	$N_2(g)$	$4H_2O(g)$	
<u>initially</u>	7.93×10^{-3}	-	-	-	mol
<u>finally</u>	0	7.93×10^{-3}	7.93×10^{-3}	$4(7.93 \times 10^{-3})$	mol

So, after reaction the total moles of gases ($N_2(g)$ and $H_2O(g)$) is $5(7.93 \times 10^{-3})$ mol = <u>0.03965 mol</u>, the temperature is (150+273) = 423 K, and the volume is 1.00 L. Thus:

$$P = \frac{nRT}{V} = \frac{(0.03965\ \text{mol})(0.0821\ \text{atm L mol}^{-1}\ \text{K}^{-1})(423\ \text{K})}{1.00\ \text{L}} = \underline{1.377\ \text{atm}}$$

(c) Since one-fifth of the gas molecules are N_2 and four-fifths are H_2O molecules,

$$p_{N_2} = \frac{1}{5}(1.377\ \text{atm}) = \underline{0.2754\ \text{atm}}\ ;\ p_{H_2O} = \frac{4}{5}(1.377\ \text{atm}) = \underline{1.102\ \text{atm}}$$

The **total pressure = 1.38 atm** and the partial pressures are **$N_2(g)$ 0.275 atm; $H_2O(g)$ 1.10 atm.**

CHAPTER 3

Periodic Table

2. Locate each of the following elements in the periodic table by group and by period. Name the element and classify it as a metal, a semimetal (metalloid), or a nonmetal: (a) He (b) P (c) K (d) Ca (e) S (f) Br (g) Al (h) F

Element	Group	Period	Type
He, helium	VIII	1	nonmetal
P, phosphorus	V	3	nonmetal
K, potassium	I	4	metal
Ca, calcium	II	4	metal
S, sulfur	VI	3	nonmetal
Br, bromine	VII	4	nonmetal
Al, aluminum	III	3	metal
F, fluorine	VII	2	nonmetal

4. The element with atomic number 22 forms crystals that melt at 1668 °C to give a liquid that boils at 3313 °C. The crystals are hard, conduct heat and electricity, can be drawn into fine wires, and emit electrons when exposed to ultraviolet light. On the basis of these properties, classify the element as a metal or a nonmetal. Which element is it?

All of the properties given are those characteristic of a metal. In the periodic table, the element with $Z = 22$ is the metal **titanium**, Ti, the second member of the first series of transition metals in period 4, all of which are metals.

5. By referring to the periodic table, classify each of the following elements as a main group element or a transition metal. For the main group elements, indicate the group to which the element belongs and whether it is a metal or a nonmetal. (a) Se (b) P (c) Mn (d) Kr (e) W (f) Al (g) Pb.

	Element	Category	Group	Type
(a)	Se, selenium	main group	VI	nonmetal
(b)	P, phosphorus	main group	V	nonmetal
(c)	Mn, manganese	**transition**	-	metal
(d)	Kr, krypton	main group	VIII	nonmetal
(e)	W, tungsten	**transition**	-	metal
(f)	Al, aluminum	main group	III	metal
(g)	Pb, lead	main group	IV	metal

7. For each of the following elements, state whether it is a metal or a nonmetal. Indicate the group to which it belongs, and give its number of valence electrons, and principal valence: (a) Li (b) Mg (c) S (d) P (e) Br (f) Ne (g) As (h) Se (i) Cl (j) Ba

---

	Symbol	Element	Group	Number of Valence Electrons	Principal Valence	Type
(a)	Li	lithium	I	1	1	metal
(b)	Mg	magnesium	II	2	2	metal
(c)	S	sulfur	VI	6	$8 - 6 = 2$	nonmetal
(d)	P	phosphorus	V	5	$8 - 5 = 3$	nonmetal
(e)	Br	bromine	VII	7	$8 - 7 = 1$	nonmetal
(f)	Ne	neon	VIII	8	$8 - 8 = 0$	nonmetal
(g)	As	arsenic	V	5	$8 - 5 = 3$	nonmetal
(h)	Se	selenium	VI	6	$8 - 6 = 2$	nonmetal
(i)	Cl	chlorine	VII	7	$8 - 7 = 1$	nonmetal
(j)	Ba	barium	II	2	2	metal

---

9. Predict the empirical formulas of the chlorides formed by the elements in group III of the periodic table.

Chlorine in group VII has a valence of **1**, and all the elements of group III, **B, Al, Ga, In, and Tℓ**, have the common valence **3**, so all have the empirical formula **ACl_3**:

$$BCl_3 \quad AlCl_3 \quad GaCl_3 \quad InCl_3 \quad TlCl_3$$

---

11. Assign each of the following elements to its appropriate group in the periodic table; classify each as a metal or a nonmetal, and write the empirical formula of its hydride. (a) C (b) Ca (c) He (d) B (e) Cl (f) Li (g) O (h) F (i) P (j) Mg

---

	Element	Group	Valence	Type	Hydride
(a)	carbon	IV	4	nonmetal	CH_4
(b)	calcium	II	2	metal	CaH_2
(c)	helium	VIII	$8 - 8 = 0$	nonmetal	none
(d)	boron	III	3	nonmetal	BH_3
(e)	chlorine	VII	$8 - 7 = 1$	nonmetal	ClH
(f)	lithium	I	1	metal	LiH
(g)	oxygen	VI	$8 - 6 = 2$	nonmetal	OH_2
(h)	fluorine	VII	$8 - 7 = 1$	nonmetal	FH

(i)	phosphorus	V	$8 - 5 = 3$	nonmetal	PH_3
(j)	magnesium	II	2	metal	MgH_2

12. Complete and balance each of the following equations: (a) $Mg(s)+Br_2(l) \rightarrow$ (b) $Ca(s)+O_2(g) \rightarrow$ (c) $Na(s)+I_2(s) \rightarrow$ (d) $Mg(s)+N_2(g) \rightarrow$

(a) Mg (Group II) has valence 2 and Br (Group VII) has valence 1, so the product is $MgBr_2$:

$$Mg(s) + Br_2(\ell) \rightarrow MgBr_2(s) \quad \textit{magnesium bromide.}$$

(b) Ca (Group II) has valence 2 and O (group VI) has valence 2, so the product is Mg:

$$Ca(s) + O_2(g) \rightarrow MgO(s) \quad \textit{magnesium oxide}$$

(c) Na (Group I) has valence 1 and I (group VII) had valence 1, so the product is NaI:

$$2Na(s) + I_2(s) \rightarrow 2NaI(s) \quad \textit{sodium iodide.}$$

(d) Mg (Group II) has valence 2 and N (Group V) has valence 3, so the product is Mg_3N_2:

$$3Mg(s) + N_2(g) \rightarrow Mg_3N_2(s) \quad \textit{magnesium nitride}$$

14. Arrange the following elements in pairs in terms of the greatest similarity of chemical and physical properties, and justify your choice: Na, Mg, C, Cl, Ca, Si, K, F.

The greatest similarities are expected between pairs of elements in the same group. Thus, we first locate the position of each element in the periodic table:

Element	Na	Mg	C	Cl	Ca	Si	K	F
Group	I	II	IV	VII	II	IV	I	VII
Period	3	3	2	3	4	3	4	2

Thus we have: Group I: Na and K; Group II: Mg and Ca; Group IV: C and Si; Group VII: F and Cl, and we note that the members of each pair are in adjacent periods, giving them closely related chemical and physical properties.

The Shell Model of the Atom

16. (a) Which among the electron shells of an atom is its valence shell? (b) What is the common feature of the valence shells of elements in the same group of the periodic table? (c) Give the electron shell structures for atoms of each of the elements in period 2 and period 3.

(a) The valence shell of an atom is its **outermost shell** of electrons.

(b) All the atoms in a specific group have the same number of valence electrons, given by the **group number.**

(c) Shell Structures In **period 2**, each of the elements has **two** shells of electrons, the first is a **filled shell** with **2** electrons, and the second (outer shell) is the valence shell where the number of valence electrons is equal to the group number.

In **period 3**, each of the elements has **three** shells of electrons, the **first** and **second** are **filled shells** with **2** and **8** electrons, respectively, and the third (outer shell) is the valence shell where the number of valence electrons is equal to the group number. Thus, we have:

Period 2	Li	Be	B	C	N	O	F	Ne
	2.1	2.2	2.3	2.4	2.5	2.6	2.7	2.8
Period 3	Na	Mg	Al	Si	P	S	Cl	Ar
	2.8.1	2.8.2	2.8.3	2.8.4	2.8.5	2.8.6	2.8.7	2.8.8

---

18. Give the electron shell structures of atoms of each of the following elements, and give their core charges: (a) Li (b) Mg (c) S (d) Br

As in Problem 17, we first locate the element by group and period; the number of the period gives the number of shells and the group number gives the number of electrons in the outer (valence) shell:

(a) Li (Group I, Period 2) **2.1** (b) Mg (Group II, Period 3) **2.8.2**

(c) S (Group VI, Period 3) **2.8.6** (d) Br (Group VII, Period 4) **2.8.18.7.**

---

20. (a) Give the electron shell structures of each of the following: (i) N (ii) N^{3-} (iii) Be (iv) Be^{2+} (v) O (vi) O^{2-}; (b) Which of the ions in (a) are isoelectronic?

(a) We locate each element in the periodic table by group and period; the period gives the number of electron shells and the group gives the number of valence (outer) electrons in a neutral atom. Starting with a neutral atom; for negatively charged species (anions) we add electrons equal in number to the magnitude of the negative charge, and for positively charged species (cations) we subtract electrons equal in number to the magnitude of the positive charge.

Atom or Ion	Period	Group	Valence Electrons	Electron Shell Structure
N	2	5	5	2.5
N^{3-}	2	5	5 + 3 = 8	2.8
Be	2	2	2	2.2
Be^{2+}	2	2	2 − 2 = 0	2
O	2	6	6	2.6
O^{2-}	2	6	6 + 2 = 8	2.8

(b) Isoelectronic ions are ions with the same number of electrons (same electron configuration), in this case:

N^{3-} and O^{2-}

---

23. What is the core charge of each of the following atoms and ions? (a) C (b) Mg (c) Mg^{2+} (d) Si (e) O (f) O^{2-} (g) S^{2-} (h) Br.

The core charge of an atom or ion is equal to the atomic number Z minus the number of electrons in inner shells (all the electrons excepting those in the outer valence shell). Note that for an ion, the outer shell is often a filled shell. We deduce the shell configuration first (from the group and period) and then subtract the number of inner shell electrons from Z:

	Atom	Atomic Number Z	Period	Group	Shell Structure	Core Charge
(a)	C	6	2	IV	2.4	+6 — 2 = +4
(b)	Mg	12	3	II	2.8.2	+12 — 10 = +2
(c)	Mg^{2+}	12	3	II	2.8	+12 —2 = +10
(d)	Si	14	3	IV	2.8.4	+14 — 10 = +4
(e)	O	8	2	VI	2.6	+8 — 2 = +6
(f)	O^{2-}	8	2	VI	2.8	+8 — 2 = +6
(g)	S^{2-}	16	3	VI	2.8.8	+16 — 10 = +6
(h)	Br	35	4	VII	2.8.18.7	+35 — 28 = +7

Chemical Bonds and Lewis Structures

25. Draw Lewis (electron dot) symbols for atoms of each of the following elements: (a) K (b) Ca (c) B (d) Sn (e) Sb (f) Te (g) Br (h) Xe (i) As (j) Ge.

To write the Lewis symbol of an element, all we need to know is the group number, which gives us the number of valence electrons to be arranged singly or as pairs, as appropriate:

Element	(a) K	(b) Ca	(c) B	(d) Sn	(e) Sb	(f) Te	(g) Br	(h) Xe	(i) As	(j) Ge
Group	I	II	III	IV	V	VI	VII	VIII	V	IV
Symbol	K·	·Ca·	·B·	·Sn·	:Sb·	:Te·	:Br·	:Xe:	:As·	·Ge·

---

26. To attain a structure in which all of the electron shells are filled, how many electrons must be (a) *lost* by each of the atoms: Li, Al, Na, Ca, Mg, Rb, and Sr; (b) *gained* by each of the atoms O, H, Cl, N, P, S, F, and I?

(a) The number of valence electrons in the neutral atom is given by the group number; cations are formed by the loss of electrons, usually to give an empty valence shell

Atom	Li	Al	Na	Ca	Mg	Rb	Sr
Group	I	III	I	II	II	I	II
Electrons Lost	1	3	1	2	2	1	2
Lewis Symbol	Li^+	Al^{3+}	Na^+	Ca^{2+}	Mg^{2+}	Rb^+	Sr^{2+}

(b) The number of valence electrons in the neutral atom is given by the group number; anions are formed by the gain of electrons to fill the valence shell:

Atom	O	H	Cl	N	P	S	F	I
Group	VI	I	VII	V	V	VI	VII	VII
Electrons Gained	2	1	2	3	3	2	1	1
Lewis Symbol	$:\ddot{\underset{..}{O}}:^{2-}$	$:H^-$	$:\ddot{\underset{..}{Cl}}:^-$	$:\ddot{\underset{..}{N}}:^{3-}$	$:\ddot{\underset{..}{P}}:^{3-}$	$:\ddot{\underset{..}{S}}:^{2-}$	$:\ddot{\underset{..}{F}}:^-$	$:\ddot{\underset{..}{I}}:^-$

Note: In all cases except H, the valence shell is filled to give a total of 8 electrons (an **octet**); the elements in period 1 are unique in having valence shells that are complete with only 2 electrons (a **duet**).

---

28. What are the empirical formulas of the ionic solids composed of each of the following pairs of ions? (a) NH_4^+, S^{2-} (b) Fe^{3+}, O^{2-} (c) Cu^+, O^{2-} (d) Al^{3+}, Cl^-

---

In each case we combine the smallest numbers of cations and anions so that their charges balance to give a neutral compound. In general for the formula M_yA_z, where the cation is M^{n+} and the anion is A^{m-}, we have for the neutral compound: **xn = ym**

(a) For a neutral compound, we obviously need twice as many NH_4^+ ions as S^{2-} ions, to give $\mathbf{(NH_4)_2S}$.

(b) To balance the charges on Fe^{3+} and O^{2-}, we need two-thirds as many Fe^{3+} ions as O^{2-} ions, or two Fe^{3+} ions for every three O^{2-} ions, to give neutral $\mathbf{Fe_2O_3}$.

(c) Obviously the charges are balanced if we have twice as many Cu^+ ions as O^{2-} ions, so the formula is $\mathbf{Cu_2O}$.

(d) The charges balance if we have three Cl^- ions for every Al^{3+} ion; the formula is $\mathbf{AlCl_3}$.

---

29. Predict the empirical formula of the ionic compound formed by each of the following pairs of elements: (a) Li,S (b) Ba,O (c) Mg,Br (d) Na,H (e) Al,I.

We first deduce the charges on the respective ions, using the method outlined in Problem 27, and then use the method of Problem 28 to deduce the formula of the compound:

(a) Li is a metal from group I and S is a nonmetal from group VI; Li loses its single electron to give Li^+, and S gains 2 electrons to give S^{2-}:

$Li\cdot \rightarrow Li^+ + e^-$; $\cdot\dot{\underset{..}{S}}: + 2e^- \rightarrow :\ddot{\underset{..}{S}}:^{2-}$; $2Li^+ + S^{2-} \rightarrow Li_2S$; Ionic structure: $[Li^+]_2\ :\ddot{\underset{..}{S}}:^{2-}$

(b) Ba (group II) loses $2e^-$ to give Ba^{2+}, and O (group VI) gains $2e^-$ to give O^{2-}:

$\cdot Be\cdot \rightarrow Be^{2+} + 2e^-$; $\cdot\ddot{\underset{..}{O}}: + 2e^- \rightarrow :\ddot{\underset{..}{O}}:^{2-}$; $Be^{2+} + O^{2-} \rightarrow BeO$; Ionic structure: $Be^{2+}\ :\ddot{\underset{..}{O}}:^{2-}$

(c) Mg (group II) loses $2e^-$ to give Mg^{2+}, and Br (group VII) gains $1e^-$ to give Br^-:

$\cdot Mg\cdot \rightarrow Mg^{2+} + 2e^-$; $:\dot{\underset{..}{Br}}: + e^- \rightarrow :\ddot{\underset{..}{Br}}:^-$; $Mg^{2+} + 2Br^- \rightarrow MgBr_2$; Ionic structure: $Mg^{2+}\ [:\ddot{\underset{..}{Br}}:^-]_2$

(d) Na (group I) loses $1e^-$ to give Na^+, and H gains $1e^-$ to form the hydride ion, H^-:

$Na\cdot \rightarrow Na^+ + e^-$; $H\cdot + e^- \rightarrow :H^-$; $Na^+ + H^- \rightarrow NaH$; Ionic structure: $Na^+\ :H^-$

(e) Al (group III) loses $3e^-$ to give Al^{3+}, and I (group VII) gains $1e^-$ to give I^-:

$\cdot\dot{Al}\cdot \rightarrow Al^{3+} + 3e^-$; $:\dot{\underset{..}{I}}: + e^- \rightarrow :I:^-$; $Al^{3+} + 3I^- \rightarrow AlI_3$ Ionic structure: $Al^{3+}\ [:\ddot{\underset{..}{I}}:^-]_3$

---

31. Write the empirical formula of each of the following compounds, and give the balanced equation for its formation from its elements: (a) barium iodide (b) aluminum chloride (c) calcium oxide (d) sodium sulfide (e) aluminum oxide.

This is essentially the same kind of problem as Problems 28 to 30: Barium iodide is a compound of barium and iodine; aluminum chloride is a compound of Al and Cl; calcium oxide a compound of Ca and O; sodium sulfide a compound of Na and S, and aluminum oxide a compound of Al and O.

(a) Ba (group II) loses $2e^-$ to give Ba^{2+}; I (group VII) gains $1e^-$ to give I^-:

$\cdot Ba \cdot \rightarrow Ba^{2+} + 2e^-$; $:\ddot{I}: + e^- \rightarrow :\ddot{I}:^-$; $Ba^{2+} + 2I^- \rightarrow BaI_2$; **$Ba(s) + I_2(s) \rightarrow BaI_2(s)$**

(b) Al (group III) loses $3e^-$ to give Al^{3+}; Cl (group VII) gains $1e^-$ to give Cl^-:

$\cdot\dot{Al}\cdot \rightarrow Al^{3+} + 3e^-$; $:\dot{Cl}: + e^- \rightarrow :\ddot{Cl}:^-$; $Al^{3+} + 3Cl^- \rightarrow AlCl_3$; **$2Al(s) + 3Cl_2(s) \rightarrow 2AlCl_3(s)$**

(c) Ca (group II) loses $2e^-$ to give Ca^{2+}; O (group VI) gains $2e^-$ to give O^{2-}:

$\cdot Ca \cdot \rightarrow Ca^{2+} + 2e^-$; $\cdot\dot{O}: + 2e^- \rightarrow :\ddot{O}:^{2-}$; $Ca^{2+} + O^{2-} \rightarrow CaO$; **$2Ca(s) + O_2(s) \rightarrow 2CaO(s)$**

(d) Na (group I) loses $1e^-$ to give Na^+; S (group VI) gains $2e^-$ to give S^{2-}:

$Na\cdot \rightarrow Na^+ + e^-$; $\cdot\dot{S}: + 2e^- \rightarrow :\ddot{S}:^{2-}$; $2Na^+ + S^{2-} \rightarrow Na_2S$; **$2Na(s) + S(s) \rightarrow Na_2S(s)$**

(e) Al (group III) loses $3e^-$ to give Al^{3+}; O (group VI) gains $2e^-$ to give O^{2-}:

$\cdot\dot{Al}\cdot \rightarrow Al^{3+} + 3e^-$; $\cdot\dot{O}: + 2e^- \rightarrow :\ddot{O}:^{2-}$; $2Al^{3+} + 3O^{2-} \rightarrow Al_2O_3$; **$4Al(s) + 3O_2(s) \rightarrow 2Al_2O_3(s)$**

---

32. Select from the following as many pairs of elements as possible that would be expected to form *ionic* binary compounds (containing two elements), and give the empirical formulas of the compounds: (a) H (b) O (c) F (d) Mg (e) Al (f) Ca.

The possible compounds will all contain a positive ion (cation) and a negative ion (anion), so the strategy here is first to deduce the ions that the elements named can form:

Element	Group	Ion	Element	Group	Ion
(a) H	I	H^+ or H^-	(d) Mg	II	Mg^{2+}
(b) O	VI	O^{2-}	(e) Al	III	Al^{3+}
(c) F	VII	F^-	(f) Ca	II	Ca^{2+}

Thus, the binary compounds in question are those formed between each of the cations Mg^{2+}, Al^{3+}, and Ca^{2+}, and each of the anions H^-, O^{2-}, and F^-. (H^+, the proton, is excluded from the cations because it forms no simple ionic binary compounds but, rather, covalent compounds).

		IONIC FORMULA
$Mg^{2+} + 2H^- \rightarrow MgH_2$	or	$Mg^{2+}\ [:H^-]_2$
$Mg^{2+} + O^{2-} \rightarrow MgO$	or	$Mg^{2+}\ :\ddot{O}:^{2-}$
$Mg^{2+} + 2F^- \rightarrow MgF_2$	or	$Mg^{2+}\ [:\ddot{F}:^-]_2$
$Al^{3+} + 3H^- \rightarrow AlH_3$	or	$Al^{3+}\ [:H^-]_3$
$2Al^{3+} + 3O^{2-} \rightarrow Al_2O_3$	or	$[Al^{3+}]_2\ [:\ddot{O}:^{2-}]_3$
$Al^{3+} + 3F^- \rightarrow AlF_3$	or	$Al^{3+}\ [:\ddot{F}:^-]_2$
$Ca^{2+} + 2H^- \rightarrow CaH_2$	or	$Ca^{2+}\ [:H^-]_2$
$Ca^{2+} + O^{2-} \rightarrow CaO$	or	$Ca^{2+}\ :\ddot{O}:^{2-}$
$Ca^{2+} + 2F^- \rightarrow CaF_2$	or	$Ca^{2+}\ [:\ddot{F}:^-]$

---

34. Draw the Lewis structures for each of the following molecules, for which all of the bonds are single covalent bonds: (a) H_2 (b) HCl (c) PH_3 (d) SiF_4 (e) F_2O (f) Cl_2

All the molecules in this problem are composed of nonmetal atoms; hence, the bonds are all covalent. We are told that they are all single (one electron pair) covalent bonds. Thus, in each case, we write the Lewis symbols for the atoms that are bonded together and arrange them so that in the molecules all of the electrons that are initially unpaired (single) form shared pairs:

(a) H· + ·H → H:H or H-H

(b) H· + ·C̈l: → H:C̈l: or H-C̈l:

(c) 3 H· + ·Ṗ: → H:P:H (with H below P) or H-P-H (with H bonded below P)

(d) 4 :F̈· + ·Ṡi· → :F:Si:F: (with :F: above and below Si) or :F-Si-F: (with :F: bonded above and below Si)

(e) 2 :F̈· + ·Ö: → :F̈:Ö:F̈: or :F̈-Ö-F̈:

(f) :C̈l· + ·C̈l: → :C̈l:C̈l: or :C̈l-C̈l:

---

35. Draw the Lewis structure for each of the following molecules and ions: (a) H_2CO (b) P_2 (c) CN^- (d) H_2NNH_2 (e) HOOH

All the atoms in these molecular species are nonmetals, so all the bonds are covalent bonds. Thus, we start with the Lewis symbol for each participating atom and combine them to give a Lewis structure: (i) where all the unpaired valence electrons form pairs with valence electrons of other atoms to form electron pair bonds, (ii) that is consistent with the valence of each atom and (iii) as far as possible, where all the atoms other than H (which forms only one bond) obey the octet rule:

(a) 2 H· + ·Ċ· + ·Ö: → H_2C=Ö: (Two of the unpaired electrons on C are used to form two C-H bonds leaving the remaining two electrons to form a double C=O bond with with the two unpaired electrons on the O atom)

(b) :Ṗ· + ·Ṗ: → :P≡P: (Three unpaired electrons on each P atom combine to form a triple bond).

(c) Here we are dealing with an ion rather than a molecule. CN^- has one more electron than neutral CN, and it is useful to take this into account first by adding an extra electron to the carbon atom, to give C^- with 5 valence electrons (one electron pair and three unpaired electrons):

:Ċ·⁻ + ·Ṅ: → ⁻:C≡N:

Note: we might have added the extra electron to the N atom, to give ·N̈:⁻ with two electron pairs and two unshared electrons, but then the C atom could form only two bonds to N, and the resulting structure would not obey the octet rule.

(d) 2 H· + ·N̈· + ·N̈· + 2 ·H → H—N̈—N̈—H (each N also bonded to an H below)

(e) H· + ·Ö· + ·Ö· + ·H → H—Ö—Ö—H

37*. In which of the following molecules is the central atom an exception to the octet rule? (a) Cl_2O (b) $BeCl_2$ (c) $AlCl_3$ (d) PCl_3

The best strategy here is to draw the respective Lewis structures and then count up the electrons on the central atoms:

(a) :Cl· + ·O· + ·Cl: → :Cl—O—Cl: (b) :Cl· + ·Be· + ·Cl: → :Cl—Be—Cl:

(c) 3 :Cl· + ·Al· → :Cl—Al—Cl: with :Cl: on Al (d) 3 :Cl· + ·P· → :Cl—P—Cl: with :Cl: on P

The central atoms in **$BeCl_2$** and **$AlCl_3$** have less than an octet of electrons in their valence shells; **$BeCl_2$ and $AlCl_3$** are **exceptions** to the octet rule.

Electronegativity

39. Without reference to Figure 3.11, select the element of higher electronegativity in each of the following pairs: (a) F, Cl; (b) F, O (c) P, S (d) C, Si (e) O, P (f) Br, Se (g) P, Al.

We make use of the relationships that (i) electronegativity increases from left to right across any period, and (ii) decreases down any group (see Problem 38). Thus, if two elements are *in the same period*, that which lies to the right has the higher electronegativity, and if two elements are *in the same group*, that which is higher in the group has the higher electronegativity:

(a) F and Cl are both in Group VII, and F (Period 2) is above Cl (Period 3), so **F is the more electronegative**; (indeed, F has the highest electronegativity of any element).

(b) F and O are both in Period 2; F (Group VII)is to the right of O (Group VI), so **F is the more electronegative.**

(c) P and S are both in Period 3, S (Group VI) is to the right of P (Group V), so **S is the more electronegative.**

(d) C is above Si in Group IV, so **C is the more electronegative.**

(e) O is in Group VI in Period 2 and P is in Group V in Period 3; Because it is both to the right and above P in the periodic table, **O is the more electronegative.**

(f) Br and Se are both in Period 4, Br (Group VII) is to the right of Se (Group VI), so Br is the more electronegative.

(g) P and Al are both in Period 3, P (Group V) is to the right of Al (Group III), so **P is the more electronegative.**

Ionic, Polar Covalent, and Covalent Bonds

40. Classify the bonds in each of the following as ionic, covalent, or or polar covalent: (a) Cl_2 (b) PCl_3 (c) LiCl (d) ClF (e) $MgCl_2$.

When two identical atoms (with the same electronegativity, χ) are bonded together, the electron pair is exactly equally shared by the two atoms and the bond is nonpolar covalent; when the two atoms have different χ values, the atom with the higher χ has a greater share of the electron pair than the atom with the smaller χ and the bond may be polar covalent or ionic. The latter is the situation in bond formation when one element is a metal and the other is a nonmetal:

Compound	Bond Type	Explanation
(a) **Cl_2**	nonpolar covalent	bond between *identical* atoms
(b) **PCl_3** and (d) **ClF**	polar covalent	bonds between *nonmetals* of different electronegativities
(c) **LiCl** and (e) **$MgCl_2$**	ionic	*metal-nonmetal* bonds

---

43. Name the compounds with each of the following formulas. State which are ionic and which are covalent. For the covalent substances, draw the Lewis structure and indicate the polarity of the bonds by writing δ+ and δ- on the appropriate atoms: (a) $MgCl_2$ (b) SCl_2 (c) PCl_3 (d) HF (e) OCl_2 (f) CS_2 (g) LiH (h) NF_3

We first decide the bond types according to the criteria given in Problem 40; then we write the Lewis structures and, finally, for the polar bonds we assign δ+ to the less electronegative, and δ- to the more electronegative of the atoms forming the bonds:

Formula	Name	Bond Type	Lewis Structure
(a) $MgCl_2$	magnesium chloride	ionic	Mg^{2+} [:C̈l:⁻]$_2$
(b) SCl_2	sulfur dichloride	polar covalent	$^{\delta-}$:C̈l—S̈$^{2\delta+}$—C̈l:$^{\delta-}$
(c) PCl_3	phosphorus trichloride	polar covalent	$^{\delta-}$:C̈l—P̈$^{3\delta+}$—C̈l:$^{\delta-}$ with :C̈l:$^{\delta-}$ below P
(d) HF	hydrogen fluoride	polar covalent	$^{\delta+}$H—F̈:$^{\delta-}$
(e) OCl_2	oxygen dichloride	polar covalent	$^{\delta+}$:C̈l—Ö$^{2\delta-}$—C̈l:$^{\delta+}$
(f) CS_2	carbon disulfide	polar covalent	$^{\delta+}$:S̈=C$^{2\delta-}$=S̈:$^{\delta+}$
(g) LiH	lithium hydride	ionic	Li^+ :H^-
(h) NF_3	nitrogen trifluoride	polar covalent	$^{\delta-}$:F̈—N̈$^{3\delta+}$—F̈:$^{\delta-}$ with :F̈:$^{\delta-}$ below N

---

45. Explain why sodium chloride forms a solid ionic compound at room temperature, rather than a gas consisting of Na^+Cl^- molecules (ion pairs).

Sodium chloride is a solid ionic compound because it is an aggregate of a large number of ions in which each Na^+ ion attracts as many Cl^- ions as possible to surround it, and vice-versa, so that in an NaCl crystal, each Na^+ ion is surrounded by an octahedral arrangement of six Cl^- ions, and each Cl^- ion is surrounded by an octahedral arrangement of six Na^+ ions. At room temperature, NaCl(s) consists of equal, very large, numbers of Na^+ and Cl^- ions that strongly attract each other by electrostatic forces and are arranged in an infinite lattice. It takes a large amount of energy to pull the lattice apart, so that NaCl(s) melts at a high temperature and boils at an even higher temperature to give NaCl vapor. In the vapor, the NaCl molecules are best regarded as highly polar covalent molecules.

VSEPR Model and Molecular Geometry

47. Draw diagrams to illustrate the geometry of each of the following types of molecule. In each case name the geometric shape: (a) AX_3E (b) AX_2E_2 (c) AXE_3 (d) AX_3 (e) AX_2.

Remember that in an AX_nE_m molecule, it is the arrangement of all of the n+m electron pairs in the valence shell of the central atom, A, that determines the shape, but the geometry describes the relative positions of the central atom A and the X atoms:

Molecule Type	Arrangement of Electron Pairs	Molecular Shape	Name of Molecular Shape
(a) AX_3E	tetrahedral	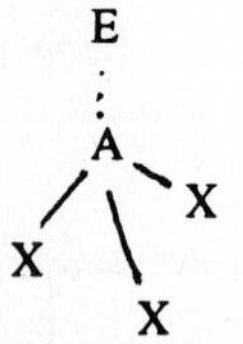 E ⋮ A X X X	triangular pyramid
(b) AX_2E_2	tetrahedral	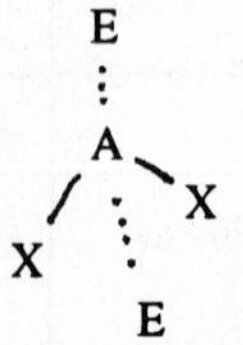 E ⋮ A X X E	angular
(c) AXE_3	tetrahedral	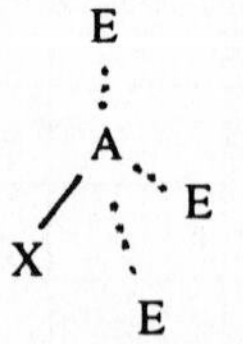 E ⋮ A X E E	linear (diatomic)
(d) AX_3	triangular planar	X | A X X	triangular planar
(e) AX_2	linear	X—A—X	linear

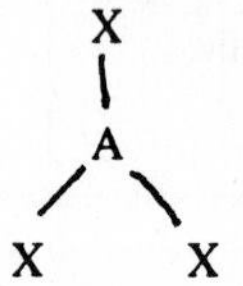

---

49. Deduce to which AX_nE_m type each of the following molecules belongs, and hence predict the shape: (a) H_2O (b) PCl_3 (c) BCl_3 (d) SiH_4.

We first write the Lewis structure for each molecule, which allows us to categorize it in the AX_nE_m nomenclature, where n is the number of bonding pairs and m is the number of unshared (lone) pairs in the valence shell of the central atom A. The arrangement of the central atom A and the n X atoms gives the molecular shape:

	Molecule	Lewis Structure	Type	Shape
(a)	H_2O	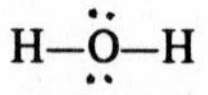 H—Ö—H	AX_2E_2	angular
(b)	PCl_3	:C̈l—P̈—C̈l: | :C̈l:	AX_3E	triangular pyramid
(c)	BF_3	:F̈—B—F̈: | :F̈:	AX_3	triangular planar

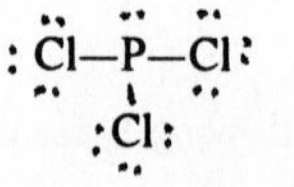

(d)	SiH_4	H–Si–H (with H above and below Si)	AX_4	tetrahedral

52. Use the electron pair domain model to predict the geometric shape of each of the following molecules: (a) CF_4 (b) CO_2 (c) H_2CO (d) ClCN (e) F_2CCF_2

The electron pair domain model is an extension of the simple VSEPR model to include double and triple bonds to a central atom A, as well as single bonds and lone pairs. To apply it to any molecule, we first write its Lewis structure and then classify it according to the usual AX_nE_m nomenclature, which gives the shape:

	Molecule	Lewis Structure	Bond Domain Description	Shape
(a)	CF_4	:F̤–C–F̤: with :F̤: above and below C	4 single (AX_4)	tetrahedral
(b)	CO_2	:Ö=C=Ö:	2 double (AX_2)	linear
(c)	H_2CO	H–C–H with Ö: double-bonded below C	2 single & 1 double (AX_3)	planar triangular
(d)	ClCN	:C̤̈l–C≡N:	1 single & 1 triple (AX_2)	linear
(e)	F_2CCF_2	:F̤–C=C–F̤: with :F̤: below each C	2 single & 1 double (AX_3)	planar triangular*

* at each C atom, and planar overall because there is no rotation about the double bond.

General Problems

54.* Draw a cube and connect the four appropriate corners to form a tetrahedron. Join the midpoint of the cube (the midpoint of the tetrahedron) to each of the four corners of the tetrahedron. These lines represent the bonds in a tetrahedral AX_4 molecule, such as CH_4. Using trigonometry, calculate the angle between the bonds.

A tetrahedron is constructed inside a cube by first drawing two *face-diagonals*, such as **AC** and **HF**. **ACHF** then defines the tetrahedron. **O** is the center of the cube (and tetrahedron). Joining **A**, **C**, **H**, and **F** to **O** gives the four bond directions of a tetrahedral molecule. The required **tetrahedral angle** is, for example, **<AOC**.

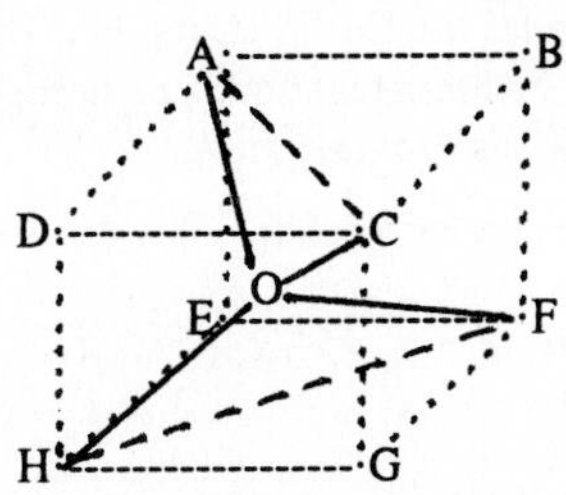

If we give the cube an edge of length **2a**, for the triangle AOC, AC is a face diagonal of the cube, and the distance OX is one-half of the length of the edge of the cube, ½(2a) = a. For the face-diagonal AC, we have from the Pythagoras theorem:

$$AC^2 = AD_2 + DC^2 = (2a)^2 + (2a)^2 = 8a^2$$

$$AC = \sqrt{8a^2} = \underline{2\sqrt{2}a}$$

and $$AX = ½(AC) = \underline{\sqrt{2}a}$$

Thus, $$\tan \Theta = AX/AO = \sqrt{2}a/a = \sqrt{2} = \underline{1.414}$$

$$\Theta = 54° 44'$$

$$<AOC = 2\Theta = \underline{109° 28'}$$

56.* Indium oxide contains 82.7 mass percent indium. It occurs naturally in ores containing zinc oxide, ZnO(s), and it was therefore originally assumed to have the empirical formula InO. On this basis, calculate the atomic mass of indium, and predict its location in the periodic table. Explain why this location is unsuitable for indium. Mendeleev suggested that the empirical formula of the oxide must be In_2O_3. Calculate the atomic mass of indium on this basis, and find its location in the periodic table. Is this location reasonable?

From the data given, indium oxide contains 82.7 mass % In and 17.4 mass % O, and we can determine the atomic mass of indium from any given empirical formula for the oxide by first calculating the number of moles of O in a given sample, which conveniently we take to be 100 g:

$$\text{100 g indium oxide contains 17.3 g O} = (17.3 \text{ g O})(\frac{1 \text{ mol O}}{16.00 \text{ g O}}) = \underline{1.08 \text{ mol O}}$$

Thus, for empirical formula InO: mol indium = mol O = $\underline{1.08 \text{ mol}}$

$$\text{Molar mass of indium} = (\frac{82.7 \text{ g indium}}{1.08 \text{ mol indium}}) = \underline{76.6 \text{ g mol}^{-1}}$$

An atomic mass of this magnitude would place indium between As and Se in Period 4, and presumably in either Group V or Group VI. The empirical formula InO for the oxide would not be suitable for a Group V element (valence 3), but would be suitable for a Group VI element (valence 2). However, In would then have to be a <u>nonmetal</u>, like Se, and its oxide InO would be covalent rather than ionic and would not resemble the metallic oxide ZnO. The other problem would be having to relocate Se elsewhere in the periodic table!! On the basis of the empirical formula In_2O_3:

$$\text{100.0 g oxide contains:} \quad (1.08 \text{ mol O})(\frac{2 \text{ mol indium}}{3 \text{ mol O}}) = \underline{0.720 \text{ mol indium}}$$

$$\text{On this basis:} \quad \text{atomic mass of indium} = (\frac{82.7 \text{ g indium}}{0.720 \text{ mol indium}}) = \underline{115 \text{ g mol}^{-1}}$$

An atomic mass of this magnitude places indium in Period 5 and Group III, between cadmium, Cd, the last of the transition metals of this period, and the metal tin, Sn, in Group IV. This position in the periodic table makes In a Group III metal with the common valence 3, consistent with an empirical formula In_2O_3 for its <u>ionic</u> oxide, and the resemblance of the oxide to ZnO, another ionic oxide.

CHAPTER 4

Chemical Reactions

1. Briefly explain three ways by which the rate of a reaction may be increased.

For molecules to react, they must collide with sufficient energy and with the correct relative orientation. The collision rate, and thus the rate of reaction, increases with (1) increasing partial pressures or concentrations of reactants and (2) increasing temperature. Another way to increase the rate of a reaction is (3) to introduce a catalyst, which lowers the energy required for the molecules to react and/or orientates molecules so that it is easier for them to react.

The Halogens

3.(a) List the halogens in order of increasing atomic number. (b) Describe the physical states and appearances of the halogens. (c) Explain why their melting points and boiling points increase, and their electronegativities decrease, with increasing atomic number.

(a) Fluorine < chlorine < bromine < iodine < astatine

(b) All occur as covalent diatomic, X_2, molecules: F_2 as a greenish-yellow gas; Cl_2 as a green gas; Br_2 as a red-brown liquid, and I_2 as a dark purple solid.

(c) Melting points and boiling points are a manifestation of the strength of intermolecular forces between the nonpolar X-X halogen molecules. These increase as the size of the halogen atoms increase, in going down Group VII, with addition of successive filled shells of electrons. Electronegativity (the relative power of a halogen to attract the electrons of a covalent bond) also increases in the same order. All the elements in Group VII have the same core charge of +7 but the valence electrons become increasingly distant from the nucleus, and are thus less strongly attracted to it, as the atom increases in size.

6. Classify the bonds in each of the following as ionic, covalent, or polar covalent, Draw the Lewis structures for (c), (d), (f), and (i): (a) NaCl (b) CaF_2 (c) OF_2 (d) PCl_3 (e) $MgBr_2$ (f) CCl_4 (g) HF (h) KI (i) ClF

We first use the respective Lewis symbols of the elements and combine them to form the requisite number of bonds. Ionic and covalent compounds are distinguished on the basis that bonds between a metal and a nonmetal are ionic, whereas those between two nonmetals are covalent (pure covalent, *nonpolar*, when the bonding pairs are equally shared between atoms of the same electronegativity, and polar covalent, *polar*, when the atoms have different electronegativities, so that their bonding pairs are unequally shared):

	Lewis Symbols	Combination	Structure	Bond Type
(a)	Na· + ·Cl: → Na:Cl:	metal-nonmetal	Na^+ :Cl:$^-$	IONIC
(b)	·Ca· + 2 ·F: → :F:Ca:F:	metal-nonmetal	Ca^{2+} [:F:$^-$]$_2$	IONIC
(c)	:O· + 2 ·F: → :F:O:F:	two nonmetals	:F-O-F:	POLAR COVALENT
(d)	:P· + 3 ·Cl: → :Cl:P:Cl: (:Cl: below P)	two nonmetals	:Cl—P—Cl: (:Cl: below P)	POLAR COVALENT
(e)	·Mg· + 2 ·Br: → :Br:Mg:Br:	metal-nonmetal	Mg^{2+} [:Br:$^-$]$_2$	IONIC

(f) ·C· + 4 ·Cl: → :Cl:C:Cl: (with :Cl: above and below C)	two nonmetals	:Cl—C—Cl: (with :Cl: above and below C)	POLAR COVALENT
(g) H· + ·F: → H:F:	two nonmetals	H—F:	POLAR COVALENT
(h) K· + ·I: → K:I:	metal-nonmetal	K^+ :I:$^-$	IONIC
(i) :Cl· + ·F: → :Cl:F:	two nonmetals	:Cl-F:	POLAR COVALENT

---

8. Write balanced equations for each of the following reactions. Name each of the products. (a) barium with chlorine, (b) aluminum with bromine, (c) potassium with iodine, (d) phosphorus with chlorine, (e) phosphorus with iodine

---

The reactions between metals and halogens (nonmetals) give ionic halides, and those between nonmetals and halogens give covalent halides:

(a) Barium is in group II: $Ba(s) + Cl_2(g) \rightarrow BaCl_2(s)$ <u>barium chloride</u>

(b) Aluminum is in group III: $2Al(s) + 3Br_2(g) \rightarrow 2AlBr_3(\ell)$ <u>aluminum bromide</u>

(c) Potassium is in group I: $2K(s) + I_2(s) \rightarrow 2KI(s)$ <u>potassium iodide</u>

(d) Phosphorus is in group V: $2P(s) + 3Cl_2(g) \rightarrow 2PCl_3(\ell)$ <u>phosphorus trichloride</u>

<u>and</u> $2P(s) + 5Cl_2(g) \rightarrow 2PCl_5(s)$ <u>phosphorus pentachloride</u>

(e) Phosphorus is in group V: $2P(s) + 3I_2(g) \rightarrow 2PI_3(s)$ <u>phosphorus triiodide</u> (note: PI_5 is unknown).

Oxidation-Reduction (Redox) Reactions

11. The reaction between potassium and bromine to give potassium bromide is classified as an oxidation-reduction reaction. (a) Write the balanced equation for this reaction. (b) Show that oxidation-reduction involves the transfer of electrons between the reactants. (c) In terms of the transfer of electrons, define the terms "oxidation," "reduction," "oxidizing agent," and "reducing "agent."

(a)

$$2K(s) + Br_2(\ell) \rightarrow 2KBr(s)$$

(b) The product of the reaction, KBr(s), is an ionic compound composed of potassium ions, K^+, and bromide ions, Br^-. In the reaction, potassium atoms lose electrons and transfer them to Br_2 molecules to give potassium ions, K^+, and bromide ions, Br^-, which we can write as the sum of the two half-reactions:

$2K \rightarrow 2K^+ + 2e^-$	oxidation
$Br_2 + 2e^- \rightarrow 2Br^-$	reduction
$2K + Br_2 \rightarrow 2K^+Br^-$	overall

(c) As exemplified in part (b): **oxidation** is a process in which a substance <u>loses</u> electrons; **reduction** is a process in which a substance <u>gains</u> electrons; an **oxidizing reagent** is a reactant that causes oxidation by accepting electrons from the substance that is oxidized, e.g., <u>bromine</u> in the above reaction, and a **reducing agent** is a reactant that causes reduction by donating electrons to the reactant that is reduced, e.g., <u>potassium</u> in the above reaction.

---

13. How are (a) chlorine and (b) fluorine prepared industrially?

(a) **Chlorine** is prepared by the electrolysis of <u>aqueous</u> (or molten) sodium chloride, NaCl(aq):

$$2Cl^-(aq) \rightarrow Cl_2(g) + 2e^-$$

The chlorine produced does not react extensively with water, and bubbles off; the other product is sodium hydroxide.

(b) **Fluorine** cannot be prepared in an analogous way by the electrolysis of NaF(aq) because fluorine reacts vigorously with water. It is prepared by the electrolysis of <u>molten</u> sodium fluoride

$$2F^-(\ell) \rightarrow F_2(g) + 2e^-$$

where the other product of the reaction is sodium metal, from the reduction of $Na^+(\ell)$ ions of molten NaF.

---

15. Complete and balance each of the following equations. If no reaction occurs, write NR. Identify the oxidizing agent and the reducing agent in each reaction: (a) $Br_2(l) + NaCl(aq) \rightarrow$ (b) $NaCl(s) + F_2(g) \rightarrow$ (c) $I_2(s) + NaF(aq) \rightarrow$ (d) $CaBr_2(s) + F_2(g) \rightarrow$ (e) $KF(aq) + Br_2(l) \rightarrow$ (f) $MgI_2(aq) + Cl_2(g) \rightarrow$

Whether a reaction occurs or not (**NR**) depends on whether the halogen used can oxidize the halide ion in question. A particular halogen, X_2, can oxidize any of the halide ions, Y^-, below it in the periodic table (but not those above):

$$2Y^- + X_2 \rightarrow Y_2 + 2X^-$$

In each case, the halogen X_2 is the <u>oxidizing agent</u> (electron acceptor), **O**, and the source of Y^-, the metal halide, is the <u>reducing agent</u> (electron donor), **R**:

(a) $Br_2(l) + NaCl(aq) \rightarrow$ **NR**, (because Cl is above Br in Group VII)

(b) $2NaCl(s) + F_2(g) \rightarrow Cl_2(g) + 2NaF(s)$
R O

(c) $I_2(s) + NaF(aq) \rightarrow$ **NR**, (because F is above I in Group VII)

(d) $CaBr_2(s) + F_2(g) \rightarrow Br_2(\ell) + CaF_2(s)$
R O

(e) $KF(aq) + Br_2(l) \rightarrow$ **NR**, (because F is above Br in Group VII)

(f) $MgI_2(aq) + Cl_2(g) \rightarrow I_2(s) + MgCl_2(aq)$*
R O

*$I_2(s)$ dissolves due to the reaction $I_2(s) + I^-(aq) \rightarrow I_3^-(aq)$

---

17. Complete and balance each of the following equations. If no reaction occurs, write NR: (a) $Cl_2(g) + KI(aq) \rightarrow$ (b) $I_2(s) + NaCl(aq) \rightarrow$ (c) $Br_2(l) + NaI(aq) \rightarrow$ (d) $F_2(g) + H_2O(l) \rightarrow$

This problem is another variation on problems 15 and 16; part (d) illustrates the fact that fluorine is such a powerful oxidizing agent that it oxidizes water to $O_2(g)$:

(a) $Cl_2(g) + 3KI(aq) \rightarrow 2KCl(aq) + KI_3(aq)$

(b) $I_2(s) + NaCl(aq) \rightarrow$ **NR**

(c) $Br_2(l) + 2NaI(aq) \rightarrow 2NaBr(aq) + I_2(s)$; $I_2(s) + NaI(aq) \rightarrow NaI_3(aq)$

(d) $F_2(g) + 2H_2O(l) \rightarrow 4HF(aq) + O_2(g)$

---

19. For each of the following reactions, identify the oxidizing agent, the reducing agent, the molecule or ion that is oxidized, and the molecule or ion that is reduced: (a) $2Rb(s) + I_2(s) \rightarrow 2RbI(s)$, (b) $4Al(s) + 3O_2(g) \rightarrow 2(Al^{3+})_2(O^{2-})_3(s)$ (c) $Zn(s) + S(s) \rightarrow Zn^{2+}S^{2-}(s)$ (d) $Mg(s) + 2HCl(aq) \rightarrow MgCl_2(aq) + H_2(g)$

The oxidizing reagent (**O**) is the electron acceptor, which is also the substance reduced, and the reducing reagent (**R**) is the electron donor, which is also the substance oxidized.

(a) $2Rb(s) + I_2(s) \rightarrow 2RbI(s)$
R O

(b) $4Al(s) + 3O_2(g) \rightarrow 2(Al^{3+})_2(O^{2-})_3(s)$
R O

(c) $Zn(s) + S(s) \rightarrow Zn^{2+}S^{2-}(s)$
R O

(d) $Mg(s) + 2HCl(aq) \rightarrow MgCl_2(aq) + H_2(g)$
R O

Acid-Base Reactions

21. (a) In terms of proton transfer explain why the reaction between HBr(g) and $NH_3(g)$ to give the salt $NH_4Br(s)$ is described as an acid-base reaction. (b) Explain why the reaction between magnesium and an aqueous solution of HCl to give $MgCl_2(aq)$ and hydrogen is not an acid-base reaction.

(a) In terms of proton transfer, as expressed in the Bronsted-Lowry concept of acids and bases, an **acid** is a proton donor and a **base** is a proton acceptor. In the reaction $HBr(g) + :NH_3(g) \rightarrow NH_4^+Br^-(s)$, HBr(g) is a proton donor, and thus an acid, while $NH_3(g)$ is a proton acceptor, and thus a base, utilizing its lone pair of electrons to form another N-H bond and thus the NH_4^+ ion.

(b) In the reaction of a metal with an aqueous acid, the metal atoms lose electrons to give metal ions, and the hydronium ion, $H_3O^+(aq)$, accepts an electron and decomposes to water and hydrogen gas:

$$Mg(s) \rightarrow Mg^{2+}(aq) + 2e^-$$

$$2H_3O^+ + 2e^- \rightarrow [2H_3O(aq)] \rightarrow 2H_2O(\ell) + H_2(g)$$

$$Mg(s) + 2H_3O^+(aq) \rightarrow Mg^{2+}(aq) + 2H_2O(\ell) + H_2(g)$$

Thus, the reaction of magnesium with an aqueous acid, to give hydrogen gas is an electron transfer reaction, rather than a proton transfer reaction, and is thus an **oxidation-reduction** reaction rather than an acid-base reaction.

---

23. Write balanced equations to show how each of the following bases is ionized in aqueous solution. Classify each base as a strong base or a weak base: (a) sodium oxide (b) potassium hydroxide (c) ammonia (d) lithium hydride

---

It is useful in any problem concerned with reactions to first consider the nature of the reactants; here, one reactant is water:

(a) **Sodium oxide,** $Na_2O(s)$, is an ionic compound, $(Na^+)_2\ O^{2-}(s)$. In solution, Na^+ is a spectator ion, and the oxide ion, O^{2-}, behaves as a strong base, $O^{2-}(aq) + H_2O(\ell) \rightarrow 2OH^-(aq)$, so the reaction is:

$$Na_2O(s) + H_2O(\ell) \rightarrow 2Na^+(aq) + 2OH^-(aq) \qquad \textbf{strong base}$$

(b) **Potassium hydroxide,** KOH(s), is ionic, $K^+\ OH^-(s)$, so in aqueous solution it simply dissociates quantitatively into its constituent ions:

$$KOH(s) \rightarrow K^+(aq) + OH^-(aq) \qquad \textbf{strong base}$$

(c) **Ammonia,** $:NH_3$, is covalent with N-H bonds insufficiently polar to be acidic in water, but it has an unshared pair of electrons on the nitrogen atom which can accept a proton from water, so NH_3 behaves as a weak base:

$$OH_2(\ell) + :NH_3(aq) \rightleftarrows OH^-(aq) + NH_4^+(aq) \qquad \textbf{weak base}$$

(d) **Lithium hydride,** LiH(s), is ionic, $Li^+:H^-(s)$. In solution in water, Li^+ is a spectator ion but $:H^-$ accepts a proton from water and behaves as a strong base: $:H^-(aq) + H_2O(\ell) \rightarrow H_2(g) + OH^-(aq)$, to give $H_2(g)$ that bubbles off, so the reaction is:

$$LiH(s) + H_2O(\ell) \rightarrow H_2(g) + OH^-(aq) + Li^+(aq) \qquad \textbf{strong base}$$

---

25. Write the balanced equation for a suitable acid-base reaction for the preparation of each of the following salts: (a) calcium sulfate (b) lithium fluoride (c) ammonium nitrate (d) magnesium perchlorate

---

Note: Salts are ionic compounds composed of cations and anions. They result from the reaction of an acid with a base, with the acid supplying the appropriate anion and the base supplying the appropriate cation. Acid-base reactions are often carried out in aqueous solutions. If the salt is insoluble, it may be simply filtered from the solution, washed, and dried. If it is soluble, then the solution has to be concentrated by evaporation and the salt allowed to separate on cooling. Thus, the strategy here is to first write the formula of the salt in its ionic form, which enables the appropriate base and acid to be determined:

(a) **Calcium sulfate,** $CaSO_4$, contains the cation Ca^{2+} and the anion SO_4^{2-} (derived from the acid H_2SO_4, sulfuric acid). An appropriate base would be $Ca(OH)_2$(s) or CaO(s), and suitable acid-base reactions are:

$$Ca(OH)_2(s) + H_2SO_4(\ell) \rightarrow CaSO_4(s) + 2H_2O(\ell)$$

$$CaO(s) + H_2SO_4(\ell) \rightarrow CaSO_4(s) + H_2O(\ell)$$

(b) **Lithium fluoride,** LiF(s), contains the cation Li^+ and the anion F^- (derived from the acid HF, hydrofluoric acid). An appropriate base would be LiOH(s) or Li_2O(s), and suitable acid-base reactions are:

$$LiOH(s) + HF(aq) \rightarrow LiF(aq) + H_2O(\ell)$$

$$Li_2O(s) + HF(aq) \rightarrow 2LiF(aq) + H_2O(\ell)$$

(c) **Ammonium nitrate,** NH_4NO_3, contains NH_4^+ (derived from the base ammonia, NH_3), and NO_3^- (derived from nitric acid, HNO_3). The appropriate reaction is:

$$NH_3(aq) + HNO_3(aq) \rightarrow NH_4^+(aq) + NO^{3-}(aq).$$

(d) **Magnesium perchlorate** is $Mg^{2+}[ClO_4^-]_2$, and appropriate reactions are:

$$Mg(OH)_2(s) + 2HClO_4(aq) \rightarrow Mg(ClO_4)_2(aq) + 2H_2O(\ell)$$

$$MgO(s) + 2HClO_4(aq) \rightarrow Mg(ClO_4)_2(aq) + H_2O(\ell)$$

---

26. (a) Classify each of the following as a strong acid, a weak acid, or as a molecule or ion with no acidic properties in aqueous solution: (i) HCl (ii) HF (iii) HNO_3 (iv) CH_4 (v) NH_4^+. (b) Classify each of the following as a strong base, a weak base, or a molecule or ion with negligible basicity, in aqueous solution: (i) Cl^- (ii) F^- (iii) NO_3^- (iv) NH_3 (v) O^{2-}.

(a) (i) HCl(aq), **strong acid**; (ii) HF(aq), **weak acid**; (iii) HNO_3(aq), **strong acid**; (iv) CH_4(g), **no basic or acidic properties**; (v) NH_4^+(aq), **weak acid** (because its conjugate base NH_3(aq) is a weak base).

(b) (i) Cl^-(aq), **negligible basicity** (because its conjugate acid HCl(aq) is a strong acid); (ii) F^-(aq), **weak base** (because its conjugate acid HF(aq) is a weak acid); (iii) NO_3^-(aq), **negligible basicity** (because its conjugate acid HNO_3(aq) is a strong acid; (iv) NH_3(aq), **weak base**, (v) O^{2-}(aq), **strong base**.

28.* (a) What is the simplest equation that represents the reaction between a solution of an acid, such as HCl(aq), and a solution of a strong base, such as NaOH(aq)? Explain. (b) When an aqueous solution of a strong base, such as NaOH(aq), is added to a solution of a weak acid, such as HF(aq), why does the reaction go to completion to form a solution of a salt?

(a) All acids, for example HCl(aq) or HNO_3(aq), transfer a proton to water to give the hydronium ion, H_3O^+(aq), and all bases, such as NaOH(aq) or $Ca(OH)_2$(aq), give hydroxide ion, OH^-(aq), in solution. Thus, in aqueous solution all acid-base reactions are simply the reaction:

$$H_3O^+(aq) + OH^-(aq) \rightarrow 2H_2O(\ell)$$

Any other ions in solution, such as Cl^-(aq), NO_3^-(aq), Na^+(aq), or Ca^{2+}(aq), behave simply as *spectator ions*.

(b) A weak acid such as HF(aq) is only partially ionized

$$HF(aq) + H_2O(\ell) \rightleftarrows H_3O^+(aq) + F^-(aq)$$

but as $Na^+OH^-(aq)$ is added the $OH^-(aq)$ formed reacts with the $H_3O^+(aq)$ from the acid to form more water and removes it. More HF(aq) then ionizes to replace the $H_3O^+(aq)$ removed, and eventually all the HF(aq) is converted to $F^-(aq)$, and all the $H_3O^+(aq)$ resulting from its ionization is neutralized, leaving only the $Na^+(aq)$ and $F^-(aq)$ ions in solution; in other words, a solution of NaF(aq):

$$Na^+(aq) + OH^-(aq) + HF(aq) \rightarrow Na^+(aq) + F^-(aq) + H_2O(\ell),$$

so the reaction goes to completion.

Polyatomic Ions: Lewis Structures

30.* Draw the Lewis structures of each of the following ions. (The first atom in the formula is the central atom in each case): (a) NH_2^- (b) H_2F^+ (c) PH_4^+ (d) BF_4^-

Ion	(a) NH_2^-	(b) H_2F^+	(c) PH_4^+	(d) BF_4^-
Lewis Structure	H—N⁻—H (two lone pairs on N)	H—F⁺—H (two lone pairs on F)	H—P⁺—H, with H above and below P	:F—B⁻—F:, with :F: above and below B (three lone pairs on each F)

*In each case, the formal charge is on the unique central atom.

Precipitation Reactions

32. (a) Write the simplest equation that describes the reaction between $AgNO_3(aq)$ and NaBr(aq) to give a precipitate of AgBr(s). (b) explain what is meant by the term "spectator ion".

(a) $AgNO_3(aq)$ is ionized in water to give $Ag^+(aq)$ and $NO_3^-(aq)$, and NaBr(aq) is ionized to give $Na^+(aq)$ and $Br^-(aq)$, but only the $Ag^+(aq)$ and $Br^-(aq)$ combine to give a precipitate. Thus, the reaction is:

$$Ag^+(aq) + Br^-(aq) \rightarrow AgBr(s)$$

(b) In solutions where precipitation reactions occur, other ions, for example, $NO_3^-(aq)$ and $Na^+(aq)$ in the reaction in (a), that are present in solution but take no part in the reaction are described as "**spectator ions**".

---

33. Use the solubility rules (Table 4.6) to predict which of the following salts are soluble and which are insoluble or sparingly soluble in water: (a) MgF_2 (b) AgI (c) K_2SO_4 (d) KF (e) $BaCl_2$ (f) $BaSO_4$ (g) NaOH (h) $Mg(OH)_2$

(a) MgF_2, **insoluble**; (b) AgI, **insoluble**; (c) K_2SO_4, **soluble**; (d) KF, **soluble**;

(e) $BaCl_2$, **soluble**; (f) $BaSO_4$, **insoluble**; (g) NaOH, **soluble**; (h) $Mg(OH)_2$, **insoluble**.

---

35. Predict whether or not a precipitate will form when aqueous solutions of each of the following pairs of substances are mixed. Where a reaction occurs, write the balanced equation for the reaction and the corresponding net ionic equation: (a) $FeCl_3$ + NaOH → (b) $BaCl_2$ + KOH → (c) $Pb(NO_3)_2$ + H_2SO_4 → (d) $AgNO_3$ + Na_2S → (e) $AgNO_3$ + HI → (f) $Pb(NO_3)_2$ + HCl →

(a) $FeCl_3(aq) + 3NaOH(aq) \rightarrow Fe(OH)_3(s) + 3NaCl(aq)$; $Fe^{3+}(aq) + 3OH^-(aq) \rightarrow Fe(OH)_3(s)$

(b) $BaCl_2(aq) + KOH(aq) \rightarrow$ **no reaction**

(c) $Pb(NO_3)_2(aq) + H_2SO_4(aq) \rightarrow PbSO_4(s) + 2HNO_3(aq)$; $Pb^{2+}(aq) + SO_4^{2-}(aq) \rightarrow PbSO_4(s)$

(d) $2AgNO_3(aq) + Na_2S(aq) \rightarrow Ag_2S(s) + 2NaNO_3(aq)$; $2Ag^+(aq) + S^{2-}(aq) \rightarrow Ag_2S(s)$

(e) $AgNO_3(aq) + HI(aq) \rightarrow AgI(s) + HNO_3(aq)$; $Ag^+(aq) + I^-(aq) \rightarrow AgI(s)$

(f) $Pb(NO_3)_2(aq) + 2HCl(aq) \rightarrow PbCl_2(s) + 2HNO_3(aq)$; $Pb^{2+}(aq) + 2Cl^-(aq) \rightarrow PbCl_2(s)$

Reaction Types

37. Balance each of the following equations for reactions in aqueous solution and classify each as an acid-base, an oxidation-reduction, or a precipitation reaction: (a) $AgNO_3 + BaCl_2 \rightarrow Ba(NO_3)_2 + AgCl$ (b) $NH_3 + H_2SO_4 \rightarrow (NH_4)_2SO_4$ (c) $Zn(s) + 2HCl \rightarrow ZnCl_2 + H_2$ (d) $NaHCO_3 + HCl \rightarrow NaCl + CO_2 + H_2O$

(a) $2AgNO_3(aq) + BaCl_2(aq) \rightarrow Ba(NO_3)_2(aq) + 2AgCl(s)$ — **precipitation**

(b) $2NH_3(aq) + H_2SO_4(aq) \rightarrow (NH_4)_2SO_4(aq)$ — **acid-base**

(c) $Zn(s) + 2HCl(aq) \rightarrow ZnCl_2(aq) + H_2(g)$ — **oxidation-reduction**

(d) $NaHCO_3(aq) + HCl(aq) \rightarrow NaCl(aq) + CO_2(g) + H_2O(\ell)$ — **acid-base**

---

39. Which of the following reactions are acid-base reactions, and which are oxidation-reduction reactions? For the acid-base reactions, identify the acid and the base. For the oxidation-reduction reactions, identify the oxidizing agent and the reducing agent: (a) $Cl_2(aq) + 2I^-(aq) \rightarrow 2Cl^-(aq) + I_2(s)$ (b) $HCl(aq) + H_2O(l) \rightarrow Cl^-(aq) + H_3O^+(aq)$ (c) $Zn(s) + 2HCl(aq) \rightarrow ZnCl_2(aq) + H_2(g)$ (d) $HCO_3^-(aq) + H_3O^+ \rightarrow CO_2(aq) + 2H_2O(l)$

(a) $Cl_2(aq) + 2I^-(aq) \rightarrow 2Cl^-(aq) + I_2(s)$ — **oxidation-reduction**
O (Cl₂) **R** (I⁻)

(b) $HCl(aq) + H_2O(\ell) \rightarrow Cl^-(aq) + H_3O^+(aq)$ — **acid-base**
acid (HCl) **base** (H_2O)

(c) $Zn(s) + 2H^+Cl^-(aq)* \rightarrow Zn^{2+}(Cl^-)_2(aq) + H_2(g)$ — **oxidation-reduction**
R (Zn) **O** (H^+Cl^-)

(d) $HCO_3^-(aq) + H_3O^+ \rightarrow CO_2(aq) + 2H_2O(l)$ — **acid-base**
base (HCO_3^-) **acid** (H_3O^+)

* **Note:** In (c), the actual reducing agent in aqueous solution is $H_3O^+(aq)$

General Problems

42. Astatine, At, is a member of the halogen family of elements. Predict the following for astatine: (a) its physical state; (b) its relative strength among the halogens as an oxidizing agent; (c) the balanced equations for its reactions with: (i) Na, (ii) Ca, (iii) P, (iv) H_2, and (v) Br_2; (d) the acid strength and reaction of its hydride with water; (e) the type of bond in the molecule it forms with bromine; (f) the type of bond in, and the physical state of, its compound with potassium.

(a) F_2 and Cl_2 are gases, Br_2 is a liquid, and I_2 is a solid; intermolecular forces increase in going down Group VII; astatine would also be expected to form At_2 molecules and to be a **solid**.

(b) F > Cl> Br > I > At; astatine would be expected to be the weakest oxidizing agent among the halogens.

(c) (i) $2Na(s) + At_2(s) \rightarrow 2Na^+At^-(s)$

(ii) $Ca(s) + At_2(s) \rightarrow Ca^{2+}(At^-)_2(s)$

(iii) $2P(s) + 3At_2(s) \rightarrow 2PAt_3(s)$

(iv) $H_2(g) + At_2(s) \rightarrow 2HAt(g)$

(v) $At_2(s) + Br_2(\ell) \rightarrow 2AtBr(s \text{ or } \ell)$.

(d) Like HCl(aq), HBr(aq), and HI(aq), but not HF (aq), HAt(aq) would be expected to be a strong acid in water;

$$HAt(g) + H_2O(\ell) \rightarrow H_3O^+(aq) + At^-(aq).$$

(e) The At—Br bond between two nonmetals with different electronegativities will be **polar covalent**.

(f) The K-At bond is between a metal and a nonmetal and would be expected to be **ionic**, K^+:At:$^-$

---

43. (a) What are the formulas of the simplest hydrides of (i) carbon (ii) nitrogen (iii) oxygen (iv) fluorine. (b) Write balanced equations for the reactions, if any, of each of the hydrides in (a) with water. Describe the acid-base properties of each of these hydrides as a strong acid, a strong base, a weak acid, a weak base, or neither an acid nor a base (more than one may apply).

(a) (i) Carbon (Group IV), **CH_4**; (ii) Nitrogen (Group III), **NH_3**; (iii) Oxygen (Group VI), **H_2O**; (iv) Fluorine (Group VII), **HF**.

(b) (i) **Methane**, $CH_4(g)$, is insoluble in water; its C—H bonds are insufficiently polar for it to behave as an acid, and it has no unshared electron pairs to accept protons. It is **neither an acid nor a base.**

(ii) **Ammonia**, $NH_3(g)$,is very soluble in water; it has an unshared electron pair on the nitrogen atom and behaves as a **weak base**:

$$NH_3(g) + H_2O(\ell) \rightleftarrows NH_4^+(aq) + OH^-(aq)$$

(iii) **Water**, $H_2O(\ell)$, behaves both as a **weak acid** and as a **weak base**:

$$H_2O(\ell) + H_2O(\ell) \rightleftarrows H_3O^+(aq) + OH^-(aq).$$

(iv) **Hydrogen fluoride**, HF(g), is very soluble in water and has a polar H—F bond, so it is a **weak acid**:

$$HF(aq) + H_2O(\ell) \rightleftarrows H_3O^+(aq) + F^-(aq).$$

45.* The salt $MX_2(s)$ is a metal halide that, when dissolved in water, gives a neutral solution. The aqueous solution does not react with bromine, but turns brown when $Cl_2(g)$ is bubbled through it. A solution containing 5.35 g of MX_2 required 710 mL of chlorine, measured at 25°C and 1.00 atm pressure, for complete reaction. Identify the salt MX_2, and explain how each of the preceding observations supports your identification.

---

1. $MX_2(s)$ is a metal halide. It must be an **ionic** compound, $M^{2+}(X^-)_2(s)$, so that M must be a metal from **Group II, and be** an alkaline earth metals.
2. $MX_2(aq)$ is not oxidized by Br_2, so X^- cannot be iodide ion, but X^- is oxidized by $Cl_2(g)$; X^- must be **bromide ion**, Br^-, which reacts to give bromine, hence the brown color:

$$MBr_2(aq) + Cl_2(g) \rightarrow M^{2+}(aq) + 2Cl^-(aq) + Br_2(aq)$$

3. From the information given, we can calculate the moles of Cl_2 that react with 5.35 g of MX_2:

$$\text{mol } Cl_2 = \frac{PV}{RT} = \frac{(1.00 \text{ atm})(710 \text{ mL } Cl_2)(\frac{1 \text{ L}}{10^3 \text{ mL}})}{(0.0821 \text{ atm L mol}^{-1} \text{ K}^{-1})(298 \text{ K})} = \underline{0.0290 \text{ mol}}$$

$$\text{mol } MBr_2 = (0.0290 \text{ mol } Cl_2)(\frac{1 \text{ mol } MBr_2}{1 \text{ mol } Cl_2}) = \underline{0.0290 \text{ mol}}$$

$$\text{Molar mass } MBr_2 = \frac{5.35 \text{ g } MBr_2}{0.0290 \text{ mol } MBr_2} = \underline{184 \text{ g mol}^{-1}}$$

Thus, molar mass of MBr_2 = [x + 2(79.90)] g = 184 g, where x is the molar mass of M. Hence, **x = 24 g mol^{-1}**, which from the table of atomic masses identifies M as **magnesium** (molar mass 24.30 g mol^{-1}).

The salt MX_2 is **magnesium bromide**, $MgBr_2$(s).

47*. The density of sodium chloride is 2.17 g cm^{-3}. How many sodium ions and chloride ions are there in a cubic NaCl crystal with sides of length 1 mm?

We first calculate the volume of the cube, then use the density to give the mass of the cube, which we convert to moles of NaCl and, finally, (using Avogadro's number) to the number of NaCl formula units:

$$\text{Volume of cube} = (1\ \text{mm})^3\left(\frac{1\ \text{cm}}{10\ \text{mm}}\right)^3$$

$$\text{Mass of cube} = (1\ \text{mm})^3\left(\frac{1\ \text{cm}}{10\ \text{mm}}\right)^3\left(\frac{2.17\ \text{g NaCl}}{1\ \text{cm}^3\ \text{NaCl}}\right) = \underline{2.17 \times 10^{-3}\ \text{g NaCl}}$$

$$\text{Number of NaCl formula units} = (2.17\text{x}10^{-3}\ \text{g NaCl})\left(\frac{1\ \text{mol NaCl}}{58.44\ \text{g NaCl}}\right)\left(\frac{6.022\text{x}10^{23}\ \text{NaCl units}}{1\ \text{mol NaCl}}\right)$$

$$= \underline{2.24 \times 10^{19}\ \text{NaCl formula units}}$$

Thus, the cube contains: **2.24 x 10^{19} Na^+ ions, and 2.24 x 10^{19} Cl^- ions.**

CHAPTER 5

Molar Masses and Moles

1. What are the molar masses of each of the following? (a) H_2O (b) CH_4 (c) NH_3 (d) CO_2 (e) NaCl

---

Molar masses are simply the sum of the masses of all the constituent atoms of a molecule in 1 mole, and usually given to four significant figures:

(a) H_2O Molar mass = [2(1.008) + 16.00] = **18.02 g mol⁻¹**
(b) CH_4 Molar mass = [12.01 + 4(1.008)] = **16.04 g mol⁻¹**
(c) NH_3 Molar mass = [14.01 + 3(1.008)] = **17.03 g mol⁻¹**
(d) CO_2 Molar mass = [12.01 + 2(16.00)] = **44.01 g mol⁻¹**
(e) NaCl Molar mass = [22.99 + 35.45] = **58.44 g mol⁻¹**

---

3. Convert each of the following to moles: (a) 100.0 g of water (b) 5.000 g of methane (c) 7.345 g of ammonia (d) 2.367 g of carbon dioxide (e) 12.50 g of sodium chloride.

Note: the molar masses were calculated in Problem 1:

$$\text{(a)}\quad \text{Mol } H_2O = (100.0\text{ g } H_2O)\left(\frac{1\text{ mol } H_2O}{18.02\text{ g } H_2O}\right) = \underline{5.549\text{ mol}}$$

$$\text{(b)}\quad \text{Mol } CH_4 = (5.000\text{ g } CH_4)\left(\frac{1\text{ mol } CH_4}{16.04\text{ g } CH_4}\right) = \underline{0.3117\text{ mol}}$$

$$\text{(c)}\quad \text{Mol } NH_3 = (7.345\text{ g } NH_3)\left(\frac{1\text{ mol } NH_3}{17.03\text{ g } NH_3}\right) = \underline{0.4313\text{ mol}}$$

$$\text{(d)}\quad \text{Mol } CO_2 = (2.367\text{ g } CO_2)\left(\frac{1\text{ mol } CO_2}{44\ 01\text{ g } CO_2}\right) = \underline{0.05378\text{ mol}}$$

$$\text{(e)}\quad \text{Mol NaCl} = (12.50\text{ g NaCl})\left(\frac{1\text{ mol NaCl}}{58.44\text{ g NaCl}}\right) = \underline{0.2139\text{ mol}}$$

Additional note: In each case the answer should be given to 4 significant figures.

---

5. Balance each of the following equations and calculate the mass of the second reactant that will react completely with 10.00 g of the first reactant: (a) $HCl + Ba(OH)_2 \rightarrow BaCl_2 + H_2O$ (b) $Cl_2 + NaBr \rightarrow NaCl + Br_2$ (c) $NaOH + H_2SO_4 \rightarrow Na_2SO_4 + H_2O$ (d) $HNO_3 + CaO \rightarrow Ca(NO_3)_2 + H_2O$

The balancing of the equations is straightforward; to calculate the mass of the second reactant we need the molar masses of both reactants. The procedure after the equation has been balanced is explained in part (a):

(a) $2HCl + Ba(OH)_2 \rightarrow BaCl_2 + 2H_2O$ <u>balanced</u>

Molar masses: HCl, 36.36 g mol⁻¹; $Ba(OH)_2$, 171.3 g mol⁻¹

To calculate the mass of $Ba(OH)_2$ that will react completely with 10.00 g of HCl, we convert the mass of HCl to moles of HCl, then moles of HCl to moles of $Ba(OH)_2$, and finally moles of $Ba(OH)_2$ to the mass of $Ba(OH)_2$:

$$\text{mass of Ba(OH)}_2 = (10.00 \text{ g HCl})\left(\frac{1 \text{ mol HCl}}{36.46 \text{ g HCl}}\right)\left(\frac{1 \text{ mol Ba(OH)}_2}{2 \text{ mol HCl}}\right)\left(\frac{171.3 \text{ g Ba(OH)}_2}{1 \text{ mol Ba(OH)}_2}\right) = \underline{23.49 \text{ g}}$$

(b) $Cl_2 + 2NaBr \rightarrow 2NaCl + Br_2$ balanced

Molar masses: Cl_2, 70.90 g mol^{-1}; NaBr, 102.9 g mol^{-1}

$$\text{mass of NaBr} = (10.00 \text{ g Cl}_2)\left(\frac{1 \text{ mol Cl}_2}{70.90 \text{ g Cl}_2}\right)\left(\frac{2 \text{ mol NaBr}}{1 \text{ mol Cl}_2}\right)\left(\frac{102.9 \text{ g NaBr}}{1 \text{ mol NaBr}}\right) = \underline{29.03 \text{ g}}$$

(c) $2NaOH + H_2SO_4 \rightarrow Na_2SO_4 + 2H_2O$ balanced

Molar masses: NaOH, 40.00 g mol^{-1}; Na_2SO_4, 142.1 g mol^{-1}

$$\text{mass of H}_2\text{SO}_4 = (10.00 \text{ g NaOH})\left(\frac{1 \text{ mol NaOH}}{40.00 \text{ g NaOH}}\right)\left(\frac{1 \text{ mol H}_2\text{SO}_4}{2 \text{ mol NaOH}}\right)\left(\frac{98.09 \text{ g H}_2\text{SO}_4}{1 \text{ mol H}_2\text{SO}_4}\right) = \underline{12.26 \text{ g}}$$

(d) $2HNO_3 + CaO \rightarrow Ca(NO_3)_2 + H_2O$ balanced

Molar masses: HNO_3, 63.02 g mol^{-1}; CaO, 56.08 g mol^{-1}

$$\text{mass of CaO} = (10.00 \text{ g HNO}_3)\left(\frac{1 \text{ mol HNO}_3}{63.02 \text{ g HNO}_3}\right)\left(\frac{1 \text{ mol CaO}}{2 \text{ mol HNO}_3}\right)\left(\frac{56.08 \text{ g CaO}}{1 \text{ mol CaO}}\right) = \underline{4.449 \text{ g}}$$

---

7. Ammonia gas, $NH_3(g)$, and hydrogen chloride gas, HCl(g),react to give the white solid ammonium chloride, $NH_4Cl(s)$. Write the balanced equation for this reaction, and calculate the mass of HCl that reacts completely with 0.200 g of $NH_3(g)$.

The balanced equation is: $NH_3(g) + HCl(g) \rightarrow NH_4Cl(s)$

Molar masses: NH_3, 17.03 g mol^{-1}, and HCl 36.46 g mol^{-1};

$$\text{mass of HCl} = (0.200 \text{ g NH}_3)\left(\frac{1 \text{ mol NH}_3}{17.03 \text{ g NH}_3}\right)\left(\frac{1 \text{ mol HCl}}{1 \text{ mol NH}_3}\right)\left(\frac{36.46 \text{ g HCl}}{1 \text{ mol HCl}}\right) = \underline{0.428 \text{ g}}$$

---

9. Metals, M(s), such as magnesium and zinc react with dilute sulfuric acid to give hydrogen, according to the equation $M(s) + H_2SO_4(aq) \rightarrow MSO_4(aq) + H_2(g)$. When 5.00 g of a mixture of finely divided (powdered) Mg and Zn were dissolved in (excess) sulfuric acid, 0.284 g of hydrogen was collected. What was the composition of the mixture of Mg and Zn expressed as mass percentages?

We first calculate moles of $H_2(g)$ and use the balanced equation

$$M(s) + H_2SO_4(aq) \rightarrow MSO_4(aq) + H_2(g)$$

to give the total moles of Mg and Zn. From the molar masses of Mg (24.31 g mol^{-1}) and Zn (65.38 g mol^{-1}) we can then find how many moles of each are in the 5.00 g sample:

$$\text{Moles of metal} = \text{moles of } H_2(g) = (0.248 \text{ g } H_2)(\frac{1 \text{ mol } H_2}{2.016 \text{ g } H_2}) = \underline{0.141 \text{ mol}}$$

Thus, if mass of Mg = $\underline{x\ g}$, mass of Zn = $\underline{(5.00-x)\ g}$, then:

$$(x \text{ g Mg})(\frac{1 \text{ mol Mg}}{24.31 \text{ g Mg}}) + [(5.00 - x) \text{ g Zn}](\frac{1 \text{ mol Zn}}{65.38 \text{ g Zn}}) = 0.141 \text{ mol}$$

Whence, $\underline{x = 2.50\ g}$

<u>Mass of Mg</u> = <u>2.50 g</u>; <u>Mass of Zn</u> = <u>2.50 g</u>

Thus, the composition of the mixture is: **50.0 mass % Mg, and 50.0 mass % Zn**

---

11. A 1.000-kg sample of impure limestone, containing 74.2% calcium carbonate, $CaCO_3(s)$, and 25.8% inert and involatile impurities, by mass, is heated until all the carbonate is decomposed to calcium oxide, $CaO(s)$ and $CO_2(g)$. What mass of CO_2 is produced?

The reaction is: $CaCO_3(s) \rightarrow CaO(s) + CO_2(g)$; we first calculate the mass of $CaCO_3(s)$ in the impure limestone:

$$\text{Mass of} CaCO_3 = (\frac{74.2}{100})(1.000 \text{ kg}) = \underline{0.742 \text{ kg}}$$

Molar masses: $CaCO_3$, 100.1 g mol^{-1}; CO_2, 44.01 g mol^{-1}

$$\text{Mass of } CO_2 = (0.742 \text{ kg})(\frac{1 \text{ mol } CaCO_3}{100.1 \text{ g } CaCO_3})(\frac{1 \text{ mol } CO_2}{1 \text{ mol } CaCO_3})(\frac{44.01 \text{ g } CO_2}{1 \text{ mol } CO_2}) = \underline{0.326 \text{ kg}}$$

---

13. How many liters of oxygen at 25°C and 740 mm Hg pressure are required to burn 5.00 g of magnesium completely?

The balanced equation for the reaction is: $2Mg(s) + O_2(g) \rightarrow 2MgO(s)$

$$\text{mol } O_2 = (5.00 \text{ g Mg})(\frac{1 \text{ mol Mg}}{24.31 \text{ g Mg}})(\frac{1 \text{ mol } O_2}{2 \text{ mol Mg}}) = \underline{0.1028 \text{ mol}}$$

$$V_{O_2} = \frac{nRT}{P} = \frac{(0.1028 \text{ mol})(0.0821 \text{ atm L mol}^{-1} \text{ K}^{-1})(298 \text{ K})}{(740 \text{ mm Hg})(\frac{1 \text{ atm}}{760 \text{ mm Hg}})} = \underline{2.58 \text{ L}}$$

<u>Limiting Reactants</u>

15. Sulfuric acid is produced when sulfur dioxide reacts with oxygen and water in the presence of a catalyst: $2SO_2(g) + O_2(g) + 2H_2O(l) \rightarrow 2H_2SO_4$. If 5.6 mol of SO_2 react with 4.8 mol of O_2 and a large excess of water, what is the maximum number of moles of H_2SO_4 that can be obtained?

We are told that water is in large excess, so it cannot be the limiting reagent. Thus, we must consider the SO_2 and O_2. From the balanced equation given, we can calculate the moles of O_2 that react with 5.6 mol SO_2, and, if required, the number of moles of SO_2 that react with 4.8 mol O_2:

$$5.6 \text{ mol } SO_2 \text{ react with: } (5.6 \text{ mol } SO_2)(\frac{1 \text{ mol } O_2}{2 \text{ mol } SO_2}) = \underline{2.8 \text{ mol } O_2}$$

Clearly, $O_2(g)$ is **in excess** and SO_2 is the **limiting reagent.**

$$\text{mol } H_2SO_4 = (5.6 \text{ mol } SO_2)(\frac{2 \text{ mol } H_2SO_4}{2 \text{ mol } SO_2}) = \underline{5.6 \text{ mol } H_2SO_4}$$

---

17. The unbalanced equation for the reaction of aluminum with hydrogen chloride (gas) is: $Al(s) + HCl(g) \rightarrow AlCl_3 + H_2(g)$. Determine the maximum mass of $AlCl_3$ produced, and the mass of Al or HCl that remains unreacted, when 2.70 g of Al reacts with 4.00 g of HCl.

The balanced equation is: $2Al(s) + 6HCl(g) \rightarrow 2AlCl_3 + 3H_2(g)$, and now we can calculate the number of moles of each reactant:

$$\text{mol Al} = (2.70 \text{ g Al})(\frac{1 \text{ mol Al}}{26.98 \text{ g Al}}) = \underline{0.1000 \text{ mol}}$$

$$\text{mol HCl} = (4.00 \text{ g HCl})(\frac{1 \text{ mol HCl}}{36.46 \text{ g HCl}}) = \underline{0.1097 \text{ mol}}$$

From the balanced equation, 1 mol Al reacts with 3 mol HCl, so that in this case the **limiting reactant** is HCl, and Al is **in excess:**

$$\text{mass of Al reacted} = (0.1097 \text{ mol HCl})(\frac{2 \text{ mol Al}}{6 \text{ mol HCl}})(\frac{26.98 \text{ g Al}}{1 \text{ mol Al}}) = \underline{0.987 \text{ g Al}}$$

$$\text{mass of } AlCl_3 \text{ formed} = (0.1097 \text{ mol HCl})(\frac{2 \text{ mol } AlCl_3}{6 \text{ mol HCl}})(\frac{133.3 \text{ g } AlCl_3}{1 \text{ mol } AlCl_3}) = \underline{4.87 \text{ g } AlCl_3}$$

$$\text{mass of unreacted Al} = (2.70-0.99)\text{g} = \underline{1.71 \text{ g}}$$

Theoretical Yield and Percent Yield

19. Carbon in the form of graphite was heated strongly with sulfur, and the resulting carbon disulfide, $CS_2(\ell)$, was distilled off and condensed to liquid. If 2.530 g of graphite gave 12.50 g of CS_2, what was the percent yield?

The balanced equation for the reaction is: $C(s) + S(s) \rightarrow CS_2(\ell)$

$$\text{Theoretical yield } CS_2 = (2.530 \text{ g C})(\frac{1 \text{ mol C}}{12.01 \text{ g C}})(\frac{1 \text{ mol } CS_2}{1 \text{ mol C}})(\frac{76.13 \text{ g } CS_2}{1 \text{ mol } CS_2}) = \underline{16.037 \text{ g } CS_2}$$

$$\% \text{ Yield} = (\frac{12.50 \text{ g } CS_2}{16.037 \text{ g } CS_2}) \times 100\% = \underline{77.94\%}$$

Determination of Empirical and Molecular Formulas

21. What are the mass percent elemental compositions of each of the following? (a) water, H_2O (b) sodium chloride, NaCl (c) ethane, C_2H_6 (d) magnesium bromide, $MgBr_2$ (e) carbon dioxide, CO_2.

In calculating the mass % elemental composition of a substance, we first find the molar mass, then:

$$\text{mass \% X} = \left(\frac{\text{mass of X per mol}}{\text{molar mass}}\right) \times 100\%$$

(a) For the H and O in H_2O (molar mass 18.02 g mol^{-1}):

$$\text{mass \% H} = \frac{\text{mass of 2 mol H}}{\text{molar mass of } H_2O} \times 100\% = \frac{2(1.008\text{ g})}{18.02\text{ g}} \times 100\% = \underline{11.19\%\text{ H}}$$

$$\text{mass \% O} = \frac{\text{mass of 1 mol O}}{\text{molar mass of } H_2O} \times 100\% = \frac{16.00\text{ g}}{18.02\text{ g}} \times 100\% = \underline{88.79\%\text{ O}}$$

(b) **NaCl**	(molar mass 58.44 g mol^{-1})	**39.34% Na; 60.66% Cl**
(c) **C_2H_6**	(molar mass 30.07 g mol^{-1})	**79.89% C; 20.11% H**
(d) **$MgBr_2$**	(molar mass 184.1 g mol^{-1})	**13.20% Mg; 86.80% Br**
(e) **CO_2**	(molar mass 44.01 g mol^{-1})	**27.29% C; 72.71% O**

---

23. Calculate the mass percentage of each element in each of the following compounds: (a) ethene, C_2H_4 (b) ethyne, C_2H_2 (c) magnesium chloride hexahydrate, $MgCl_2{\cdot}6H_2O$ (d) iron sulfate heptahydrate, $FeSO_4.7H_2O$

For the method see Problem 21.

(a) C_2H_4 **85.63% C, 14.37% H** (b) C_2H_2 **92.26% C, 7.74% H**

(c) $MgCl_2 \cdot 6H_2O$ **11.96% Mg, 34.87% Cl, 5.95% H, 47.22% O**

(d) $FeSO_4 \cdot 7H_2O$ **20.09% Fe, 11.53% S, 5.08% H, 63.31% O**

---

25. Ammonium sulfate, $(NH_4)_2SO_4$, is a common agricultural fertilizer. (a) What is the percentage of nitrogen by mass in this compound? (b) What mass of ammonium sulfate contains 100 g of nitrogen?

See problem 23 for the method:

(a) Molar mass $(NH_4)_2SO_4$ = 132.2 g mol^{-1}; **% N by mass = 21.20%**

(b) **Mass of $(NH_4)_2SO_4$(s) containing 100 g Nitrogen** $= (100\text{ g N})\left(\frac{100\text{ g }(NH_4)_2SO_4}{21.20\text{ g N}}\right) = \underline{471.7\text{ g}}$

EMPIRICAL FORMULAS

Note: When we are given the percent elemental composition by mass of a compound:

1. We express this as the **mass of each element** in a 100 g sample
2. We convert **grams** of each element to **moles**
3. The **ratio of moles** of each element is **the same as** the **ratio of atoms**
4. We divide by the smallest number of moles, and
5. Express the answer as a ratio of **whole numbers of atoms.**

27. The hydrocarbon anthracene has the composition 94.33% C and 5.67% H, by mass. What are its empirical formula and empirical formula mass?

	C	H
Mass %	94.33	5.67
Grams in 100 g	94.33	5.67
Moles 100 g	$\frac{94.33}{12.01}$ = 7.854	$\frac{5.67}{1.008}$ = 5.625
Ratio of moles = ratio of atoms	$\frac{7.584}{5.625}$ = 1.40 = 7.00	$\frac{5.625}{5.625}$ = 1.00 = 5.00

Thus the **empirical formula** of anthracene is $\mathbf{C_7H_5}$

Empirical formula mass = [2(12.01)+5(1.008)] = 89.11 u

29. When 3.10 g of a compound containing only carbon, hydrogen, and oxygen were completely burned in oxygen, 4.40 g CO_2 and 2.70 g H_2O were produced. (a) What is the empirical formula of the compound? (b) If its molecular mass is 62.1 u, what is its molecular formula?

(a) In this type of problem, the aim is to calculate the ratio of the number of moles of each atom in the compound, since this is the same as the ratio of atoms, which gives the empirical formula. The mass of the sample is the total mass of all of its atoms. The mass of CO_2 collected after burning contains all of the **carbon** atoms in the original sample. The mass of CO_2 can be converted to moles of CO_2, and hence to moles of C, to give the moles of carbon in the original sample:

$$\text{mol C} = (4.40\ \text{g CO}_2)\left(\frac{1\ \text{mol CO}_2}{44.01\ \text{g CO}_2}\right)\left(\frac{1\ \text{mol C}}{1\ \text{mol CO}_2}\right) = \underline{0.100\ \text{mol C}}$$

Similarly, the mass of H_2O from complete combustion contains all of the **hydrogen** in the original sample:

$$\text{mol H} = (2.70\ \text{g H}_2\text{O})\left(\frac{1\ \text{mol H2O}}{18.02\ \text{g H}_2\text{O}}\right)\left(\frac{2\ \text{mol H}}{1\ \text{mol H}_2\text{O}}\right) = \underline{0.300\ \text{mol H}}$$

(**Note**: the unit conversion factor here is **1 mol H_2O = 2 mol H**, because 1 mol H_2 contains 2 mol H).

We have obtained moles of C and moles of H in the original sample; we now need to obtain moles of O, which can be achieved by converting the **moles of C into the corresponding mass of C**, and the **moles of H into the corresponding mass of H**, and **subtracting the sum of these masses from the mass of the original sample** (which contains only C, H, and O):

mass of C = (0.100 mol)(12.01 g mol^{-1}) = **1.201 g of carbon**

mass of H = (0.300 mol)(1.008 g mol^{-1}) = **0.302 g of hydrogen**

Thus, **mass of O in sample** = (3.10-1.20-0.30) = **1.60 g**, which we can now convert to **moles of oxygen atoms:**

$$\text{mol O} = (1.60 \text{ g } CO_2)(\frac{1 \text{ mol O}}{16.00 \text{ g O}}) = \underline{0.100 \text{ mol}}$$

Ratio of moles C : H : O = 0.100 : 0.300 : 0.100 = **1 : 3 : 1**

and the ratio of atoms is also 1 : 3 : 1, so the **empirical formula is CH_3O.**

The empirical formula mass is 31.03 u, which is one-half of the given molecular mass of 62.1 u, giving the **molecular formula $C_2H_6O_2$**

---

31. When 0.100 mol of a compound of carbon, hydrogen, and nitrogen was burned completely in oxygen, 26.4 g of CO_2, 6.30 g H_2O, and 4.60 g of NO_2 were produced. What is the empirical formula of the compound?

Here we can convert the masses of CO_2, H_2O, and NO_2 directly to moles of C, H, and N, resepectively, since the compound contains only these elements:

$$\text{mol C} = (26.4 \text{ g } CO_2)(\frac{1 \text{ mol } CO_2}{44.01 \text{ g } CO_2})(\frac{1 \text{ mol C}}{1 \text{ mol } CO_2}) = \underline{0.600 \text{ mol C}}$$

$$\text{mol H} = (6.30 \text{ g } H_2O)(\frac{1 \text{ mol } H_2O}{18.02 \text{ g } H_2O})(\frac{2 \text{ mol H}}{1 \text{ mol } H_2O}) = \underline{0.700 \text{ mol H}}$$

$$\text{mol N} = (4.60 \text{ g } NO_2)(\frac{1 \text{ mol } NO_2}{46.01 \text{ g } NO_2})(\frac{1 \text{ mol N}}{1 \text{ mol } NO_2}) = \underline{0.100 \text{ mol N}}$$

Thus, ratio of moles of C : H : N = ratio of atoms = 0.600 : 0.700 : 0.100 = $\underline{6 : 7 :1}$

Thus, the **empirical formula** is **C_6H_7N**

---

33. A sample of mass 6.20 g of a compound containing only sulfur, hydrogen, and carbon reacted completely with chlorine and gave 21.9 g of hydrogen chloride, HCl, and 30.8 g of tetrachloromethane, CCl_4. What is the empirical formula of the compound?

Assuming that all the carbon in the sample reacts to give CCl_4 (molar mass 153.8 g mol^{-1}), and all the hydrogen reacts to give HCl (molar mass 36.46 g mol^{-1}):

$$\text{mol C} = (30.8 \text{ g } CCl_4)(\frac{1 \text{ mol } CCl_4}{153.8 \text{ g } CCl_4})(\frac{1 \text{ mol C}}{1 \text{ mol } CCl_4}) = \underline{0.200 \text{ mol C}}$$

$$\text{mol H} = (21.9 \text{ g HCl})(\frac{1 \text{ mol HCl}}{36.46 \text{ g HCl}})(1 \text{ mol H1 mol HCl}) = \underline{0.600 \text{ mol H}}$$

so in the 6.20 g sample of the compound, we have:

$$(0.200 \text{ mol C})(\frac{12.01 \text{ g C}}{1 \text{ mol C}}) = \underline{2.40 \text{ g C}}\ ; \qquad (0.600 \text{ mol H})(\frac{1.008 \text{ g H}}{1 \text{ mol H}}) = \underline{0.605 \text{ g H}}$$

By difference: mass of S = (6.20 − 2.40 − 0.60)g = $\underline{3.20 \text{ g}}$ = $(3.20 \text{ g S})(\frac{1 \text{ mol S}}{32.07 \text{ g S}}) = \underline{0.100 \text{ mol S}}$

Ratio of moles C : H : S = 0.200 : 0.600 : 0.100 = ratio of atoms = **2 : 6 : 1**

Thus, the **empirical formula** is **C_2H_6S**

35. Upon heating in air, 2.862 g of a red copper oxide gave 3.182 g of a black copper oxide. When the latter was heated strongly in hydrogen, it gave 2.542 g of pure copper. What are the empirical formulas of the two copper oxides?

Both samples contain the same amount of copper (2.542 g)

Black oxide: 3.182 g contains 2.542 g Cu and 0.640 g O.

Red Oxide: 2.862 g contains 2.542 g Cu and 0.320 g O.

	Red Oxide		Black Oxide	
	Cu	O	Cu	O
grams	2.542	0.320	2.542	0.640
moles	$\frac{2.542}{63.55}$ = 0.040	$\frac{0.320}{16.00}$ = 0.020	$\frac{2.542}{63.55}$ = 0.040	$\frac{0.640}{16.00}$ = 0.040
ratio of moles (atoms)	2.0 : 1.0		1.0 : 1.0	
Empirical formula	Cu_2O		CuO	

---

37. A volatile compound has the composition 62.04% carbon, 10.41% hydrogen, and 27.55% oxygen by mass. At 100°C and 1.00 atm pressure, 440 mL of the gaseous compound had a mass of 1.673 g. What are the compound's molar mass and molecular formula?

We use the gas data and the ideal gas law, PV = nRT, to calculate the **moles** of gas. From the given mass, we can then calculate the **molar mass**:

$$n = \frac{PV}{RT} = \frac{(1.00\text{ atm})(440\text{mL})(\frac{1\text{ L}}{10^3\text{ mL}})}{(0.0821\text{ atm L mol}^{-1}\text{ K}^{-1})(373\text{ K})} = \underline{1.44 \times 10^{-2}\text{ mol}}\ ; \quad \text{molar mass} = \frac{1.673\text{ g}}{1.44\text{x}10^{-2}\text{ mol}} = \underline{116\text{ g mol}^{-1}}$$

The empirical formula comes from the elemental composition:

	C	H	O
Mass %	62.04	10.41	27.55
grams in 100 g	62.04	10.41	27.55
moles in 100 g	$\frac{62.04}{12.01}$	$\frac{10.41}{1.008}$	$\frac{27.55}{16.00}$
	5.166	10.33	1.722
mole (atom) ratio	$\frac{5.166}{1.722}$	$\frac{10.33}{1.722}$	$\frac{1.722}{1.722}$
	3.00 :	6.00 :	1.00

Thus, the **empirical formula** is C_3H_6O with an empirical formula mass of 58.08 u. From the molar mass calculation, **molecular mass** = **116 u**. Thus, **molecular formula** = 2(empirical formula), i.e., $C_6H_{12}O$

Stoichiometry of Gas Reactions

38. What volumes $N_2(g)$ and $H_2(g)$, measured at STP, react completely to give 1.00 kg of $NH_3(g)$?

Molar mass of $NH_3(g)$ = 17.03 g mol^{-1} and we first calculate the moles of NH_3 in 1.00 kg, and its volume at STP, using the relationship that 1 mol of an ideal gas has a volume of 22.41 L at STP:

$$V_{NH_3} = (1 \text{ kg } NH_3)(\frac{10^3 \text{ g}}{1.00 \text{ kg}})(\frac{1 \text{ mol } NH_3}{17.03 \text{ g } NH_3})(22.41 \text{ L}) = \underline{1.316 \times 10^3 \text{ L}} \text{ at STP}$$

and from the balanced equation and Avogadro's law:

$$N_2(g) + 3H_2(g) \rightarrow 2NH_3(g)$$
$$1 \text{ volume} + 3 \text{ volumes} \rightarrow 2 \text{ volumes}$$

$$V_{N_2} = (1.316 \times 10^3 \text{ L } NH_3)(\frac{1 \text{ mol } N_2}{2 \text{ mol } NH_3}) = \underline{6.58 \times 10^2 \text{ L}}; \quad V_{H_2} = (1.316 \times 10^3 \text{ L } NH_3)(\frac{3 \text{ mol } H_2}{2 \text{ mol } NH_3}) = \underline{1.97 \times 10^3 \text{ L}}$$

---

40. $CH_4(g)$, methane (natural gas), reacts with steam to give *synthesis gas*, a mixture of $CO(g)$ and $H_2(g)$. What total volume of synthesis gas, and what volumes of $CO(g)$ and $H_2(g)$ gas result from the complete reaction of 100 L of methane (all measured at STP)?

$$CH_4(g) + H_2O(g) \rightarrow CO(g) + 3H_2(g)$$
$$1 \text{ vol} + 1 \text{ vol} \rightarrow 1 \text{ vol} + 3 \text{ vol}$$
$$100 \text{ L} + 100 \text{ L} \rightarrow 100 \text{ L} + 300 \text{ L}$$

Total volume of synthesis gas = 400 L; or 100 L CO(g) and 300 L $H_2(g)$

Stoichiometry of Reactions in Solution

42. How many moles of sodium hydroxide are contained in each of the following? (a) 1.00 L of 0.0100-M NaOH(aq); (b) 250 mL of 0.0100-M NaOH(aq); (c) 25.15 mL of 0.0100-M NaOH(aq); (d) 25.15 mL of 0.0134-M NaOH(aq).

(a) $$\text{mol NaOH} = (1.00\text{L})(\frac{0.0100 \text{ mol}}{1 \text{ L}}) = \underline{0.0100 \text{ mol}}$$

(b) $$\text{mol NaOH} = (250 \text{ mL})(\frac{1 \text{ L}}{1000 \text{ mL}})(\frac{0.0100 \text{ mol}}{1 \text{ L}}) = \underline{2.50 \times 10^{-3} \text{ mol}}$$

(c) $$\text{mol NaOH} = (25.15 \text{ mL})(\frac{1 \text{ L}}{1000 \text{ mL}})(\frac{0.0100 \text{ mol}}{1 \text{ L}} = \underline{2.52 \times 10^{-4} \text{ mol}}$$

(d) $$\text{mol NaOH} = (25.15 \text{ mL})(\frac{1 \text{ L}}{1000 \text{ mL}})(\frac{0.0134 \text{ mol}}{1 \text{ L}}) = \underline{3.37 \times 10^{-4} \text{ mol}}$$

---

44. A 12.00-g sample of potassium permanganate, $KMnO_4(s)$, was dissolved in sufficient distilled water to give 2.00 L of solution. What was the molarity of potassium permanganate in the solution?

$$\text{mol KMnO}_4 = (12.00 \text{ g KMnO}_4)(\frac{1 \text{ mol KMnO}_4}{158.0 \text{ g KMnO}_4}) = \underline{7.595\text{x}10^{-2} \text{ mol}}$$

$$\text{Molarity of KMnO}_4 = \frac{\text{mol KMnO}_4}{\text{volume of solution}} = \frac{7.595\text{x}10^{-2} \text{ mol}}{2.00 \text{ L}} = \underline{0.0380 \text{ M}}$$

---

46. How many milliliters of concentrated sulfuric acid, which is 98.0% H_2SO_4 by mass and has a density of 1.842 g L^{-1}, are needed to prepare 500 mL of a 0.175-M solution of H_2SO_4(aq)?

We first calculate moles of H_2SO_4 in 500 mL of 0.175-M H_2SO_4,convert this to grams of H_2SO_4, and then, using the density of the H_2SO_4, convert mass to volume:

$$\text{mass H}_2\text{SO}_4 = (500 \text{ mL})(\frac{1 \text{ L}}{1000 \text{ mL}})(\frac{0.175 \text{ mol H}_2\text{SO}_4}{1 \text{ L}})(\frac{98.08 \text{ g H}_2\text{SO}_4}{1 \text{ mol H}_2\text{SO}_4}) = \underline{8.582 \text{ g}}$$

Using the density: mass of 1 mL H_2SO_4 = 1.842 g, which contains

$$(1.842 \text{ g})(\frac{98.0 \text{ g H}_2\text{SO}_4}{100 \text{ g}}) = 1.805 \text{ g H}_2\text{SO}_4 \text{ mL}^{-1}$$

Thus, for 8.582 g H_2SO_4, we need: $(8.582 \text{ g H}_2\text{SO}_4)(\frac{1 \text{ mL}}{1.805 \text{ g H}_2\text{SO}_4}) = \underline{4.75 \text{ mL } 98.0\% \text{ H}_2\text{SO}_4}$

---

48. How could each of the following be prepared? (a) 6.30 L of 0.00300-M $Ba(OH)_2$(aq) from a 0.100-M $Ba(OH)_2$(aq) solution; (b) 750 mL of 0.0250 M $Cr_2(SO_4)_3$(aq) from a solution containing 35.0% $Cr_2(SO_4)_3$ by mass (density 1.412 g cm^{-3}).

(a) To prepare a more dilute solution, we dilute the solution of greater concentration by adding distilled water. Suppose the volume of the concentrated solution required is V mL, then, since both solutions must contain the same number of moles of $Ba(OH)_2$:

$$(\text{V mL})(\frac{0.100 \text{ mol}}{1 \text{ L}}) = (6.3 \text{ L})(\frac{10^3 \text{ mL}}{1 \text{ L}})(\frac{3.00\text{x}10^{-3} \text{ mol}}{1 \text{ L}}): \ \underline{\text{V} = 189 \text{ mL}}$$

Thus, to prepare the $3.00\text{x}10^{-3}$ M solution, take **189 mL** of the 0.100 M $Ba(OH)_2$(aq) solution and dilute it with distilled water to **6.30 L**.

(b) We first calculate the molarity of the more concentrated solution, and then proceed as in part (a).

For 1 L of 35 mass % $Cr_2(SO_4)_3$(aq):

$$\text{Molarity} = (\frac{35.0 \text{ g Cr}_2(\text{SO}_4)_3}{100 \text{ g sol'n}})(\frac{1 \text{ mol Cr}_2(\text{SO}_4)_3}{392.2 \text{ g Cr}_2(\text{SO}_4)_3})(\frac{1.412 \text{ g } 35\% \text{ sol'n}}{1 \text{ cm}^3 \ 35\% \text{ sol'n}})(\frac{1 \text{ mL}}{1 \text{ cm}^3})(\frac{1000 \text{ mL}}{1 \text{ L}}) = \underline{1.260 \text{ mol L}^{-1}}$$

$$\text{Thus:} \quad (\text{V mL})(\frac{1.260 \text{ mol}}{1 \text{ L}}) = (750 \text{ mL})(\frac{0.025 \text{ mol}}{1 \text{ L}}); \quad \underline{\text{V} = 14.9 \text{ mL of } 35.0\% \text{ Cr}_2(\text{SO}_4)_3(\text{aq})}$$

---

50. Write the balanced equation that describes the neutralization of any acid by any base in aqueous solution. Suppose that 250 mL of 1.00 M HCl(aq) is neutralized by 500 mL of 0.500 M NaOH(aq). (a) What is the concentration of NaCl(aq) in the resulting solution? (b) When it is evaporated to dryness, what mass of sodium chloride is obtained?

All acids give H_3O^+ and all bases give OH^- in water. The common reaction in all acid-base reactions is:

$$H_3O^+(aq) + OH^-(aq) \rightarrow 2H_2O(\ell)$$

For all acid-base reactions, we first write the balanced equation for the reaction. In this case:

$$HCl(aq) + NaOH(aq) \rightarrow NaCl(aq) + H_2O(\ell)$$

and then we can calculate the initial moles of reactants:

$$\text{mol HCl(aq)} = (250 \text{ mL})(\frac{1 \text{ L}}{1000 \text{ mL}})(\frac{1.00 \text{ mol HCl}}{1 \text{ L}}) = \underline{0.250 \text{ mol}}$$

$$\text{mol NaOH(aq)} = (500 \text{ mL})(\frac{1 \text{ L}}{1000 \text{ mL}})(0.500 \text{ mol NaOH1 L}) = \underline{0.250 \text{ mol}}$$

Thus, we have:

	HCl(aq) +	NaOH(aq) →	NaCl(aq) +	$H_2O(\ell)$	
initially	0.250	0.250	0	-	mol
finally	0	0	0.250	-	mol

(a) In the solution after reaction, we have 0.250 mol NaCl in a total volume of solution of (250+500) = 750 mL, and the concentration of NaCl is:

$$\frac{(0.250 \text{ mol NaCl})}{(750 \text{ mL})(\frac{1 \text{ L}}{1000 \text{ mL}})} = \underline{0.333 \text{ M}}$$

(b) And the mass of NaCl obtained is given by:

$$(0.250 \text{ mol NaCl})(\frac{58.44 \text{ g NaCl}}{1 \text{ mol NaCl}}) = \underline{14.6 \text{ g NaCl(s)}}$$

---

52. (a) What volume of 0.124-M HBr(aq) would neutralize 25.00 mL of 0.107-M NaOH(aq) solution? (b) What volume of 0.115-M KOH(aq) solution would react completely with 100.0 mL of 0.211-M HF(aq)?

(a) From the balanced equation for the reaction: $HBr(aq) + NaOH(aq) \rightarrow NaBr(aq) + H_2O(\ell)$

1 mol HBr reacts with 1 mol NaBr, so if the volume of HBr(aq) needed is V mL :

$$\text{mol HBr reacted} = \text{mol NaOH reacted}$$

$$(V \text{ mL})(0.124 \text{ mol L}^{-1}) = (25.00 \text{ mL})(0.107 \text{ mol L}^{-1})$$

$$\mathbf{V = 21.6 \text{ mL}}$$

(b) From the balanced equation for the reaction: $HF(aq) + KOH(aq) \rightarrow KF(aq) + H_2O(\ell)$

1 mol KOH reacts with 1 mol HF, so if the volume of KOH(aq) needed is V mL:

$$\text{mol KOH reacted} = \text{mol HF reacted}$$

$$(V \text{ mL})(0.115 \text{ mol L}^{-1}) = (100.0 \text{ mL})(0.211 \text{ mol L}^{-1})$$

$$\mathbf{V = 183 \text{ mL}}$$

---

54. (a) How much bromine is liberated when 25.00 mL of 0.102-M NaBr(aq) reacts with excess $Cl_2(g)$? (b) How much iodine is liberated when 10.00 mL of 0.120-M $Br_2(aq)$ is mixed with 10.00 mL of 0.100-M KI(aq)?

(a) Initial moles NaBr = (25.00 mL)(0.102 mol L^{-1}) = 2.55 mmol (**2.55 x 10^{-3} mol**), and from the equation:

$$NaBr(s) + Cl_2(g) \rightarrow 2NaCl(s) + Br_2(\ell)$$

$$\text{Mass of } Br_2 = (2.55\text{x}10^{-3} \text{ mol NaBr})(\frac{1 \text{ mol } Br_2}{2 \text{ mol NaBr}})(\frac{159.8 \text{ g } Br_2}{1 \text{ mol } Br_2}) = \underline{0.204 \text{ g}}$$

(b)

$$Br_2 + 2KI \rightarrow 2KBr + I_2$$

and the **limiting reactant** is KI(aq), since we have the same volume of each initial solution but the concentration of KI(aq) is the smaller:

$$\text{Mass of } I_2 = (10.00 \text{ mL})(\frac{0.100 \text{ mol KI}}{1 \text{ L}})(\frac{1 \text{ L}}{10^3 \text{ mL}})(\frac{1 \text{ mol } I_2}{2 \text{ mol KI}})(\frac{253.8 \text{ g } I_2}{1 \text{ mol } I_2}) = \underline{127 \text{ mg}} \quad (0.127 \text{ g})$$

General Problems

56. An impure sample of limestone contains calcium carbonate, $CaCO_3(s)$, and inert and involatile impurities. Upon heating, the reaction $CaCO_3(s) \rightarrow CaO(s) + CO_2(g)$ occurs. When 1.000 g of the impure limestone was strongly heated to constant mass, 215 mL of gas was evolved at 25°C and a pressure of 755 torr. What was the mass percentage of $CaCO_3(s)$ in the limestone?

We first calculate the theoretical yield of $CO_2(g)$ from 1 g of pure calcium carbonate in moles, then convert the 215 mL of gas into moles, and finally compare the two:

$$\text{Theoretical yield} = (1 \text{ g } CaCO_3)(\frac{1 \text{ mol } CaCO_3}{100.1 \text{ g } CaCO_3}) = \underline{9.99 \text{ x } 10^{-3} \text{ mol}}$$

$$\text{Actual yield} = \frac{PV}{RT} = \frac{(755 \text{ torr})(\frac{1 \text{ atm}}{760 \text{ torr}})(215 \text{ mL})(\frac{1 \text{ L}}{10^3 \text{ mL}})}{(0.0821 \text{ atm L mol}^{-1} \text{ K}^{-1})(298 \text{ K})} = \underline{8.73 \text{ x } 10^{-3} \text{ mol}}$$

$$\text{Mass \% } CaCO_3 = 100(\frac{8.73 \text{ x } 10^{-3} \text{ mol}}{9.99 \text{ x } 10^{-3} \text{ mol}}) = \underline{87.4\%}$$

---

58. When heated, 0.2800 g of blue hydrated copper sulfate, $CuSO_4{\cdot}xH_2O$, gave 0.1789 g of colorless anhydrous copper sulfate, $CuSO_4$. What is the empirical formula of hydrated copper sulfate?

On heating, n moles of $CuSO_4.xH_2O$ gives n moles of $CuSO_4$ (molar mass 159.6 g mol^{-1}) and and y moles of H_2O (molar mass 18.02 g mol^{-1}):

	$CuSO_4.xH_2O(s) \rightarrow$	$CuSO_4(s)$ +	$xH_2O(\ell)$
initially	0.2800 g	0	0
after heating	0	0.1789 g	0.1011 g

$$\text{mol CuSO}_4 = (0.1789 \text{ g CuSO}_4)\left(\frac{1 \text{ mol CuSO}_4}{159.6 \text{ g CuSO}_4}\right) = \underline{1.121\text{x}10^{-3} \text{ mol}}$$

$$\text{mol H}_2\text{O} = (0.1011 \text{ g H}_2\text{O})\left(\frac{1 \text{ mol H}_2\text{O}}{18.02 \text{ g H}_2\text{O}}\right) = \underline{5.610\text{x}10^{-3} \text{ mol}}$$

Thus: Ratio of moles $CuSO_4$: H_2O = 1.121 : 5.610 = <u>1.00 : 5.00</u>

i.e., empirical formula = $CuSO_4{\cdot}5H_2O$

---

60. One of the CFC (chloroflurocarbon) gases with the composition 11.50% C, 54.57% F, and 33.94% Cl by mass has a density of 4.66 g L^{-1} at STP. What are its empirical formula mass, molecular mass, and molecular formula?

From the gas density and PV = nRT we can find the molar mass:

$$n = \frac{PV}{RT} = \frac{(1.00 \text{ atm})(1 \text{ L})}{(0.0821 \text{ atm L mol}^{-1} \text{ K}^{-1})(273 \text{ K})} = \underline{0.0446 \text{ mol}} \; ; \quad \text{Molar mass} = \frac{4.66 \text{ g}}{0.0446 \text{ mol}} = 104 \text{ g mol}^{-1}$$

and we calculate the empirical formula from the elemental composition in the normal way:

	C	F	Cl
Mass %	11.50	54.57	33.94
grams in 100 g	11.50	54.57	33.94
moles in 100 g	$\frac{11.50}{12.01}$	$\frac{54.57}{19.00}$	$\frac{33.94}{35.45}$
	0.958	2.872	0.957
mole (atom) ratio	$\frac{0.958}{0.957}$	$\frac{2.872}{0.957}$	$\frac{0.957}{0.957}$
	1.00 :	3.00 :	1.00

Thus the **empirical formula** is CF_3Cl (empirical formula mass 104.5 u)

molecular formula = empirical formula = CF_3Cl (molecular mass 104.5 u)

---

62. The primary active ingredient in a common antacid is the salt $NaAl(OH)_2CO_3(s)$, which reacts with stomach acid according to the equation: $NaAl(OH)_2CO_3(s) + 4HCl(aq) \rightarrow NaCl(aq) + AlCl_3(aq) + 3H_2O(\ell) + CO_2(g)$. What mass of this salt would react completely with 2.00 L of 0.120-M HCl(aq)?

$$\text{Mass of antacid} = (2.00 \text{ L})\left(\frac{0.120 \text{ mol HCl}}{1 \text{ L}}\right)\left(\frac{1 \text{ mol antacid}}{4 \text{ mol HCl}}\right)\left(\frac{144.0 \text{ g antacid}}{1 \text{ mol antacid}}\right) = \underline{8.64 \text{ g}}$$

CHAPTER 6

Light and Other Electromagnetic Radiation

1. The light that is scattered from molecules in the atmosphere and gives the sky its blue color has a frequency of 7.5×10^{14} Hz. What is the corresponding wavelength?

Conversion of a frequency, ν, to the corresponding wavelength, λ, and vice versa, is often required, and is accomplished using the relationship: $\lambda\nu = c = 3.00 \times 10^8$ m s^{-1}

$$\nu\lambda = c\,; \quad \lambda = \frac{c}{\nu} = \frac{3.00 \times 10^8 \text{ m s}^{-1}}{7.5 \times 10^{14} \text{ s}^{-1}} = \underline{4.0 \times 10^{-7} \text{ m}} \quad (400 \text{ nm})$$

---

3. Citizens' band (CB) radio operates at a frequency of 27.3 MHz. What is the wavelength of this radio wave?

This problem is similar to Problem 1:

$$\nu\lambda = c; \quad \lambda = \frac{c}{\nu} = \left(\frac{3.00 \times 10^8 \text{ m s}^{-1}}{27.3 \text{ MHz}}\right)\left(\frac{1 \text{ MHz}}{10^6 \text{ s}^{-1}}\right) = \underline{11.0 \text{ m}}$$

---

5. A helium-neon laser produces light of wavelength 633 nm. What are the color and frequency of this light?

$$\nu\lambda = c; \quad \nu = \frac{c}{\lambda} = \left(\frac{3.00 \times 10^8 \text{ m s}^{-1}}{633 \text{ nm}}\right)\left(\frac{10^9 \text{ nm}}{1 \text{ m}}\right) = \underline{4.74 \times 10^{14} \text{ s}^{-1}} \quad (\text{Hz})$$

which corresponds to light in the red part of the visible spectrum.

---

6. The atomic spectrum of lithium has a strong red line at 670.8 nm. (a) What is the energy of each photon of this wavelength? (b) What is the energy of 1 mol of these photons?

The energy of a photon, frequency ν, is given by **E** = hν, (h = Planck's constant = 6.63×10^{-34} J s), and $\nu = c/\lambda$:

(a) For 1 photon: $$E = h\nu = \frac{hc}{\lambda} = \frac{(6.63 \times 10^{-34} \text{ J s})(3.00 \times 10^8 \text{ m s}^{-1})}{(670.8 \text{ nm})\left(\frac{1 \text{ m}}{10^9 \text{ nm}}\right)} = \underline{2.97 \times 10^{-19} \text{ J}}$$

(b) For 1 mol photons: $$E = (2.97 \times 10^{-19} \text{ J photon}^{-1})\left(\frac{6.022 \times 10^{23} \text{ photons}}{1 \text{ mol photons}}\right)\left(\frac{1 \text{ kJ}}{10^3 \text{ J}}\right)$$

$$= \underline{1.79 \times 10^2 \text{ kJ mol}^{-1}} \quad (179 \text{ kJ})$$

---

8. By the use of a suitable filter, a green mercury emission line of wavelength 546.1 nm can be isolated. Calculate (a) the energy of one photon of light of this wavelength; (b) 1 mol of photons of light of this wavelength.

This problem is similar to Problem 7:

(a) For <u>1 photon</u>: $E = h\nu = \frac{hc}{\lambda} = \frac{(6.63 \times 10^{-34}\ J\ s)(3.00 \times 10^{8}\ m\ s^{-1})}{(546.1\ nm)(\frac{1\ m}{10^{9}\ nm})} = \underline{3.64 \times 10^{-19}\ J}$

(b) For <u>1 mol photons</u>: $E = (3.64 \times 10^{-19}\ J\ photon^{-1})(\frac{6.022 \times 10^{23}\ photons}{1\ mol\ photons})(\frac{1\ kJ}{10^{3}\ J})$

$= \underline{2.19 \times 10^{2}\ kJ\ mol^{-1}}$ (219 kJ)

---

10. Photons of minimum energy 496 $kJ{\cdot}mol^{-1}$ are needed to ionize sodium atoms. (a) Calculate the lowest frequency of light that will ionize a sodium atom. What is the color of this light? (b) If light of energy 600 $kJ{\cdot}mol^{-1}$ is used, what is the kinetic energy of each emitted electron?

(a) $Na(g) + 496\ kJ\ mol^{-1} \rightarrow Na^{+}(g)$

For <u>one</u> photon: $E = \frac{496\ kJ\ mol^{-1}}{N_A\ mol^{-1}} = h\nu;\quad \nu = \frac{(496\ kJ\ mol^{-1})(\frac{10^{3}\ J}{1\ kJ})}{(6.63 \times 10^{-34}\ J\ s)(6.022 \times 10^{23}\ mol^{-1})} = \underline{1.24 \times 10^{15}\ s^{-1}}$ (Hz)

Light of this frequency is in the <u>ultraviolet</u> at higher frequency than visible light, which has insufficient energy to ionize sodium atoms.

(b) If light of energy 600 kJ mol^{-1} is used, **496 kJ mol^{-1}** is needed to ionize one mole of Na(g) atoms and the excess energy, (600 - 496) = **104 kJ mol^{-1}** appears as the kinetic energy of 1 mol of emitted electrons.

Thus, for <u>one</u> electron: kinetic energy $= (104\ kJ\ mol^{-1})(\frac{10^{3}\ J}{1\ kJ})(\frac{1\ mol\ electrons}{6.022 \times 10^{23}\ electrons}) = \underline{1.727 \times 10^{-19}\ J}$

---

12. One type of burglar alarm uses the photoelectric effect. Provided that visible light falling on a metal plate causes the emission of photoelectrons, the alarm is inactive. When the light beam is blocked by an intruder, the alarm is set off. Would magnesium metal be a suitable material for the metal plate, given that the lowest frequency that can cause the emission of an electron from magnesium is 8.95×10^{14} Hz?

From the lowest frequency that causes the emission of an electron, we can calculate the corresponding maximum wavelength:

$$\lambda\nu = c\ ;\quad \lambda = \frac{c}{\nu} = \frac{(3.00 \times 10^{8}\ m\ s^{-1})(\frac{10^{9}\ nm}{1\ m})}{8.95 \times 10^{14}\ s^{-1}} = \underline{335\ nm}$$

This corresponds to ultraviolet light with a wavelength less than that of visible light (400 to 750 nm); only ultraviolet light or radiation of smaller wavelength would set off the alarm. Magnesium would be an unsuitable metal except when the alarm operated in bright sunlight.

---

13. Light of minimum wavelength 493 nm dissociates chlorine molecules into chlorine atoms. (a) Will light of the same wavelength dissociate bromine molecules into bromine atoms, given that the dissociation energy of $Br_2(g)$ is 190 $kJ{\cdot}mol^{-1}$? (b) What is the minimum wavelength of light that will dissociate $Br_2(g)$ molecules into bromine atoms?

(a) First we should calculate the energy of 1 mole of photons of wavelength 493 nm:

$$E = N_A h\nu = \frac{N_A hc}{\lambda} = \frac{(6.022 \times 10^{23}\ \text{mol}^{-1})(6.63 \times 10^{-34}\ \text{J s})(3.00 \times 10^{8}\ \text{m s}^{-1})}{(493\ \text{nm})(\frac{1\ \text{m}}{10^{9}\ \text{nm}})} = \underline{2.43 \times 10^{5}\ \text{J}} \quad (243\ \text{kJ})$$

An energy of 243 kJ mol^{-1} is greater than the 190 kJ mol^{-1} needed to dissociate Br_2 molecules into Br atoms, so light of this wavelength will dissociate Br_2 molecules into Br atoms.

(b) To calculate the minimum wavelength of light that will just dissociate Br_2 molecules into Br atoms, we need to calculate the wavelength of photons of energy 190 kJ mol^{-1}:

$$E = N_A h\nu = \frac{N_A hc}{\lambda}$$

$$\lambda = \frac{N_A hc}{E} = \frac{(6.022 \times 10^{23}\ \text{mol}^{-1})(6.63 \times 10^{-34}\ \text{J s})(3.00 \times 10^{8}\ \text{m s}^{-1})}{(190\ \text{kJ mol}^{-1})(\frac{10^{3}\ \text{J}}{1\ \text{kJ}})} = \underline{6.30 \times 10^{-7}\ \text{m}} \quad (630\ \text{nm})$$

Atomic Spectra

15. What wavelength of light is emitted when an electron moves from the n = 6 to the n = 2 energy level of a hydrogen atom? In what region of the electromagnetic spectrum is the corresponding spectral line found?

The energy emitted when an electron moves from an energy level n_2 to an energy level n_1 is given by:

$$\text{For } n_1 = 6,\ n_2 = 2:\quad \Delta E = -1312(\frac{1}{n_1^2} - \frac{1}{n_2^2})\ \text{kJ mol}^{-1} = -1312(\frac{1}{(6)^2} - \frac{1}{(2)^2})\ \text{kJ mol}^{-1} = \underline{-291.6\ \text{kJ mol}^{-1}}$$

and we now need to calculate the wavelength corresponding to this energy:

$$\frac{\Delta E}{N_A} = \frac{hc}{\lambda};\quad \lambda = \frac{N_A hc}{\Delta E} = \frac{(6.022 \times 10^{23}\ \text{mol}^{-1})(6.63 \times 10^{-34}\ \text{J s})(3.00 \times 10^{8}\ \text{m s}^{-1})}{(291.6\ \text{kJ mol}^{-1})(\frac{10^{3}\ \text{J}}{1\ \text{kJ}})}$$

$$= (4.11 \times 10^{-7}\ \text{m})(\frac{10^{9}\ \text{nm}}{1\ \text{m}}) = \underline{411\ \text{nm}}$$

A wavelength of 411 nm corresponds to a line at the **blue** end of the visible spectrum.

---

17. Lines in the ultraviolet region of the emission spectrum of atomic hydrogen arise from transitions to the n = 1 level. One of these lines has a wavelength of 103 nm. What is the n quantum number of the electrons in the excited atoms that give rise to this line?

We first calculate the energy per mole corresponding to photons of wavelength of 103 nm, for which:

$$\Delta E = \frac{N_A hc}{\lambda} = \frac{(6.022 \times 10^{23}\ \text{mol}^{-1})(6.63 \times 10^{-34}\ \text{J s})(3.00 \times 10^{8}\ \text{m s}^{-1})}{(103\ \text{nm})(\frac{1\ \text{m}}{10^{9}\ \text{nm}})} = \underline{1.162 \times 10^{6}\ \text{J mol}^{-1}}$$

Then, because the spectral lines are due to transitions from energy levels n > 1 to n = 1

$$\Delta E = (1312 \text{ kJ mol}^{-1})(\frac{1}{(1)^2} - \frac{1}{n^2}) = (1.162 \times 10^6 \text{ J mol}^{-1})(\frac{1 \text{ kJ}}{10^3 \text{ J}}) = \underline{1162 \text{ kJ mol}^{-1}}$$

$$\text{Thus:} \quad n^2 = \frac{1312}{150} = 8.75 \text{ ; } \quad \text{Hence, } \underline{n = 3} \text{ (since n must be an integer)}$$

The excited electrons that give rise to the 103 nm line in the spectrum are in the **n = 3 energy level.**

---

19. The visible series of emission lines from atomic hydrogen end in the n = 2 energy level. What is the wavelength of the longest wavelength line in this series?

For transitions from levels with n > 2 to n = 2, and the longest wavelength spectral line corresponds to the transition of lowest energy (λ inversely proportional to energy), i.e., that from n = 3 to n = 2, for which:

$$\Delta E = -1312(\frac{1}{(2)^2} - \frac{1}{(3)^2}) \text{ kJ mol}^{-1} = \underline{182.2 \text{ kJ mol}^{-1}}$$

and the corresponding wavelength is given by:

$$\lambda = \frac{(6.022 \times 10^{23} \text{ mol}^{-1})(6.63 \times 10^{-34} \text{ J s})(3.00 \times 10^8 \text{ m s}^{-1})}{(182.2 \text{ kJ mol}^{-1})(\frac{10^3 \text{ J}}{1 \text{ kJ}})} = (6.57 \times 10^{-7} \text{ m})(\frac{10^9 \text{ nm}}{1 \text{ m}}) = \underline{657 \text{ nm}}$$

---

21. The infrared series of emission lines from atomic hydrogen arise from transitions to the n = 3 level. One of these lines has a wavelength of 1094 nm. Determine the value of the n quantum number for the upper level involved in this transition.

Using the method used in for example Problem 20, we can show that a wavelength of 1094 nm corresponds to a difference between energy levels, ΔE, of 109.5 kJ mol^{-1}, so

$$\Delta E = 109.5 \text{ kJ mol}^{-1} = -1312(\frac{1}{n^2} - \frac{1}{(3)^2}) \text{ kJ mol}^{-1} \text{ ; } \quad \underline{n = 6}$$

Electron Configurations

23. (a) Explain briefly how ionization energies are obtained from a photoelectron spectrometer. (b) Argon has ionization energies of -1.52, -2.82, -24.1, -31.5, and -309 MJ·mol^{-1}. Interpret these ionization energies in terms of the electron configuration of argon.

(a) In *photoelectron spectroscopy*, gaseous samples are irradiated with high energy photons of known energy, $h\nu$, which eject photoelectrons with kinetic energies $\frac{1}{2}mv^2$ from the atoms. These energies are then measured by deflection in a magnetic field. The photon provides the energy to remove an electron from a particular energy level and to give it kinetic energy, so that $h\nu = IE + \frac{1}{2}mv^2$, where IE is the ionization energy of the electron. Since for the large collection of atoms in any sample an ejected electron can come from any of their many energy levels, in principle all of the energy levels of an element may be obtained provided suitably energetic photons are used.

(b) The photoelectron spectrum of argon (Z = 18) shows that its 18 electrons are in five different energy levels. The peaks in the photoelectron spectrum are identified with each of the five energy levels given by the electron configuration of argon, $1s^2\ 2s^2\ 2p^6\ 3s^2\ 3p^6$, as follows:

Peak (MJ mol^{-1})	-1.52	-2.82	-24.1	-31.5	-309
Energy Level	3p	3s	2p	2s	1s

The 1s electrons in the n = 1 shell are closest to the nucleus and are the most difficult to remove. They have the largest ionization energy of 309 MJ mol^{-1}. There are two fairly closely spaced ionization energies for the electrons of the n=2 shell next closest to the nucleus, indicating two energy levels, designated 2s and 2p, and the most easily removed electrons are those of the outer n=3 shell, which are also found in two closely spaced energy levels designated 3s and 3p.

---

25. The ionization energy of helium is 2.37 MJ·mol^{-1}. What is the kinetic energy of the electrons produced when helium gas is irradiated with radiation of wavelength 40.0 nm?

We make use of the relationship that the energy of the photon equals the ionization energy of the photoelectron plus its kinetic energy, E_{photon} = IE + KE, but first we must find the energy per mole of photons:

$$E = N_A E_{photon} = N_A h\nu = \frac{N_A hc}{\lambda}$$

$$= \frac{(6.022 \times 10^{23}\ mol^{-1})(6.63 \times 10^{-34}\ J\ s)(3.00 \times 10^{8}\ m\ s^{-1})}{(40.0\ nm)(\frac{1\ m}{10^9\ nm})} = \underline{2.99 \times 10^6\ J} \quad (2.99\ MJ)$$

Thus, the kinetic energy of the electrons is:

$$KE = E_{photon} - IE = (2.99\text{-}2.37)\ MJ\ mol^{-1} = \mathbf{0.62\ MJ\ mol^{-1}}\ (620\ kJ\ mol^{-1})$$

---

27. (a) What is the maximum number of electrons associated with each of the n = 1, n = 2, and n = 3 energy levels? (b) How many subshells are associated with each of these levels, and how is each labeled? (c) What is the maximum number of electrons associated with each subshell?

(a) The total number of electrons in a shell designated **n** is **$2n^2$**.

n = 1, **2**; n = 2, **8**; n = 3, **18**

(b) The subshells are labelled ns, np, nd, nf, ... in order of ascending energy levels, where the number of subshells for a given value of n is also n:

n = 1, **1s**; n = 2, **2s** and **2p**; n = 3, **3s**, **3p**, and **3d**; n = 4, **4s**, **4p**, **4d**, and **4f**.

(c) The number of electrons associated with each subshell is: ns, **2**: np, **6**; nd, **10**, and nf, **14**.

---

28. Which of the following energy level designations are not allowed? (a) 6s (b) 1p (c) 4d (d) 2d (e) 3p (f) 4d (g) 5p (h) 2s.

For a given value of n, there are only n energy levels, designated in order: s, p, d, f, ...

(a) **n = 6**. Six energy levels are allowed, of which the first is 6s.

(b) **n = 1**. Only one energy level is allowed, designated 1s; 1 p is **not allowed**.

(c) **n = 4**. Four energy levels are allowed (4s, 4p, 4d, and 4f, of which the third is 4d.

(d) **n = 2** Two energy levels are allowed (2s and 2p); **2d is not allowed.**
(e) **n = 3** Three energy levels are allowed (3s, 3p, and 3d), of which the second is 3p.
(f) **n = 4** Four energy levels are allowed (4s, 4p, 4d, and 4f), of which the third is 4d.
(g) **n = 5** Five energy levels are allowed (5s, 5p, 5d, 5f, and 5e), of which the second is 5p.
(h) **n = 2** Two energy levels are allowed (2s and 2p), of which the first is 2s.

Thus, only **(b)** and **(d)** are **not allowed.**

---

30. In terms of the Pauli exclusion principle and/or Hund's rule, explain the following: (a) Beryllium cannot have the electron configuration $1s^4$; (b) the ground state of nitrogen has three unpaired electrons.

(a) Beryllium (Z = 4) with a total of four electrons has the ground state electron configurations $1s^2\ 2s^2$. The configuration $1s^4$, with 4 electrons, is not possible because any single orbital can accommodate only two electrons, and then only provided they have opposite spins.

(b) Nitrogen (Z = 7) has the ground state electron configuration $1s^2\ 2s^2\ 2p^3$. As a consequence of Hund's rule, the three 2p electrons each occupy one of the three separate 2p orbitals and have the same spin. These three electrons thereby minimize their electrostatic repulsions by occupying orbitals with the same energy level in different regions of space (in different **domains**).

---

32. Using the orbital box notation, write the ground state electron configurations of each of the following atoms and ions: (a) Be (b) N (c) F (d) Mg (e) Cl^+ (f) Ne^+ (g) Al^{3+}

We first locate each atom in the periodic table by group and period, from which the basic shell structure follows. The order of the energy levels and thus the sequence in which they are filled is:

1s < 2s < 2p < 3s < 3p < 4s < 3d < 4p ....

Note that for negative ions we have to add the requisite number of electrons to the configuration for the neutral atom, and for positive ions we have to subtract the requisite number of electrons, to those of the neutral atom, ensuring that the Pauli principle and Hund's rule are followed:

	Group	Period	Z	Orbital boxes				
				1s	2s	2p	3s	3p
(a) **Be**	II	2	4	[↑↓]	[↑↓]			
(b) **N**	V	2	7	[↑↓]	[↑↓]	[↑ | ↑ | ↑]		
(c) **F**	VII	2	9	[↑↓]	[↑↓]	[↑↓ | ↑↓ | ↑]		
(d) **Mg**	II	3	12	[↑↓]	[↑↓]	[↑↓ | ↑↓ | ↑↓]	[↑↓]	[| |]
(e) **Cl^+**	VII	3	17	[↑↓]	[↑↓]	[↑↓ | ↑↓ | ↑↓]	[↑↓]	[↑↓ | ↑ | ↑]
(f) **Ne^+**	VIII	2	18	[↑↓]	[↑↓]	[↑↓ | ↑↓ | ↑]		
(g) **Al^{3+}**	III	3	13	[↑↓]	[↑↓]	[↑↓ | ↑↓ | ↑↓]		

---

34. Identify the elements with each of the following ground-state configurations: (a) $1s^22s^1$ (b) $1s^22s^22p^3$ (c) $[Ar]4s^2$ (d)$[Ar]3d^{10}4s^24p^3$.

There are two possible approaches; (1) Count up the electrons and thus obtain the atomic number Z, or, *preferably*, (2) Count up the number of shells, to give the *Period*, and the number of valence electrons, to give the *Group*.

(a) $1s^2 2s^1$ — The final (valence) shell is n = 2 (**Period 2**) and contains 1 electron (**Group I**). Thus the element is the first alkali metal, **lithium.**

(b) $1s^2 2s^2 2p^3$ — The final (valence) shell is n = 2 (**Period 2**) and contains 5 electrons (**Group V**). Thus the element is the first element in Group V, **nitrogen.**

(c) $[Ar]4s^2$ — The final (valence) shell is n = 4 (**Period 4**) and contains 2 electrons (**Group II**). Thus the element is the alkaline earth metal in Period 4, **calcium.**

(d) $[Ar]3d^{10}4s^2 4p^3$. — The 3d shell is filled, so this is not a transition metal. The final (valence) shell is n = 4 (**Period 4**) and contains 5 electrons (**Group V**). Thus the element is the Group V element in Period IV, **arsenic.**

---

35. How many unpaired electrons are there in the ground states of each of the following? (a) O (b) O^- (c) O^{2-} (d) S (e) F (f) Ar

---

We first locate each element in the Periodic Table by Period and Group. The latter gives us the number of valence shell electrons. After allowing for any charges on ions, we arrange these valence electrons in valence-shell orbitals according to Hund's rule and the Pauli exclusion principle.

(a) O in Period 2 and Group VI has 6 valence electrons arranged as $2s^2 2p_x^2 2p_y^1 2p_z^1$, giving **2 unpaired** electrons

(b) O^- has 7 valence electrons (1 more than neutral O) arranged as $2s^2 2p_x^2 2p_y^2 2p_z^1$, giving **1 unpaired** electron

(c) O^{2-} has 8 valence electrons (2 more than neutral O) arranged as $2s^2 2p_x^2 2p_y^2 2p_z^2$, giving **0 unpaired** electrons

(d) S in Period 3 and Group VI has 6 valence electrons arranged as $3s^2 3p_x^2 3p_y^1 3p_z^1$, giving **2 unpaired** electrons

(e) F in Period 2 and Group VII has 7 valence electrons arranged as $2s^2 2p_x^2 2p_y^2 2p_z^1$, giving **1 unpaired** electron

(f) Ar in Period 3 and Group VIII has 8 valence electrons arranged as $3s^2 3p_x^2 3p_y^2 3p_z^2$, giving **0 unpaired** electrons

---

37. Without reference to Table 6.4, draw orbital box diagrams for the ground-state valence shells of each of the following atoms: (a) P (b) Ca (c) O (d) Br (Z = 35)

(a) P, group V, period 3 — 3s [↑↓] 3p [↑ | ↑ | ↑]

(b) Ca, group II, period 4 — 4s [↑↓]

(c) O, group VI, period 2 — 2s [↑↓] 2p [↑↓ | ↑ | ↑]

(d) Br, group VII, period 4 — 4s [↑↓] 4p [↑↓ | ↑↓ | ↑]

---

39. By drawing orbital box diagrams, determine how many unpaired electrons there are for the ground-state electron configurations of each of the following atoms: (a) P (b) Si (c) I (d) Se

---

We first identify the position of the element in the periodic table. The group number gives the number of valence electrons, which for the ground state are arranged in orbitals according to Hund's rule and the Pauli exclusion principle. All electrons in inner shells are paired, so we need to consider only the valence shells.

(a) **Phosphorus, P** (Group V, Period 3) — 3s [↑↓] 3p [↑ | ↑ | ↑] — three unpaired electrons

	3s	3p	
(b) **Silicon, Si** (Group IV, Period 3)	[↑↓]	[↑][↑][ ]	**two unpaired electrons**
	5s	5p	
(c) **Iodine, I** (Group VII, Period 5)	[↑↓]	[↑↓][↑↓][↑]	**one unpaired electron**
	4s	4p	
(d) **Selenium, Se** (Group VI, Period 4)	[↑↓]	[↑↓][↑][↑]	**two unpaired electrons**

---

41. (a) To what group and period of the periodic table does arsenic belong? (b) How many energy levels are occupied in the ground state of arsenic? (c) How many of these energy levels are only singly occupied? (d) Name the elements in the same group as arsenic in Periods 2 and 3, and write their ground-state electron configurations.

(a) **Arsenic** is in Group V, Period 4.

(b) The electron configuration is $1s^22s^22p^63s^23p^64s^23d^{10}4p^3$, with **1s, 2s, 2p, 3s, 3p, 4s, 3d,** and **4p** energy levels occupied, for a total of **8 energy levels.**

(c) There are **5** electrons in the n = 4 valence shell; **2** electrons occupy a 4s orbital and the other **3** occupy each of the three 4p orbitals **singly**, $4s^24p^3$.

(d) The Group V elements in periods 2 and 3 are **nitrogen** and **phosphorus**, respectively, with the electron configurations: **N, [He]$2s^22p^3$** and **P, [Ne]$3s^23p^3$**.

---

42. Write the electron configurations of the elements with each of the following atomic numbers. Without reference to any other information, name each element and classify it as an s-block, p-block, or d-block element:
(a) Z = 5 (b) Z = 11 (c) Z = 19 (d) Z = 22 (e) Z = 23 (f) Z = 29 (g) Z = 33 (h) Z = 36

Given an **atomic number**, Z, we have to remember the order of filling of the energy levels **1s < 2s < 2p < 3s < 3p < 4s < 3d < 4p ....** and that **s-levels** contain a maximum of **2** electrons, **p** a total of **6** electrons, and **d** a total of **10** electrons. We can then write the **ground state electron configuration.** The shell last filled is the **valence shell,** which gives the **period,** and the **group.**

(a) **Z = 5,** $1s^22s^22p^1$	Period 2, Group III; **boron, p-block.**
(b) **Z = 11,** $1s^22s^22p^63s^1$	Period 3, Group I; **sodium, s-block.**
(c) **Z = 19,** $1s^22s^22p^63s^23p^64s^1$	Period 4, Group I; **potassium, s-block.**
(d) **Z = 22,** $1s^22s^22p^63s^23p^64s^23d^2$	Period 4, second transition metal; **titanium, d-block.**
(e) **Z = 23,** $1s^22s^22p^63s^23p^64s^23d^3$	Period 4, third transition metal; **vanadium, d-block.**
(f) **Z = 29,** $1s^22s^22p^63s^23p^64s^13d^{10}$*	Period 4, ninth transition metal; **copper, d-block.**
(g) **Z = 33,** $1s^22s^22p^63s^23p^64s^23d^{10}4p^3$	Period 4, Group V; **arsenic, p-block.**
(h) **Z = 36,** $1s^22s^22p^63s^23p^64s^23d^{10}4p^6$	Period 4, Group VIII; **krypton, p-block.**

Note: You may have written $4s^23d^9$, but for reasons we cannot go into the ground state configuration is $4s^13d^{10}$

The Hydrogen Molecule and the Covalent Bond

45. (a) Write the ground-state electron configuration of silicon. (b) How many unpaired electrons are there in the ground state? (c) What valence would you predict for silicon? (d) How do you explain that silicon forms the compounds $SiCl_4$ and SiH_4?

(a) **Silicon** is in Period 3 and Group IV, so its ground state electron configuration is **[Ne]$3s^23p^2$.**

(b) The **ground state** electron configuration has **two** 3p electrons in separate atomic orbitals (Hund's rule):

3s 3p
Si [Ne] [↑↓] [↑ | ↑ |]

(c) In its **ground state** electron configuration, the anticipated valence of **silicon** is **2**.

(d) The observed valence of **4** in SiH_4 and $SiCl_4$ means that they must both be formed from an **excited state** of silicon, with **four unpaired valence electrons** and the electron configuration

3s 3p
Si [Ne] [↑] [↑ | ↑ | ↑]

---

47. (a) Explain why the bonding in the CH_4 molecule cannot be conveniently described in terms of the atomic 2s and 2p orbitals on carbon. (b) How are the bonds in this molecule usually described in terms of hybrid orbitals?

(a) The **ground state** of carbon is $[He]2s^22p^2$, so that consistent with the Pauli exclusion principle and Hund's rule the atomic orbital description is:

2s 2p
C [He] [↑↓] [↑ | ↑ |]

which with **two** unpaired electrons gives carbon a valence of **two**, so that with hydrogen, for example, it might be expected to form only *angular* AX_2E $:CH_2$. In order to form *tetrahedral* AX_4 methane, $\mathbf{CH_4}$, the C atom must start with **four singly-occupied** atomic orbitals, which can be achieved by starting with the **excited state**

2s 2p
C [He] [↑] [↑ | ↑ | ↑]

which with **four** unpaired electrons gives carbon a valence of **four**, as required to form CH_4. It would not, however, give a molecule with the correct **geometry** because the 2s and 2p atomic orbitals do not have the geometry required to form four tetrahedral orbitals by overlap with hydrogen 1s orbitals.

For **tetrahedral geometry**, we must start with *four equivalent* singly occupied orbitals on carbon, which is achieved by mixing together the *one* 2s and *three* 2p orbitals to give **four equivalent** orbitals, described as **a set of sp^3 hybrid orbitals**, each of which can overlap with a hydrogen 1s orbital to form four tetrahedral bonding orbitals.

---

49. Give hybrid orbital descriptions of the bonding in the C_2H_2, HCN, and CO_2 molecules.

Each of these molecules is linear, so in each case we start with one 2s and three 2p atomic orbitals on each atom and first combine the 2s orbital with *one* of the 2p orbitals to form **two linear** sp hybrid orbitals.

Acetylene, C_2H_2, H—C≡C—H We start with each C atom having its 4 valence electrons in two singly-occupied sp hybrid orbitals and two 2p orbitals. According to the **σ-π model**, each C atom utilizes its two sp hybrid orbitals to form two colinear σ bonds (one to a H atom and the other to the other C atom); the remaining two singly-occupied 2p orbitals on each C atom overlap sideways to form two π bonds.

Hydrogen cyanide, HCN, **H—C≡N:** We start with the C atom having its 4 valence electrons in two singly-occupied sp hybrid orbitals and two 2p orbitals, and the N atom having its 5 valence electrons distributed as one lone pair and a single electrons in two sp hybrid orbitals, and single electrons in each of the two 2p orbitals. The C atom utilizes its two sp hybrid orbitals to form two colinear σ bonds (one to a H atom and the other to the N atom); the remaining two singly-occupied 2p orbitals on the C atom, and the two singly-occupied 2p orbitals on the N atom overlap sideways to form two π bonds.

Carbon dioxide, CO_2, :O=C=O: We start with the C atom having its 4 valence electrons in two singly-occupied sp hybrid orbitals and two 2p orbitals, and each O atom having its 6 valence electrons distributed as one lone pair and a single electron in two sp hybrid orbitals, and a lone pair and a single electron in two 2p orbitals. The C atom utilizes its two singly-occupied sp hybrid orbitals to form two colinear σ bonds with the singly-occupied sp hybrid orbital of each O atom; the two singly-occupied 2p orbitals on C overlap, respectively, with the singly-occupied 2p orbital of each O atom to form two π bonds.

---

51. For each of the following molecules use the VSEPR model to find the approximate value of each of the marked angles:

(a) 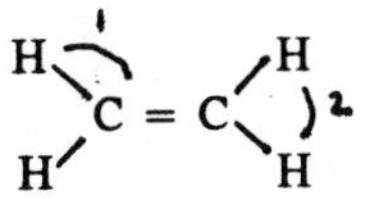 Ethene (b) 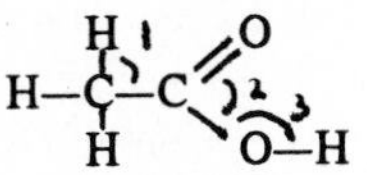 Acetic acid (c) 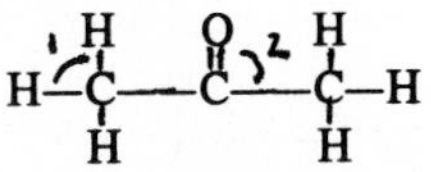 Acetone

---

The approximate values of each of the marked bond angles comes from considering the arrangement of the number of bond and unshared (lone) pairs in the valence shell of the atom at the centre of each angle:

Molecule	Bond Angle	Arrangement of Bond and Unshared Electron Pair Domains on Atom at Centre of Bond Angle	Expected Bond Angle
(a) Ethene	1. <HCC 2. <HCH	AX_3 triangular planar AX_3 triangular planar	120° 120°
(b) Acetic acid	1. <HCC 2. <OCO 3. <COH	AX_4 tetrahedral AX_3 triangular planar AX_2E_2 tetrahedral	109.5° 120° 109.5°
(c) Acetone	1. <HCH 2. <OCC	AX_4 tetrahedral AX_3 triangular planar	109.5° 120°

---

53. For each of the molecules in Problem 51, describe the bonding in terms of hybrid orbitals and the σ-π model. How many σ bonds and how many π bonds are there in each molecule?

(a) Ethene Each C atom is AX_3 triangular planar and forms σ bonds to two H atoms and to the other C atom, utilizing three sp^2 hybrid orbitals, leaving a single electron in a 2p orbital on each C atom. The parallel p orbitals then overlap sideways to form the π bond component of the C=C double bond. **Total bonds = five σ bonds** and **one π bond.**

(b) Acetic acid The C atom of the CH_3 group is AX_4 tetrahedral and utilizes four sp^3 hybrid orbitals to form σ bonds to three H atoms and the adjacent C atom, which is AX_3 triangular planar and forms three sp^2 hybrid orbitals. One of these is utilized in forming the σ bond to the CH_3 group, the second forms a σ bond to the O atom of the C=O group, and the third forms a σ bond to the O atom of the OH group. The O atom of the C=O group has two unshared pairs in sp^2 hybrid orbitals and utilizes the third sp^2 hybrid orbital to form the σ bond to carbon. This leaves singly-occupied 2p orbitals on the C and O atoms of the C=O group that overlap sideways to form a π bond. The O atom of the —OH group has four sp^3 hybrid orbitals. Two of these are occupied by unshared pairs, and the remaining two form σ bonds to the adjacent C and H atoms. In summary: the C atom of the —CH_3 group forms four σ bonds (three to H atoms and one to the adjacent C atom) and the C atom of the —CO_2H group forms one σ bond to the —OH group, and one σ bond and one π bond to the other O atom, and in addition to the σ bond to C the O atom of —OH forms another σ bond to the H atom. Thus, acetic acid contains a total of **seven σ bonds** and **one π bond.**

(c) Acetone Each of the two —CH_3 groups is AX_4 tetrahedral and utilizes four sp^3 hybrid orbitals to form σ bonds to each H atom and the C atom of the C=O group, for a total of eight σ bonds. The AX_3 triangular planar C atom of the C=O group has three sp^2 hybrid orbitals; two are used to form the σ bonds to the adjacent —CH_3 groups and the third forms a σ bond to the O atom. This leaves singly occupied 2p orbitals on the C and O atoms of the C=O group that overlap sideways to form the π bond component of the C=O double bond. **Total bonds = nine σ bonds** and **one π bond.**

General Problems

55. Excited barium atoms in a bunsen burner flame can return to their ground state by emitting photons of energy 3.62×10^{-19} J. What color is imparted to the flame by light of this wavelength?

The photons emitted with an energy of 3.62×10^{-19} J have a wavelength, λ, given by:

$$E_{photon} = h\nu = \frac{hc}{\lambda};\ \lambda = \frac{hc}{E} = \frac{(6.63 \times 10^{-34}\ \text{J s})(3.00 \times 10^{8}\ \text{m s}^{-1})}{3.62 \times 10^{-19}\ \text{J}} = \underline{5.49 \times 10^{-7}\ \text{m}}\ (549\ \text{nm})$$

Light of this wavelength is in the <u>green</u> region of the visible spectrum. The green color imparted by barium to hot flames is utilized in fireworks and to identifying barium compounds in the laboratory by emission spectroscopy, or by <u>flame tests</u> in which an inert platinum wire moistened with a solution of a barium compound is heated in a hot bunsen burner flame.

---

57. When light of wavelength 470.0 nm falls on the surface of potassium metal, electrons are emitted with a velocity of 6.4×10^4 m·s^{-1}. What is (a) the kinetic energy of the emitted electrons? (b) the energy of a 470.0-nm photon? (c) the minimum energy required to remove an electron from potassium metal?

(a) $$KE = \frac{1}{2}mv^2 = \frac{1}{2}(9.11 \times 10^{-31}\ \text{kg})(6.4 \times 10^{4}\ \text{m s}^{-1})^2\left(\frac{1\ \text{J}}{1\ \text{kg m}^2\ \text{s}^{-2}}\right)\left(\frac{6.022 \times 10^{23}}{1\ \text{mol}}\right) = \underline{1.124 \times 10^{3}\ \text{J mol}^{-1}}$$

(b) For one 470 nm photon: $$E = h\nu = \frac{hc}{\lambda} = \frac{(6.63 \times 10^{-34}\ \text{J s})(3.00 \times 10^{8}\ \text{m s}^{-1})}{(470.0\ \text{nm})\left(\frac{1\ \text{m}}{10^{9}\ \text{nm}}\right)} = \underline{4.23 \times 10^{-19}\ \text{J}}$$

and for 1 mole of 470.0 nm photons:

$$E = (4.232 \times 10^{-19}\ \text{J})\left(\frac{1\ \text{kJ}}{10^{3}\ \text{J}}\right)\left(\frac{6.022 \times 10^{23}}{1\ \text{mol}}\right) = \underline{254.9\ \text{kJ mol}^{-1}}$$

(c) The minimum energy required to remove an electron from potassium metal is the difference between the energy of a 470.0 nm photon and the kinetic energy of an emitted electron, which is given by:

$$(254.9 - 1.1)\ \text{kJ mol}^{-1} = \mathbf{253.8\ kJ\ mol^{-1}}$$

---

60. Explain why two nitrogen atoms attract each other when brought together to form N_2, two oxygen atoms attract each other to form O_2, and two fluorine atoms attract each other to form F_2, while two neon atoms do not attract each other to form Ne_2.

The atoms in question have the following valence shell electron configurations:

	2s	2p		2s	2p		2s	2p		2s	2p
N	[↑↓]	[↑][↑][↑]	O	[↑↓]	[↑↓][↑][↑]	F	[↑↓]	[↑↓][↑↓][↑]	Ne	[↑↓]	[↑↓][↑↓][↑↓]

In principle, any atom that has unpaired electrons in one or more atomic orbitals can attract unpaired electrons in atomic orbitals of another atom to share electron pairs and form covalent bonds.

Nitrogen, N, with three singly occupied atomic orbitals and a lone pair in its valence shell, can attract another N atom to form **three** bonding molecular orbitals each containing two electrons, and in each of which there is a small increase in the electron density in the internuclear region, which pulls the two atoms together by electrostatic attraction. The bond in N_2 is therefore described as a triple bond:

$$:N{\equiv}N:$$

Oxygen, O, with two singly occupied atomic orbitals and two lone pairs in its valence shell, can similarly combine with another O atom to form two bonding molecular orbitals, so that the bond between the two O atoms in O_2 is described as a double bond:

$$:\ddot{O}{=}\ddot{O}:$$

Fluorine, with one one singly occupied atomic orbital and three lone pairs in its valence shell, can similarly combine with another F atom to form only one bonding molecular orbital, so that the bond between the two F atoms in F_2 is described as a single bond:

$$:\underset{..}{\ddot{F}}{-}\underset{..}{\ddot{F}}:$$

In contrast, **neon,** Ne, with four lone pairs in its valence shell has no orbitals containing unpaired electrons and cannot form any bonds.

CHAPTER 7

Sulfur

2. Give an example and write a balanced equation for each of the following types of reaction of sulfuric acid: (a) acid-base, (b) oxidation-reduction; (c) dehydration.

(a) Reactions in which sulfuric acid acts as an acid (transfers a proton to another substance) are numerous, and include, for example:
Sulfuric acid is a strong acid in water; it reacts completely to give the hydrogensulfate ion HSO_4^-:

$$H_2SO_4(\ell) + H_2O(\ell) \rightarrow H_3O^+(aq) + HSO_4^-(aq)$$

The HSO_4^- ion is also an acid although it is a weak acid as its reaction with water is incomplete:

$$HSO_4^-(aq) + H_2O(\ell) \rightleftarrows H_3O^+(aq) + SO_4^{2-}(aq)$$

Sulfuric acid is a diprotic acid with two ionizable hydrogen atoms; it reacts with bases to form two series of salts, the hydrogensulfates and the sulfates, for example:

$$H_2SO_4(aq) + KOH(aq) \rightarrow \underset{\text{potassium hydrogensulfate}}{KHSO_4(aq)} + H_2O(\ell)$$

$$H_2SO_4(aq) + 2KOH(aq) \rightarrow \underset{\text{potassium sulfate}}{K_2SO_4(aq)} + 2H_2O(\ell)$$

(b) Reactions in which sulfuric acid acts as an oxidizing agent are those in which a reactant is oxidized and H_2SO_4 is reduced to $SO_2(g)$. Examples include:

$$2Br^- + 5H_2SO_4(\ell) \rightarrow Br_2 + SO_2(g) + 2H_3O^+ + 4HSO_4^-$$

$$2I^- + 5H_2SO_4(\ell) \rightarrow I_2 + SO_2(g) + 2H_3O^+ + 4HSO_4^-$$

$$Cu(s) + 5H_2SO_4(\ell) \rightarrow Cu^{2+} + SO_2(g) + 2H_3O^+ + 4HSO_4^-$$

$$2Ag(s) + 5H_2SO_4(\ell) \rightarrow 2Ag^+ + SO_2(g) + 2H_3O^+ + 4HSO_4^-$$

(c) Water is completely ionized in solution in sulfuric acid. It behaves as a strong base:

$$H_2O(\ell) + H_2SO_4(\ell) \rightarrow H_3O^+ + HSO_4^-$$

This property makes concentrated sulfuric acid a very good **dehydrating agent**; it removes water (or the elements of water) from many substances, for example, **hydrated salts**, such as $CuSO_4 \cdot 5H_2O$, lose their water of crystallization when stored in a closed container (desiccator) with concentrated sulfuric acid. A small amount of water vapor is normally in equilibrium with blue hydrated copper sulfate but when this water is absorbed by sulfuric acid equilibrium is never established and the hydrated salt gradually loses water and is eventually completely converted to white anhydrous copper sulfate:

$$\underset{\text{blue}}{CuSO_4 \cdot 5H_2O} \rightarrow \underset{\text{white}}{CuSO_4} + \underset{\text{absorbed by } H_2SO_4}{5H_2O}$$

The tendency of sulfuric acid to combine with water is so strong that it will remove hydrogen and oxygen as water from many compounds that contain no H_2O molecules. For example, from carbohydrates and many other organic compounds, leaving a residue of carbon. Wood, paper, starch, cotton, and sugar (sucrose) are all dehydrated in this way. For example:

$$\underset{\text{sucrose}}{C_{12}H_{22}O_{11}(s)} + 11H_2SO_4(\ell) \rightarrow 12C(s) + 11H_3O^+ + 11HSO_4^-$$

---

4. Write a balanced equation for the reaction of (a) zinc with dilute sulfuric acid, (b) concentrated sulfuric acid with sodium iodide; (c) concentrated sulfuric acid with copper; (d) dilute sulfuric acid with solid magnesium hydroxide.

(a) Zn reduces H_3O^+ in aqueous solution to give H_2O and $H_2(g)$

$$Zn + 2H_3O^+ \rightarrow Zn^{2+} + [2H_3O] \downarrow 2H_2O + H_2(g)$$

For the overall reaction: $Zn(s) + H_2SO_4(aq) \rightarrow Zn^{2+}(aq) + SO_4^{2-}(aq) + H_2(g)$

(b) Concentrated sulfuric acid oxidizes iodide ion, I^-, to iodine, and is itself reduced to SO_2, for the overall reaction:

$$2I^- + 5H_2SO_4(\ell) \rightarrow I_2 + SO_2(g) + 2H_3O^+ + 4HSO_4^-$$

or $$2NaI(s) + 5H_2SO_4(\ell) \rightarrow I_2(s) + SO_2(g) + 2Na^+ + 2H_3O^+ + 4HSO_4^-$$

(c) Concentrated sulfuric acid oxidizes metallic Cu to Cu^{2+} ions, and is itself reduced to SO_2:

$$Cu(s) + 5H_2SO_4(\ell) \rightarrow Cu^{2+} + SO_2(g) + 2H_3O^+ + 4HSO_4^-$$

(d) Aqueous sulfuric acid reacts with the base $Mg(OH)_2(s)$ to give the soluble salt $MgSO_4$:

$$Mg(OH)_2(s) + H_2SO_4(aq) \rightarrow Mg^{2+}(aq) + SO_4^{2-}(aq) + 2H_2O(\ell)$$

---

6. An aqueous solution is known to contain either $SO_4^{2-}(aq)$ or $SO_3^{2-}(aq)$, but not both. Suggest two tests that could be used to identify the anion in the solution.

The characteristic test for $SO_4^{2-}(aq)$ in solution is to add a solution of a soluble barium salt, such as $BaCl_2(aq)$. A white precipitate of $BaSO_4(s)$ indicates the presence of sulfate ion. (All sulfites, including $BaSO_3(s)$, are soluble).

$$Ba^{2+}(aq) + SO_4^{2-}(aq) \rightarrow BaSO_4(s)$$

A number of tests could detect the presence of $SO_3^{2-}(aq)$, sulfite ion, containing S(IV), a good reducing agent. It is readily oxidized to $SO_4^{2-}(aq)$ containing S(VI). For example, it readily reduces $Br_2(aq)$ to bromide ion, $Br^-(aq)$:

$$Br_2(aq) + SO_3^{2-}(aq) + 3H_2O(\ell) \rightarrow 2Br^-(aq) + SO_4^{2-}(aq) + 2H_3O^+(aq)$$

Also, addition of a dilute acid, such as HCl(aq), to a solution containing sulfite ion liberates $SO_2(g)$, because sulfite ion is a weak base:

$$SO_3^{2-}(aq) + 2H_3O^+(aq) \rightarrow 2H_2SO_3(aq)$$

and sulfurous acid, H_2SO_3, is unstable in aqueous solution:

$$H_2SO_3(aq) \rightarrow SO_2(g) + H_2O(\ell)$$

The $SO_2(g)$ is readily detected by its characteristic odor, and its bleaching properties; for example, it bleaches moistened litmus paper.

---

8. Water from springs and wells is typically contaminated with small concentrations of hydrogen sulfide, which gives it a bad smell. The H_2S can be removed by treating the water with chlorine, which oxidizes the H_2S to sulfur. (a) Write a balanced equation for this reaction. (b) If the H_2S content of the water from a particular source is 2 parts per million (ppm) by mass, how much chlorine will be needed to remove the H_2S from 5000 L of water?

(a) H_2S is oxidized to S(s), so $Cl_2(aq)$ must be reduced to $Cl^-(aq)$, so we can write:

$$Cl_2(aq) + H_2S(aq) \rightarrow 2Cl^-(aq) + 2H^+(aq) + S(s)$$

(b) The H_2S content of 5000 L of spring water is:

$$(5000 \text{ L water})\left(\frac{1000 \text{ g water}}{1 \text{ L water}}\right)\left(\frac{22 \text{ g } H_2S}{10^6 \text{ g water}}\right) = \underline{1.10 \times 10^2 \text{ g } H_2S}$$

We now convert this mass to moles of H_2S, then, using the balanced equation, to moles, and finally to grams of chlorine:

$$\text{Mass } Cl_2 = (1.10 \times 10^2 \text{ g } H_2S)\left(\frac{1 \text{ mol } H_2S}{34.09 \text{ g } H_2S}\right)\left(\frac{1 \text{ mol } Cl_2}{1 \text{ mol } H_2S}\right)\left(\frac{70.90 \text{ g } Cl_2}{1 \text{ mol } Cl_2}\right) = \underline{2 \times 10^2 \text{ g } Cl_2}$$

---

10. Write the formula for each of the following compounds: (a) potassium sulfate (b) calcium hydrogensulfate (c) calcium sulfite (d) potassium diphosphate (e) aluminum sulfate

---

All salts are composed of ions. Thus, the formula of a salt comes from combining the appropriate ions to give a neutral compound:

(a) $(K^+)_2SO_4^{2-}$ (b) $Ca^{2+}(HSO_4^-)_2$ (c) $Ca^{2+}SO_3^{2-}$ (d) $(K^+)_4\ ^{2-}O_3P\text{-}O\text{-}PO_3^{2-}$ (e) $(Al^{3+})_2(SO_4^{2-})_3$,

that is **(a) K_2SO_4 (b) $Ca(HSO_4)_2$ (c) $CaSO_3$ (d) $K_4P_2O_7$ (e) $Al_2(SO_4)_3$**

---

12. Name each of the following and classify each as ionic or covalent. Draw Lewis structures for (c) through (g). (a) $Na_2S(s)$ (b) $MgS(s)$ (c) $S_8(s)$ (d) $CS_2(\ell)$ (e) $SO_2(g)$ (f) $SO_3(g)$ (g) $H_2SO_4(\ell)$.

Formula	Name	Type	Structure
(a) Na_2S	sodium sulfide	ionic	$(Na^+)_2\ S^{2-}$
(b) MgS	magnesium sulfide	ionic	$Mg^{2+}\ S^{2-}$
(c) S_8	orthorhombic sulfur	covalent	eight-membered ring: S–S–S–S–S–S–S–S (each S with two lone pairs)
(d) CS_2	carbon disulfide	covalent	$:\ddot{S}=C=\ddot{S}:$
(e) SO_2	sulfur dioxide	covalent	$:\ddot{O}=S=\ddot{O}:$
(f) SO_3	sulfur trioxide	covalent	$:\ddot{O}=S(=\ddot{O}:)=\ddot{O}:$
(g) H_2SO_4	sulfuric acid	covalent	$(H\text{–}\ddot{O}:)_2S(=\ddot{O}:)_2$

---

14. Complete and balance each of the following equations:
(a) $H_2SO_4(conc) + Cu(s) \rightarrow Cu^{2+} +$ (b) $H_2SO_4(aq) + CuS(s) \rightarrow$ (c) $Na_2SO_3(aq) + H_2SO_4(aq) \rightarrow$
(d) $H_2S(aq) + Cl_2(aq) \rightarrow$ (e) $H_2S(aq) + Pb^{2+}(aq) \rightarrow$ (f) $H_2SO_4(conc) + NaI(s) \rightarrow$

(a) Cu(s) is oxidized to Cu^{2+}, and H_2SO_4(conc) is reduced to $SO_2(g)$:

$$5H_2SO_4 + Cu(s) \rightarrow Cu^{2+} + SO_2(g) + 2H_3O^+ + 4HSO_4^-$$

(b) $H_3O^+(aq)$ in $H_2SO_4(aq)$ protonates S^{2-} ion in CuS(s) to give $H_2S(g)$:

$$H_2SO_4(aq) + CuS(s) \rightarrow Cu^{2+}(aq) + SO_4^{2-}(aq) + H_2S(g)$$

(c) $H_3O^+(aq)$ in $H_2SO_4(aq)$ protonates SO_3^{2-} ion in $Na_2SO_3(aq)$ to give $H_2SO_3(aq)$ (which is unstable and decomposes to $SO_2(g)$ and water:

$$Na_2SO_3(aq) + H_2SO_4(aq) \rightarrow 2Na^+(aq) + SO_4^{2-} + H_2O(\ell) + SO_2(g)$$

(d) $Cl_2(g)$ oxidizes S^{2-} ion to sulfur and is reduced to chloride ion, Cl^-:

$$H_2S(aq) + Cl_2(aq) \rightarrow 2HCl(aq) + S(s)$$

(e) $Pb^{2+}(aq)$ and $S^{2-}(aq)$ combine to give a precipitate of lead sulfide, PbS(s):

$$H_2S(aq) + Pb^{2+}(aq) + 2H_2O(\ell) \rightarrow PbS(s) + 2H_3O^+(aq)$$

(f) Concentrated sulfuric acid oxidizes $I^-(aq)$ to iodine, and is reduced to $SO_2(g)$:

$$2I^- + 5H_2SO_4 \rightarrow I_2 + SO_2(g) + 2H_3O^+ + 4HSO_4^-$$

16*. A compound contains 23.7% S, 52.6% Cl, and 23.7% O by mass and has a boiling point of 69°C. The volume occupied by 0.337 g of the gaseous compound at 100°C and 770 mm Hg pressure is 75.6 mL. (a) Find the empirical and molecular formulas. (b) Draw a Lewis structure for the molecule.

(a) The first step is to calculate the **empirical formula** of the compound from the given analytical data, assuming a 100 g sample:

Element:	S	Cl	O
100 g contains:	23.7 g	52.6 g	23.7 g
Moles:	$\frac{23.7}{32.06}$	$\frac{42.6}{35.45}$	$\frac{23.7}{16.00}$
Ratio of moles (atoms):	0.739 :	1.48 :	1.48
=	$\frac{0.739}{0.739}$:	$\frac{1.48}{0.739}$:	$\frac{1.48}{0.739}$
=	1.00 :	2.01 :	2.01

Thus, the **empirical formula** is SCl_2O_2 (formula mass 135.0 u).

From the gas data we can calculate the moles of the compound in the 0.337-g sample, and hence the molar mass:

$$\text{Moles} = n = \frac{PV}{RT} = \frac{(770\ \text{mm Hg})(\frac{1\ \text{atm}}{760\ \text{mm Hg}})((75.6\ \text{mL})(\frac{1\ \text{L}}{1000\ \text{mL}})}{(0.0821\ \text{atm L mol}^{-1}\ \text{K}^{-1})(373\ \text{K})} = \underline{2.50 \times 10^{-3}\ \text{mol}}$$

$$\text{Thus, molar mass} = \frac{0.337\ \text{g}}{2.50 \times 10^{-3}\ \text{mol}} = \underline{135\ \text{g mol}^{-1}}$$

The molar mass is the same as the formula mass, so the **molecular formula** is SO_2Cl_2.

(b) The relatively low boiling point of 69°C suggests that the compound is a covalent substance. The oxidation state of sulfur in the compound is S(VI), suggesting the **Lewis structure**:

```
       ..
       O:
  ..   ||  ..
 :Cl—S—Cl:      (with AX4 tetrahedral geometry)
  ..   ||  ..
       O:
       ..
```

Phosphorus

18. Explain the following properties of the allotropes of phosphorus in terms of their structures: (a) White phosphorus has a lower melting point than red phosphorus and is more volatile. (b) White phosphorus is soluble in carbon disulfide, $CS_2(l)$, whereas red phosphorus is insoluble.

(a) White phosphorus consists of P_4 molecules and is a molecular solid. To melt it, energy is only required to overcome to some extent the rather weak intermolecular forces between the P_4 molecules to allow them to slide past each other. In contrast, red phosphorus is an infinite covalent network solid. To melt red phosphorus, many relatively strong P—P covalent bonds have to be broken. Similarly, $P_4(s)$ is much more volatile than red phosphorus because it is relatively easy for P_4 molecules to break away from the solid since no covalent bonds have to be broken.

(b) The relatively small P_4 molecules of white phosphorus go into solution in the covalent carbon disulfide, $CS_2(\ell)$, solvent. In contrast red phosphorus is insoluble in carbon disulfide, because in order for it to go into solution many covalent P-P bonds would have to be broken.

---

20. Striking a match involves the combustion of P_4S_3 to produce a white smoke of $P_4O_{10}(s)$ and $SO_2(g)$. Write a balanced equation for this reaction. Calculate the maximum volume of $SO_2(g)$ at 20°C and 772 mm Hg pressure that results from the combustion of 0.157 g of P_4S_3.

The balanced equation for the reaction is:

$$P_4S_3(s) + 8O_2(g) \rightarrow P_4O_{10}(s) + 3SO_2(g)$$

We first calculate the moles of SO_2 produced by the complete combustion of 0.157 g of $P_4S_3(s)$, and then use the ideal gas law to calculate the maximum possible volume of SO_2:

$$\text{moles } SO_2 = (0.157 \text{ g } P_4S_3)\left(\frac{1 \text{ mol } P_4S_3}{220.1 \text{ g } P_4S_3}\right)\left(\frac{3 \text{ mol } SO_2}{1 \text{ mol } P_4S_3}\right) = \underline{2.140 \times 10^{-3} \text{ mol}}$$

$$V_{SO_2} = \frac{nRT}{P} = \frac{(2.140 \times 10^{-3} \text{ mol})(0.0821 \text{ atm L mol}^{-1} \text{ K}^{-1})(293 \text{ K})}{(772 \text{ mm Hg})\left(\frac{1 \text{ atm}}{760 \text{ mm Hg}}\right)} = \underline{5.07 \times 10^{-2} \text{ L}} \text{ (50.7 mL)}$$

---

22. (a) What reaction is likely between $PH_3(g)$ and $HCl(g)$? (b) Write an equation for the reaction. (c) Draw a Lewis structure for the compound formed and suggest a name for it. (d) Would you predict it to be a solid, a liquid, or a gas?

(a) Like ammonia, NH_3, the common hydride of nitrogen in group V, PH_3, the hydride of phosphorous, also in group V, would be expected to behave as a base. In the same way that $NH_3(g)$ and $HCl(g)$ react to give a white cloud of ammonium chloride, $NH_4^+Cl^-(s)$, $PH_3(g)$ and $HCl(g)$ would also be expected to react:

$$PH_3(g) + HCl(g) \rightarrow PH_4Cl(s)$$

to give the **salt**, $PH_4^+Cl^-(s)$, a **solid** composed of PH_4^+ and Cl^- ions.

(c) The Lewis structure is:

```
      H
      |        ..
   H—P⁺—H    :Cl:⁻
      |        ..
      H
```

which by analogy with ammonium chloride is called **phosphonium chloride.**

(d) Since it is an ionic compound, phosphonium chloride is a **solid.**

---

24. Write a formula for each of the following compounds: (a) tetraphosphorus hexaoxide (b) calcium dihydrogenphosphate (c) calcium phosphide (d) phosphorous acid (e) sodium triphosphate

---

(a) P_4O_6 (b) $Ca(H_2PO_4)_2$ (c) Ca_3P_2 (d) H_3PO_3 (e) $Na_5P_3O_{10}$

---

26. Starting with white phosphorus, write balanced equations for the preparation of (a) tetraphosphorus hexaoxide; (b) phosphorus pentachloride; (c) phosphoric acid; (d) phosphine.

(a) When phosphorus is burned in a limited supply of air, the principal product is **tetraphosphorus hexaoxide:**

$$P_4(s) + 3O_2(g) \rightarrow P_4O_6(s)$$

(b) **Phosphorus pentachloride** is the product of the reaction of phosphorus with excess chlorine gas:

$$P_4(s) + 10Cl_2(g) \rightarrow 4PCl_5$$

(c) When phosphorus is burned in excess oxygen, the principal product is **tetraphosphorus pentaoxide,** $P_4O_{10}(s)$, the anhydride of phosphoric acid.

$$P_4(s) + 5O_2(g) \rightarrow P_4O_{10}(s)$$

Thus, burning white (or red) phosphorus in oxygen and dissolving the $P_4O_{10}(s)$ formed in the exactly calculated amount of water from the balanced equation

$$P_4O_{10}(s) + 6H_2O(\ell) \rightarrow 4H_3PO_4(\ell)$$

gives **phosphoric acid.**

(d) **Phosphine** can be made by the direct reaction of red or white phosphorus with hydrogen

$$2P(s) + 3H_2(g) \rightarrow 2PH_3(g)$$

or by first forming a metal phosphide, for example by heating calcium with phosphorus

$$3Ca(s) + 2P(s) \rightarrow Ca_3P_2(s)$$

followed by the reaction of the phosphide with dilute aqueous acid

$$[Ca^{2+}]_3[P^{3-}]_2(s) + 6H_3O^+(aq) \rightarrow 3Ca^{2+}(aq) + 2PH_3(g) + 6H_2O(\ell)$$

28.* Dry hydrogen fluoride, HF(g), reacts with $P_4O_{10}(s)$ to give a gas containing 29.8% P, 54.8% F, and 15.4% O by mass. This gas has a density of 4.64 $g \cdot L^{-1}$ at STP. (a) What is its molecular formula? (b) Suggest a possible Lewis structure.

(a) We first calculate the empirical formula of the phosphorus oxofluoride from its mass % composition:

Empirical Formula		P		O		F
100 g compound contains:		29.8 g		15.4 g		54.8 g
moles	=	$\frac{28.9\ g}{30.97\ g}$		$\frac{15.4\ g}{16.00\ g}$		$\frac{54.8\ g}{19.00\ g}$
mole (atom) ratio	=	0.933	:	0.963	:	2.88

$$\text{i.e.,} \quad \frac{0.933}{0.933} : \frac{0.963}{0.933} : \frac{2.88}{0.933} = \underline{1.00 : 1.03 : 3.09}$$

Giving the **empirical formula** POF_3 (formula mass 104.0 g mol^{-1})

Using the density and the relationship that 1 mol of an ideal gas at STP has a volume of 22.4 L:

$$\text{molar mass} = \left(\frac{4.64\ \text{g}}{1\ \text{L}}\right)\left(\frac{22.4\ \text{L}}{1\ \text{mol}}\right) = \underline{104\ \text{g mol}^{-1}}$$

and this molar mass is equal to the formula mass, so the **molecular formula is POF_3**.

(b) The oxidation number of P is +5, so the likely Lewis structure has all of the atoms bonded to a central phosphorus atom with a total of <u>five</u> covalent bonds (3 single P—F bonds and 1 double P=O bond):

```
      ..
     :F:
  ..  |  ..
 :F—P=O:
  ..  |
     :F:
      ..
```

<u>Oxoacids</u>

30. Explain why KOH and $Ca(OH)_2$ are strong bases in aqueous solution, while $Si(OH)_4$ is a very weak acid.

Potassium (Group I) and calcium (Group II) have low electronegativities compared to oxygen and therefore their X-O bonds are highly polar (ionic) in the sense $X^{\delta+}$-$O^{\delta-}$, so K^+OH^- and $Ca^{2+}(OH^-)_2$ dissociate fully into their cations and OH^- ions in aqueous solution. Thus they behave as **strong bases**.

In contrast, silicon has a higher electronegativity than K or Ca and is a nonmetal, and silicic acid, $Si(OH)_4$, is a covalent molecule. The Si-O bond is correspondingly less polar and both the O atom and the Si atom pull electrons away from the hydrogen atoms, so that the O—H bonds becomes polar in the sense Si—$O^{\delta-}$—$H^{\delta+}$; the hydrogen atom can be donated as H^+ to a base and silicic acid is a **weak acid.**

---

32. Write the formulas of the following ions and classify each as a strong acid or a weak acid:
(a) hydrogensulfate ion (b) hydrogenphosphate ion (c) dihydrogenphosphate ion (d) hydrogensulfite ion.

All anions are weaker acids than the corresponding parent acids, because it is more difficult to remove a proton from a negative ion than it is from a neutral acid, and the greater the negative charge on the anion the weaker the anion is as an acid in aqueous solution relative to its parent acid:

Anion	Formula	Acid Strength
(a) hydrogensulfate ion	$HOSO_3^-$	weak
(b) hydrogenphosphate ion	$HOPO_3^{2-}$	weak
(c) dihydrogenphosphate ion	$(HO)_2PO_2^-$	weak
(d) hydrogensulfite ion	$HOSO_2^-$	weak

---

34. Write balanced equations for all of the possible reactions of each of the following oxoacids with sodium hydroxide in aqueous solution, and name each of the possible salts that could be obtained from the solutions:
(a) nitric acid (b) sulfuric acid (c) phosphoric acid (d) perchloric acid.

The number of salts obtainable from an oxoacid depends on its number of ionizable H^+ ions).

(a) **Nitric acid,** $HONO_2$, is **a monoprotic** acid, and gives only **nitrates:**

$$HNO_3(aq) + NaOH(aq) \rightarrow NaNO_3(aq) + H_2O(\ell)$$

sodium nitrate

(b) **Sulfuric acid,** $(HO)_2SO_2$, is a **diprotic** acid and gives **hydrogensulfates** and **sulfates**

$$H_2SO_4(aq) + NaOH(aq) \rightarrow NaHSO_4(aq) + H_2O(\ell)$$

sodium hydrogensulfate

$$H_2SO_4(aq) + 2NaOH(aq) \rightarrow Na_2SO_4(aq) + 2H_2O(\ell)$$

sodium sulfate

(c) **Phosphoric acid,** $(HO)_3PO$, is a **triprotic** acid and gives **dihydrogenphosphates, hydrogenphosphates,** and **phosphates:**

$$H_3PO_4(aq) + NaOH(aq) \rightarrow NaH_2PO_4(aq) + H_2O(\ell)$$

sodium dihydrogenphosphate

$$H_3PO_4(aq) + 2NaOH(aq) \rightarrow Na_2HPO_4(aq) + 2H_2O(\ell)$$

sodium hydrogenphosphate

$$H_3PO_4(aq) + 3NaOH(aq) \rightarrow Na_3PO_4(aq) + 3H_2O(\ell)$$

sodium phosphate

(d) **Perchloric acid,** $HOClO_3$, is a **monprotic** acid and gives only **perchlorates:**

$$HClO_4(aq) + NaOH(aq) \rightarrow NaClO_4(aq) + H_2O(\ell)$$

sodium perchlorate

---

36. What molecules or ions are present in aqueous solutions of each of the following? Give both names and formulas: (a) sulfuric acid (b) sulfur dioxide (c) phosphoric acid (d) carbonic acid

---

(a) **Sulfuric acid,** $(HO)_2SO_2$, is a **diprotic acid:**

$$H_2SO_4(aq) + H_2O(\ell) \rightarrow H_3O^+(aq) + HSO_4^-(aq)$$

hydronium ion hydrogensulfate ion

$$HSO_4^-(aq) + H_2O(\square) \rightleftarrows H_3O^+(aq) + SO_4^{2-}(aq)$$

hydrogensulfate ion hydronium ion sulfate ion

(b) **Sulfur dioxide,** $SO_2(g)$, dissolves to form (to a small extent) **sulfurous acid,** a diprotic acid:

$$SO_2(aq) + H_2O(\ell) \rightleftarrows H_2SO_3(aq), \text{ the}$$

$$H_2SO_3(aq) + H_2O(\ell) \rightleftarrows H_3O^+(aq) + HSO_3^-(aq)$$

sulfurous acid hydronium ion hydrogensulfite ion

$$HSO_3^-(aq) + H_2O(\ell) \rightleftarrows H_3O^+(aq) + SO_3^{2-}(aq)$$

hydrogensulfite ion hydronium ion sulfite ion

(c) **Phosphoric acid,** $(HO)_3PO$, a weak **triprotic acid:**

$$H_3PO_4(aq) + H_2O(\ell) \rightleftarrows H_3O^+(aq) + H_2PO_4^-(aq)$$

phosphoric acid hydronium ion dihydrogenphosphate ion

$$H_2PO_4^-(aq) + H_2O(\ell) \rightleftarrows H_3O^+(aq) + HPO_4^{2-}(aq)$$

dihydrogenphosphate ion hydronium ion hydrogenphosphate ion

$$HPO_4^{2-}(aq) + H_2O(\ell) \rightleftarrows H_3O^+(aq) + PO_4^{3-}(aq)$$

hydrogenphosphate ion hydronium ion phosphate ion

(d) **Carbonic acid,** $(HO)_2CO$, a **diprotic acid,** which is only present in water in a low concentration:

$$H_2CO_3(aq) \rightleftarrows CO_2(aq) + H_2O(\ell)$$

carbonic acid carbon dioxide

$$H_2CO_3(aq) + H_2O(\ell) \rightleftarrows H_3O^+(aq) + HCO_3^-(aq)$$

carbonic acid **hydronium ion** **hydrogen carbonate ion**

$$HCO_3^-(aq) + H_2O(\ell) \rightleftarrows H_3O^+(aq) + CO_3^{2-}(aq)$$

hydrogencarbonate ion **hydronium ion** **carbonate ion**

Chlorine

38. Write equations for the half-reactions that describe the behavior of each of ClO_3^-, ClO_2^-, and ClO^- when they are reduced to chloride ion in acidic aqueous solution.

$$ClO_3^- + 6e^- + 6H^+ \rightarrow Cl^- + 3H_2O$$

$$ClO_2^- + 4e^- + 4H^+ \rightarrow Cl^- + 2H_2O$$

$$ClO^- + 2e^- + 2H^+ \rightarrow Cl^- + H_2O$$

40*. Bleaching powder is obtained by reacting slaked lime, $Ca(OH)_2(s)$, with chlorine. It has the composition 36.6% Ca, 43.2% Cl, 19.5% O, and 0.6% H by mass. When excess $AgNO_3(aq)$ was added to 1.000 g of bleaching powder dissolved in water, 0.874 g of AgCl(s) precipitated. When an acidified aqueous solution containing 1.000 g of bleaching powder was titrated with 0.100-M KI(aq), 121.9 mL of the KI(aq) was needed to reduce all of the hypochlorite ion, $OCl^-(aq)$, to chloride ion, $Cl^-(aq)$. (a) What is the empirical formula of bleaching powder? (b) Write a balanced equation for the preparation of bleaching powder.

Empirical Formula:	Ca	O	Cl	H
100 g compound contains:	36.6 g	19.5 g	43.2 g	0.6 g

$$\text{Moles} = \frac{36.6\ g}{40.08\ g} \quad \frac{19.5\ g}{16.00\ g} \quad \frac{43.2\ g}{35.45\ g} \quad \frac{0.6\ g}{1.008\ g}$$

$$\text{mole (atom) ratio} = 0.913 : 1.219 : 1.219 : 0.595$$

$$= \frac{0.913}{0.595} : \frac{1.219}{0.595} : \frac{1.219}{0.595} : \frac{0.595}{0.595} = \underline{1.53 : 2.05 : 2.05 : 1.00} = \underline{3 : 4 : 4 : 2}$$

The **empirical formula** is $Ca_3O_4Cl_4H_2$, and the description of the chemistry of its aqueous solution suggests that it is an ionic compound consisting of Ca^{2+}, Cl^-, OCl^-, and OH^- ions.

$$\text{Moles of bleaching powder} = (1.000\ g)\left(\frac{1\ mol}{327.9\ g}\right) = \underline{3.050 \times 10^{-3}\ mol}$$

and the reaction with $AgNO_3(aq)$ gives the amount of $Cl^-(aq)$ ion in the solution from the precipitation reaction:

$$Ag^+(aq) + Cl^-(aq) \rightarrow AgCl(s)$$

For the reaction of $OCl^-(aq)$ with $I^-(aq)$, the balanced equation is:

$$OCl^-(aq) + 2I^-(aq) + 2H^+(aq) \rightarrow Cl^-(aq) + I_2(s) + H_2O(\ell)$$

Thus, moles of OCl^- in 1.000 g of bleaching powder is given by:

$$\text{mol AgCl(s)} = (0.874 \text{ g AgCl})(\frac{1 \text{ mol AgCl}}{143.4 \text{ g AgCl}}) = \underline{6.095 \times 10^{-3} \text{ mol}}$$

Thus, for 1.000 g bleaching powder, X: $\frac{\text{mol } Cl^-}{\text{mol X}} = \frac{6.095 \times 10^{-3} \text{ mol}}{3.050 \times 10^{-3} \text{ mol}} = \underline{2.0}$

$$(121.9 \text{ mL})(\frac{1 \text{ L}}{10^3 \text{ mL}})(\frac{0.100 \text{ mol } I^-}{1 \text{ L}})(\frac{1 \text{ mol } OCl^-}{2 \text{ mol } I^-}) = \underline{6.10 \times 10^{-3} \text{ mol}}$$

Thus, for 1.000 g of X: $\frac{\text{mol } OCl^-}{\text{mol X}} = \frac{6.10 \times 10^{-3}}{3.05 \times 10^{-3}} = \underline{2.0}$

and we can rewrite the empirical formula $Ca_3O_4Cl_4H_2$ as:

$$(Ca^{2+})_3(Cl^-)_2(OCl^-)_2(OH^-)_2$$

(b) For the formation of OCl^- and Cl^- in <u>basic</u> solution:

$$Cl_2(aq) + 4OH^- \rightarrow 2OCl^-(aq) + 2e^- + H_2O(\ell)$$
$$Cl_2(aq) + 2e^- \rightarrow 2Cl^-(aq)$$
$$\overline{2Cl_2(aq) + 4OH^-(aq) \rightarrow 2OCl^-(aq) + 2Cl^-(aq) + 2H_2O(\ell)}$$

Thus, for the formation of bleaching powder from $Ca(OH)_2(s)$ and $Cl_2(g)$:

$$3Ca(OH)_2 + 2Cl_2 \rightarrow Ca_3[Cl_2(OCl)_2(OH)_2]$$

42*. (a) Using the VSEPR model, predict the geometry of the molecules ClF_3 and ClF_5. How many different possible geometries can you find for ClF_3 and Cl_5? (b) Given that lone pairs always occupy the three equivalent equatorial sites of a trigonal bipyramid, draw the predicted shape of ClF_3. Can you explain why this is often called a T-shaped structure?

(a) First we draw the Lewis structures for ClF_3 and ClF_5:

ClF_3 1. Connecting the F atoms to Cl by single bonds gives

```
  F
  |
F—Cl—F
```

2. All the atoms are in group VII with 7 valence electrons, so the total number of valence electrons is 4x7 = <u>28</u>.

3. 3x2 = <u>6</u> electrons are needed to form three Cl-F bonds, leaving 22 electrons; the octet around each F atom is completed by assigning to each atom 3 unshared pairs (6 electrons), for a total of 3x6 = <u>18</u> electrons, so that only 22-18 = <u>4</u> electrons remain to be assigned as <u>two</u> unshared pairs to the Cl atom, giving the Lewis structure:

```
   ..
   :F:
 ..  |  ..
:F—Cl—F:
 ..  ..  ..
```

ClF_5 1. Connecting the F atoms to Cl by single bonds gives

```
   F
   |
F—Cl—F
  /  \
 F    F
```

2. All the atoms are in group VII with 7 valence electrons, so the total number of valence electrons is 6x7 = <u>42</u>.

3. 5x2 = 10 electrons are needed to form the five Cl-F bonds, leaving 32 electrons; the octet around each F atom is completed by assigning to each atom 3 unshared pairs (6 electrons), for a total of 5x6 = 30 electrons, so that only 32-30 = 2 electrons remain to be assigned as an unshared pair to the Cl atom, giving the Lewis structure:

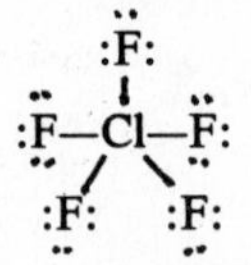

:F:
:F—Cl—F:
:F: :F:

Geometries:
According to the VSEPR model, ClF_3 is an AX_3E_2 type molecule; its structure is based on the **trigonal bipyramidal** arrangement of five electron pairs in the valence shell of Cl. The *geometry* is defined by the position of the Cl atom and the three F atoms, which can be arranged in *three* different ways:

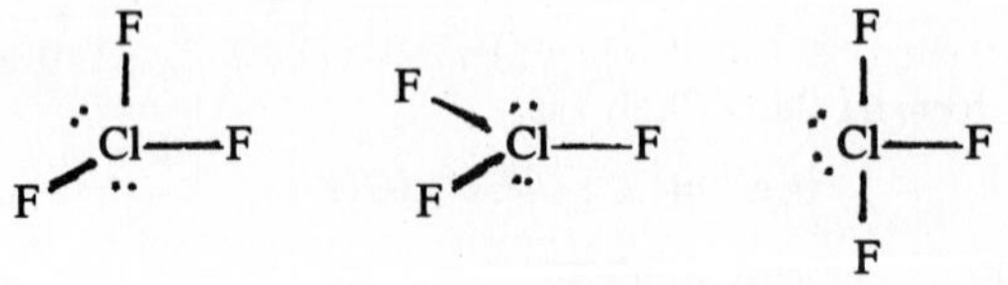

I II III

According to the VSEPR model, ClF_5 is an AX_5E type molecule; its structure is based on the **octahedral** arrangement of six electron pairs in the valence shell of Cl. The **geometry** is defined by the position of the Cl atom and the five F atoms, and is described as a **square pyramid**, and there is only one possible structure because all the corners of an octahedron are equivalent.

(b) Since in the trigonal bipyramidal arrangement of five electron pairs, lone pairs preferably occupy equatorial positions (around the central atom) rather than the axial positions above or below the central atom, of the three possible structures structure III is preferred. This has a linear F-Cl-F arrangement of atoms with the third F atom perpendicular to this linear arrangement (the shape of a "T"), so ClF_3 is conveniently described as a **"T-shaped" molecule.**

Oxidation Numbers and Oxidation States

44. What are the common oxidation states of (a) sulfur and (b) phosphorus in their compounds? Give an example of a compound for each oxidation state.

(a) **Sulfur**

Oxidation State	Examples
-II	H_2S, HS^-, S^{2-}
0	S_8, S_n
+II	SO, SF_2, SCl_2
+IV	SO_2, SO_3^{2-}, HSO_3^-, SF_4,
+VI	SO_3, H_2SO_4, HSO_4^-, SO_4^{2-}, SF_6

(b) **Phosphorus**

Oxidation State	Examples
-III	PH_4^+, PH_3, P^{3-}
0	P_4, S_n
+III	P_4O_6, PCl_3, PF_3, H_3PO_3, $H_2PO_3^-$, HPO_3^{2-}
+V	P_4O_{10}, H_3PO_4, $H_2PO_4^-$, HPO_4^{2-}, PO_4^{3-}, PCl_5, PF_5, $POCl_3$, POF_3

---

46. What are the oxidation numbers of each of the following? (a) C in CO (b) S in SO_2 (c) P in $Ca_5(PO_4)_3(OH)$ (d) S in SF_6 (e) S in S_2^{2-} (f) P in $H_2PO_3^-$

(a) CO	ON(0) = —2, so	**ON(C) = +2**
(b) SO_2	ON(O) = —2, so ON(S) + 2(—2) = 0;	**ON(S) = +4**
(c) $Ca_5(PO_4)_3(OH)$	contains P in the PO_4^{3-} ions, for which ON(P) + 4ON(O) = —3; ON(P) + 4(—2) = —3;	**ON(P) = +5**
(d) SF_6	ON(S) + 6ON(F) = 0; ON(S) + 6(—1) = 0;	**ON(S) = +6**
(e) S_2^{2-}	2ON(S) = —2;	**ON(S) = —1**
(f) $H_2PO_3^-$	2ON(H) + ON(P) + 3ON(O) = -1; 2(+1) + ON(P) + 3(-2) = -1;	**ON(P) = +3**

---

48. Assign oxidation numbers to each element in each of the following: (a) N_2 (b) H_2CO_3 (c) NH_3 (d) PH_3 (e) MnO_2 (f) HNO_3

(a) N_2	**N(0)**	(b) H_2CO_3	**H(+1) C(+4) O(—2)**	(c) NH_3	**N(—3) H(+1)**
(d) PH_3	**P(—3) H(+1)**	(e) MnO_2	**Mn(+4) O(—2)**	(f) HNO_3	**H(+1) N(+5) O(—2)**

---

50. (a) If an atom in a molecule is oxidized, does its oxidation number increase or decrease? (b) If an atom in a molecule is reduced, does its oxidation number increase or decrease? (c) Write an equation for each of the following half-reactions in aqueous acid: (i) the oxidation of Cu^+ to Cu^{2+}; (ii) the reduction of S to S^{2-}; (iii) the reduction of NO_3^- to NO.

(a) **Oxidation** is **electron loss**, which makes the oxidation number of an atom more *positive*, and therefore **increases** the oxidation number of the atom.

(b) **Reduction** is **electron gain**, which makes the oxidation number of an atom more *negative*, and therefore **decreases** the oxidation number of the atom.

(c) (i) The ON. of Cu increases from +1 to +2: $Cu^+ \rightarrow Cu^{2+} + e^-$

(ii) The O.N. of S decreases from 0 to —2: $S + 2e^- \rightarrow S^{2-}$

(iii) The O.N. of N decreases from +5 to +2: $NO_3^- + 3e^- + 4H^+ \rightarrow NO + 2H_2O$

---

52. (a) When an ionic bromide is heated with concentrated sulfuric acid, bromide ion is oxidized to bromine, and the sulfuric acid is reduced to sulfur dioxide. Write the balanced equation for this reaction. (b) When sulfur dioxide is bubbled through an aqueous solution of bromine, the bromine is reduced to bromide ion, and the sulfur dioxide is oxidized to sulfate ion. Write the balanced equation for this reaction.

(a) For the reaction of Br^- ion with concentrated H_2SO_4:

$$2Br^- \rightarrow Br_2 + 2e^- \quad \text{(oxidation)}$$
$$H_2SO_4 + 2e^- + 2H^+ \rightarrow SO_2 + 2H_2O \quad \text{(reduction)}$$
$$2Br^- + H_2SO_4 + 2H^+ \rightarrow Br_2 + SO_2 + 2H_2O \quad \text{(balanced)}$$

Since the reaction is in concentrated H_2SO_4, this must be the source of the $2H^+$, and the H_2O formed must react with the acid to give H_3O^+ and HSO_4^-, so the reaction is more realistically written as:

$$2Br^- + 5H_2SO_4 \rightarrow Br_2 + SO_2 + 2H_3O^+ + 4HSO_4^-$$

(b) For the reaction of bromine in aqueous solution:

$$Br_2 + 2e^- \rightarrow 2Br^- \quad \text{(reduction)}$$
$$SO_2 + 2H_2O \rightarrow SO_4^{2-} + 2e^- + 4H^+ \quad \text{(oxidation)}$$
$$Br_2 + SO_2 + 2H_2O \rightarrow 2Br^- + SO_4^{2-} + 4H^+ \quad \text{(balanced)}$$

Note: The position of equilibrium lies far to the right in the above reaction because in dilute aqueous solution all the sulfuric acid is ionized to SO_4^{2-}(aq) and H^+(aq), so that there is no unionized molecular H_2SO_4. Only molecular H_2SO_4 is the strong oxidizing agent needed to oxidize Br^- to Br_2.

Lewis Structures

53. Draw Lewis structures for the following molecules and ions, and then use the VSEPR model to predict the geometry around the S atoms: (a) SO_2 (b) SO_3 (c) H_2SO_4 (d) SO_3^{2-}

(a) **SO_2** 1. Connecting the atoms by single bonds gives O—S—O

2. S and O are both in group VI, so there is a total of 3(6) = **18** valence electrons.

3. 2x2 = <u>4</u> electrons are needed to form the two S-O bonds, leaving (18-4) = **14** electrons, and three pairs of electrons (6) are needed to complete the octets of electrons around each O atom, leaving 14-2(6) = **2** electrons, which are assigned as an **unshared pair** to the central S atom, to give: :Ö—S̈—Ö:

4. The S atom has an incomplete octet, so a lone pair is moved from one of the O atoms to form an additional S—O bond, thus converting an S—O bond to an S=O bond, and giving: :Ö—S̈=Ö:
Finally, calculating formal charges gives: ⁻:Ö—S̈⁺=Ö:

5. Sulfur is from period 3, so the formal charges on the —O:⁻ and S atoms can be removed by transferring a lone pair from this O atom into the SO bonding region, to finally give the most appropriate Lewis structure as:

:Ö=S̈=Ö: **AX_2E** angular

(b) **SO_3** 1. Connecting the atoms by single bonds gives O—S—O (with a third O bonded to S below)

2. S and O are both in group VI, so there is a total of 4(6) = **24** valence electrons.

3. 3x2 = **6** electrons are needed to form the three S—O single bonds, leaving (24-6) = **18** electrons; three pairs of electrons (6 electrons) are needed to complete the octets around each O atom, leaving 18-3(6) = **0** electrons. All the electrons have now been assigned to give, after assigning formal charges:

:Ö:⁻ (bonded to S)
⁻:Ö—S—Ö:⁻ (S bearing 3+)

5. Sulfur is from period 3, so the formal charges on the —O:⁻ and S^{3+} atoms can be removed by transferring a lone pair from each of the O atoms to the bonding region, to give finally:

$$:\ddot{O}{=}\ddot{S}{=}\ddot{O}: \text{ (with } \ddot{O}: \text{ double-bonded above S)}$$ **AX_3 triangular planar**

(c) **H_2SO_4** 1. Sulfuric acid is best written as $(HO)_2SO_2$, and connecting the atoms by single bonds gives:

```
O     O—H
 \   /
   S
 /   \
O     O—H
```

2. S and O are both in group VI, and H is in group I, so there is a total of 5(6)+2(1) = 32 valence electrons.

3. 6x2 = **12** electrons are needed to form the four S—O bonds and two O—H bonds, leaving (32-12) = **20** electrons; three pairs of electrons (6 electrons) are needed to complete the octets around each of two —O atoms, and two pairs to complete the octets around each of the O atoms of the —O—H groups, for a total of 2(6)+2(4) = **20** electrons, leaving (20-20) = **0** electrons. All the electrons have been assigned to give, after assigning formal charges:

```
⁻:Ö:     Ö—H
    \   /
     S²⁺
    /   \
⁻:Ö:     Ö—H
```

5. Sulfur is from period 3, so the formal charges on the -O:⁻ and S^{2+} atoms can be removed by transferring a lone pair from each of the O atoms to the bonding region, to give finally:

```
:Ö     Ö—H
   \\  /
    S
   // \
:Ö     Ö—H
```

AX_4 tetrahedral

(d) **SO_3^{2-}** 1. Connecting the atoms by single bonds gives O—S—O (with a third O bonded below S)

2. S and O are both in group VI. There is a total of 4(6)+2 = **26** valence electrons, allowing for the 2- charge.

3. 3x2 = **6** electrons are needed to form the three S—O bonds, leaving (26-6) = **20** electrons; three pairs of electrons (6 electrons) are needed to complete the octets around each O atom, leaving 20-3(6) = **2** electrons to be assigned as an unshared pair to the S atom, to give, after assigning formal charges:

```
       :Ö:⁻
        |
⁻:Ö—S⁺—Ö:⁻
```

5. Sulfur is from period 3, so the formal charges on the —O:⁻ and S^+ atoms can be removed by transferring a lone pair from *one* of the O atoms to the bonding region, to give finally:

```
     :Ö:⁻
      |
:Ö=S—Ö:⁻
```

AX_3E triangular pyramidal

---

55. (a) On the basis of the VSEPR model, what are the expected shapes of AX_5 and AX_6 molecules? (b) Give examples of an AX_5 and an AX_6 molecule.

(a) The geometry of an **AX_6 type molecule** is that of **six** electron pairs arranged around a central atom A so that they maximize their distances apart; the six electron pairs (domains) are arranged at the corners of **a regular octahedron** (square bipyramid).
The geometry of an **AX_5 type molecule** is that of **five** electron pairs arranged around a central atom A so that they maximize their distances apart; the five electron pairs (domains) are arranged at the corners of **a regular triangular bipyramid.**

(b) **AX_6 octahedral molecules:** SF_6, PF_6^-, PCl_6^-
AX_5 triangular bipyramidal molecules: PCl_5, PF_5, AsF_5

57*. Draw Lewis structures for each of the following: (a) phosphoric acid (b) diphosphoric acid (c) fluorosulfate ion, FSO_3^- (d) fluorophosphate ion, FPO_3^{2-}.

For the method, see Problem 54.

```
(a)        :Ö—H      (b)   H—Ö:    :Ö—H        (c)      Ö:         (d)      :Ö:⁻
            |                 |       |                   ||                   |
       H—Ö—P=Ö:          H—Ö—P—Ö—P—Ö—H             :F—S=Ö:              :F—P=Ö:
            |                 ||      ||                  |                    |
           :Ö—H               Ö:      Ö:                 :Ö:⁻                 :Ö:⁻
```

phosphoric acid **diphosphoric acid** **fluorosulfate ion** **fluorophosphate ion**

General Problems

58. A sample of coal contains 5.00% sulfur by mass. When burned in a power plant, it gives sulfur dioxide in the stack gas, which is converted in the atmosphere to sulfuric acid and deposited as acid rain. (a) Write the simplest equations to show how sulfur dioxide is converted to sulfuric acid in the atmosphere. (b) Calculate the mass of sulfuric acid that could result from burning 1000 kg of this coal. (c) How much $Ca(OH)_2(s)$ would be required to neutralize this sulfuric acid?

(a) The $SO_2(g)$ produced by burning the sulfur in the coal reacts with $O_2(g)$ from the air to give sulfur trioxide, $SO_3(g)$, which in turn reacts with water vapor or rain in the atmosphere to give sulfuric acid:

$$2SO_2(g) + O_2(g) \rightarrow 2SO_3(g)$$

$$SO_3(g) + H_2O(g/\ell) \rightarrow H_2SO_4(aq)$$

(b) $$10^3 \text{ kg coal contains } \left(\frac{5.00 \text{ g S}}{100 \text{ g coal}}\right)(10^3 \text{ kg coal}) = \underline{50.0 \text{ kg S}}$$

$$\text{Mass of } H_2SO_4 = (50.0 \text{ kg S})\left(\frac{1 \text{ mol S}}{32.07 \text{ g S}}\right)\left(\frac{1 \text{ mol } H_2SO_4}{1 \text{ mol S}}\right)\left(\frac{98.08 \text{ g } H_2SO_4}{1 \text{ mol } H_2SO_4}\right) = \underline{153 \text{ kg } H_2SO_4}$$

(c) For neutralization of H_2SO_4 with $Ca(OH)_2$, we have:

$$H_2SO_4 + Ca(OH)_2 \rightarrow CaSO_4 + 2H_2O$$

Thus, the mass of $Ca(OH)_2(s)$ to neutralize 153 kg H_2SO_4 is given by:

$$(153 \text{ kg } H_2SO_4)\left(\frac{1 \text{ mol } H_2SO_4}{98.08 \text{ g } H_2SO_4}\right)\left(\frac{1 \text{ mol } Ca(OH)_2}{1 \text{ mol } H_2SO_4}\right)\left(\frac{74.10 \text{ g } Ca(OH)_2}{1 \text{ mol } Ca(OH)_2}\right) = \underline{116 \text{ kg } Ca(OH)_2}$$

60*. A yellow solid A was heated in air and gave a colorless gas B, which when dissolved in water gave a colorless solution C. Another sample of the yellow solid was mixed with iron powder and heated strongly to produce a black solid D. When dilute sulfuric acid was added to D, a colorless gas E formed. When E was bubbled through C, a yellow precipitate formed. It was found to be identical to A in its chemical and physical properties. Identify A, B, C, D, and E, and write balanced equations for all of the reactions described.

Among the yellow solids encountered in this chapter, the most important was **sulfur**, whose reactions fit the observations described.

A is **sulfur** which when heated in air burns to give **sulfur dioxide gas**, **B**, according to the equation:

$$S(s) + O_2(g) \rightarrow SO_2(g)$$

and $SO_2(g)$ is soluble in water to give a solution of **sulfurous acid**, $H_2SO_3(aq)$, **C**:

$$SO_2(g) + H_2O(\ell) \rightleftarrows H_2SO_3(aq)$$

Sulfur, **A**, combines with iron when heated strongly, to give **iron(II)sulfide**, FeS(s), **D**, which is a black solid:

$$Fe(s) + S(s) \rightarrow FeS(s)$$

and FeS(s), **D**, reacts with dilute sulfuric acid to give **hydrogen sulfide gas**, $H_2S(g)$, **E**:

$$FeS(aq) + H_2SO_4(aq) \rightarrow FeSO_4(aq) + H_2S(g)$$

$H_2S(g)$, **E**, is oxidized in aqueous solution to yellow sulfur, **A**, by $H_2SO_3(aq)$, **C**:

$$2H_2S(aq) + H_2SO_3(aq) \rightarrow 3S(s) + 3H_2O(\ell)$$

62*. An element X is a reactive white solid that is insoluble in water but soluble in organic solvents, such as carbon disulfide, from which it can be recovered by evaporating the solvent. It melts at 44°C and boils at 280°C. As a vapor X has a density of 2.242 $g{\cdot}L^{-1}$ at 400$_o$C and 1 atm pressure. When 25.0 mL of X(g) reacts completely with 150 mL of chlorine at a given temperature and pressure, it gives 100 mL of a gaseous chloride at the same temperature and pressure. This chloride contains 77.4% Cl by mass, boils below 100$_o$C, is liquid at room temperature, and has a molar mass of 137.3 $g{\cdot}mol^{-1}$. (a) Identify X. (b) Draw its Lewis structure and that of its chloride. (c) Write the balanced equation for the reaction of X(g) with $Cl_2(g)$.

(a) X has the properties described for white phosphorus, P_4, which is confirmed by calculating the molar mass from the gas data:

$$\text{mol X} = \frac{PV}{RT} = \frac{(1\ \text{atm})(1\ \text{L})}{(0.0821\ \text{atm L mol}^{-1}\ \text{K}^{-1})(673\ \text{K})} = \underline{1.81 \times 10^{-2}\ \text{mol}}$$

$$\text{molar mass} = \frac{2.242\ \text{g}}{1.81\text{x}10^{-2}\ \text{mol}} = \underline{124\ \text{g mol}^{-1}}$$

which is to be compared to **123.9 g mol^{-1}**, the molar mass of P_4

For the gaseous reaction of $P_4(g)$ with $Cl_2(g)$, we deduce from Avogadro's law of combining volumes:

$$P_4(g) + 6Cl_2 \rightarrow 4PCl_3(g)$$

25 mL 150 mL 100 mL

1 vol 6 vol 4 vol

and the formation of PCl_3 (molar mass 137.3 g mol^{-1}) is confirmed by the observed molar mass of the product (137.3 g mol^{-1}), and by its elemental composition:

$$\text{mass \% Cl for PCl}_3 = \left(\frac{106.4\ \text{g Cl mol}^{-1}}{137.3\ \text{g PCl}_3\ \text{mol}^{-1}}\right) \times 100\% = \underline{77.5\%}$$

Which is to be compared with the experimental value of **77.4%**.

(b) Lewis structures: **P_4** **PCl_3**

```
        ..
        P
      / | \
  :P---+---P:
      \ | /
        P
        ..
```

```
  ..   ..   ..
 :Cl—P—Cl:
  ..   |    ..
      :Cl:
       ..
```

(c) $P_4(g) + 6Cl_2 \rightarrow 4PCl_3(g)$

CHAPTER 8

1. Account for the rarity of diamond and the very different physical properties of diamond and graphite.

The position of equilibrium between diamond and graphite lies to the right at ordinary temperature and pressure, but this change is infinitessimally slow at ordinary temperatures because many strong carbon-carbon bonds in diamond have to be broken for the change to occur. At high pressure and temperature, the position of the equilibrium shifts to the left, and small industrial diamonds are made from graphite in this way using a catalyst. This is also what apparently must also have occurred rarely in places like South Africa in the hot earth's mantle 200-300 km below the surface, where the pressure is sufficiently high. The diamonds were then carried to the surface by volcanic activity. Diamond has an infinite three-dimensional covalent lattice structure in which each carbon atom is joined by four strong covalent bonds to a tetrahedral arrangement of four other surrounding carbon atoms. This structure makes it very hard and chemically unreactive. In contrast, graphite consists of infinite two-dimensional sheets of carbon atoms stacked upon each other at relatively large distances, with the layers held together only by weak intermolecular forces. Graphite can be thought of as an almost infinitely large polycyclic arene. Its layers can slide over each other relatively easily, so that graphite is a soft substance that feels slippery and is a good lubricant.

---

3. In terms of their molecular structures, explain why the bond length in diamond is 154 pm, whereas all the bonds in the planar layers of graphite have equal lengths of 142 pm, intermediate in length between that of a single CC bond (154 pm) and a double CC bond (134 pm).

The valence of carbon (Group IV) is 4, and diamond has an infinite covalent network structure in which each C atom is bonded to four others and the bonds to each carbon atom have an AX_4 tetrahedral arrangement. Thus, all the bonds are single bonds, consistent with the observed bond length of 154 pm. **Graphite** consists of infinite sheets of carbon atoms in which each carbon atom is bonded to **three** other carbon atoms, and the sheets are bonded only by weak intermolecular forces, allowing them to slip easily over each other. Thus, the arrangement of bonds to each carbon atom is AX_3 triangular planar, and many Lewis structures can be written for graphite in which each carbon atom is joined to three neighboring C atoms by **two** C-C single bonds and **one** C=C double bond, and in which each C atom forms part of a regular hexagon. Averaging these structures, gives each CC bond in graphite a bond order of $\frac{1}{3}(1+1+2)$, or $1\frac{1}{3}$, which accounts for the observed bond length of 142 pm, shorter than in diamond (bond order 1) and intermediate in length between the length of a single bond and a double bond of bond order 2.

Inorganic Compounds of Carbon

5. Write balanced equations for the reaction of carbon monoxide with each of the following, and state the conditions under which each reaction occurs: (a) $H_2(g)$ (b) $O_2(g)$ (c) $H_2O(g)$ (d) $Fe_2O_3(s)$.

(a) $CO(g) + H_2(g) \xrightarrow[\text{high T}]{\text{catalyst}} H_2CO(g) \xrightarrow{H_2} H_3COH$

(b) $2CO(g) + O_2(g) \xrightarrow[\text{high T}]{\text{combustion}} 2CO_2(g)$

(c) $CO(g) + H_2O(g) \xrightarrow[\text{high T}]{\text{catalyst}} CO_2(g) + H_2(g)$

(d) $Fe_2O_3(s) + 3CO(g) \xrightarrow{\text{high T}} 2Fe(s) + 3CO_2(g)$

---

6. Write balanced equations for each of the following: (a) two ways of preparing carbon monoxide; (b) two ways of preparing carbon dioxide; (c) one way of preparing carbon disulfide.

(a) **Carbon monoxide** is made in the laboratory by burning carbon (charcoal) in a limited supply of air:

$$2C(s) + O_2(g) \rightarrow 2CO(g)$$

Industrial methods include the reaction of natural gas with steam at high temperature in the presence of a catalyst:

$$CH_4(g) + H_2O(g) \rightarrow CO(g) + 3H_2(g) \text{ (synthesis gas)}$$

(b) **Carbon dioxide** is the combustion product of complete reaction of carbon (charcoal or coke) in oxygen (or air):

$$C(s) + O_2(g) \rightarrow CO_2(g)$$

Industrially it results from heating calcium carbonate:

$$CaCO_3(s) \rightarrow CaO(s) + CO_2(g)$$

or from the complete combustion of any hydrocarbon:

$$2C_4H_{10}(g) + 13O_2(g) \rightarrow 8CO_2(g) + 10H_2O(g)$$

In the laboratory it is conveniently made by reacting a metal carbonate with an acid such as sulfuric acid:

$$Na_2CO_3(s) + H_2SO_4(\ell) \rightarrow Na_2SO_4(s) + CO_2(g) + H_2O(\ell)$$

(c) **Carbon disulfide** results from the reaction of methane gas with sulfur:

$$CH_4(g) + 4S(s) \rightarrow CS_2(\ell) + 2H_2S(g)$$

---

8. What is the chemical composition of each of the following? (a) lime (b) soda water (c) natural gas (d) coke (e) chalk

---

(a) **Lime** is calcium oxide, $CaO(s)$; (b) **Soda water** is an aqueous solution of carbon dioxide, $CO_2(aq)$; also known as carbonated water. (c) **Natural gas** is a mixture of hydrocarbon gases, mainly methane, $CH_4(g)$, with some ethane, $C_2H_6(g)$, and propane, $C_3H_8(g)$. (d) **Coke** is an impure form of carbon obtained by strongly heating coal in the absence of air. (e) **Chalk** is a form of calcium carbonate, $CaCO_3(s)$.

---

10. Write a balanced equation to describe the reaction that occurs when methane is strongly heated with each of the following: (a) $H_2O(g)$ (b) $NH_3(g)$ (c) $O_2(g)$

(a) This is the reaction for producing *synthesis gas*:

$$CH_4(g) + H_2O(g) \rightarrow CO(g) + 3H_2(g)$$

(b) Methane and ammonia react at high temperature, using a catalyst:

$$CH_4(g) + NH_3(g) \rightarrow HCN(g) + 3H_2(g)$$

(c) With limited $O_2(g)$, carbon monoxide is formed:

$$2CH_4(g) + 3O_2(g) \rightarrow 2CO(g) + 4H_2O(g)$$

and with excess $O_2(g)$, combustion to give $CO_2(g)$ is complete:

$$CH_4(g) + 2O_2(g) \rightarrow CO_2(g) + 2H_2O(g)$$

---

12. (a) Write balanced equations for the preparation of ethyne, starting with calcium oxide, coke and water. (b) Why is the carbide ion, C_2^{2-}, completely converted to acetylene, $C_2H_2(g)$ in water?

(a) First **calcium carbide** has to be synthesized from calcium oxide and coke; this can then be reacted with water to give ethyne (acetylene):

$$CaO(s) + 3C(s) \rightarrow CO(g) + CaC_2(s)$$

$$CaC_2(s) + 2H_2O(\ell) \rightarrow Ca(OH)_2(s) + C_2H_2(g)$$

(b) Acetylene, H—C≡C—H, is a very weak acid. so carbide ion, $^-$:C≡C:$^-$, is a strong base and the reaction

$$C_2^{2-}(aq) + 2H_2O(\ell) \rightleftarrows H_2C_2(g) + 2OH^-(aq)$$

is driven to completion as the acetylene, which is insoluble in water, escapes.

14*. When concentrated sulfuric acid is dripped onto carbon tetrabromide at 160°C, a (gaseous) compound containing 6.4% C, 85.0% Br, and 8.6% O by mass, is obtained. At 25°C and a pressure of 1.00 atm, 0.940 g of this compound has a volume of 122 mL. (a) What is the molecular formula of the compound? (b) Draw its Lewis structure, and deduce its molecular shape.

From the mass % composition we can calulate the empirical formula:

$$\text{Moles of C per 100 g compound} = (6.4\text{ g C})\left(\frac{1\text{ mol C}}{12.01\text{ g C}}\right) = 0.53\text{ mol C}$$

$$\text{Moles of Br per 100 g compound} = (85.0\text{ g Br})\left(\frac{1\text{ mol Br}}{79.90\text{ g Br}}\right) = 1.06\text{ mol Br}$$

$$\text{Mol O per 100 g compound} = (8.6\text{ g O})\left(\frac{1\text{ mol O}}{16.00\text{ g O}}\right) = 0.54\text{ mol O}$$

$$\text{Ratio of moles (atoms) C : Br : O} = 0.53 : 1.06 : 0.54 = \frac{0.53}{0.53} : \frac{1.06}{0.53} : \frac{0.54}{0.53} = \underline{1.00 : 2.00 : 1.02}$$

Therefore, the **empirical formula** is CBr_2O (formula mass 187.8 u), and from the other data:

$$\text{moles of gas} = n = \frac{PV}{RT} = \frac{(1.00\text{atm})(122\text{ mL})\left(\frac{1\text{ L}}{10^3\text{ mL}}\right)}{(0.0821\text{ atm L mol}^{-1}\text{ K}^{-1})(298\text{ K})} = \underline{4.99 \times 10^{-3}\text{ mol}}$$

$$\text{Thus:} \quad \text{molar mass} = \frac{0.940\text{ g}}{4.99 \times 10^{-3}\text{ mol}} = \underline{188\text{ g mol}^{-1}}$$

Thus, the **molecular formula** is CBr_2O, the same as the empirical formula, and since the compound was prepared from CBr_4 it must retain two C—Br bonds and have the Lewis structure

```
 ..     ..
:Br—C—Br:
 ..  ||  ..
     O:
     ..
```

with AX_3, **triangular planar**, geometry.

15.* When we heat aluminum oxide with coke in an electric furnace, we obtain a yellow carbide of aluminum that is stable up to 1400°C and reacts with water to give methane. We react a sample of the carbide of mass 0.500 g with excess water and collect the methane in a 250-mL bulb at 25 °C. The pressure in the bulb is 1.02 atm. (a) What is the empirical formula of the carbide? (b) What mass of aluminum carbide is needed to produce 20.0 L of methane at 25°C and 1.00 atm pressure?

(a) A carbide contains only a metal and carbon, in this case Al and C, and the mass of the $CH_4(g)$ from a known mass of carbide gives the amount of C it contains and, by difference, the amount of Al:

$$\text{Mol CH}_4 = n = \frac{PV}{RT} = \frac{(1.02\text{ atm})(250\text{ mL})(\frac{1\text{ L}}{10^3\text{ mL}})}{(0.0821\text{ atm L mol}^{-1}\text{ K}^{-1})(298\text{ K})} = 0.817\text{ mol}$$

$$\text{Mass of C} = (0.817\text{ mol CH}_4)(\frac{1\text{ mol C}}{1\text{ mol CH}_4})(\frac{12.01\text{ g C}}{1\text{ mol C}}) = 0.125\text{ g C}$$

and, by difference: **Mass of Al** in the 0.500 g sample = (0.500-0.125) = **0.375 g**

$$\text{moles of Al} = (0.375\text{ g Al})(\frac{1\text{ mol Al}}{26.98\text{ g Al}}) = \underline{0.0139\text{ mol}}\ ; \quad \text{moles of C} = (0.125\text{ g C})(\frac{1\text{ mol C}}{12.01\text{ g C}}) = \underline{0.0104\text{ mol}}$$

$$\text{Thus, ratio mol Al : mol C} = \text{ratio of atoms} = 0.0139 : 0.0104 = \frac{0.139}{1.04} : \frac{0.104}{0.104} = 1.34 : 1.00 \quad \text{i.e., } \underline{4 : 3}$$

The **empirical formula** of the carbide is Al_4C_3 - which is consistent with the valences (3 for Al and 4 for C). Thus, the balanced equation for the reaction with water is:

$$\mathbf{Al_4C_3(s) + 6H_2O(\ell) \rightarrow 2Al_2O_3(s) + 3CH_4(g)}$$

(b) To calculate the mass of carbide required to produce 20.0 L of methane at 25°C and 1.00 atm pressure, we first calculate the equivalent moles of CH_4, convert this to moles of Al_4C_3, and then to grams of Al_4C_3:

$$n_{CH_4} = \frac{PV}{RT} = \frac{(1.00\text{ atm})(20.0\text{ L})}{(0.0821\text{ atm L mol}^{-1}\text{ K}^{-1})(298\text{ K})} = 0.8175\text{ mol CH}_4$$

$$\text{Mass Al}_4\text{C}_3 = (0.8175\text{ mol CH}_4)(\frac{1\text{ mol Al}_4\text{C}_3}{3\text{ mol CH}_4})(\frac{144.0\text{ g Al}_4\text{C}_3}{1\text{ mol Al}_4\text{C}_3}) = \underline{39.2\text{ g}}$$

Alkanes, Alkenes, and Alkynes; Benzene and the Arenes: Aromatic Hydrocarbons

Hydrocarbons: Names, Formulas, and Isomers

16. Give the names and draw the Lewis structures of the alkanes with (a) one, (b) two, and (c) three carbon atoms each.

```
   H           H  H           H  H  H
   |           |  |           |  |  |
H—C—H      H—C—C—H       H—C—C—C—H
   |           |  |           |  |  |
   H           H  H           H  H  H
```

methane **ethane** **propane**

---

18. Give the molecular formula of each of the following: (a) an alkane with eight carbon atoms; (b) an alkene with six carbon atoms and one double bond; (c) an alkyne with five carbon atoms and one triple bond; (d) a cycloalkane containing a six-membered ring of carbon atoms.

The general formulas are C_nH_{2n+2} for an **alkane**, C_nH_{2n} for an **alkene** with one double bond, or a **cycloalkane**, and C_nH_{2n-2} for an **alkyne** with one triple bond:

Thus: (a) C_nH_{2n+2} with n = 8 gives $\mathbf{C_8H_{18}}$ (b) C_nH_{2n} with n = 6 gives $\mathbf{C_6H_{12}}$

(c) C_nH_{2n-2} with n = 5 gives C_5H_8 (d) C_nH_{2n} with n = 6 gives C_6H_{12}

---

20. Which of the following names do not conform to IUPAC rules for naming organic compounds? Rename those that are incorrect: (a) 2-ethylbutane (b) 3,3-dimethylbutane (c) 1-ethylpropane (d) 2,2-dimethylpropane (e) 1,1-dimethylpropane

	Name Given	Structural Formula	Comment
(a)	2-ethylbutane	CH_3—CH—CH_2—CH_3 (with CH_2—CH_3 on the second C)	Compound has **five** C atoms in a continuous chain and should be renamed **3-methylpentane**
(b)	3,3-dimethylbutane	CH_3—CH_2—C—CH_3 (with two CH_3 on the third C)	The substituent CH_3 groups should have the lowest possible number in the continuous chain of 4 C atoms; it should be renamed **2,2-dimethylbutane**
(c)	1-ethylpropane	CH_3—CH_2 attached to CH_2—CH_2—CH_3	This has a continuous chain of 5 C atoms and should be renamed **pentane**
(d)	2,2-dimethylpropane	CH_3—C—CH_3 (with two CH_3 on the central C)	This has a continous chain of 3 C atoms and is **correctly named**
(e)	1,2-dimethylpropane	H_2C—CH—CH_3 (with H_3C on the first C and CH_3 on the second C)	This has a continous chain of 4 C atoms and should be renamed **2-methylbutane**

---

21. Draw the structure and give the IUPAC names of each of the isomers with the molecular formula $C_2H_2Cl_2$.

There are three possible substituted dichloroethenes with the molecular formula $C_2H_2Cl_2$:

```
Cl      H          Cl      H          Cl      Cl
  \    /             \    /             \    /
   C=C                C=C                C=C
  /    \             /    \             /    \
Cl      H           H      Cl          H      H
```

1,1-dichloroethene **trans-1,2-dichloroethene** **cis-1,2-dichloroethene**

---

23. Draw the structural formula and give the IUPAC name of each of the nine isomers of heptane.

To deduce the structures of all the isomers of heptane, C_7H_{16}, we start with n-heptane with a continuous chain of 7 C atoms, consider next all possible isomers with a continuous chain of 6 C atoms, then those with a continuous chain of 5 C atoms, and so on until the structures of all nine isomers have been deduced:

I. CH_3-CH_2-CH_2-CH_2-CH_2-CH_2-CH_3 **n-heptane**

II. CH_3-CH-CH_2-CH_2-CH_2-CH_3 (with CH_3 on C-2) **2-methylhexane**

III. CH_3-CH_2-CH-CH_2-CH_2-CH_3 (with CH_3 on C-3) **3-methylhexane**

IV. 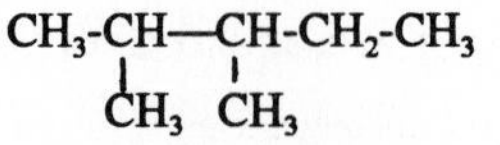 CH_3-CH—CH-CH_2-CH_3 (with CH_3 on C2 and CH_3 on C3)

2,3-dimethylpentane

V. 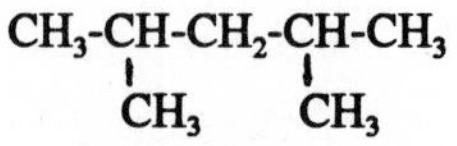 CH_3-CH-CH_2-CH-CH_3 (with CH_3 on C2 and CH_3 on C4)

2,4-dimethylpentane

VI. 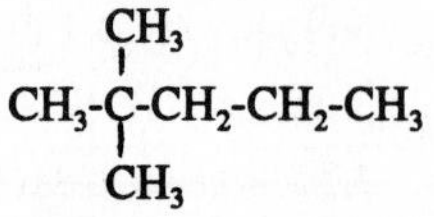 CH_3-C-CH_2-CH_2-CH_3 (with two CH_3 on C2)

2,2-dimethylpentane

VII. 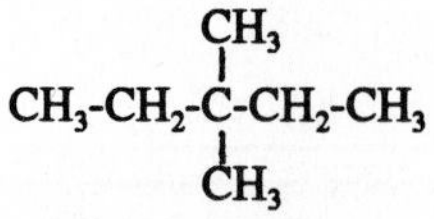 CH_3-CH_2-C-CH_2-CH_3 (with two CH_3 on C3)

3,3-dimethylpentane

VIII. CH_3-CH_2-CH-CH_2-CH_3 (with CH_2-CH_3 on C3)

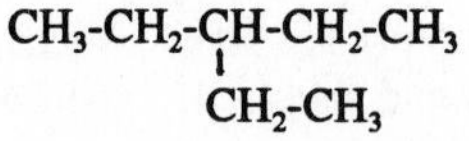

3-ethylpentane

IX. 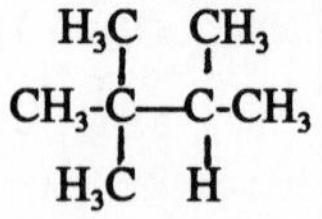 CH_3-C—C-CH_3 (with H_3C above and H_3C below C2; CH_3 above and H below C3)

2,2,3-trimethylbutane

---

24. Draw the structure and give the IUPAC name of each of the six isomers with the molecular formula C_4H_8.

C_4H_8 is the molecular formula of an **alkene** with 4 C atoms, or a **cycloalkane**:

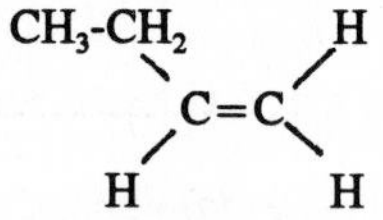

I. CH_3-CH_2(H)C=C(H)(H) — **1-butene**

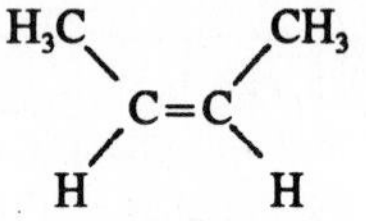

II. H_3C(H)C=C(H)CH_3 — **cis-2-butene**

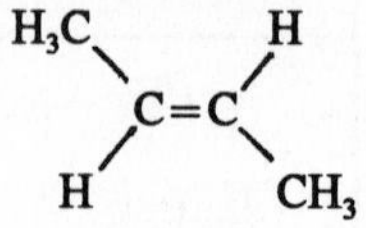

III. H_3C(H)C=C(H)CH_3 — **trans-2-butene**

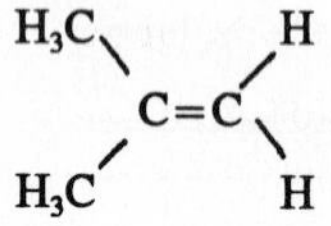

IV. $(H_3C)_2$C=C(H)(H) — **2-methylpropene**

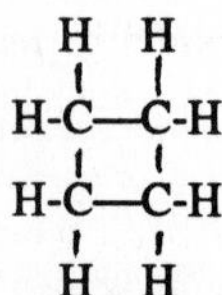

V. **cyclobutane**

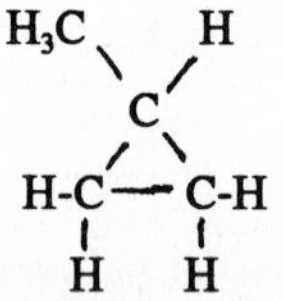

VI. H_3C(H)C in a ring with H-C—C-H (each C bearing H) — **methylcyclopropane**

---

27. Deduce the IUPAC name of each of the following:

(a) 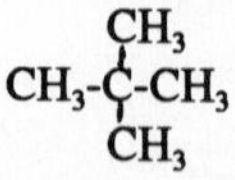 CH_3-C-CH_3 (with CH_3 above and CH_3 below) (b) 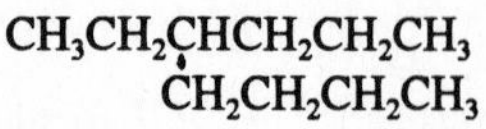 $CH_3CH_2CHCH_2CH_2CH_3$ (with $CH_2CH_2CH_2CH_3$ on the CH) (c) CH_3-CH=CH_2 (d) CH_3CH=$C(CH_3)_2$

(e) $CH_3CH_2CH_2CH_2CH$=$CHCH_3$ (f) 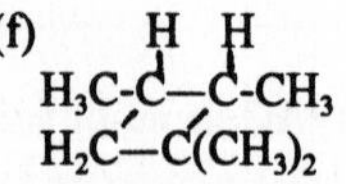 H_3C-C(H)—C(H)-CH_3 / H_2C—$C(CH_3)_2$ (g) CH_2=$CHCH_2CH$=$CHCH_3$

(a) **2,2-dimethylpropane**; (b) **4-ethyloctane**; (c) **propene**; (d) **2-methyl-2-butene**; (e) **2-heptene**; (f) **1,1,2,3-tetramethylcyclobutane**; (g) **1,4-hexadiene.**

---

29. Which of the following is not an isomer of heptane? (a) 2-methylhexane (b) 2,2-dimethylpentane (c) 2,3-dimethylbutane (d) 2,3-dimethylpentane

---

All of the formulas given are those of saturated hydrocarbons, and all isomers of heptane must have the molecular formula $\mathbf{C_7H_{16}}$, and therefore contain **seven C atoms**, which is true of (a), (b), and (d), but **not (c)**, which contains only six C atoms, has the molecular formula C_6H_{14}, and is an isomer of hexane.

2,3-dimethylbutane is not an **isomer of heptane.**

---

30. Draw the structures of as many isomers of C_5H_{10} as you can, and give the IUPAC name of each.

Alkenes or cycloalkanes have the general formula C_nH_{2n}, so isomers of formula C_5H_{10} are either **alkenes** or **cycloalkanes** with five C atoms; there are **five** isomeric alkenes and **four** isomeric cycloalkanes with this formula:

CH_3-CH_2-CH_2-$CH=CH_2$	CH_3-CH_2-$C(H)=CH(CH_3)$	$(CH_3)_2C=C(H)CH_3$	$H_2C=C(CH_3)CH_2$-CH_3
1-pentene	2-pentene (cis & trans)	2-methyl-2-butene	2-methyl-2-butene

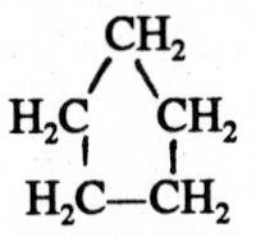

CH_2 / H_2C CH_2 / H_2C—CH_2

H_3C / H-C—CH_2 / H_2C—CH_2

H_3C CH_3 / C / H_2C—CH_2

H CH_3 / C / H_2C—C-CH_3 / H

$H_2C=CH$—$CH(CH_3)_2$

cyclopentane	methylcyclobutane	1,1-dimethylcyclopropane	1,2-dimethylcyclopropane (cis and trans)	3-methyl-1-butene

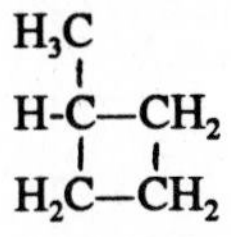

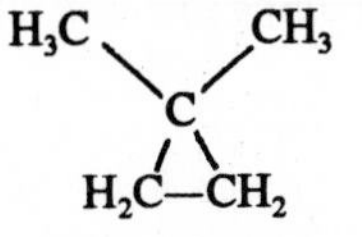

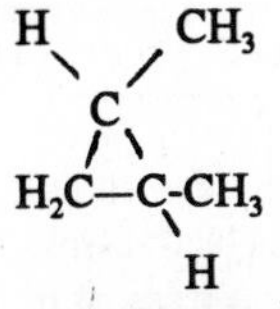

---

32. Draw the structures and name the isomers of the substituted benzenes with the formulas: (a) $C_6H_4(CH_3)_2$ and (b) $C_6H_3(CH_3)_3$

This kind of problem is best answered by considering the relative positions of the groups attached to the benzene ring and then naming each likely isomer **systematically** using the IUPAC rules, giving the substituents the lowest possible numbers. Only structures that have different names are distinguishable isomers:

(a) **Dimethylbenzenes**

Putting one -CH_3 at any of the C atoms of the benzene ring, gives the same compound because all six C atoms in the ring are equivalent; there is only one methylbenzene. We next consider where to place the second -CH_3 group: With the first -CH_3 group at C_1, the second can go at C_2, at C_3, or at C_4, giving:

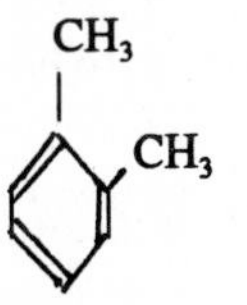

CH_3 CH_3 CH_3 CH_3 CH_3 CH_3

1,2-dimethylbenzene	**1,3-dimethylbenzene**	**1,4-dimethylbenzene**

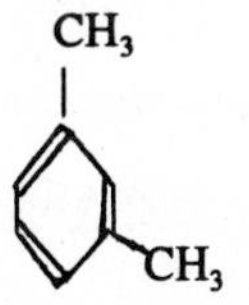

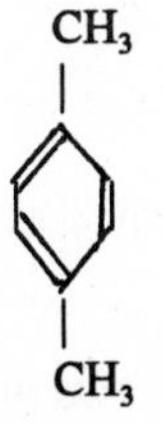

which are also called **ortho-**, **meta-**, and **para-** dimethylbenzene, respectively (or o-, m-, and p-xylene).

Note: we number here in a clockwise direction, with position -1- at the top of the benzene ring, to give the substituents the lowest possible numbers. We could also consider the following possibilities:

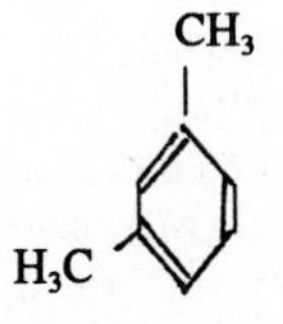

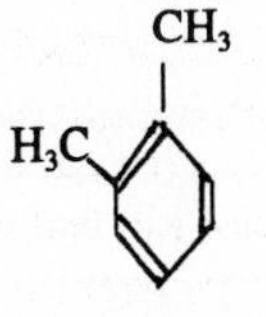

CH_3 H_3C CH_3 H_3C

but these are also **1,3-dimethylbenzene** and **1,2-dimethylbenzene**, respectively, because to give the substituents the lowest possible numbers we now have to number the six ring C atoms anti-clockwise from the top carbon atom.

(b) **Trimethylbenzenes**

The **four** possible isomers are shown below:

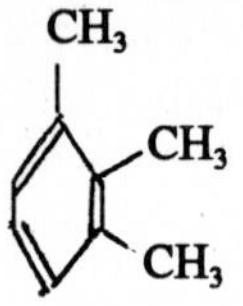

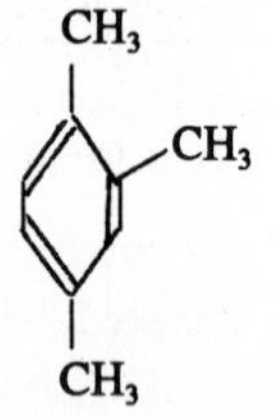

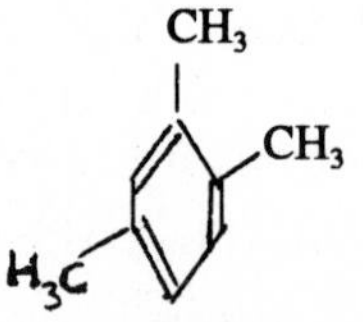

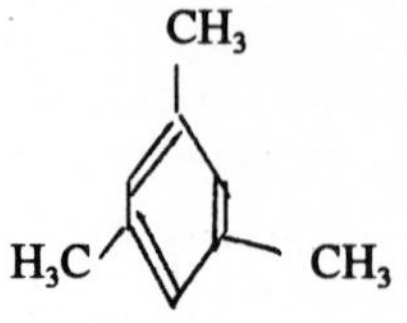

CH_3 CH_3 CH_3 | CH_3 CH_3 CH_3 CH_3 H_3C CH_3 CH_3 H_3C CH_3 CH_3

1,2,3-trimethylbenzene **1,2,4-trimethylbenzene** **1,2,5-trimethylbenzene** **1,3,5-trimethylbenzene**

---

34. Complete combustion of a sample of a hydrocarbon gave 0.318 g CO_2 and 0.163 g H_2O. The mass of the hydrocarbon that occupied a 250-mL flask at 100°C and 1.00 atm pressure was 0.4743 g. (a) Determine the empirical and molecular formulas of the hydrocarbon. (b) Draw structural formulas for, and name, all the isomeric hydrocarbons with this molecular formula.

The mass of CO_2 from the combustion of the sample gives the moles of C atoms, and the mass of H_2O gives the moles of H atoms, and thus the ratio mol C : mol H in the sample and the empirical formula:

$$\text{mol C} = (0.318 \text{ g } CO_2)\left(\frac{1 \text{ mol } CO_2}{44.01 \text{ g } CO_2}\right)\left(\frac{1 \text{ mol C}}{1 \text{ mol } CO_2}\right) = 7.23\text{x}10^{-3} \text{ mol}$$

$$\text{mol H} = (0.163 \text{ g } H_2O)\left(\frac{1 \text{ mol} H_2O}{18.02 \text{ g } H_2O}\right)\left(\frac{2 \text{ mol H}}{1 \text{ mol } H_2O}\right) = 1.81\text{x}10^{-2} \text{ mol}$$

Ratio of moles (atoms) = $7.23\text{x}10^{-3}$ mol C : $1.81\text{x}10^{-2}$ mol H = 1.00 : 2.51 <u>or</u> **2 : 5**

Thus, the **empirical formula** is C_2H_5 (formula mass 29.06 u), and from the gas data:

$$\text{mol gas} = n = \frac{PV}{RT} = \frac{(1.00 \text{ atm})(250 \text{ mL})\left(\frac{1 \text{ L}}{10^3 \text{ mL}}\right)}{(0.0821 \text{ atm L mol}^{-1} \text{ K}^{-1})(373 \text{ K})} = 8.16\text{x}10^{-3} \text{ mol}$$

$$\text{Molar mass} = \frac{0.4743 \text{ g}}{8.16\text{x}10^{-3} \text{ mol}} = \underline{58.1 \text{ g mol}^{-1}}$$

The empirical formula mass of 29.06 u and the experimental molecular mass of 58.1 u, give:

the **molecular formula**: $(C_2H_5)_2$, or $\mathbf{C_4H_{10}}$.

This is the molecular formula of an **alkane**, C_nH_{2n+2}, which for n = 4 has just **two** isomers:

$H_3C\text{-}CH_2\text{-}CH_2\text{-}CH_3$ — **butane**

$H_3C\text{-}CH(CH_3)\text{-}CH_3$ — **2-methylpropane**

---

35. Complete combustion of 0.1540 g of a hydrocarbon gave 0.4832 g of CO_2. The mass of hydrocarbon that filled a 250-mL flask at 100°C and 1.00 atm was 0.4580 g. (a) Determine the empirical and molecular formulas of the hydrocarbon. (b) Draw structural formulas for, and name, at least four isomers that have this molecular formula.

This problem is very similar to Problem 34. We first calculate the moles of C and the moles of H in the sample:

$$\text{mol C} = (0.4832\ \text{g } CO_2)\left(\frac{1\ \text{mol } CO_2}{44.01\ \text{g } CO_2}\right)\left(\frac{1\ \text{mol C}}{1\ \text{mol } CO_2}\right) = 1.098 \times 10^{-2}\ \text{mol;}$$

$$\text{Thus, sample contains } (1.098 \times 10^{-2}\ \text{mol C})\left(\frac{12.01\ \text{g C}}{1\ \text{mol C}}\right) = \underline{0.1319\ \text{g C}}$$

and $$(0.1540\ \text{g} - 0.1319\ \text{g}) = \mathbf{0.0221\ g\ H}$$

$$\text{i.e., mol H} = (0.0221\text{g H})\left(\frac{1\ \text{mol H}}{1.008\ \text{g H}}\right) = 2.19 \times 10^{-2}\ \text{mol}$$

$$\text{Thus, ratio mol C : mol H (= atom ratio)} = 1.098 \times 10^{-2} : 2.19 \times 10^{-2} = \underline{1.0 : 2.0}$$

Therefore, the **empirical formula** is CH_2 (formula mass 14.03 u), and from the gas data:

$$\text{mol gas} = n = \frac{PV}{RT} = \frac{(1.00\ \text{atm})(250\ \text{mL})\left(\frac{1\ \text{L}}{10^3\ \text{mL}}\right)}{(0.0821\ \text{atm L mol}^{-1}\ \text{K}^{-1})(373\ \text{K})} = 8.16 \times 10^{-3}\ \text{mol}$$

$$\text{Molar mass} = \frac{0.4580\ \text{g}}{8.16 \times 10^{-3}\ \text{mol}} = \underline{56.1\ \text{g mol}^{-1}}$$

Thus, from the empirical formula mass of 14.03 u and the experimental molecular mass of 56.1 u:

the **molecular formula** is C_4H_8.

This is the molecular formula of an alkene or a cycloalkane with 4 C atoms, for which the structural formulas and names of the **six** isomers were given in the answer to Problem 24. The particular isomer could be identified using mass, nmr, and infrared spectroscopy (see Chapter 14).

Reactions of Hydrocarbons

37. By writing an appropriate balanced equation, give an example of each of the following reaction types: (a) the combustion of an alkane; (b) a cracking reaction of an alkane; (c) an addition reaction of an alkene (d) an addition reaction of an alkyne; (e) a polymerization reaction of an alkene

---

(a) The products of complete **combustion** of any hydrocarbon are carbon dioxide and water, for example:

$$CH_4(g) + 2O_2(g) \rightarrow CO_2(g) + H_2O(\ell)$$

(b) In a **cracking reaction**, substances decomposes at high temperature into two simpler substances, for example:

$$H_3C{-}CH_3(g) \rightarrow H_2C{=}CH_2(g) + H_2(g)$$

$$H_3C{-}CH_2{-}CH_3(g) \rightarrow H_2C{=}CH_2(g) + CH_4(g)$$

(c) In an **addition reaction of an alkene**, a molecule adds across a C=C double bond to give a substituted alkane:

$$H_2C{=}CH_2(g) + H_2(g) \rightarrow H_3C{-}CH_3(g)$$

$$H_2C{=}CH_2(g) + HBr(g) \rightarrow H_3C{-}CH_2Br(\ell)$$

$$H_2C{=}CH_2(g) + H_2O(g) \rightarrow H_3C{-}CH_2OH(\ell)$$

(d) In an **addition reaction of an alkyne**, a molecule adds across a C≡C triple bond to give a substituted alkene:

$$H{-}C{\equiv}C{-}H(g) + Br_2(g) \rightarrow H{-}\underset{Br}{\underset{|}{C}}{=}\underset{Br}{\underset{|}{C}}{-}H\ (g)$$

which can then undergo further addition:

$$\underset{\text{Br Br}}{\text{H—C=C—H}}\ (g) + Br_2(g) \rightarrow \text{H—}\underset{\text{Br Br}}{\overset{\text{Br Br}}{\text{C—C}}}\text{—H}\ (g)$$

(e) In a **polymerization reaction**, a large number of small molecules (monomers) take part in addition reactions and add together to form a long chain (polymer), for example, ethene polymerizes to polyethene (polyethylene):

$$n\ CH_2{=}CH_2(g) \rightarrow [-CH_2\text{-}CH_2-]_n\ (s)$$

---

39. Describe a simple chemical test you might use to distinguish between each of the following pairs of gases and what you would observe experimentally: (a) ethane and ethyne (b) carbon dioxide and propane

---

(a) **Ethane**, $H_3C\text{-}CH_3$, is a gaseous saturated hydrocarbon while ethyne gas, $HC{\equiv}CH$, is unsaturated; thus, for instance, Br_2 will add to the triple bond in ethyne but will not react with ethane under ordinary conditions. Experimentally, you could bubble the gases through bromine water, $Br_2(aq)$, the red-brown color of which would be decolorized by ethyne but not by ethane:

$$H-C{\equiv}C-H + Br_2 \rightarrow H(Br)C{=}CH(Br) \rightarrow HBr_2C-CBr_2H$$

(b) Propane burns in air with a bright flame whereas carbon dioxide does not support combustion:

$$C_3H_8(g) + 5O_2(g) \rightarrow 3CO_2(g) + 4H_2O(g)$$

Alternatively, when the gases are bubbled through an aqueous solution of $Ca(OH)_2$ (limewater), only $CO_2(g)$ reacts to give a white precipitate of calcium carbonate, $CaCO_3(s)$:

$$Ca(OH)_2(aq) + CO_2(g) \rightarrow CaCO_3(s) + H_2O(\ell)$$

---

41. Ethene can be made in the laboratory by heating ethanol, $C_2H_5OH(\ell)$, with concentrated sulfuric acid. (a) Write the balanced equation for this reaction, and (b) suggest how 2-methylpropene might be prepared.

(a) In this reaction sulfuric acid behaves as a dehydrating agent, removing H and OH as H_2O from ethanol:

$$C_2H_5OH + H_2SO_4 \rightarrow H_2C{=}CH_2 + H_3O^+ + HSO_4^-$$

(b) 2-methylpropene, with 4 C atoms, could be prepared by dehydration of an appropriate alcohol, also with 4 C atoms. Since addition of H—OH to the double bond of 2-methylpropene gives *two* possible products:

(1) $$H_2C{=}\underset{CH_3}{\underset{|}{C}}-CH_3 + H-OH \rightarrow H_3C-\overset{OH}{\overset{|}{\underset{CH_3}{\underset{|}{C}}}}-CH_3$$ **2-methyl-2-propanol (tertiary butanol)**

(2) $$H_2C{=}\underset{CH_3}{\underset{|}{C}}-CH_3 + H-OH \rightarrow HO-\underset{CH_3}{\underset{|}{CH_2}}-\overset{H}{\overset{|}{C}}-CH_3$$ **2-methyl-1-propanol**

Thus, the dehydration of *either* **2-methyl-2-propanol** *or* **2-methyl-1-propanol** gives **2-methylpropene.**

---

43. Each of the following compounds may be synthesized from an alkene, or an alkyne, and another reactant. In each case give the name and structure of the alkene or alkyne, and the other reactant:
(a) 2-propanol (b) 2,2,3,3-tetrabromobutane (c) 1-butene (d) 2-bromopropene

---

Each product results from the appropriate addition reaction:

(a) $H_3C{-}\overset{H}{C}{=}\overset{H}{C}{-}H + H_2O \xrightarrow{\text{acid catalyst}} H_3C{-}\underset{OH}{\overset{H}{C}}{-}CH_3$ (b) $H_3C{-}C{\equiv}C{-}CH_3 + 2Br_2 \rightarrow H_3C{-}\underset{Br}{\overset{Br}{C}}{-}\underset{Br}{\overset{Br}{C}}{-}CH_3$

propene **water** 2-propanol **2-butyne** **bromine** 2,2,3,3-tetrabromobutane

(c) $H{-}C{\equiv}C{-}CH_2{-}CH_3 + H_2 \xrightarrow{\text{catalyst}} H_2C{=}\overset{H}{C}{-}CH_2{-}CH_3$ (d) $H_3C{-}C{\equiv}C{-}H + HBr \rightarrow H_3C{-}\overset{Br}{C}{=}CH_2$

1-butyne **hydrogen** 1-butene **propyne** **hydrogen bromide** 2-bromopropene

---

Resonance Structures and Bond Order

45. Draw Lewis structures for each of the following molecules and ions, and name each: (a) CO_2 (b) CO (c) CN^- (d) C_2^{2-} (e) HCN

Here, we count up the total number of electrons and use the argument that the difference between the number needed to complete the valence shells of all the atoms in the molecule and the number available must be the number of electrons that are shared, in other words, the number involved in bonding. For example in **CO_2 in part (a)**, **C** (Group IV) contributes 4 valence electrons, and two O atoms (Group VI) contribute 2(6) = 12 electrons, **for a total** of **4+2(6) = 16** valence electrons. In the Lewis structure of CO_2, all three atoms obey the *octet rule*; the number of electrons to complete octets on 3 atoms = **3(8) = 24**. Thus, in CO_2 the number of shared (bonding) electrons is **3(8) - 16 = 8**. Eight electrons form four electron-pair bonds that are arranged as **two double bonds** between the atoms, and completing octets around each atom gives:

$$:\ddot{O}{=}C{=}\ddot{O}:$$

for the Lewis structure of CO_2. Calculations of this kind for each molecule or ion are shown below:

Species	**Name**	**Valence Electrons**	**Bonding Electrons**	**Total Bonds**	**Lewis Structure**
(a) CO_2	**Carbon dioxide**	4+2(6) = 16	3(8)-16 = 8	4	$:\ddot{O}{=}C{=}\ddot{O}:$
(b) CO	**Carbon monoxide**	4+6 = 10	2(8)-10 = 6	3	$^-:C{\equiv}O:^+$
(c) CN^-	**Cyanide ion**	4+5+1 = 10	2(8)-10 = 6	3	$^-:C{\equiv}N:$
(d) C_2^{2-}	**Carbide ion**	2(4)+2 = 10	2(8)-10 = 6	3	$^-:C{\equiv}C:^-$
(e) HCN	**Hydrogen cyanide**	1+4+5 = 10	[2(8)+2]-10 = 8	4	$H{-}C{\equiv}N:$

---

47. Explain why a single Lewis structure is insufficient to describe the observed structure of CO_3^{2-}, which is triangular planar with all the OCO angles equal to 120° and all the CO bonds with the same length of 131 pm.

(a) The Lewis (localized electron pair) structure of CO_3^{2-}, with a total of 4+3(6)+2 = 24 valence electrons, contains a total of 4(8)-24 = 8 bonding electrons, or **4 pairs**, and is thus:

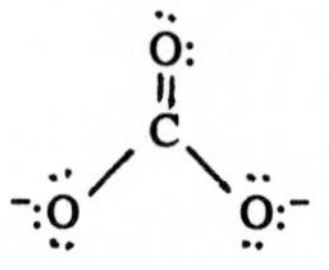

:Ö:
||
C
$^-$:Ö/ \Ö:$^-$

which, with two single C-O bonds and one C=O double bond, does not represent the true structure, because it implies that one CO bond should be shorter than the other two, and the bond angles would not all be equal to 120°, because <O=C-O would be greater than <O-C-O (due to the domain of a double bond being larger than that of a single bond).

(b) The actual structure is represented by mixing together the **three** possible Lewis structures:

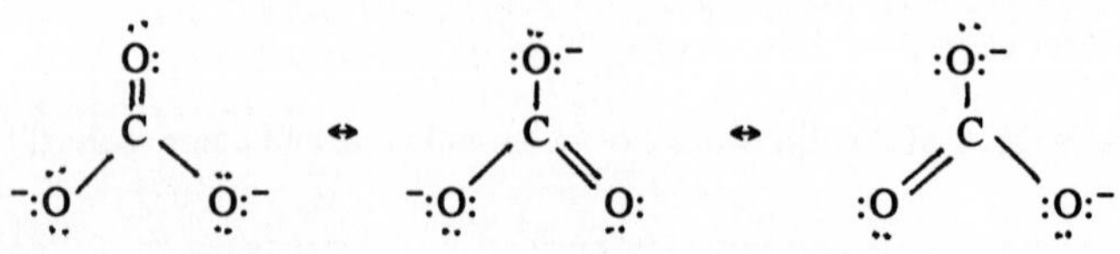

:Ö: || C $^-$:Ö/ \Ö:$^-$ ↔ :Ö:$^-$ | C $^-$:Ö/ \\Ö: ↔ :Ö:$^-$ | C :O// \Ö:$^-$

corresponding to a structure in which 3 of the total of 4 bonding electron pairs are localized in 3 C-O single bonds and the additional electron pair is delocalized over all four atoms, giving each bond the **same C-O bond order of 1⅓**, which is consistent with the observed regular trigonal planar geometry and all CO bonds of equal length, intermediate in length between those of a single C—O and a double C=O bond.

---

49. Draw Lewis structures, including possible resonance structures where relevant, for each of the following. Give the sulfur-oxygen bond orders in each, and use the VSEPR model to predict the expected molecular geometries: (a) H_2SO_4 (b) SO_2 (c) SO_4^{2-} (d) HSO_3^- (e) SO_3

(a) H_2SO_4

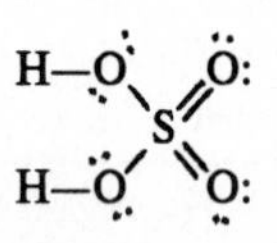

H—Ö Ö: / S \\ H—Ö Ö:

S-OH bond order = 1
S=O bond order = 2

(b) **SO_2**

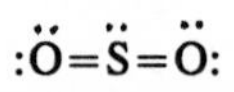

:Ö=S̈=Ö:

S=O bond order = 2

(c) **SO_4^{2-}**

:Ö:, $^-$:Ö—S=Ö:, :Ö:$^-$ ↔ $^-$:Ö—S—Ö:$^-$ ↔ :Ö=S—Ö:$^-$ ↔ $^-$:Ö—S=Ö: ↔ :Ö=S=Ö: ↔ :Ö=S—Ö:$^-$

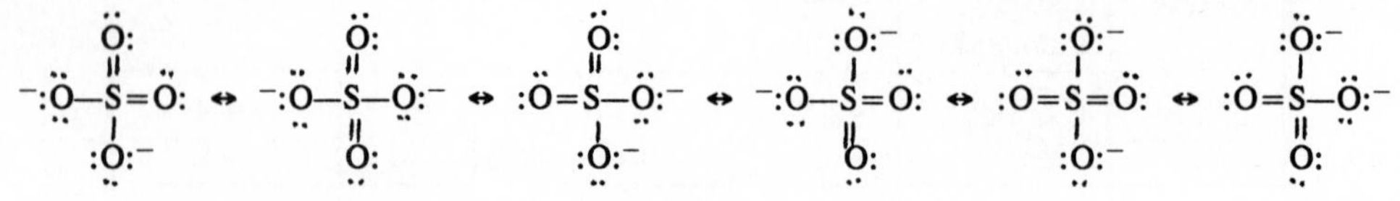

S-O bond order = $\frac{(2+2+2+1+1+1)}{6}$ = **1½**

(d) **HSO_3^{2-}**

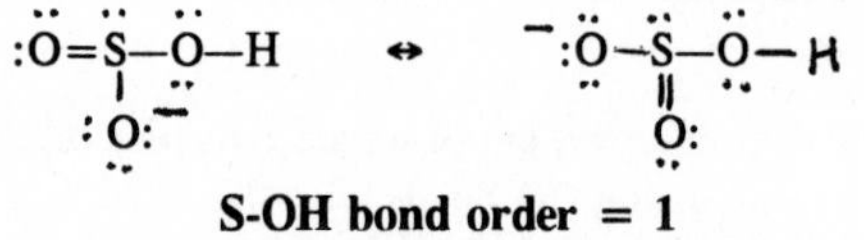

:Ö=S—Ö—H, :Ö:$^-$ ↔ $^-$:Ö—S—Ö—H, Ö:

S-OH bond order = 1
S-O bond order = ½(2+1) = 1½

(e) SO_3

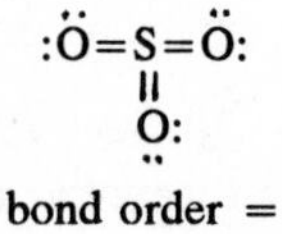

:Ö=S=Ö:, || Ö:

SO bond order = 2

We can now categorize each molecule or ion in terms of the AX_nE_m nomenclature of the VSEPR model:

	Species	Bond Order	Description	Geometry
(a)	H_2SO_4	SO(H) 1 SO 2	AX_4	tetrahedral
(b)	SO_2	SO 2	AX_2E	angular
(c)	SO_4^{2-}	SO 1½	AX_4	tetrahedral
(d)	HSO_3^-	SO(H) 1 SO 1½	AX_3E	trigonal pyramidal
(e)	SO_3	SO 2	AX_3	triangular planar

51.* The nitrate ion, NO_3^-, is a triangular planar molecule with all the nitrogen-oxygen bonds of equal length, and all the oxygen-nitrogen-oxygen angles equal to 120°. Draw the possible resonance structures and deduce the expected nitrogen-oxygen bond orders.

NO_3^- ion has a total of 5+3(6)+1 = 24 valence electrons. Since 4x8 = 32 electrons are required to complete octets around each atom, the number of shared (bonding) electrons in NO_3^- is 32-24 = 8 electrons, or 4 bonding pairs, giving the Lewis structure:

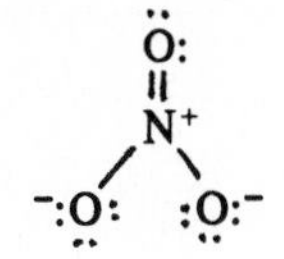

Ö:
‖
N^+
/ \
⁻:Ö: :Ö:⁻

However, to depict the true structure of NO_3^-, we must write all the possible Lewis structures and average them:

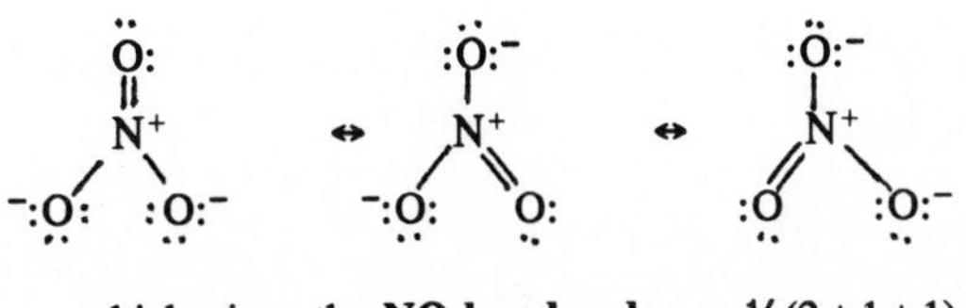

Ö: ‖ N^+ / \ ⁻:Ö: :Ö:⁻ ↔ :Ö:⁻ | N^+ / ∖∖ ⁻:Ö: Ö: ↔ :Ö:⁻ | N^+ // \ :Ö :Ö:⁻

which gives the **NO bond order** as ⅓(2+1+1) = **1⅓.**

General Problems

52. A hydrocarbon contains 82.6 mass % carbon. 0.470 g of the hydrocarbon filled a 200-mL flask at 25 °C and a pressure of 750 mm Hg. (a) What are its empirical and molecular formulas? (b) Can you write a unique structural formula for this hydrocarbon?

$$\text{Mol C per 100 g compound} = (82.6\text{ g C})(\frac{1\text{ mol C}}{12.01\text{ g C}}) = 6.88\text{ mol}$$

$$\text{Mol H per 100g compound} = (17.4\text{ g H})(\frac{1\text{ mol H}}{1.008\text{ g H}}) = 17.3\text{ mol}$$

Thus, ratio mol C : mol H = ratio of atoms = 6.88 : 17.3 = 1.00 : 2.51, or **2 : 5** and the **empirical formula** is C_2H_5 (formula mass = 29.06 u).

From the gas data:

$$\text{mol gas} = n = \frac{PV}{RT} = \frac{(750 \text{ mm Hg})(\frac{1 \text{ atm}}{760 \text{ mm Hg}})(200 \text{ mL})(\frac{1 \text{ L}}{10^3 \text{ mL}})}{(0.0821 \text{ atm L mol}^{-1} \text{ K}^{-1})(298 \text{ K})} = \underline{8.07 \times 10^{-3} \text{ mol}}$$

$$\text{Thus,} \quad \text{molar mass} = \frac{0.470 \text{ g}}{8.07 \times 10^{-3} \text{ mol}} = \underline{58.2 \text{ g mol}^{-1}}$$

and the **molecular formula** is C_4H_{10} (molecular mass 58.12 u).

(b) For the molecular formula C_4H_{10} there is no unique molecular structure; there are two isomers with this formula:

Butane $CH_3\text{-}CH_2\text{-}CH_2\text{-}CH_3$ or 2-Methylpropane $CH_3\text{-}CH(CH_3)\text{-}CH_3$

54.* A 0.200-g sample of a hydrocarbon containing 85.71% carbon by mass occupies a volume of 95.3 mL at 0.921 atm pressure and 27°C. Draw possible structures for the hydrocarbon and name them.

This problem is solved in the same way as Problems 52 and 53. The **empirical formula** is CH_2 (formula mass 14.03 u) and the experimental molar mass of 56.1 g mol^{-1} is consistent with the **molecular formula** C_4H_8, the structures of the six isomers of which were given in the solution to Problem 24.

56.* A gaseous mixture of methane and an alkene of volume 1.00 L has a mass of 0.882 g at 25°C and 744 mm Hg pressure. When burned incompletely in excess oxygen, 2.641 g of carbon dioxide and 1.442 g of water resulted. Identify the alkene and calculate the mass percentage composition of the mixture.

From the gas data we can calculate the total moles of gas comprising the mixture:

$$n = \frac{PV}{RT} = \frac{(744 \text{ mm Hg})(\frac{1 \text{ atm}}{760 \text{ mm Hg}})(1.00 \text{ L})}{(0.0821 \text{ atm L mol}^{-1} \text{ K}^{-1})(298 \text{ K})} = \underline{0.0400 \text{ mol}}$$

and for the combustion reactions, we can write:

$$CH_4 + 2O_2 \rightarrow CO_2 + 2H_2O \quad \text{and} \quad C_nH_{2n} + 2nO_2 \rightarrow nCO_2 + nH_2O$$

Now we calculate the number of moles of C and the number of moles of H in the mixture:

$$\text{Mol C} = (2.641 \text{ g } CO_2)(\frac{1 \text{ mol } CO_2}{44.01 \text{ g } CO_2})(\frac{1 \text{ mol C}}{1 \text{ mol } CO_2}) = 0.0600 \text{ mol C}$$

$$\text{Mol H} = (1.442 \text{ g } H_2O)(\frac{1 \text{ mol } H_2O}{18.02 \text{ g } H_2O})(\frac{2 \text{ mol H}}{1 \text{ mol } H_2O}) = 0.1600 \text{ mol H}$$

Thus, for a mixture containing x mol CH_4 and y mol C_nH_{2n}, we have:

From the total moles of gas: $x+y = 0.0400$ .... (1)

From the moles of C: $x+ny = 0.0600$ .... (2)

From the moles of H: $4x+2ny = 0.1600$ .... (3)

and solving equations (1), (2), and (3) for x, y, and n gives:

$x = \text{mol } CH_4 = \underline{0.0200 \text{ mol}}$, $y = \text{mol } C_nH_{2n} = \underline{0.0200 \text{ mol}}$, and $\underline{n = 2}$.

i.e., the alkene is **C_2H_4**, ethene.

We can now use the given mass of the mixture to confirm these results:

$$\text{Total mass} = (0.0200 \text{ mol } CH_4)\left(\frac{16.04 \text{ g}}{1\text{mol}}\right) + (0.0200 \text{ mol } C_2H_4)\left(\frac{28.05 \text{ g}}{1 \text{ mol}}\right) = (0.321 + 0.561) \text{ g} = \underline{0.882 \text{ g}}$$

$$\text{Thus:} \quad \text{Mass \% } CH_4 = (0.321 \text{ g})\left(\frac{100\%}{0.882 \text{ g}}\right) = \underline{36.4\%}; \quad \text{mass \% } C_2H_4 = \underline{63.6\%}$$

57.* When mercury(II) cyanide, $Hg(CN)_2(s)$, is heated, a gaseous compound X containing only carbon and nitrogen is obtained. Analysis gives 46.2% C by mass. At 100°C and 0.950 atm pressure, 0.208 g of X has a volume of 126 mL: (a) what are the empirical and molecular formulas of X? (b) Draw a possible Lewis structure for X, and deduce its molecular shape.

(a) Firstly, we calculate the empirical formula, then the empirical formula mass, and finally the molar mass: 100 g of X contains 46.2 g C and 53.8 g N, thus:

$$100 \text{ g X contains } (46.2 \text{ g C})\left(\frac{1 \text{ mol C}}{12.01 \text{ g C}}\right) = 3.85 \text{ mol C}$$

$$100 \text{ g X contains } (53.8 \text{ g N})\left(\frac{1 \text{ mol N}}{14.01 \text{ g N}}\right) = 3.85 \text{ mol N}$$

$$\text{Thus:} \quad \text{ratio mol C : mol N} = \text{ratio of atoms} = 3.85 : 3.85 = \underline{1 : 1}$$

and the **empirical formula** is CN (formula mass 26.02 u).

From the gas data:

$$n = \frac{PV}{RT} = \frac{(0.950 \text{ atm})(126 \text{ mL})\left(\frac{1 \text{ L}}{10^3 \text{ mL}}\right)}{(0.0821 \text{ atm L mol}^{-1} \text{ K}^{-1})(373 \text{ K})} = 3.908 \times 10^{-3} \text{ mol}$$

$$\text{Molar mass} = \frac{0.208 \text{ g}}{3.908 \times 10^{-3} \text{ mol}} = \underline{53.2 \text{ g mol}^{-1}}$$

Hence, the molecular mass of 53.2 u is **twice** the empirical formula mass, and the **molecular formula is C_2N_2**.

(b) All the atoms should obey the octet rule and there are 2(4)+2(5) = 18 valence electrons. Hence the number of bonding electrons is 4(8)-18 = 14, or **seven bonding pairs**, consistent with the Lewis structure

:N≡C—C≡N:

where each of the C atoms has AX_2 linear geometry, and thus all the atoms are arranged linearly; the entire molecule is **linear**.

CHAPTER 9

Thermochemistry and the First Law of Thermodynamics

1. When 0.150 g of liquid octane, $C_8H_{18}(\ell)$, was burned in a flame calorimeter containing 1.500 kg of water, the temperature of the water rose from 25.246 to 26.386 °C. What is the standard reaction ethalpy of combustion of octane at 25 °C?

Assuming that the heat capacity of the flame calorimeter is negligible, all of the heat produced by the reaction is transferred to the 1.500 kg of water, at constant pressure. Thus:

Amount of heat evolved = (mass of water)(heat capacity)(temperature change)

$$= (1.500 \text{ kg } H_2O)\left(\frac{10^3 \text{ g}}{1 \text{ kg}}\right)(4.184 \text{ J g}^{-1})([26.386-25.246] \text{ K}) = \underline{7155 \text{ J}}$$

Heat is evolved, so the reaction is **exothermic**, and q = -7155 J,(**-7.155 kJ**), which is the ΔH for the combustion of 0.150 g of liquid octane. Thus, for 1 mol of liquid octane, $C_8H_{18}(\ell)$,

$$\Delta H^\circ = \left(\frac{-7.155 \text{ kJ}}{0.150 \text{ g octane}}\right)\left(\frac{114.2 \text{ g octane}}{1 \text{ mol octane}}\right) = \underline{-5450 \text{ kJ mol}^{-1}}$$

The standard combustion reaction enthalpy of octane is —5450 kJ mol^{-1}

---

3. A 25.00 mL sample of HCl(aq) was mixed with 25.00 mL of KOH(aq) of the same concentration in a calorimeter. As a result the temperature rose from 25.00 to 26.60°C. Given that ΔH° = -56.02 kJ for the reaction $H_3O^+(aq) + OH^-(aq) \rightarrow 2H_2O(\ell)$, and that the heat capacity of dilute aqueous solutions is 75.4 $J\cdot K^{-1}\cdot mol^{-1}$, determine the concentration of the HCl(aq). (Assume the heat capacity of the calorimeter to be negligible).

Since we have equal volumes of acid and base of the same concentration, the reaction is the complete neutralization of HCl(aq) with KOH(aq), for which ΔH° = -56.02 kJ mol^{-1}. The amount of heat evolved in the experiment is:

(mass of solution)(heat capacity)(temperature change)

$$= (50 \text{ mL sol'n})\left(\frac{1 \text{ g}}{1 \text{ mL}}\right)\left(\frac{1 \text{ mol } H_2O}{18.02 \text{ g } H_2O}\right)\left(\frac{75.4 \text{ J}}{\text{K mol}}\right)(1.60 \text{ K}) = \underline{334.7 \text{ J}}$$

Thus, the initial HCl(aq) solution must have contained

$$(334.7 \text{ J})\left(\frac{1 \text{ mol HCl}}{56.02 \text{ kJ}}\right)\left(\frac{1 \text{ kJ}}{10^3 \text{ J}}\right) = \underline{5.97\text{x}10^{-3} \text{ mol HCl}}$$

and since the initial volume of the HCl(aq) was 25.00 mL, its concentration must have been:

$$\left(\frac{5.97\text{x}10^{-3} \text{ mol HCl}}{25.00 \text{ mL HCl}}\right)\left(\frac{10^3 \text{ mL}}{1 \text{ L}}\right) = \underline{0.239 \text{ mol L}^{-1}}$$

---

5. A small well-insulated hydrogenation apparatus had a heat capacity of 1.500 $kJ\cdot K^{-1}$. When 1.500 g of ethene are hydrogenated completely to ethane in the apparatus, what temperature rise should be observed? (Assume that the heat capacities of the gases are negligible compared to that of the apparatus and that the pressure remains constant.)

The reaction is: $C_2H_4(g) + H_2(g) \rightarrow C_2H_6(g)$, for which

$$\Delta H° = \Delta H_f°(C_2H_6,g) - [\Delta H_f°(C_2H_4,g) + \Delta H_f°(H_2,g)] = [-84.7 - (52.3 + 0)]\ kJ = \mathbf{-137\ kJ}$$

and the heat given out on hydrogenation of 1.500 g of $C_2H_4(g)$ is

$$\left(\frac{-137\ kJ}{1\ mol}\right)(1.500\ g)\left(\frac{1\ mol}{28.05\ g}\right) = \underline{-7.33\ kJ}$$

and for a temperature rise ΔT: $(\Delta T\ K)(1.500\ kJ\ K^{-1}) = \mathbf{7.33\ kJ}$

$$\Delta T = \frac{7.33\ kJ}{1.500\ kJ\ K^{-1}} = \underline{4.89\ K}$$, (which is the expected temperature rise).

---

6. Calculate ΔH° for the reaction $2F_2(g) + 2H_2O(\ell) \rightarrow 4HF(g) + O_2(g)$, given that $H_2(g) + F_2(g) \rightarrow 2HF(g)$ ΔH° = -542 kJ ..... (1); $2H_2(g) + O_2(g) \rightarrow 2H_2O(\ell)$ ΔH° = -572 kJ ..... (2)

In general, to apply Hess's law we arrange the equations for which the ΔH° values are given in such a way that when added together they give the equation for the reaction for which ΔH° is to be found. In this case, $F_2(g)$ appears in the final equation as 2 mol, and in equation (1) as 1 mol. The required equation is obtained by multiplying equation (1) by 2 and adding it to the reverse of equation (2):

$2[H_2(g) + F_2(g) \rightarrow 2HF(g)]$	ΔH° = 2(-542 kJ) = -1084 kJ
$2H_2O(\ell) \rightarrow 2H_2(g) + O_2(g)$	ΔH° = -(-572 kJ) = 572 kJ
$2F_2(g) + 2H_2O(\ell) \rightarrow 4HF(g) + O_2(g)$	ΔH° = (-1082+572) = **—512 kJ**

---

8. Calculate the standard enthalpy change for the reaction $2C(s) + H_2(g) \rightarrow C_2H_2(g)$, given that

$2C_2H_2(g) + 5O_2(g) \rightarrow 4CO_2(g) + 2H_2O(\ell)$	ΔH° = -2600 kJ .... (1)
$C(s) + O_2(g) \rightarrow CO_2(g)$	ΔH° = -394 kJ .... (2)
$2H_2(g) + O_2(g) \rightarrow 2H_2O(\ell)$	ΔH° = -572 kJ .... (3)

Reverse equation (1) and divide it by 2; multiply equation (2) by 2, and add to equation (3) divided by 2:

$\frac{1}{2}[4CO_2(g) + 2H_2O(\ell) \rightarrow 2C_2H_2(g) + 5O_2(g)]$	ΔH° = -½(-2600 kJ) = 1300 kJ
$2[C(s) + O_2(g) \rightarrow CO_2(g)]$	ΔH° = 2(-394 kJ) = -788 kJ
$\frac{1}{2}[2H_2(g) + O_2(g) \rightarrow 2H_2O(\ell)]$	ΔH° = ½(-572 kJ) = -286 kJ
$2C(s) + H_2(g) \rightarrow C_2H_2(g)$	ΔH° = (+1300-780-286) = **+226 kJ**

---

10. It has been proposed that the following reaction occurs in the stratosphere: $HO(g) + Cl_2(g) \rightarrow HOCl(g) + Cl(g)$. Calculate the standard reaction enthalpy for the reaction from the following data:

$Cl_2(g) \rightarrow 2Cl(g)$	ΔH° = 242 kJ ..... (1)
$H_2O_2(g) \rightarrow 2OH(g)$	ΔH° = 134 kJ ..... (2)
$H_2O_2(g) + 2Cl(g) \rightarrow 2HOCl(g)$	ΔH° = -209 kJ ..... (3)

Halve the reverse of equation (2), add it to equation (1), and then add one-half of equation (3):

$\frac{1}{2}[2OH(g) \rightarrow H_2O_2(g)]$	ΔH° = -½(134 kJ) = -67 kJ
$Cl_2(g) \rightarrow 2Cl(g)$	ΔH° = 242 kJ
$\frac{1}{2}[H_2O_2(g) + 2Cl(g) \rightarrow 2HOCl(g)]$	ΔH° = ½(-209 kJ) = -105 kJ
$HO(g) + Cl_2(g) \rightarrow HOCl(g) + Cl(g)$	ΔH° = (-67+242-105) = **70 kJ**

---

12. What are the enthalpies of formation of each of $H_2O(\ell)$, $H_2O(g)$, and $NH_3(g)$, given that

$$H_2(g) + \tfrac{1}{2}O_2(g) \rightarrow H_2O(\ell) \quad \Delta H° = -285.8 \text{ kJ} \ldots\ldots (1)$$
$$H_2O(g) \rightarrow H_2O(\ell) \quad \Delta H° = -44.0 \text{ kJ} \ldots\ldots (2)$$
$$2NH_3(g) \rightarrow N_2(g) + 3H_2(g) \quad \Delta H° = 92.4 \text{ kJ} \ldots\ldots (3)$$

By definition, the standard enthalpy of formation of any substance is the ΔH° for the reaction where 1 mol of the substance is formed from its elements in their standard states. Thus, for $H_2O(\ell)$ the appropriate equation is in fact equation (1) above, for which ΔH° = **$\Delta H_f°(H_2O,\ell)$ = —285.8 kJ.**

For the formation of $H_2O(g)$ from its elements in their standard states, $\Delta H_f°(H_2O,g)$ is the ΔH° for the reaction: $H_2(g) + \tfrac{1}{2}O_2(g) \rightarrow H_2O(g)$, which results from adding equation (1) and the reverse of equation (2):

$$H_2(g) + \tfrac{1}{2}O_2(g) \rightarrow H_2O(\ell) \quad \Delta H° = -285.8 \text{ kJ}$$
$$H_2O(\ell) \rightarrow H_2O(g) \quad \Delta H° = -(-44.0 \text{ kJ}) = +44.0 \text{ kJ}$$
$$H_2(g) + \tfrac{1}{2}O_2(g) \rightarrow H_2O(g) \quad \mathbf{\Delta H°(H_2O,g)} = (-285.8+44.0) = \mathbf{-241.8 \text{ kJ}}$$

The balanced equation for the formation of $NH_3(g)$ from its elements is one-half of the reverse of equation (3); i.e.,

$$\tfrac{1}{2}[2NH_3(g) \rightarrow N_2(g) + 3H_2(g)] \quad \Delta H_f°(NH_3,g) = -\tfrac{1}{2}(92.4) = \mathbf{-46.2 \text{ kJ}}$$

In summary: **$\Delta H_f°(H_2O,\ell)$ = —285.8 kJ; $\Delta H_f°(H_2O,g)$ = —241.8 kJ; $\Delta H_f°(NH_3,g)$ = —46.2 kJ**

---

13. The standard combustion reaction enthalpy of liquid n-heptane, $C_7H_{16}(\ell)$, is -4816.9 kJ. The products of this combustion are liquid water and carbon dioxide gas. Calculate the standard reaction enthalpy of formation of liquid n-heptane.

We first write the balanced equation for the reaction at 25°C:

$$C_7H_{16}(\ell) + 11O_2(g) \rightarrow 7CO_2(g) + 8H_2O(\ell)$$

The $\Delta H_f°$ value for any reactant or product can be calculated provided the $\Delta H_f°$ values for the remainder of the reactants and products, and ΔH° for the reaction are known, since for **standard enthalpies of formation**

$$\mathbf{\Delta H° = \Sigma\, [n_p(\Delta H_f°)_p] - \Sigma\, [n_r(\Delta H_f°)_r)}$$

In this case we have:

$$\Delta H° = [7\Delta H_f°(CO_2,g) + 8\Delta H_f°(H_2O,\ell)] - [\Delta H_f°(C_7H_{14},\ell) + 11\Delta H_f°(O_2,g)]$$

Using the given ΔH° value for the reaction and $\Delta H_f°$ values from Table 9.1

$$-4816.9 \text{ kJ} = [7(-393.5) + 8(-285.8)] - [\Delta H_f°(C_7H_{14},\ell) + 11(0)] \text{ kJ}$$

Hence: $$\mathbf{\Delta H_f°(C_7H_{14},\ell)} = +4816.9 - 2754.5 - 2286.4 = \mathbf{-224 \text{ kJ}}$$

---

15. Calculate $\Delta H°_f$ for ethyne, $C_2H_2(g)$, from the standard reaction enthalpy of -312 kJ for the reaction $C_2H_2(g) + 2H_2(g) \rightarrow C_2H_6(g)$ and the standard enthalpy of formation of ethane, $C_2H_6(g)$, given in Table 9.1.

Using the method of Problem 13, we have for: $C_2H_2(g) + 2H_2(g) \rightarrow C_2H_6(g)$; ΔH° = —312 kJ

$$\Delta H° = [\Delta H_f°(C_2H_6,g)] - [\Delta H_f°(C_2H_2,g) + 2\Delta H_f°(H_2,g)]; \quad -312 \text{ kJ} = [-84.7] - [\Delta H_f°(C_2H_2,g) + 2(0)]$$

$$\mathbf{\Delta H_f°(C_2H_2,g) = +227 \text{ kJ}}$$

---

17. Calculate $\Delta H°_f$ for propane, $C_3H_8(g)$, from its standard reaction enthalpy of combustion (-2220 kJ mol^{-1}) and from other data in Table 9.1.

The reaction is: $C_3H_8(g) + 5O_2(g) \rightarrow 3CO_2(g) + 4H_2O(\ell)$, for which:

$$\Delta H° = [3\Delta H_f°(CO_2,g) + 4\Delta H_f°(H_2O,\ell)]-[\Delta H_f°(C_3H_8,g) + 5\Delta H_f°(O_2,g)] = -2220 \text{ kJ}$$
$$= [3(-393.5) + 4(-285.8]-[\Delta H_f°(C_3H_8,g) + 5(0)] \text{ kJ}$$
$$\mathbf{\Delta H_f°(C_3H_8,g) = -103.7 \text{ kJ mol}^{-1}}$$

---

19. When 2 mol of gaseous hydrogen iodide forms 1 mol gaseous hydrogen and 1 mol solid iodine under standard conditions, 52.8 kJ of heat is absorbed. What is the standard enthalpy of formation of HI(g)?

For H_2 and I_2 are in their standard states, $\Delta H°$ for the reaction $H_2(g) + I_2(s) \rightarrow 2HI(g)$, is $2\Delta H_f°(HI,g)$

and we are given: $H_2(g) + I_2(s) \rightarrow 2HI(g)$; $\Delta H° = +52.8$ kJ.

Thus, $2\Delta H_f° = \Delta H°(HI,g) = -52.8$ kJ; $\mathbf{\Delta H_f°(HI,g) = +26.4\ kJ\ mol^{-1}}$

---

21. The standard reaction enthalpies of combustion of graphite and diamond to $CO_2(g)$ are -393.5 kJ mol^{-1} and -395.6 kJ mol^{-1} respectively. What is the enthalpy of formation of C(diamond) from C(graphite)? Which is more stable, diamond or graphite?

The reaction is C(graphite,s) → C(diamond,s), and writing the balanced equations for the combustion reactions, and reversing that for the combustion of diamond and adding the equations, we have:

$$C(graphite,s) + O_2(g) \rightarrow CO_2(g) \quad \Delta H° = -393.5\ kJ$$
$$CO_2(g) \rightarrow C(diamond,s) + O_2(g) \quad \Delta H° = -(-395.6) = +395.6\ kJ$$
$$C(graphite,s) \rightarrow C(diamond,s) \quad \Delta H° = +395.6-393.5 = \mathbf{+1.9\ kJ\ mol^{-1}}$$

The reaction is endothermic; more energy is required to break the bonds in graphite than to form the bonds in diamond, so **graphite is more stable than diamond.**

Bond Energies

23. The standard enthalpy of formation of ClF(g) is -55.7 kJ mol^{-1}, and the dissociation energies of $F_2(g)$ and $Cl_2(g)$ are 155 kJ mol^{-1} and 242 kJ mol^{-1}, respectively. What is the bond dissociation energy of ClF(g)?

The bond dissociation energy of ClF(g) is $\Delta H°$ for the reaction ClF(g) → Cl(g) + F(g), for which:

$$\Delta H° = [\Delta H_f°(Cl,g) + \Delta H_f°(F,g)] - [\Delta H_f°(ClF,g)]$$

As we saw in Problem 22, $\Delta H_f°(X,g)$ is one-half of the dissociation energy of $X_2(g)$, so that:

$$\Delta H_f°(Cl,g) = ½(242\ kJ) = 121\ kJ\ mol^{-1}; \quad \Delta H_f°(F,g) = ½(155\ kJ) = 77.5\ kJ\ mol^{-1}$$

Thus: $\Delta H° = [\Delta H_f°(Cl,g) + \Delta H_f°(F,g)] - [\Delta H_f°(ClF,g)] = [121 + 77.5] - [-55.7] = \mathbf{254\ kJ\ mol^{-1}}$

---

25. Given the $\Delta H°_f$ values (in kJ mol^{-1}) H(g), 218.0; O(g), 247.5; $H_2O_2(g)$, -136.4; $H_2O(g)$, -241.8: (a) Calculate the average O-H bond energy in water. (b) Using the result of (a), calculate the O-O bond energy in hydrogen peroxide. (c) Is the O-O bond in H_2O_2 a strong or a weak bond, compared, for example, to the C-C bond?

(a) For the reaction $H_2O(g) \rightarrow 2H(g) + O(g)$, $\Delta H° = 2BE(O\text{-}H)$, and

$$\Delta H° = [2\Delta H_f°(H,g) + \Delta H_f°(O,g)] - [\Delta H_f°(H_2O,g)] = [2(218) + (247.5)] - [-(241.8)]\ kJ = \underline{925.3\ kJ\ mol^{-1}}$$
$$2BE(O\text{-}H) = 925.3\ kJ\ mol^{-1};\ BE(O\text{-}H) = \mathbf{463\ kJ\ mol^{-1}}$$

(b) For the reaction $H_2O_2(g) \rightarrow 2H(g) + 2O(g)$

$$\Delta H° = 2BE(O\text{-}H) + BE(O\text{-}O) = [2\Delta H_f°(H,g) + 2\Delta H_f°(0,g)] - [\Delta H_f°(H_2O_2,g)] = [2(218) + 2(247.5)] - [-(-136.4)]$$
$$= \underline{1067.4\ kJ\ mol^{-1}}$$

and assuming BE(O-H) = 463 kJ mol^{-1}, from part (a),

$$\Delta H° = 1070.6\ kJ\ mol^{-1} = 2(463\ kJ\ mol^{-1}) + BE(O\text{-}O);\ BE(O\text{-}O) = \mathbf{141\ kJ\ mol^{-1}}$$

(c) The O-O single bond in H_2O_2 is **much weaker** than the average C-C bond energy of 348 kJ mol^{-1}.

---

27. Using the average C-C and C-H bond energies given in Table 9.2, estimate the standard enthalpy of formation of ethane, $C_2H_6(g)$.

To solve this type of problem, we use the relationship:

$$\Delta H^\circ = \Sigma[\text{bond energies reactants}] - \Sigma[\text{bond energies products}]$$

The standard enthalpy of formation, $\Delta H_f^\circ(C_2H_6,g)$, is ΔH° for the reaction:

$$2C(\text{graphite,s}) + 3H_2(g) \rightarrow C_2H_6(g)$$

The reaction can be achieved by first converting C(graphite,s) to C(atomic,g), and $H_2(g)$ to 2H(atomic,g), and then combining all the atoms to give $C_2H_6(g)$.

Thus: $\Delta H^\circ = \Sigma[\text{bond energies reactants}] - \Sigma[\text{bond energies reactants}]$

```
  H H
  | |
H—C—C—H
  | |
  H H
```

$$\Delta H_f^\circ(C_2H_6,g) = [2BE(C,s) + 3BE(H\text{-}H)] - [BE(C\text{-}C) + 6BE(C\text{-}H)]$$

$$= [2(716.7) + 3(436)] - [(348) + 6(413)] \text{ kJ mol}^{-1} = \mathbf{-84.6\ kJ\ mol^{-1}}$$

---

29. Repeat Problem 28 for each of the following gas-phase reactions: (a) $C_2H_2 + C_2H_6 \rightarrow 2C_2H_4$ (b) $2H_2O_2 \rightarrow 2H_2O + O_2$ (c) $C_2H_6 \rightarrow C_2H_4 + H_2$

(a) $C_2H_2(g) + C_2H_6(g) \rightarrow 2C_2H_4(g)$

bonds:

```
              H H        H H
              | |        | |
H—C≡C—H     H—C—C—H    2 C=C
              | |        | |
              H H        H H
```

i.e., **reactants**: 8(C—H); 1(C≡C); 1(C—C) **products**: 8(C—H); 2(C=C)

$$\Delta H^\circ = [8BE(C\text{-}H) + BE(C{\equiv}C) + BE(C\text{-}C)] - [8BE(C\text{-}H) + 2BE(C{=}C)]$$

$$= BE(C{\equiv}C) + BE(C\text{-}C) - 2BE(C{=}C) = (812)+(348)-2(619) = \mathbf{-78\ kJ}$$

(b) $2H_2O_2(g) \rightarrow 2H_2O(g) + O_2(g)$

bonds: 2 H—O—O—H 2 H—O—H O=O

i.e., **reactants**: 4(O-H); 2(O-O) **products**: 4(O-H); 1(O=O)

$$\Delta H_o = [4BE(O\text{-}H)+2BE(O\text{-}O)]-[4BE(O\text{-}H)+BE(O{=}O)] = 2BE(O\text{-}O)-BE(O{=}O) = 2(138)-494 = -218\text{ kJ}$$

(c) $C_2H_6(g) \rightarrow C_2H_4(g) + H_2(g)$

bonds:

```
  H H      H H
  | |      | |
H—C—C—H    C=C     H—H
  | |      | |
  H H      H H
```

i.e., **reactants**: 6(C-H); 1(C-C) **products**: 4(C-H); 1(C=C); 1(H-H)

$$\Delta H^\circ = [6BE(C\text{-}H) + BE(C\text{-}C)] - [4BE(C\text{-}H) + BE(C{=}C) + BE(H\text{-}H)]$$

$$= 2BE(C\text{-}H) + BE(C\text{-}C) - BE(C{=}C) - BE(H\text{-}H) = 2(413) + (348) - (619) - (436) = \mathbf{+119\ kJ}$$

---

31. Given that the standard enthalpy of dissociation of $H_2(g)$, $Cl_2(g)$, and HCl(g) are 436, 242, and 431 kJ mol^{-1}, respectively, what is the standard enthalpy of formation of HCl(g)?

The standard enthalpy of dissociation of HCl(g) is BE(H-Cl), so we can use the bond energy method to calculate $\Delta H_f^\circ(HCl,g)$, which is the standard reaction enthalpy for the reaction:

$$\tfrac{1}{2}H_2(g) + \tfrac{1}{2}Cl_2(g) \rightarrow HCl(g)$$

$$\mathbf{\Delta H_f^\circ(HCl,g)} = [\tfrac{1}{2}BE(H\text{-}H) + \tfrac{1}{2}BE(Cl\text{-}Cl)] - [BE(H\text{-}Cl)] = [\tfrac{1}{2}(436) + \tfrac{1}{2}(242)] - [(431)] = \mathbf{-92\ kJ}$$

Alternative Energy Sources

32. (a) Assuming that coke (graphite) and natural gas (methane) have identical costs per gram, which of these fuels is the more economical for heating a home? (b) If the price per gram of the more economical fuel of (a) is doubled, would that fuel still be the more economical?

Assuming complete combustion, we have:

$$C(s,gr) + O_2(g) \rightarrow CO_2(g); \quad \Delta H^\circ = \Delta H_f^\circ(CO_2,g) = \mathbf{-393.5\ kJ\ mol^{-1}}$$

and $$CH_4(g) + 2O_2(g) \rightarrow CO_2(g) + 2H_2O(g)$$

for which $$\mathbf{\Delta H^\circ} = [\Delta H_f^\circ(CO_2,g) + 2\Delta H_f^\circ(H_2O,g)] - [\Delta H_f^\circ(CH_4,g) + 2\Delta H_f^\circ(O_2,g)]$$
$$= [(-393.5) + 2(-241.8)] - [(-74.5) + 2(0)] = \mathbf{-802.6\ kJ\ mol^{-1}}$$

(a) The heat evolved per gram is:

Coke: $$\Delta H^\circ = \left(\frac{-393.5\ kJ}{1\ mol\ C}\right)\left(\frac{1\ mol\ C}{12.01\ g\ C}\right) = \underline{-32.8\ kJ\ g^{-1}}$$

Methane: $$\Delta H^\circ = \left(\frac{-802.6\ kJ}{1\ mol\ CH_4}\right)\left(\frac{1\ mol\ CH_4}{16.04\ g\ CH_4}\right) = \underline{-50.0\ kJ\ g^{-1}}$$

For identical costs per gram, natural gas (methane) is the more economical.
(b) For double the cost for methane, only ½(50.0 kJ) = 25.0 kJ of heat would be produced per unit cost; coke would then be the more economical.

---

34. Using Table 9.1, calculate ΔH° for the photosynthesis reaction $6CO_2(g) + 6H_2O(\ell) \rightarrow C_6H_{12}O_6(s) + 6O_2(g)$, (a) per mole, and (b) per gram of glucose. (In nature, light from the sun, rather than heat provides the large energy required for this reaction.)

(a) For the photosynthesis reaction: $6CO_2(g) + 6H_2O(\ell) \rightarrow C_6H_{12}O_6(s) + 6O_2(g)$

$$\mathbf{\Delta H^\circ} = [\Delta H_f^\circ(C_6H_{12}O_6,s) + 6\Delta H_f^\circ(O_2,g)] - [6\Delta H_f^\circ(CO_2,g) + 6\Delta H_f^\circ(H_2O,g)]$$
$$= [(-1273) + 6(0)] - [6(-393.5) + 6(-285.8)] = \mathbf{+2803\ kJ\ mol^{-1}}$$

(b) And for 1 gram of glucose:

$$q = \left(\frac{+2803\ kJ}{1\ mol\ glucose}\right)\left(\frac{1\ mol\ glucose}{180.2\ g\ glucose}\right) = \underline{15.55\ kJ\ g^{-1}}$$

Entropy and the Second Law of Thermodynamics

37. For each of the following, use qualitative reasoning to decide which system will have the greater entropy: (a) 1 mole of ice at 0°C, or 1 mole of water at the same temperature? (b) A pack of cards arranged in suits, or a pack of cards randomly shuffled? (c) A collection of jigsaw pices, or a completed puzzle? (d) Solid ammonium chloride, or an aqueous solution of ammonium chloride?

The answers here are all related to the concept that *greater* entropy is associated with greater disorder.

(a) The water molecules of liquid water are relatively free to move around each other and have a less ordered arrangement than the regular array of water molecules in crystalline ice; **water has the greater entropy.**

(b) A pack of cards randomly shuffled has a greater disorder and thus **a greater entropy** than the orderly arrangement of a pack of cards arranged in suits.

(c) A random collection of jigsaw pieces has **a greater disorder** and thus a greater entropy than the completed puzzle.

(d) Solid NH_4Cl is a crystalline solid with a structure consisting of a regular repeating pattern of NH_4^+ and Cl^- ions, while in an aqueous solution these ions move about randomly. **The solution has the greater entropy.**

---

39. Predict the sign of the entropy change, ΔS, for each of the following reactions: (a) $CaCO_3(s) \rightarrow CaO(s) + CO_2(g)$, (b) $NH_3(g) + HCl(g) \rightarrow NH_4Cl(s)$, (c) $BaO(s) + CO_2(g) \rightarrow BaCO_3(s)$

(a) A solid is replaced by a solid and a gas, resulting in increased disorder and an increase in entropy; $\boldsymbol{\Delta S > 0}$.

(b) 2 mol of gas are replaced by 1 mol of solid, giving a large decrease in disorder; entropy decreases; $\boldsymbol{\Delta S < 0}$.

(c) One mole of gas and one mole of solid are replaced by one mole of solid; giving a decrease in disorder (decreased entropy); $\boldsymbol{\Delta S < 0}$.

---

41. Repeat Problem 39 for each of the following reactions: (a) $2CO(g) + O_2(g) \rightarrow 2CO_2(g)$; (b) $Mg(s) + Cl_2(g) \rightarrow MgCl_2(s)$; (c) $2C_2H_6(g) + 7O_2(g) \rightarrow 4CO_2(g) + 6H_2O(g)$; (d) $CH_4(g) + 2O_2(g) \rightarrow CO_2(g) + 2H_2O(\ell)$

(a) 3 moles of gas gives 2 moles of gas; the disorder of the system decreases, $\boldsymbol{\Delta S < 0}$.

(b) 1 mole of solid and 1 mole of gas are replaced by 1 mole of solid; disorder in the system decreases, $\boldsymbol{\Delta S < 0}$.

(c) 9 moles of gas gives 10 moles of gas; the disorder of the system increases, $\boldsymbol{\Delta S > 0}$.

(d) 3 moles of gas give 1 mole of gas and 2 moles of liquid; the disorder in the system increases, $\boldsymbol{\Delta S < 0}$.

---

42. Repeat Problem 39 for each of the following reactions: (a) $H_2(g) + Br_2(\ell) \rightarrow 2HBr(g)$; (b) $ZnO(s) + H_2S(g) \rightarrow ZnS(s) + H_2O(\ell)$; (c) $2H_2(g) + O_2(g) \rightarrow 2H_2O(\ell)$; (d) $2C_2H_6(g) + 7O_2(g) \rightarrow 4CO_2(g) + 6H_2O(\ell)$

(a) 1 mole of gas and 1 mole of liquid are replaced by 2 moles of gas; disorder in the system increases; $\boldsymbol{\Delta S > 0}$.

(b) 1 mole of solid and 1 mole of gas are replaced by 1 mole of solid and 1 mole of liquid; disorder in the system decreases, $\boldsymbol{\Delta S < 0}$.

(c) 3 moles of gas are replaced by two moles of liquid; disorder in the system decreases, $\boldsymbol{\Delta S < 0}$.

(d) 9 moles of gas are replaced by 4 moles of gas and 6 moles of liquid; the system's disorder decreases, $\boldsymbol{\Delta S < 0}$.

---

44. Calculate the standard entropy change, $\Delta S°$, associated with each of the following reactions at 298 K: (a) $C(s,\text{graphite}) + O_2(g) \rightarrow CO_2(g)$; (b) $C_2H_5OH(\ell) + 3O_2(g) \rightarrow 2CO_2(g) + 3H_2O(\ell)$; (c) $C_6H_{12}O_6(s) + 6O_2(g) \rightarrow 6CO_2(g) + 6H_2O(\ell)$; (d) $H_2(g) + I_2(s) \rightarrow 2HI(g)$.

To calculate the standard entropy change, ΔS, for a reaction we use S° values from Table 9.3 and the formula:

$$\boldsymbol{\Delta S° = \Sigma[S°(\text{products})] - \Sigma[S°(\text{reactants})]}$$

(a) $\Delta S° = [S°(CO_2,g)] - [S°(C,\text{s graphite}) + S°(O_2,g)] = [(213.7)] - [(5.8) + (205.0)] = \mathbf{2.9\ J\ K^{-1}\ mol^{-1}}$

(b) $\Delta S° = [2S°(CO_2,g) + 3S°(H_2O,\ell)] - [S°(C_2H_5OH,\ell) + 3S°(O_2,g)]$
$= [2(213.7) + 3(70.0)] - [(160.7) + 3(205.0)] = \mathbf{-138.3\ J\ K^{-1}\ mol^{-1}}$

(c) $\Delta S° = [6S°(CO_2,g) + 6S°(H_2O,\ell)] - [S°(C_6H_{12}O_6,s) + 6S°(O_2,g)]$
$= [6(213.7) + 6(70.0)] - [(182.4) + 6(205.0)] = \mathbf{289.8\ J\ K^{-1}\ mol^{-1}}$

(d) $\Delta S° = [2S°(HI,g)] - [S°(H_2,g) + S°(I_2,s)] = [2(206.0)] - [(130.6) + (116.1)] = \mathbf{166.3\ J\ K^{-1}\ mol^{-1}}$

---

46. Using Table 9.3, calculate the standard entropy change, $\Delta S°$, associated with each of the following reactions at 298 K: (a) $C_2H_2(g) + 2H_2(g) \rightarrow C_2H_6(g)$; (b) $2C(s,graphite) + O_2(g) \rightarrow 2CO(g)$; (c) $H_2(g) + Br_2(\ell) \rightarrow 2HBr(g)$

The method is the same as that used in Problem 44.

(a) $\Delta S° = [S°(C_2H_6,g)] - [S°(C_2H_2,g) + 2S°(H_2,g)] = [(229.5)] - [(200.8) + 2(130.6)] =$ **—232.5 J K^{-1}**

(b) $\Delta S° = [2S°(CO,g)] - [2S°(C,s\ gr) + S°(O_2,g)] = [2(197.6)] - [2(5.8) + (205.0)] =$ **178.6 J K^{-1}**

(c) $\Delta S_o = [2S°(HBr,g)] - [S°(H_2,g) + S°(Br_2,\ell)] = [2(198.6)] - [(130.6) + (152.2)] =$ **114.4 J K^{-1}**

Gibbs Free Energy

48. Calculate $\Delta G°$ for each of the following reactions and state whether each is spontaneous or not under standard conditions: (a) $C_3H_8(g) + 5O_2(g) \rightarrow 3CO_2(g) + 4H_2O(g)$; (b) $2H_2(g) + O_2(g) \rightarrow 2H_2O(g)$; (c) $2SO_2(g) + O_2(g) \rightarrow 2SO_3(g)$.

The criterion here is that a reaction is spontaneous as written at standard conditions when its $\Delta G°$ value is negative.

(a) $\Delta G° = [3\Delta G_f°(CO_2,g) + 4\Delta G_f°(H_2O,g)] - [\Delta G_f°(C_3H_8,g) + 5\Delta G_f°(O_2,g)]$

$= [3(-137.2) + 4(-228.6)] - [(-23.4)+5(0)] =$ **—1302.6 kJ**

$\Delta G°$ is negative; the reaction is **spontaneous**

(b) $\Delta G° = [2\Delta G_f°(H_2O,g)] - [2\Delta G_f°(H_2,g) + \Delta G_f°(O_2,g)] = [2(-228.6)] - [2(0) + 1(0)] =$ **—457.2 kJ**

$\Delta G°$ is negative; the reaction is **spontaneous**

(c) $\Delta G° = [2\Delta G_f°(SO_3,g)] - [2\Delta G_f°(SO_2,g) + \Delta G_f°(O_2,g)] = [2(-371.1)] - [2(300.1) + 1(0)] =$ **—142.0 kJ**

$\Delta G°$ is negative; the reaction is **spontaneous**

---

49. Predict the sign of the reaction Gibbs free energy change at *low* temperature for reactions where: (a) ΔH is positive and ΔS is positive; (b) ΔH is negative and ΔS is positive; (c) ΔH is negative and ΔS is negative; (d) ΔH is positive and ΔS is negative

---

In solving this problem we use **$\Delta G = \Delta H - T\Delta S$**

(a) $\Delta H > 0$, $-T\Delta S < 0$; ΔG is either positive or negative depending on the values of ΔH and $T\Delta S$.

At *low temperature*, T is small, so ΔG will most often be **positive.**

(b) $\Delta H < 0$, $-T\Delta S < 0$; ΔG is **negative** under all circumstances.

(c) $\Delta H < 0$, $-T\Delta S > 0$; ΔG is either negative or positive depending on the values of ΔH and $T\Delta S$.

At *low temperature*, T is small, so ΔG will most often be **negative.**

(d) $\Delta H < 0$, $-T\Delta S < 0$; ΔG is **positive** under all circumstances.

---

51. (a) Calculate $\Delta G°$ for the formation of 1 mol of $CO_2(g)$ from C(s,diamond) and $O_2(g)$ at 298 K and 1 atm pressure. (b) What does the sign of this $\Delta G°$ signify? (c) Should the owners of diamonds be concerned about the spontaneous conversion of diamond to carbon dioxide? Explain.

(a) For $C(s,diamond) + O_2(g) \rightarrow CO_2(g)$ under standard conditions:

$\Delta G° = [\Delta G_f°(CO_2,g)] - [\Delta G_f°(C,s,diamond) + \Delta G_f°(O_2,g)] = [(-394.4)] - [(2.9) + (0)] =$ **—397.3 kJ mol^{-1}**

(b) The negative value of $\Delta G°$ indicates that the reaction is spontaneous as written.

(c) Although the reaction is spontaneous, it must be in fact a *very slow* reaction under standard conditions. Although thermodynamics tells us about the spontaneity of a reaction but nothing about its **rate**.

General Problems

54. Natural gas typically contains unwanted $H_2S(g)$, which can be removed by reaction with $SO_2(g)$, according to the equation $2H_2S(g) + SO_2(g) \rightarrow 3S(s) + 2H_2O(g)$. (a) Identify the oxidizing agent and the reducing agent in this reaction. (b) Calculate ΔH° at 25 °C for the reaction, assuming that the sulfur formed is orthorhombic sulfur consisting of S_8 molecules (its most stable form at room temperature).

(a) In this reaction, S(+IV) in $SO_2(g)$ is reduced to S(0) in S(s), and S(-II) in H_2S is oxidized to S(0) in S(s); $SO_2(g)$ is the **oxidizing agent** and H_2S is the **reducing agent.**

(b) From the data in Table 9.1 and the balanced equation:

$$\Delta H° = [3\Delta H_f°(S,s) + 2\Delta H_f°(H_2O,g)] - [2\Delta H_f°(H_2S,g) + \Delta H_f°(SO_2,g))]$$

$$= [3(0) + 2(-241.8)] - [2(-20.6) + (-296.8)] = \mathbf{-145.6\ kJ}$$

---

55. Calculate the standard reaction free energy for each of the following reactions, and use the results to comment on the relative powers of the elements F_2, Cl_2 and Br_2 as oxidizing agents:

(a) $2NaF(s) + Cl_2(g) \rightarrow 2NaCl(s) + F_2(g)$; (b) $2NaBr(s) + Cl_2(g) \rightarrow 2NaCl(s) + Br_2(\ell)$

(a)
$$\Delta G° = [2\Delta G_f°(NaCl,s) + \Delta G_f°(F_2,g)] - [2\Delta G_f°(NaF,s) + \Delta G_f°(Cl_2,g)]$$

$$= [2(-384.3) + (0)] - [2(-546.3) + (0)] = \mathbf{+324.0\ kJ}$$

(b)
$$\Delta G° = [2\Delta G_f°(NaCl,s) + \Delta G_f°(Br_2,\ell)] - [2\Delta G_f°(NaBr,s) + \Delta G_f°(Cl_2,g)]$$

$$= [2(-384.3) + (0)] - [2(-349.1) + (0)] = \mathbf{-70.4\ kJ}$$

Reaction (a) is not spontaneous as written, but the reverse reaction is spontaneous, so **$F_2(g)$ is a stronger oxidizing agent than $Cl_2(g)$.** Reaction (b) is spontaneous as written, showing that **$Cl_2(g)$ is a stronger oxidizing agent than $Br_2(\ell)$**, confirming the order of oxidizing power of the halogens as **$F_2(g) > Cl_2(g) > Br_2(\ell)$.**

56*. (a) Define standard enthalpy of formation, with reference to $MgCO_3(s)$. (b) When 0.203 g of magnesium were dissolved in an excess of dilute HCl(aq) in a vacuum flask, the temperature rose by 8.61 K. In another experiment, 506 J were required to raise the temperature of the vacuum flask and its contents 1.02 K. Calculate the heat released in the first experiment, and hence find the reaction enthalpy per mole of magnesium. (c) In a similar experiment with the same apparatus, $MgCO_3(s)$ reacted with excess HCl(aq) and the standard reaction enthalpy was found to be -90.4 kJ mol^{-1}. Use the result from (b), and the standard enthalpies of formation of $H_2O(\ell)$ and $CO_2(g)$ to find the standard enthalpy of formation of $MgCO_3(s)$.

(a) The "standard enthalpy of formation" of magnesium carbonate, **$\Delta H_f°(MgCO_3,s)$,** is the ΔH° for the reaction in which 1 mole of $MgCO_3(s)$ is formed from Mg(s), C(s,graphite), and $O_2(g)$ (in their standard states) under standard conditions, according to the equation:

$$Mg(s) + C(s,graphite) + 1\tfrac{1}{2}O_2(g) \rightarrow MgCO_3(s) \quad .... (1)$$

(b) The heat capacity of the calorimeter and its contents is found from the results of the second experiment:

$$\text{Heat capacity of calorimeter} = \frac{506\ J}{1.02\ K} = \underline{496\ J\ K^{-1}}$$

Thus, the heat *given out* in the first experiment, q, is given by:

$$q = -(496\ J\ K^{-1})(8.61\ K) = \mathbf{-4270\ J}$$

Thus, for 1 mole Mg reacting with HCl(aq): $\Delta H = \left(\frac{-4270\ J}{0.203\ g\ Mg}\right)\left(\frac{24.30\ g\ Mg}{1\ mol\ Mg}\right)\left(\frac{1\ kJ}{1000\ J}\right) = \underline{\mathbf{-511\ kJ\ mol^{-1}}}$

(c) For the reaction of $MgCO_3(s)$ with HCl(aq):

$$MgCO_3(s) + 2HCl(aq) \rightarrow MgCl_2(aq) + CO_2(g) + H_2O(\ell);\ \Delta H = -90.4\ kJ\ mol^{-1}\ ....\ (2)$$

and from part (b), we have for the reaction:

$$Mg(s) + 2HCl(aq) \rightarrow MgCl_2(aq) + H_2(g);\ \Delta H = -511\ kJ\ mol^{-1}\ ....\ (3)$$

Now, using these equations, we can calculate ΔH for reaction (1) given in part (a).

Reversing equation (2) and adding it to equation (3) gives:

$$Mg(s) + CO_2(g) + H_2O(\ell) \rightarrow MgCO_3(s) + H_2(g)\ ....\ (4)$$

and the corresponding ΔH value is therefore: $[(-511) - (-90.4)]$ kJ = **—421 kJ**

Equation (4) contains reactants and products whose $\Delta H_f°$ values (except for $\Delta H_f°(MgCO_3,s)$), are given in Table 9.1. Thus, assuming conditions close to standard, $\Delta H = \Delta H°$, and is given by:

$$\Delta H° = [\Delta H_f°(MgCO_3,s) + \Delta H_f°(H_2,g)] - [\Delta H_f°(Mg,s) + \Delta H_f°(CO_2,g) + \Delta H_f°(H_2O,\ell)]$$

$$= [\Delta H_f°(MgCO_3,s) + (0)] - [(0) + (-393.5) + (-285.8)] = -421\ kJ$$

$$\mathbf{\Delta H_f°(MgCO_3,s) = -1100\ kJ\ mol^{-1}}$$

58.* The standard reaction enthalpies of combustion for C(s,graphite), $H_2(g)$, $C_2H_6(g)$, and $C_3H_8(g)$, are -393.5, -285.8, -1559.8, and -2219.9 kJ mol^{-1}, respectively. (a) Calculate the enthalpies of formation of $C_2H_6(g)$ and $C_3H_8(g)$. (b) Use bond energies to predict approximate $\Delta H°_f$ and $\Delta H°$(combustion) values for butane, $C_4H_{10}(g)$.

(a) We first write the balanced equations for each combustion reaction:

$$C(s,graphite) + O_2(g) \rightarrow CO_2(g) \qquad \Delta H° = -393.4\ kJ\ ....\ (1)$$

$$H_2(g) + \tfrac{1}{2}O_2(g) \rightarrow H_2O(\ell) \qquad \Delta H° = -285.8\ kJ\ ....\ (2)$$

$$C_2H_6(g) + 3\tfrac{1}{2}O_2(g) \rightarrow 2CO_2(g) + 3H_2O(\ell) \qquad \Delta H° = -1559.8\ kJ\ ....\ (3)$$

$$C_3H_8(g) + 5O_2(g) \rightarrow 3CO_2(g) + 4H_2O(\ell) \qquad \Delta H° = -2219.9\ kJ\ ....\ (4)$$

And, by definition, $\Delta H_f°(C_2H_6,g)$ is the $\Delta H°$ for the reaction: $2C(s,graphite) + 3H_2(g) \rightarrow C_2H_6(g)$

From Hess's law, twice equation (1) plus three times equation (2), plus the reverse of equation (3) gives this equation. Thus:

$$\mathbf{\Delta H_f°(C_2H_6,g)} = 2(-393.5) + 3(-285.8) - (-1559.8) = \mathbf{-84.6\ kJ\ mol^{-1}}$$

Similarly, $\Delta H_f°(C_3H_8,g)$ is the $\Delta H°$ for the reaction: $3C(s,graphite) + 4H_2(g) \rightarrow C_3H_8(g)$

From Hess's law, three times equation (1) plus four times equation (2), plus the reverse of equation (4) gives this equation. Thus:

$$\mathbf{\Delta H_f°(C_3H_8,g)} = 3(-393.5) + 4(-285.8) - (-2219.9) = \mathbf{-103.8\ kJ}$$

(b) The structures of each of the first four alkanes are

```
   H          H  H           H  H  H           H  H  H  H
   |          |  |           |  |  |           |  |  |  |
 H—C—H     H—C—C—H      H—C—C—C—H      H—C—C—C—C—H
   |          |  |           |  |  |           |  |  |  |
   H          H  H           H  H  H           H  H  H  H
```

<u>C-H bonds</u>	4	6	8	10
<u>C-C bonds</u>	0	1	2	3

Going from one alkane to the next, there are *two* additional C—H bonds and *one* additional C—C bond. Thus, approximately, we would expect the enthalpies of formation to differ by about the same amount, and we can write:

$$\Delta H_f^\circ(C_4H_{10},g) - \Delta H_f^\circ(C_3H_8,g) = \Delta H_f^\circ(C_3H_8,g) - \Delta H_f^\circ(C_2H_6,g)$$

Hence:

$\mathbf{\Delta H_f^\circ(C_4H_{10},g)} = 2\Delta H_f^\circ(C_3H_8,g) - \Delta H_f^\circ(C_2H_6,g) = 2(-103.8) - (-84.6) = \mathbf{-123\ kJ}$ (c.f., -126.1 kJ) (Table 9.1)

For the combustion of these hydrocarbons, the general balanced equation is of the form:

$$C_nH_{2n+2}(g) + \tfrac{1}{2}(3n+1)O_2(g) \rightarrow nCO_2(g) + (n+1)H_2O(\ell)$$

Bonds:	C-H	C-C	O=O	C=O	O-H
n = n	2n+2	n-1	½(3n+1)	2n	2(n+1)
n = n+1	2n+4	n	½(3n+4)	2n+2	2(n+2)
Difference:	2	1	1½	2	2

which tells us that the change in ΔH°(combustion) in going from $C_3H_8(g)$ to $C_4H_{10}(g)$ should be approximately the same as that obtained in going from $C_2H_6(g)$ to $C_3H_8(g)$, which is:

$$(-2219.9) - (-1559.8) = \mathbf{-660.0\ kJ}$$

i.e. $\mathbf{\Delta H^\circ(combustion, C_4H_{10},g)} = [(-2219.9) + (-660.0)] = \mathbf{-2879.9\ kJ}$

This can be compared to the result from standard enthalpies of formation:

$$\mathbf{\Delta H^\circ} = [4\Delta H_f^\circ(CO_2,g) + 5\Delta H_f^\circ(H_2O,\ell)] - [\Delta H_f^\circ(C_4H_{10},g) + 6\tfrac{1}{2}\Delta H_f^\circ(O_2,g)]$$
$$= [4(-393.5) + 5(-285.8)] - [(-123) - 6\tfrac{1}{2}(0)] = \mathbf{-2880\ kJ}$$

---

60. Methanol, $CH_3OH(\ell)$, is a potentially useful fuel. Although it provides only one-half as much energy per liter as does gasoline, burns cleanly and has a high octane number (that is, it has a low tendency to knock in an engine). It is made industrially from synthesis gas at high pressure, using a catalyst at about 300 °C:

$$2H_2(g) + CO(g) \rightarrow CH_3OH(\ell).$$

Use the following standard reaction enthalpies of combustion to determine ΔH° for the synthesis of methanol from the previous reaction:

$CH_3OH(\ell) + 1\tfrac{1}{2}(g) \rightarrow CO_2(g) + 2H_2O(\ell)$	ΔH° = -726.6 kJ ... (1)
$C(graphite,s) + \tfrac{1}{2}O_2(g) \rightarrow CO(g)$	ΔH° = -110.5 kJ ... (2)
$C(graphite,s) + O_2(g) \rightarrow CO_2(g)$	ΔH° = -393.5 kJ ... (3)
$H_2(g) + \tfrac{1}{2}O_2(g) \rightarrow H_2O(\ell)$	ΔH° = -285.8 kJ ... (4)

This is a Hess's law problem, and ΔH° for the reaction in question is obtained as follows:

First double equation (4) and then add it to the reverse of equation (2); then add the reverse of equation (1) and, finally, add equation (3):

$2[H_2(g) + \tfrac{1}{2}O_2(g) \rightarrow H_2O(\ell)]$	ΔH° = 2(-285.8) = -571.6 kJ
$CO(g) \rightarrow C(graphite,s) + \tfrac{1}{2}O_2(g)$	ΔH° = -(-110.5) = +110.5 kJ
$CO_2(g) + 2H_2O(\ell) \rightarrow CH_3OH(\ell) + 1\tfrac{1}{2}O_2(g)$	ΔH° = -(-726.6) = +726.6 kJ
$C(graphite,s) + O_2(g) \rightarrow CO_2(g)$	ΔH° = -393.5 kJ
$2H_2(g) + CO(g) \rightarrow CH_3OH(\ell)$	ΔH° = **—128.0 kJ**

CHAPTER 10

Properties and Uses of Metals

2. The metals gold, silver, copper, and iron are mentioned in the Bible, but sodium, potassium, calcium, and aluminum, for example, were not isolated as pure elements until the nineteenth century. What differences in properties between the earliest-known elements and the Group I and II metals and aluminum account for the late discovery of the latter?

Metals are produced from their ores by the reduction of metal cations, M^{n+}. The activity series of common metals gives the order $\mathbf{Au^+ > Ag^+ > Cu^{2+} > Fe^{2+} > Al^{3+} > Ca^{2+} > K^+ > Na^+}$ for the ease of reduction of the ions in question, which is the *reverse* of the order of the ease of oxidation of the metals themselves. Given appropriate conditions of the kind found in the earth's crust, the "biblical" metals could easily form by reaction with suitable reducing agents, such as charcoal, C(s), as in the smelting of iron ores, or S^{2-} ion, as in the preparation of Cu(s) from Cu_2S(s). However, none of the natural reducing agents are strong enough to reduce alkali metal, alkaline earth, or aluminum cations, so once formed these remained as their compounds. Only very strong reducing agents, such as the electrons used in electrolysis reactions or a more reactive metal, are suitable to form the "non-biblical" metals from their compounds. In the 19th century after the first batteries were invented, it was possible for the first time to use an electric current and sodium and aluminum were both first made by Humphrey Davy in 1807-1808 by electrolyzing molten salts. Potassium or calcium could then be made, either by electrolysis or by reducing a potassium or calcium compound with sodium.

Structure and Bonding

4. Explain: (a) what is a metallic bond, and (b) how it differs from a covalent bond.

(a) **A metallic bond** is the type of bond that holds the atoms together in a metal. According to the *electron gas model* the metallic bonds in metals are described as the attraction between the *positively charged* **metal cations** and the *negatively charged* **electron cloud** formed by their valence electrons, which is dispersed throughout the metal.

(b) In a **covalent bond** one or more pairs of electrons are *localized* between the two atoms forming the bond, whereas the **metallic bond** is the consequence of the attraction between the positively charged atomic cores and a cloud of electrons that is *delocalized* over all the atoms.

---

6. Explain why (a) a metal is a good conductor of heat and electricity as both a liquid and a solid, and (b) the electrical conductivity of a metal increases as temperature decreases.

(a) According to the **electron-gas model**, the delocalized electrons in a metal, both as solid and liquid, are free to move between the positively charged ions and they move freely under the influence of an electrical potential. When a source of current (a battery) is connected between two parts of a metal, electrons from one terminal of the battery replace electrons in the metal as they are removed from the metal at the other battery terminal and a current flows through the metal.

(b) In **electrical conduction**, electrons move freely through a metal and their movement is only impeded by collisions with the vibrating metal ions. With decrease in temperature, the amplitude of the vibrations of the ions decrease, fewer collisions of electrons with metal ions occur, and the electrical conductivity increases.

---

7. (a) Explain the following increase in melting point: Na 98 °C $<$ Mg 650 °C $<$ Al 660°C. (b) Suggest a possible reason why iron melts at 1535 °C, at a temperature so much higher than that for the three metals in (a).

(a) The melting points of metals are related to the strength of the **metallic bonding**, which depends on the number of electrons per metal available for metallic bonding and the sizes of the ions to which the electrons are attracted.

Na (Group I), Mg (Group II), and Al (Group III) have, respectively, 1, 2, and 3 easily ionized valence electrons and form Na^+, Mg^{2+}, and Al^{3+} ions, and the sizes of these isoelectronic ions decrease in the order $Na^+ > Mg^{2+}$ Al^{3+} (with increasing core charge). Thus, the strength of attraction for an ever increasing number of delocalized metallic bonding electrons is expected to increase in the order Na < Mg < Al, which is the observed order of increasing melting point.

(b) Iron has the valence shell electron configuration $4s^23d^6$ and in all probability has available a greater number of valence electrons per iron atom than aluminum, with 3 electrons per Al atom, which accounts for its much greater melting point than Na, Mg, or Al.

Reactions and Compounds of Group I and II Metals; Reactions and Compounds of Aluminum

11. (a) Why does the reactivity of the Group I and II metals increase as we descend each group? (b) Describe the relative reactivities of the alkali metals and of the alkaline earth metals with water.

(a) The reactivities of metals depend on the ease with which they form positive ions (cations), which is measured by their *first* ionization energies. The **alkali metals** of Group I, with 1 valence electron, all easily lose this electron to form M^+ ions. They all have the same core charge of +1 but the reactivity *increases* down the group because the first ionization energies decrease in descending the group, with the increasing distance of the valence shell from the nucleus. Similarly, the **alkaline-earth metals** of Group II, with two valence electrons, all form M^{2+} ions. They all have core charges of +2 and their reactivity increases in descending the group for the same reason it does in Group I.

(b) Since it takes more energy to form a M^{2+} ion than to form an M^+ ion, the alkaline-earth metals are less reactive than the alkali metals, as exemplified by their relative ease of reaction with water. All the alkali metals react readily with water, ($2M(s) + 2H_2O(\ell) \rightarrow 2M^+OH^-(aq) + H_2(g)$). Lithium and sodium react relatively slowly at room temperature, potassium reacts sufficiently energetically to ignite the hydrogen as it is formed, and rubidium and cesium react explosively. In Group II, magnesium reacts only with steam, calcium reacts slowly with water, and strontium and barium react relatively vigorously but not as strongly as any of the alkali metals, according to the equation $M(s) + 2H_2O(\ell) \rightarrow M^{2+}(OH^-)_2(s) + H_2(g)$; (only $Ba(OH)_2(s)$ is appreciably soluble.)

---

13. (a) Place the metals Al, Cu, Mg, and Na in order of decreasing reactivity. (b) Cite some of the experimental observations that provide the evidence for this order of reactivity.

(a) According to the **activity series** for the common metals, the order of reactivity is: **Na > Mg > Al > Cu.**

(b) In terms of their familiar reaction chemistry, they can be distinguished in terms of the ease with which they react with oxidizing agents of increasing strength; for instance water ($H_2O(\ell)$), weak aqueous acids ($H_3O^+(aq)$), or oxidizing molecular acids such as HNO_3(conc) or H_2SO_4(conc). Among these metals, only **Na** is oxidized by water at room temperature, only **Na** and **Mg** are oxidized at higher temperature by steam, $H_2O(g)$, **Na**, **Mg**, and **Al** are oxidized by $H_3O^+(aq)$ in dilute acid, and **Na**, **Mg**, **Al**, and **Cu** are all oxidized by concentrated sulfuric acid. (Cu only on heating). The products of oxidation are Na^+, Mg^{2+}, Al^{3+}, and Cu^{2+}. Examples of balanced equations for the reactions described are:

$2Na(s) + H_2O(\ell) \rightarrow 2NaOH(aq) + H_2(g)$ $\qquad$ $Mg(s) + H_2O(g) \rightarrow MgO(s) + H_2O(g)$

$2Al(s) + 6H_3O^+(aq) \rightarrow 2Al^{3+}(aq) + 3H_2(g)$ $\qquad$ $Cu(s) + 2H_2SO_4(conc) \rightarrow Cu^{2+} + SO_2(g) + SO_4^{2-} + 2H_2O$

---

16. (a) The alkali and alkaline earth metals react with hydrogen on heating to give ionic hydrides. Write balanced equations for the reactions of potassium and calcium with hydrogen; (b) Magnesium reacts on heating with nitrogen to give a nitride containing the N^{3-} ion. Write the balanced equation for this reaction.

(a) $2K(s) + H_2(g) \rightarrow 2KH(s)$ $[K^+ H^-]$ $\qquad$ $Ca(s) + H_2(g) \rightarrow CaH_2(s)$ $[Ca^{2+}(H^-)_2]$

(b) $3Mg(s) + N_2(g) \rightarrow Mg_3N_2$ $[(Mg^{2+})_3(N^{3-})_2]$

---

18. From the metals discussed in this chapter: (a) name two metals that react with dilute hydrochloric acid, and write the balanced equations for the reactions that occur; (b) name two metals that *do not* react with dilute hydrochloric acid but react with dilute nitric acid, and explain why.

(a) All metals above H_2 in the activity series for common metals (Table 10.4) - **Na, K, Ca, Mg, Zn, Al, and Fe** - react with the $H_3O^+(aq)$ ion present in dilute HCl(aq) (or any aqueous acid).

For example:

$$Zn(s) + 2H_3O^+(aq) \rightarrow Zn^{2+}(aq) + 2H_2O(\ell) + H_2(g);$$

$$Fe(s) + 2H_3O^+(aq) \rightarrow Fe^{2+}(aq) + 2H_2O(\ell) + H_2(g).$$

(b) Metals below H_2 in the activity series (Cu, Ag, and Au in Table 10.4) do not react with $H_3O^+(aq)$. Dilute nitric acid, $HNO_3(aq)$, is fully ionized in water and contains $H_3O^+(aq)$ and $NO_3^-(aq)$:

$$HNO_3(aq) + H_2O(\ell) \rightarrow H_3O^+(aq) + NO_3^-(aq)$$

Nitrate ion, $NO_3^-(aq)$, is a stronger oxidizing agent than $H_3O^+(aq)$; thus, most metals react with dilute $HNO_3(aq)$, even metals such as **copper** and **silver** that are too weak to reduce $H_3O^+(aq)$. Nitrate ion, $NO_3^-(aq)$, is usually reduced to $NO_2(g)$or NO(g), or even $NH_3(g)$ when the metal is a very strong reducing agent, and depending on the concentration of the nitric acid.

---

20. (a) What volume of hydrogen gas at STP would result from the reaction of 2.00 g of pure calcium hydride with excess water? (b) What mass of calcium would be required to give the same amount of hydrogen as in (a) in its reaction with water?

(a) Calcium hydride, $CaH_2(s)$, contains the hydride ion, H^-, a strong base, which reacts with water according to:

$$H{:}^-(aq) + H_2O(\ell) \rightarrow H_2(g) + OH^-(aq)$$

Thus, the reaction is:

$$CaH_2(s) + 2H_2O(\ell) \rightarrow Ca(OH)_2(s) + 2H_2(g).$$

The formula mass of CaH_2 is [40.08 + 2(1.008)] = 42.10 g mol^{-1}, which we use to convert grams of $CaH_2(s)$ to moles of $CaH_2(s)$, then use the balanced equation to convert moles of CaH_2 reactant to moles of $H_2(g)$ product, and finally use the ideal gas equation (P = 1 atm, T = 273.2 K) to calculate the volume of $H_2(g)$ at STP:

$$PV = nRT \quad ; \quad V = \frac{nRT}{P}$$

$$= \frac{(2.00\ g\ CaH_2)\left(\frac{1\ mol\ CaH_2}{42.10\ g\ CaH_2}\right)\left(\frac{2\ mol\ H_2}{1\ mol\ CaH_2}\right)(0.0821\ atm\ L\ mol^{-1}\ K^{-1})(273.2\ K)}{1\ atm} = \underline{2.13\ L}$$

Alternatively, instead of using PV = nRT, we could have used the fact that 1 mol of any gas at STP occupies a volume of 22.41 L, and then converted mol $H_2(g)$ directly to its volume at STP:

$$V = (2.00\ g\ CaH_2)\left(\frac{1\ mol\ CaH_2}{42.10\ g\ CaH_2}\right)\left(\frac{2\ mol\ H_2}{1\ mol\ CaH_2}\right)\left(\frac{22.41\ L}{1\ mol}\right) = \underline{2.13\ L}$$

(b) The reaction is:

$$Ca(s) + 2H_2O(\ell) \rightarrow Ca(OH)_2(s) + H_2(g)$$

and comparison with the equation for the reaction of $CaH_2(s)$, above, shows that for production of the same amount of $H_2(g)$:

$$\textbf{moles of Ca} = \textbf{2(moles of CaH}_2\textbf{)}$$

Thus:

$$\text{Equivalent mass of Ca} = (2.00\ g\ CaH_2)\left(\frac{1\ mol\ CaH_2}{42.10\ g\ CaH_2}\right)\left(\frac{2\ mol\ Ca}{1\ mol\ CaH_2}\right)\left(\frac{40.08\ g\ Ca}{1\ mol\ Ca}\right) = \underline{3.81\ g}$$

3.81 g of calcium would give 2.13 L $H_2(g)$ at STP

---

22. (a) Write a balanced equation for the production of sodium carbonate by heating the mineral trona. (b) What is the maximum amount of hydrated sodium carbonate that can be obtained from 1 metric ton (10^3 kg) of trona ore?

(a) **Trona** has the formula $Na_5(CO_3)_2(HCO_3)\cdot 2H_2O(s)$ and on heating gives anhydrous sodium carbonate, $Na_2CO_3(s)$, carbon dioxide, $CO_2(g)$, and water:

$$2Na_5(CO_3)_2(HCO_3)\cdot 2H_2O(s) \rightarrow 5Na_2CO_3(s) + CO_2(g) + 5H_2O(g)$$

(b) **Hydrated sodium carbonate** is $Na_2CO_3\cdot 10H_2O(s)$, molar mass 286.2 g mol^{-1}; **trona** = 332.0 g mol^{-1}. Thus, from the balanced equation, maximum mass of washing soda from 10^3 kg **trona** is given by:

$$(10^3 \text{ kg trona})\left(\frac{10^3 \text{ g}}{1 \text{ kg}}\right)\left(\frac{1 \text{ mol trona}}{332.0 \text{ g trona}}\right)\left(\frac{5 \text{ mol soda}}{2 \text{ mol trona}}\right)\left(\frac{286.2 \text{ g soda}}{1 \text{ mol soda}}\right)$$

$$= (2.16 \times 10^6 \text{ g soda})\left(\frac{1 \text{ kg}}{10^3 \text{ g}}\right)\left(\frac{1 \text{ ton}}{10^3 \text{ kg}}\right) = \underline{2.2 \text{ metric ton soda}}$$

---

26. Write an equation for the reaction of each of the following oxides with water, and classify each as an acidic or basic oxide: (a) K_2O (b) SrO (c) SO_2 (d) SO_3 (e) CO_2 (f) P_4O_6

Metal oxides are *ionic* **basic oxides** (all containing the oxide ion, O^{2-}), and **nonmetal oxides** are *covalent* **and are acidic oxides.** The acids formed by the reaction of acidic oxides with water will all be dissociated into $H_3O^+(aq)$ and the appropriate anions:

Oxide	Type	Reaction with Water	
(a) K_2O	ionic	$K_2O(s) + H_2O(\ell) \rightarrow 2K^+(aq) + 2OH^-(aq)$	**basic**
(b) SrO	ionic	$SrO(s) + H_2O(\ell) \rightarrow Sr^{2+}(aq) + 2OH^-(aq)$	**basic**
(c) SO_2	covalent	$SO_2(g) + H_2O(\ell) \rightarrow H_2SO_3(aq)$	**acidic**
(d) SO_3	covalent	$SO_3(g) + H_2O(\ell) \rightarrow H_2SO_4(aq)$	**acidic**
(e) CO_2	covalent	$CO_2(g) + H_2O(\ell) \rightarrow H_2CO_3(aq)$	**acidic**
(f) P_4O_6	covalent	$P_4O_6(s) + 6H_2O(\ell) \rightarrow 4H_3PO_3(aq)$	**acidic**

Note: $H_2SO_3(aq)$ and $H_2CO_3(aq)$ solutions contain high concentrations of $SO_2(aq)$ and $CO_2(aq)$, respectively.

---

27. Write balanced equations for each of the following reactions: (a) sodium oxide with sulfur dioxide; (b) calcium oxide with sulfur trioxide; (c) sodium oxide with phosphorus(V) oxide; (d) aluminum oxide with sulfur trioxide.

These are all reactions between **basic (metal) oxides and acidic (nonmetal) oxides** to give **salts:**

(a) $Na_2O(s) + SO_2(g) \rightarrow Na_2SO_3$ (b) $CaO(s) + SO_3(g) \rightarrow CaSO_4(s)$

(c) $6Na_2O(s) + P_4O_{10}(s) \rightarrow Na_3PO_4(s)$ (d) $Al_2O_3(s) + 3SO_3(g) \rightarrow Al_2(SO_4)_3(s)$

---

29. Write balanced equations for the reactions that occur when each of the following compounds is heated and name the products: (a) calcium carbonate (b) calcium hydroxide (c) sodium hydrogencarbonate (d) iron(III) hydroxide.

(a) $CaCO_3(s) \rightarrow CaO(s) + CO_2(g)$ **calcium oxide and carbon dioxide**

(b) $Ca(OH)_2(s) \rightarrow CaO(s) + H_2O(g)$ **calcium oxide and water**

(c) $Ca(HCO_3)_2(s) \rightarrow CaCO_3(s) + CO_2(g) + H_2O(g)$ **calcium carbonate, carbon dioxide, and water**

(d) $2Al(OH)_3(s) \rightarrow Al_2O_3(s) + 3H_2O(g)$ **aluminum oxide and water**

31.* Except at very high temperature, both aluminum chloride and aluminum bromide as gases are composed of the dimeric molecules Al_2Cl_6 and Al_2Br_6, respectively, rather than monomeric $AlCl_3$ and $AlBr_3$. They have the following bridged structures. Suggest a reason why dimeric molecules form.

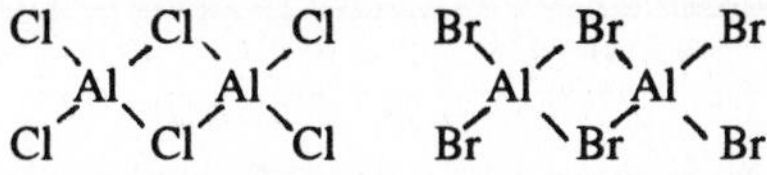

We saw in Problem 30 that Al forms the **complex ion** $AlCl_4^-$ in which Al has an **octet** of electrons in its valence shell. Although $AlCl_3$ can be regarded to a first approximation as consisting of Al^{3+} and Cl^- ions in the solid, $\mathbf{AlCl_3}$ and $\mathbf{AlBr_3}$ both form **covalent** molecules with *highly polar* bonds in the gas phase. These have only **three pairs** of electrons (an incomplete octet) in the valence shell of Al, but complete their octets of electrons by **forming dimers** in which unshared electron pairs of halogen atoms of two different molecules are donated to the Al atom of the other molecule, thus forming a dimer.

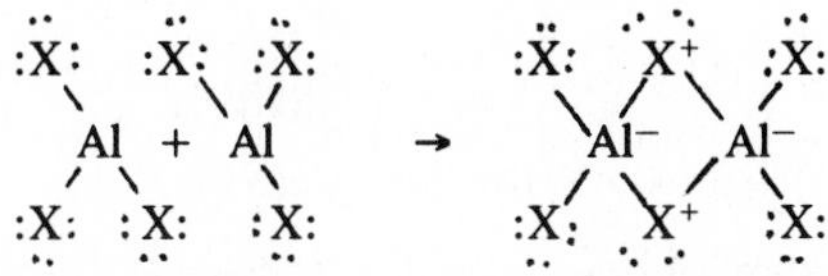

33.* Write balanced equations for each of the following reactions: (a) powdered aluminum with NaOH(aq) to give sodium aluminate, $NaAl(OH)_4(aq)$, and hydrogen; (b) aluminum oxide with carbon and chlorine to give aluminum trichloride and carbon monoxide, (c) the decomposition, upon heating, of ammonium alum, $NH_4Al(SO_4)_2.12H_2O(s)$, to give ammonia, sulfuric acid, aluminum oxide, and water.

(a) $2Al(s) + 2NaOH(aq) + 6H_2O(\ell) \rightarrow 2Na^+(aq) + 2Al(OH)_4^-(aq) + 3H_2(g)$

(b) $Al_2O_3(s) + 3C(s) + 3Cl_2(g) \rightarrow 2AlCl_3(s) + 3CO(g)$

(c) $2NH_4Al(SO_4)_2.12H_2O(s) \rightarrow 2NH_3(g) + 4H_2SO_4(\ell) + Al_2O_3(s) + 9H_2O(g)$

<u>The Transition Metals</u>

34. Write balanced equations for each of the following reactions: (a) calcium with water; (b) iron with steam at high temperature; (c) aluminum with dilute sulfuric acid; (d) copper with hot concentrated sulfuric acid.

(a) $Ca(s) + 2H_2O(\ell) \rightarrow Ca(OH)_2(s) + H_2(g)$

(b) $2Fe(s) + 3H_2O \rightarrow Fe_2O_3(s) + 3H_2(g)$

(c) $2Al(s) + 3H_2SO_4(aq) \rightarrow 2Al^{3+}(aq) + 3SO_4^{2-}(aq) + 3H_2(g)$

(d) $Cu(s) + 2H_2SO_4(conc) \rightarrow Cu^{2+}(aq) + SO_4^{2-}(aq) + SO_2(g) + 2H_2O(\ell)$

---

36. Assign oxidation numbers to each of the atoms in each of the following: (a) Al_2O_3 (b) $AlCl_4^-$ (c) $Fe_2(SO_4)_3$ (d) $CuSO_4 \cdot 5H_2O$ (e) $Cu(NH_3)_4 \cdot SO_4$

	Formulation	Oxidation Numbers
(a) $Al_2O_3.xH_2O$	$(Al^{3+})_2\ (O^{2-})_3$	Al +3 O -2
(b) $AlCl_4^-$	$Al^{3+}\ (Cl^-)_4$	Al +3 Cl -1
(c) $Fe_2(SO_4)_3$	$(Fe^{3+})_2\ (SO_4^{2-})_3$	Fe +3 S +6 O -2
(d) $CuSO_4.5H_2O$	$Cu^{2+}\ SO_4^{2-}\ (H_2O)_5$	Cu +2 S +6 O -2 H +1
(e) $Cu(NH_3)_4.SO_4$	$Cu^{2+}\ (NH_3)_4\ SO_4^{2-}$	Cu +2 N -3 H +1 S +6 O -2

---

38. Classify each of the following as a Lewis acid or a Lewis base, and give a balanced equation for a reaction of each to support your choice: (a) NH_3 (b) Cu^{2+} (c) Fe^{3+} (d) Cl^- (e) $AlCl_3$

A **Lewis acid** is an **electron pair acceptor** and a **Lewis base** is an **electron pair donor**. For a molecule or ion to be a Lewis acid it must contain an atom with an incomplete valence shell which can accept another electron pair; for a molecule or ion to behave as a Lewis base it must contain an atom with one or more unshared (lone) pairs of electrons.

(a) **Ammonia,** $:NH_3$, can behave as a **Lewis base** because its N atom has a lone pair of electrons that can be donated. For example:

$$Ni^{2+}(aq) + 6NH_3(aq) \rightarrow Ni(NH_3)_6^{2+}$$

$$Cu^{2+}(aq) + 4NH_3(aq) \rightarrow Cu(NH_3)_4^{2+}$$

(b) **Copper(II) ion** can behave as a **Lewis acid.** Cu^{2+} has the electron configuration $[Ar]3d^9$; it has unfilled 4s and 4p orbitals that can accept lone pairs from other species:

$$Cu^{2+}(aq) + 4H_2O(\ell) \rightarrow Cu(H_2O)_4^{2+}$$

$$Cu^{2+}(aq) + 4NH_3(aq) \rightarrow Cu(NH_3)_4^{2+}$$

$$Cu^{2+}(aq) + 4Cl^-(aq) \rightarrow CuCl_4^{2-}(aq)$$

(c) **Iron(III) ion** can behave as a **Lewis acid.** Fe^{3+} has the electron configuration $[Ar]3d^5$; it has unfilled 4s and 4p orbitals that can accept lone pairs from other species.

$$Fe^{3+}(aq) + 6H_2O(\ell) \rightarrow Fe(H_2O)_6^{3+}(aq)$$

$$Fe^{3+}(aq) + 6CN^-(aq) \rightarrow Fe(CN)_6^{3-}(aq)$$

(d) **Chloride ion,** $:Cl:^-$, can behave as a **Lewis base** because it has lone pairs of electrons that can be donated. An example is given in the answer to part (b).

(e) **Aluminum chloride,** $AlCl_3$, can behave as a **Lewis acid** because when Al forms only three electron-pair bonds it has an incomplete valence shell which can accept lone pairs. For example:

$$AlCl_3 + Cl^- \rightarrow AlCl_4^-$$

---

40. (a) Give three examples taken from this chapter of metal hydroxides that decompose on heating to give the corresponding oxides. (b) Write a balanced equation for each reaction.

In general, alkali and alkaline earth metal hydroxides and are stable to heat but other metal hydroxides, including most transition metal hydroxides, decompose on heating to give the corresponding oxide. For example:

(a) **Aluminum(III) hydroxide**, $Al(OH)_3$, **iron(III) hydroxide**, $Fe(OH)_3(s)$; **copper(II) hydroxide**, $Cu(OH)_2(s)$; **cobalt(II) hydroxide**, $Co(OH)_2(s)$, and **nickel(II) hydroxide**, $Ni(OH)_2(s)$.

(b) $2Al(OH)_3(s) \rightarrow Al_2O_3(s) + 3H_2O(g)$ $2Fe(OH)_3(s) \rightarrow Fe_2O_3(s) + 3H_2O(g)$

$$Cu(OH)_2(s) \rightarrow CuO(s) + H_2O(g) \qquad Co(OH)_2(s) \rightarrow CoO(s) + H_2O(g)$$

$$Ni(OH)_2(s) \rightarrow NiO(s) + H_2O(g)$$

---

42. Copper(II) ammonium sulfate was found to contain 27.03% water of crystallization by mass. Upon strongly heating it gave copper(II) oxide corresponding to 19.89% of the starting mass. What is the empirical formula of hydrated copper(II) ammonium sulfate?

The salt contains Cu^{2+}, NH_4^+, and SO_4^{2-} ions and water, H_2O. Its empirical formula must therefore be of the form $(Cu^{2+})_x\ (NH_4^+)_y\ (SO_4^{2-}]_{x+½y}.zH_2O$, where we have to determine x, y, and z, and this is conveniently rewritten as:

$$[(Cu^{2+})_x(SO_4^{2-})_x]\ [(NH_4^+)_y\ (SO_4^{2-}]_{½y}] \cdot zH_2O$$

Now we can consider the composition of a 100.0 g sample of the compound:

$$100.0 \text{ g compound contains } 27.03 \text{ g } H_2O = (27.03 \text{ g } H_2O)\left(\frac{1 \text{ mol } H_2O}{18.02 \text{ g } H_2O}\right) = \underline{1.500 \text{ mol } H_2O}$$

$$\text{Similarly, } 100.0 \text{ g gives } 19.89 \text{ g CuO(s)} = (19.89 \text{ g CuO})\left(\frac{1 \text{ mol CuO}}{79.55 \text{ g CuO}}\right)\left(\frac{1 \text{ mol Cu}}{1 \text{ mol CuO}}\right) = \underline{0.2500 \text{ mol}}$$

$$\text{Thus, } 100.0 \text{ g salt contains: } (0.2500 \text{ mol CuO})\left(\frac{1 \text{ mol } CuSO_4}{1 \text{ mol CuO}}\right)\left(\frac{159.6 \text{ g } CuSO_4}{1 \text{ mol } CuSO_4}\right) = \underline{39.90 \text{ g } Cu^{2+}SO_4^{2-}}$$

Thus, 100.0 g of compound contains **39.90 g $Cu^{2+}SO_4^{2-}$** and **27.03 g H_2O**, and the remainder of the 100.0 g must have the composition $(NH_4^+)_2SO_4^{2-}$, in the amount of (100.0 - 39.90 - 27.03) = **33.07 g, corresponding to:**

$$(33.07 \text{ g } (NH_4)_2SO_4)\left(\frac{1 \text{ mol } (NH_4)_2SO_4}{132.1 \text{ g } (NH_4)_2SO_4}\right) = \underline{0.2503 \text{ mol } (NH_4)_2SO_4}$$

	$CuSO_4$	$(NH_4)_2SO_4$	H_2O
Ratio of moles (molecules)	$\frac{0.2500}{0.2503}$	$\frac{0.2503}{0.2503}$	$\frac{1.500}{0.2503}$
	0.999	**1.00**	**5.99**

Thus, the **empirical formula** is $CuSO_4 \cdot (NH_4)_2SO_4 \cdot 6H_2O$ <u>or</u> $(NH_4)_2Cu(SO_4)_2 \cdot 6H_2O$

---

43. Describe and explain the reactions that could be used to prepare each of the following substances from copper(II) sulfate pentahydrate (more than one step may be required): (a) copper (b) copper(II) chloride (c) copper(I) oxide (d) copper(II)tetramminesulfate monohydrate, $Cu(NH_3)_4SO_4 \cdot H_2O(s)$.

Copper sulfate pentahydrate has the formula $CuSO_4 \cdot 5H_2O$, and a solution in water contains $Cu^{2+}(aq)$ and $SO_4^{2-}(aq)$ ions, as well as water molecules.

(a) **Copper metal** Addition of NaOH(aq) to $CuSO_4(aq)$ gives a pale-blue precipitate of $Cu(OH)_2(s)$, which is filtered off and heated strongly to give black copper(II) oxide, CuO(s). Copper(II) oxide may then be reduced with a reducing agent such as hydrogen or carbon:

$$Cu^{2+}(aq) + 2OH^-(aq) \rightarrow Cu(OH)_2(s)$$

$$Cu(OH)_2(s) \rightarrow CuO(s) + H_2O(g)$$

$$CuO(s) + H_2(g) \rightarrow Cu(s) + H_2O(g) \quad \textbf{or} \quad CuO(s) + C(s) \rightarrow Cu(s) + CO(g)$$

(b) Copper(II) chloride $Cu(OH)_2(s)$ or $CuO(s)$, prepared in part (a), could be dissolved in hydrochloric acid, $HCl(aq)$, to give a solution from which $CuCl_2 \cdot 2H_2O(s)$ crystallizes after concentrating the solution:

$$Cu(OH)_2(s) + 2HCl(aq) \rightarrow CuCl_2(aq) + 2H_2O(\ell)$$

$$CuO(s) + 2HCl(aq) \rightarrow CuCl_2(aq) + H_2O(\ell)$$

Alternatively, the copper from part (a) could be heated with chlorine gas.

$$Cu(s) + Cl_2(g) \rightarrow CuCl_2(aq)$$

(c) **Copper(I) oxide** This is made by reducing the Cu^{2+} ions in **copper(II) oxide**, $CuO(s)$, prepared in part (a), with copper metal, by heating the $CuO(s)$ with $Cu(s)$.

$$CuO(s) + Cu(s) \rightarrow Cu_2O(s)$$

(d) **Copper(II)tetrammine sulfate monohydrate** contains the dark-blue $Cu(NH_3)_4^{2+}$ ion, which is formed when the appropriate amount of concentrated ammonia solution, $NH_3(aq)$, is added to a solution of $CuSO_4{\cdot}5H_2O(aq)$:

$$Cu^{2+}(aq) + 4NH_3(aq) \rightarrow Cu(NH_3)_4^{2+}(aq)$$

The complex salt crystallizes from solution when it is concentrated and cooled.

---

45. For the transition metals in Period 4, what are (a) the highest oxidation states of each of the metals from Sc to Mn? (b) (i) the commonest, and (ii) the highest oxidation states for each of the metals from Co to Zn?

(a) The **highest** oxidation state for the *transition metals* from scandium to manganese correspond to those of the ions obtained by removing (or utilizing) all of their valence electrons:

Sc, $[Ar]3d^14s^2$, **+III**; **Ti**, $[Ar]3d^24s^2$, **+IV**; **V**, $[Ar]3d^34s^2$, **+V**; **Cr**, $[Ar]3d^44s^1$, **+VI**; **Mn**, $[Ar]3d^54s^2$, **+VII**;

(b) The transition metals after **manganese** are: **iron, cobalt, nickel, copper,** and **zinc**:

Oxidation State	**Co** $[Ar]3d^74s^2$	**Ni** $[Ar]3d^84s^2$	**Cu** $[Ar]3d^{10}4s^1$	**Zn** $[Ar]3d^{10}4s^2$
(i) Commonest	+II	+II	+II	+II
(ii) Highest	+III	+III	+III	+II

46.* A dark red solid resulted from heating 1.50 g of iron in excess $Cl_2(g)$. When the solid was dissolved in water and excess $NaOH(aq)$ was added, a gelatinous brown precipitate resulted, which when strongly heated formed a red-brown powder. (a) Write balanced equations for each of the reactions described and (b) identify each product. (c) What is the maximum mass of red-brown powder that could be obtained?

(a)

$$2Fe(s) + 3Cl_2(g) \rightarrow 2FeCl_3(s)$$

$$FeCl_3(aq) + 3OH^-(aq) \rightarrow Fe(OH)_3(s) + 3Cl^-(aq)$$

$$2Fe(OH)_3(s) \rightarrow Fe_2O_3(s) + 3H_2O(g)$$

(b) Iron is oxidized by $Cl_2(g)$ to $FeCl_3(s)$, a dark red solid, that dissolves in water to give $Fe^{3+}(aq)$ and $Cl^-(aq)$ in solution. Treatment of this solution with $NaOH(aq)$ gives a red-brown precipitate of $Fe(OH)_3(s)$. When this is strongly heated, red-brown $Fe_2O_3(s)$ forms.

(c) Maximum mass of $Fe_2O_3(s)$ = $(1.50 \text{ g Fe})(\frac{1 \text{ mol Fe}}{55.85 \text{ g Fe}})(\frac{1 \text{ mol } Fe_2O_3}{2 \text{ mol Fe}})(\frac{159.7 \text{ g } Fe_2O_3}{1 \text{ mol } Fe_2O_3}) = \underline{2.14 \text{ g}}$

47.* Aluminum brass contains copper, zinc, and aluminum. When 1.00 g of brass was reacted with 0.100-M $H_2SO_4(aq)$, 149.3 mL of $H_2(g)$, measured at 25°C and 1.00 atm pressure, was evolved. When an identical sample was dissolved in hot concentrated sulfuric acid, 411.1 mL of $SO_2(g)$ was obtained at 25°C and 1.00 atm pressure. What is the composition of aluminum brass?

Of the metals in the alloy, only Zn and Al react with the $H_3O^+(aq)$ in dilute $H_2SO_4(aq)$, but all the metals react with the stronger oxidizing agent H_2SO_4 in H_2SO_4(conc) to give $SO_2(g)$. For the reaction with dilute $H_2SO_4(aq)$:

$$\text{mol } H_2(g) = n_{H_2} = \frac{PV}{RT} = \frac{(1 \text{ atm})(149.3 \text{ mL})(\frac{1 \text{ L}}{10^3 \text{ mL}})}{(0.0821 \text{ atm L mol}^{-1} \text{ K}^{-1})(298 \text{ K})} = \underline{6.102 \times 10^{-3} \text{ mol}}$$

and for the reaction with concentrated H_2SO_4:

$$\text{mol } SO_2(g) = n_{SO_2} = \frac{PV}{RT} = \frac{(1 \text{ atm})(411.1 \text{ mL})(\frac{1 \text{ L}}{10^3 \text{ mL}})}{(0.0821 \text{ atm L mol}^{-1} \text{ K}^{-1})(298 \text{ K})} = \underline{1.608 \times 10^{-2} \text{ mol}}$$

In each case, the initial moles of each metal are related to the moles of gas produced, as follows:

Dilute acid:

$$Zn(s) + H_2SO_4(aq) \rightarrow Zn^{2+}(aq) + SO_4^{2-}(aq) + H_2(g)$$

$$2Al(s) + 3H_2SO_4(aq) \rightarrow 2Al^{3+}(aq) + 3SO_4^{2-}(aq) + 3H_2(g)$$

Concentrated acid:

$$Cu(s) + 2H_2SO_4(conc) \rightarrow Cu^{2+} + SO_4^{2-} + 2H_2O + SO_2$$

$$Zn(s) + 2H_2SO_4(conc) \rightarrow Zn^{2+} + SO_4^{2-} + 2H_2O + SO_2$$

$$2Al(s) + 6H_2SO_4(conc) \rightarrow 2Al^{3+} + 3SO_4^{2-} + 6H_2O + 3SO_2$$

Thus, for n_{Zn} **moles of zinc,** n_{Al} moles of aluminum, and n_{Cu} **moles** of copper initially:

$$n_{H_2} = n_{Zn} + \frac{3}{2}n_{Al} = 6.102 \times 10^{-3} \text{ mol} \quad \ldots\ldots (1) \quad ; \quad n_{SO_2} = n_{Cu} + n_{Zn} + \frac{3}{2}n_{Al} = 1.680 \times 10^{-2} \text{ mol} \quad \ldots\ldots (2)$$

Subtracting equation (1) from equation (2) gives the moles of copper:

$$n_{Cu} = (0.01680 - 0.00610) = \underline{0.01070 \text{ mol}} \ ; \quad \text{mass of Cu} = (0.01070 \text{ mol Cu})(\frac{63.55 \text{ g Cu}}{1 \text{ mol Cu}}) = \underline{0.6800 \text{ g}}$$

And, by difference, mass of Zn + mass of Al = (1.000 - 0.6800) = **0.320 g**

If the mass of Zn is **x g**, then the mass of Al is (**0.320** — x) g, and from equation (1):

$$n_{Zn} + \frac{3}{2}n_{Al} = (x \text{ g Zn})(\frac{1 \text{ mol Zn}}{65.38 \text{ g Zn}}) + \frac{3}{2}[(0.320 - x)\text{g Al}](\frac{1 \text{ mol Al}}{26.98 \text{ g Al}}) = 6.102 \times 10^{-3} \text{ mol}$$

Whence: $(1.530 \times 10^{-2})x + (1.799 \times 10^{-2}) + (5.560 \times 10^{-2})\, x = (1.779 \times 10^{-2}) - (4.03 \times 10^{-2})x = 6.102 \times 10^{-3}$

$4.03\ x = 1.169$; $\underline{x = 0.2901\ g}$;   Mass of Al = $(0.320 - x)\ g = (0.320 - 0.290)g = \underline{0.030\ g}$

In summary: 1.000 g brass contains **0.680 g Cu, 0.290 g Zn,** and **0.030 g Al**

COMPOSITION OF ALUMINUM BRASS BY MASS: **68.0% Cu, 29.0% Zn, 3.0% Al.**

Metallurgy: Extraction of Metals

50. (a) What mass of an ore containing 30% Fe_2O_3 by mass is required to produce 10^6 metric tons of steel? (b) How many kilograms of iron are present in one metric ton of this ore? (c) What mass of coke (assuming that it is pure carbon and only CO(g) forms) is needed to reduce one metric ton of the ore to iron?

(a) 100 g of ore contains 30 g of $Fe_2O_3(s)$, or in terms of moles of iron:

$$\text{100 g ore contains } (30.0\ g\ Fe_2O_3)\left(\frac{1\ mol\ Fe_2O_3}{159.7\ g\ Fe_2O_3}\right)\left(\frac{2\ mol\ Fe}{1\ mol\ Fe_2O_3}\right)\left(\frac{55.85\ g\ Fe}{1\ mol\ Fe}\right) = \underline{20.98\ g\ Fe}$$

$$\text{Thus, for production of } 10^6 \text{ ton steel: mass of ore} = (10^6\ ton\ Fe)\left(\frac{100\ g\ ore}{20.98\ g\ Fe}\right) = \underline{4.8 \times 10^6\ ton}$$

(b) 100 g ore contains 20.98 g iron, so 1 metric ton of ore contains

$$(1\ ton\ ore)\left(\frac{10^6\ g}{1\ ton}\right)\left(\frac{20.98\ g\ Fe}{100\ g\ ore}\right)\left(\frac{1\ kg}{10^3\ g}\right) = \underline{210\ kg\ Fe}$$

(c) The balanced equation for the reaction is:

$$Fe_2O_3(s) + 3CO(g) \rightarrow 2Fe(\ell) + 3CO_2(g)$$

and 1 metric ton of ore produces a maximum of 209.8 kg of iron. Thus:

$$\text{mass of carbon (coke) required} = (209.8\ kg\ Fe)\left(\frac{1\ mol\ Fe}{55.85\ g\ Fe}\right)\left(\frac{3\ mol\ C}{1\ mol\ Fe}\right)\left(\frac{12.01\ g\ C}{1\ mol\ C}\right) = \underline{135\ kg\ coke}$$

---

52. Calculate $\Delta H°$ for the reduction of FeO(s) to Fe(s) by CO(g) given:

$$Fe_2O_3(s) + CO(g) \rightarrow 2FeO(s) + CO_2(g) \qquad \Delta H° = 38\ kJ$$
$$Fe_2O_3(s) + 3CO(g) \rightarrow 2Fe(s) + 3CO_2(g) \qquad \Delta H° = -28\ kJ$$

The required reaction is:

$$FeO(s) + CO(g) \rightarrow Fe(s) + CO_2(g)$$

which comes from reversing the equation for the first reaction and halving it, and adding to it one-half of the equation for the second reaction:

$$\tfrac{1}{2}[2FeO(s) + CO_2(g) \rightarrow Fe_2O_3(s) + CO(g)] \qquad \Delta H° = -\tfrac{1}{2}(38\ kJ) = -19\ kJ$$
$$\tfrac{1}{2}[Fe_2O_3(s) + 3CO(g) \rightarrow 2Fe(s) + 3CO_2(g)] \qquad \Delta H° = \tfrac{1}{2}(-28\ kJ) = -14\ kJ$$

$$FeO(s) + CO(g) \rightarrow Fe(s) + CO_2(g) \qquad \mathbf{\Delta H° = -33\ kJ}$$

General Problems

54. The mineral atacamite has the formula $Cu_2Cl(OH)_3 \cdot xH_2O$, where x is an integer. In a titration experiment, it was found that 21.45 mL of 0.4071-M HCl(aq) were required to react completely with 0.6217 g of atacamite. What is x in this empirical formula?

The balanced equation for the reaction with HCl(aq) is:

$$Cu_2Cl(OH)_3 \cdot xH_2O(s) + 3HCl(aq) \rightarrow 2CuCl_2(aq) + (x+3)H_2O$$

First we calculate the moles of HCl(aq) used up, which gives us the moles, and then the mass, of $Cu_2Cl(OH)_3$ in the sample of mass 0.6217 g, and the remainder of this mass must be due to the water of hydration:

$$\text{moles HCl(aq) used up} = (21.45\text{ mL})\left(\frac{0.4071\text{ mol}}{1\text{ L}}\right)\left(\frac{1\text{ L}}{10^3\text{ mL}}\right) = \underline{8.732 \times 10^{-3}\text{ mol}}$$

$$\text{mass of } Cu_2Cl(OH)_3 \text{ consumed} = (8.732 \times 10^{-3}\text{ mol HCl})\left(\frac{1\text{ mol } Cu_2Cl(OH)_3}{3\text{ mol HCl}}\right)\left(\frac{213.5\text{ g } Cu_2Cl(OH)_3}{1\text{ mol } Cu_2Cl(OH)_3}\right) = \underline{0.6217\text{ g}}$$

Thus, in fact **x = 0**; *Atacamite* contains no water of hydration.

---

56. Explain qualitatively why the endothermic decomposition of calcium carbonate to calcium oxide and carbon dioxide is spontaneous at a high temperature but not at room temperature.

$\Delta G = \Delta H - T\Delta S$ has to be negative for the reaction to be spontaneous, which for an **endothermic** reaction (ΔH positive) occurs only when $T\Delta S$ is positive and sufficiently large in magnitude. For the reaction

$$CaCO_3(s) \rightarrow CaO(s) + CO_2(g)$$

ΔS is expected to be positive, since the products include a gas and a solid that will be more disordered than the solid reactant alone. Thus, $T\Delta S$ is positive, and relatively small in magnitude at room temperature, but indeed becomes greater than ΔH at a sufficiently large value of T (in fact, at about 1100°C).

---

58. (a) Give the equation for the half-reaction in which dichromate ion, $Cr_2O_7^{2-}$, is reduced to Cr^{3+} in acid solution (b) Write the complete equation for the reduction of $Cr_2O_7^{2-}(aq)$ in acidic aqueous acid solution by $SO_2(g)$.

(a) Cr(VI) in $Cr_2O_7^{2-}(aq)$ is reduced to Cr(III) in $Cr^{3+}(aq)$ in acid solution according to the equation:

$$Cr_2O_7^{2-} + 6e^- + 14H^+ \rightarrow 2Cr^{3+} + 7H_2O$$

(b) S(IV) in SO_2 will be oxidized to S(VI) to give nominally give SO_3, which will be present in aqueous solution as sulfate ion, SO_4^{2-}, so we can write:

$3[SO_2 + 2H_2O + \rightarrow SO_4^{2-} + 2e^- + 4H^+]$	oxidation
$Cr_2O_7^{2-} + 6e^- + 14H^+ \rightarrow 2Cr^{3+} + 7H_2O$	reduction
$3SO_2 + Cr_2O_7^{2-} + 2H^+ \rightarrow 3SO_4^{2-} + 2Cr^{3+} + H_2O$	overall

---

60.* A sample of hydrated iron(II) sulfate of mass 6.673 g was dissolved in water to give 250 mL of solution. Upon titration with 0.0200-M $KMnO_4(aq)$, 25.00 mL of this solution required 24.00 mL of acidic $KMnO_4(aq)$ solution to react completely. Calculate (a) the mass percentage of $FeSO_4$ in hydrated iron(II) sulfate, and hence (b) x in its formula, $FeSO_4.xH_2O$.

(a) The balanced equation for the oxidation of Fe(II) to Fe(III) by permanganate ion is given by:

$5[Fe^{2+} \rightarrow Fe^{3+} + e^-]$ oxidation

$MnO_4^- + 5e^- + 8H^+ \rightarrow Mn^{2+} + 4H_2O$ reduction

$5Fe^{2+} + MnO_4^- + 8H^+ \rightarrow 5Fe^{3+} + Mn^{2+} + 4H_2O$ overall

and from the results of the titration, we can calculate the moles of Fe^{2+} in the sample:

$$\text{mol } Fe^{2+} = \left(\frac{250 \text{ mL } FeSO_4(aq)}{25.00 \text{ mL } FeSO_4(aq)}\right)(24.00 \text{ mL})\left(\frac{0.0200 \text{ mol } KMnO_4}{1 \text{ L}}\right)\left(\frac{1 \text{ L}}{10^3 \text{ mL}}\right)\left(\frac{5 \text{ mol } FeSO_4}{1 \text{ mol } KMnO_4}\right) = \underline{2.40 \times 10^{-2} \text{ mol}}$$

Thus: $$\text{molar mass of } FeSO_4\cdot xH_2O = \frac{6.673 \text{ g}}{2.40 \times 10^{-2} \text{ mol}} = \underline{278 \text{ g mol}^{-1}}$$

$$\text{mass\% } FeSO_4 = \frac{151.9 \text{ g } FeSO_4}{278 \text{ g } FeSO_4\cdot xH_2O} \times 100\% = \underline{54.6\%}$$

(b) The difference between the molar mass of $FeSO_4\cdot xH_2O$ and the molar mass of $FeSO_4$ gives the mass of water in 1 mole of the hydrate:

$$\text{mass of water} = (278 - 151.9) \text{ g mol}^{-1} = 126.1 \text{ g}$$

$$\text{moles of water per mole of } FeSO_4\cdot xH_2O = (121.6 \text{ g } H_2O)\left(\frac{18.02 \text{ g } H_2O}{1 \text{ mol } H_2O}\right) = \underline{7.00}$$

Thus, the **empirical formula** of hydrated iron(II) sulfate is $\mathbf{FeSO_4\cdot 7H_2O}$

61.* The booster rockets of the space shuttle use a mixture of aluminum and ammonium perchlorate as fuel. The balanced equation for the reactions is $3Al(s) + 3NH_4ClO_4 \rightarrow Al_2O_3(s) + AlCl_3(s) + 3NO(g) + 6H_2O(g)$. Using data from Table 9.1, and $\Delta H_f°(NH_4Cl,s)$ —295, $\Delta H_f°(Al_2O_3,s)$ —1676, and $\Delta H_f°(AlCl_3,s)$ —704 (all in kJ mol^{-1}), calculate the standard reaction enthalpy, $\Delta H°$.

$$\Delta H° = \Sigma\, n_p(\Delta H_f°)_p - \Sigma\, n_r(\Delta H_f°)_r$$

$$\Delta H° = [\Delta H_f°(Al_2O_3,s) + \Delta H_f°(AlCl_3,s) + 3\Delta H_f°(NO,g) + 6\Delta H_f°(H_2O,g)] - [3\Delta H_f°(Al,s) + 3\Delta H_f°(NH_4ClO_4,s)]$$

$$= [(-1676) + (-704) + 3(+90.3) + 6(-241.8)] - [3(0) + 3(-295)] \text{ kJ} = \mathbf{-2675 \text{ kJ}}$$

CHAPTER 11

Solids and Liquids

1. Which of the following are molecular solids, and which are network solids? (a) C (b) S_8 (c) CO_2 (d) P_4O_6 (e) NaCl (f) MgO (g) Al

---

Molecular solids are composed of discrete molecules, while **network solids** consist of an essentially infinite array of atoms, bonded together by *metallic bonds* in **metals** and *covalent bonds* in **covalent network solids**, or ions bonded by *ionic bonds* in **ionic solids**.

(a) **Carbon, C**, both as *diamond* and as *graphite*, is a **covalent network solid.**

(b) S_8 molecules are found both in the *orthorhombic* and in the *monoclinic* allotropes of elemental sulfur, which are **molecular solids.**

(c) Solid **carbon dioxide,** $CO_2(s)$, consists of discrete CO_2 molecules; it is a **molecular solid.**

(d) **Phosphorus(III) oxide,** $P_4O_6(s)$, is composed of covalent P_4O_6 molecules; it is a **molecular solid.**

(e) **Sodium chloride,** NaCl(s), is composed of an infinite array of Na^+ and Cl^- ions; it is an **ionic network solid.**

(f) **Magnesium oxide,** MgO(s), is composed of an infinite array of Mg^{2+} and O^{2-} ions; it is an **ionic network solid.**

(g) **Aluminum,** Al(s), consists of an infinite array of Al atoms bonded together by meallic bonds; it is a **metallic network solid.**

---

3. In terms of their structures, explain why each of the following pairs of substances have very different chemical and physical properties: (a) silicon and aluminum (b) oxygen and sulfur (c) diamond and graphite

---

(a) **Silicon and Aluminum** Silicon is a covalent network solid with the diamond structure, in which each Si atom has AX_4 tetrahedral geometry and is bonded to four other Si atoms by Si—Si covalent bonds. Because Si—Si bonds have to be broken for a change in phase to occur, it has a high melting point (1410°C) and an even higher boiling point (2680°C). Pure silicon is a nonconductor of electricity. Chemically, it reacts with other nonmetals at high temperature but is insoluble in water and acids. In contrast, aluminum is a metallic network solid in which the Al atoms are bonded together by metallic bonds, and is a good conductor of heat and electricity. It melts at 660°C and boils at 2450°C, at a much higher temperature than any Group I and II metal, because its strong metallic bonding involving *three* valence electrons per Al atom. It readily loses these electrons to form Al^{3+}, so it reacts readily with many oxidizing agents, reacting, for example, with nonmetals, such as oxygen and the halogens, to give aluminum(III) compounds. Although it is expected to be a very reactive metal, its behavior is modified to some extent by the ease with which it reacts with the oxygen in air to form a hard, resistant coating of aluminum oxide, $Al_2O_3(s)$.

(b) **Oxygen and Sulfur** Oxygen is a very reactive gas, whereas sulfur is a moderately reactive solid. Both exist as allotropes (oxygen as *dioxygen*, $O_2(g)$, and *ozone*, $O_3(g)$; and sulfur as *orthorhombic* and *monoclinic* sulfur, both containing S_8 molecules, and as polymeric *plastic sulfur*). Both are typical nonmetals but oxygen has a much greater electronegativity than sulfur, which is evident in structural differences between many oxygen compounds and sulfur compounds. These differences are associated with oxygen's ability to readily form *multiple bonds*, whereas sulfur usually forms single bonds, except with highly electronegative elements (such as oxygen). For example, the two O atoms in O_2 are joined by a double bond O=O, whereas the S atoms in the S_8 ring are all joined by S—S single bonds. Both oxygen and sulfur are reactive chemically, but oxygen the more so.

(c) **Diamond and Graphite** These *allotropes* of carbon are solids that have quite different chemical reactivities as a consequence of their different structures. *Diamond* has a three dimensional covalent network structure with each carbon atom bonded to four others by C—C single bonds in an AX_4 tetrahedral arrangement. As a consequence, diamond is a very hard substance that is chemically not very reactive and is an electrical insulator. In contrast, *graphite* is relatively soft, quite reactive, and an electrical conductor. It consists of infinite sheets of carbon atoms stacked one on another. Each C atom is bonded to three others in an AX_3 planar arrangement, as part of an infinite two dimensional planar lattice consisting of interlocking hexagons of carbon atoms. The Lewis structure of graphite has each C atom forming two C—C single bonds and a C=C double bond, but the actual structure is better represented by a large number of *different resonance structures*; all the CC bonds are identical and have a bond order of 1⅓. Each C atom can be described as forming three σ bonds and a π bond, with the π electrons delocalized over an entire graphite sheet, which accounts for the electrical conductivity of graphite. Since the parallel sheets of C atoms attract each other by only weak London forces, graphite cleaves easily to form flakes and is a good lubricant. Both of these allotropes of carbon burn in oxygen to give CO(g) or CO_2(g). Because of its greater reactivity, graphite is a useful reducing agent.

---

4. Which of the following substances form network solids and which form molecular solids? (a) carbon monoxide (b) silicon carbide (c) chlorine (d) magnesium (e) magnesium chloride

---

(a) **Carbon monoxide** Carbon monoxide consists of covalent CO molecules and is a gas under ordinary conditions. Thus, as expected, it forms a low melting **molecular solid.**

(b) **Silicon carbide** Silicon carbide, *carborundum*, has the empirical formula SiC(s) and is a solid under ordinary conditions. It is a **covalent network solid** with a structure related to that of *diamond.* Each Si atom is surrounded by a tetrahedral arrangement of four C atoms and each C atom by a tetrahedral arrangement of Si atoms, forming strong covalent Si-C bonds.

(c) **Chlorine** Chlorine consists of covalent Cl_2 molecules and is a gas under ordinary conditions. Thus, as expected, it forms a low melting **molecular solid.**

(d) **Magnesium** Magnesium, Mg(s), is a metal in which Mg atoms in an infinite lattice are bonded by metallic bonds. Thus, it is a **metallic network solid.**

(e) **Magnesium chloride** Magnesium chloride, $MgCl_2$(s), consists of Mg^{2+} and Cl^- ions bonded by ionic bonds in an infinite lattice. Thus, it is an **ionic network solid.**

---

6. Sketch each of the following and deduce how many atoms each contains: (a) the body-centered cubic unit cell of barium; (b) the face-centered cubic unit cell of solid neon.

(a) **Body-centered cubic unit cell**

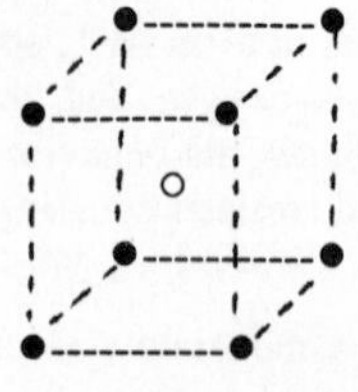

body-centered cubic cell

(b) **Face-centered cubic unit cell**

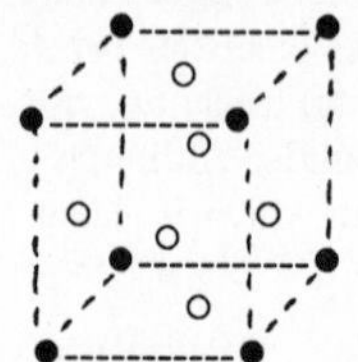

face-centered cubic cell

(a) Eight cubic unit cells intersect at any corner of this unit cell, so each of the **eight** corner Ba atoms ● contribute ⅛th of an atom to the unit cell, and the **central** Ba atom ○ belongs entirely to the unit cell.

Thus, **number of Ba atoms in the unit cell** = 8(⅛) + 1 = **2**

(b) Eight cubic unit cells intersect at any corner of this unit cell, so each of the **eight** corner Ne atoms ● contribute **⅛th** of an atom to the unit cell, and each of the Ne atoms ○ at the centers of the **six** faces are shared with **one** adjacent cell, so each of the six face-center atoms contribute ½th of a Ne atom to the unit cell.

Thus, **number of Ne atoms in the unit cell** = 8(⅛) + 6(½) = **4**

---

7. The structure of aluminum is based on a cubic lattice. At 25°C, the edge-length of the unit cell is 405 pm, and the density is 2.70 g·cm^{-3}: (a) how many Al atoms are there in the unit cell? (b) On what type of cubic lattice must the structure of aluminum be based?

(a) From the length of the cell-edge we can calculate its volume in cm^3:

$$\text{Volume of unit cell} = [(405\ \text{pm})(\frac{1\ \text{m}}{10^{12}\ \text{pm}})(\frac{10^2\ \text{cm}}{1\ \text{m}})]^3 = \underline{6.64 \times 10^{-23}\ \text{cm}^3}$$

Then: $$\text{mass of cell} = \text{volume} \times \text{density} = (6.64 \times 10^{-23}\ \text{cm}^3)(\frac{2.70\ \text{g}}{1\ \text{cm}^3}) = \underline{1.79 \times 10^{-22}\ \text{g}}$$

which must be the mass of the Al atoms in **one** unit cell.

$$\text{mass of 1 Al atom} = (\frac{26.98\ \text{g Al}}{1\ \text{mol Al}})(\frac{1\ \text{mol Al}}{6.022 \times 10^{23}\ \text{atoms}}) = \underline{4.480 \times 10^{-23}\ \text{g}}$$

$$\text{Thus, atoms per unit cell} = \frac{\text{mass of unit cell}}{1\ \text{atom}} = \frac{1.799 \times 10^{-22}\ \text{g}}{4.480 \times 10^{-23}\ \text{g}} = \underline{4.00}$$

(b) The cubic lattice with **four** atoms per unit cell is the **face-centered cubic lattice** (see Problem 6).

---

9. At 24 K solid neon has a cubic lattice with a unit cell edge-length of 450 pm and a density of 1.45 g·cm^{-3} (for solid neon). (a) How many Ne atoms are in the unit cell? (b) What type of cubic lattice does crystalline neon have? (c) What is the radius of the Ne atom?

This Problem incorporates and extends Problems 7 and 8.

(a) We first calculate the volume of one unit cell, then use the density to calculate its mass, and from the mass of one Ne atom we deduce the number of Ne atoms per unit cell.

$$\text{Volume of unit cell} = [(450\ \text{pm})(\frac{1\ \text{m}}{10^{12}\ \text{pm}})(\frac{10^2\ \text{cm}}{1\ \text{m}})]^3 = \underline{9.11 \times 10^{-23}\ \text{cm}^3}$$

$$\text{Mass of unit cell} = (9.11 \times 10^{-23}\ \text{cm}^3)(\frac{1.45\ \text{g}}{1\ \text{cm}^3}) = \underline{1.32 \times 10^{-22}\ \text{g}}$$

$$\text{Number of atoms per unit cell} = (1.32 \times 10^{-22}\ \text{g})(\frac{1\ \text{mol}}{20.18\ \text{g}})(\frac{6.022 \times 10^{-23}\ \text{atoms}}{1\ \text{mole}}) = \underline{3.94\ \text{atoms}}$$

The number of atoms must be an integral number. Clearly it is **4** and the discrepancy presumably has to do with the accuracy of determining the density of neon which melts at -249°C.

(b) The cubic lattice with **four** atoms per unit cell is the **face-centered cubic lattice** (see Problem 6).

(c) When we examine any *face* of the face-centered cubic lattice, we see that there are four atoms with their centers at the corners of a square and one with its center at the center of the square, all touching. Thus, the **face-diagonal, b, is four times the length of the radius of a Ne atom, r_{Ne}.**

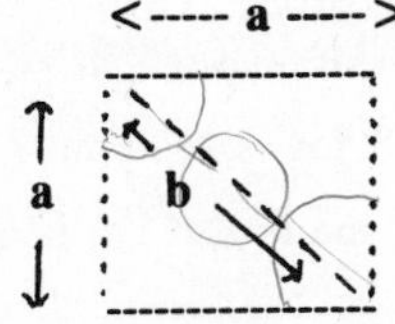

From Pythagoras' theorem

$b = (a^2 + a^2)^{1/2}$

$b = \sqrt{2}a$

In solid neon, a = 450 pm, so $b = 4r_{Ne} = \sqrt{2}(450\text{ pm}) = \underline{636\text{ pm}}$;

$$r_{Ne} = \textbf{159 pm.}$$

---

11. In crystalline sodium chloride at 25°C, the distance between the centers of neighboring Na^+ and Cl^- ions is 281 pm and the density is 2.165 $g{\cdot}cm^{-3}$. Calculate a value for Avogadro's number.

Since NaCl has the cubic close-packed (face-centered cubic) lattice, there are **four NaCl formula units per unit cell.** Along any cell-edge, we have Na^+---Cl^----Na^+ (or Cl^----Na^+---Cl^-) ions touching, so the **length of the cell-edge is twice the** Na^+---Cl^- distance, or 2(281 pm) = **562 pm.** As in Problem 10, we first calculate the mass of four NaCl formula units from the volume of the unit cell and the density of NaCl. Avogadro's number, which is the number of NaCl formula units in one mole of NaCl, or (22.99 + 35.45) = 58.44 g NaCl, can then be obtained:

Mass of 4 NaCl formula units = (density of NaCl)(volume of unit cell)

$$= (2.165\text{ g cm}^{-3})[(562\text{ pm})(\frac{1\text{ m}}{10^{12}\text{ pm}})(\frac{10^2\text{ cm}}{1\text{ m}})]^3 = \underline{3.843 \times 10^{-22}\text{ g}}$$

$$\text{Number of NaCl formula units per mole} = (\frac{58.44\text{ g NaCl}}{1\text{ mol NaCl}})(\frac{4\text{ NaCl units}}{3.843 \times 10^{-22}\text{ g}})$$

$$= \underline{6.08 \times 10^{23}\text{ NaCl units mol}^{-1}}$$

---

13. Copper(I) chloride, CuCl(s), of density 3.41 $g{\cdot}cm^{-3}$, has the zinc sulfide structure: (a) What is the edge-length of the unit cell? (b) What is the shortest distance between the centers of a Cu^+ ion and a Cl^- ion? (c) The radius of the Cl^- ion is 180 pm. What is the radius of the Cu^+ ion?

CuCl(s) has the ZnS(s) (face-centered cubic) structure with four CuCl formula units per unit cell.
(a) Four CuCl formula units occupy the unit cell; thus the density of CuCl is given by:

$$\text{density} = \frac{\text{mass of 4 CuCl formula units}}{\text{volume of unit cell}}$$

$$\text{1 mol of CuCl formula units has a mass of } (63.55 + 35.45) = \underline{99.00\text{ g}}$$

$$\text{Thus,}\quad V_{\text{unit cell}} = (4\text{ units})(\frac{99.00\text{ g CuCl}}{1\text{ mol CuCl units}})(\frac{1\text{ mol CuCl units}}{6.022 \times 10^{23}\text{ CuCl units}})(\frac{1\text{ cm}^3}{3.41\text{ g}}) = \underline{1.93 \times 10^{-22}\text{ cm}^3}$$

$$\text{If } a \text{ is the length of the cell-edge:}\quad V_{\text{unit cell}} = a^3 = 1.93 \times 10^{-22}\text{ cm}^3$$

$$a = (1.93 \times 10^{-22}\text{ cm}^3)^{\frac{1}{3}}(\frac{1\text{ m}}{10^2\text{ cm}})(\frac{10^{12}\text{ pm}}{1\text{ m}}) = \underline{578\text{ pm}}$$

(b) In the unit cell, the Cu^+ ions are at the lattice points and the Cl^- ions are one-quarter of the length of a **body-diagonal** away. We can use Pythagoras' theorem (see below) that the length of the face diagonal **b** is $\sqrt{2}$**a, and that** of the body-diagonal **c** is $\sqrt{3}$a:

$$\text{For } a = 578\text{ pm; } c = \sqrt{3}(578\text{ pm}) = 1000\text{ pm; } \tfrac{1}{4}c = \tfrac{1}{4}(1000\text{ pm}) = \textbf{250 pm}$$

(c) The distance between the centers of a Cu^+ and a Cl^- ion is 250 pm; $r_+ + r_- = r_+ + 181$ pm; $\mathbf{r_+}$ **= 69 pm.**

The radius of the Cu^+ ion in CuCl(s) is **69 pm.**

---

15. Which unit cell has the higher mass, that of cesium chloride or that of sodium chloride? Explain.

Cesium chloride has the body-centered cubic unit cell, with one Cl^- (or Cs^+) ion at the center of a cube and eight Cs^+ (or Cl^-) ions at each of the eight corners. Thus, the number of CsCl formula units per unit cell is **one**. **Sodium chloride** has the face-centered unit cell, with a Na^+ (or Cl^-) ion at each lattice point and an equal number of Cl^- (or Na^+) ions situated halfway between closest pairs of Na^+ (or Cl^-) ions. Thus, the number of NaCl formula units per unit cell is **four**. Now we compare the mass of one CsCl formula unit with that of four NaCl formula units:

$$\frac{\text{mass of 1 CsCl formula unit}}{\text{mass of 4 NaCl formula units}} = \frac{(132.9\ \text{u} + 35.45\ \text{u})}{4(22.99\ \text{u} + 35.45\ \text{u})} = \frac{168.4}{233.8} = \underline{0.720}$$

The NaCl unit cell has the higher mass

---

16. Cesium bromide, CsBr(s), crystallizes in the cesium chloride structure. If the closest distance between the centers of oppositely charged ions is 371 pm, what is the density of cesium bromide?

Cesium bromide has a body-centered cubic unit cell with a Cs^+ (or Br^-) ion at the center and Br^- (or Cs^+) ions at each of the 8 corners. There is one CsBr formula unit per unit cell (see Problem 6). Along any body diagonal we have Br^----Cs^+---Br^- (or Cs^+---Br^----Cs^+) ions touching. Thus, the length of the body diagonal is $2(r_+ + r_-)$, twice $r_+ + r_- = 371$ pm. We can calculate the length of the body diagonal **c** from the length of the cell edge **a and** the length of the face-diagonal $\mathbf{b} = \sqrt{2}a$, as we did in Problem 13: $\mathbf{c} = \sqrt{3}\mathbf{a} = \sqrt{3}(371\ \text{pm}) = \mathbf{428\ pm}$.

$$\text{Thus, density of CsBr} = \frac{\text{mass of 1 CsBr formula unit}}{\text{volume of unit cell}} = \frac{\left(\frac{212.8\ \text{g CsBr}}{1\ \text{mol CsBr}}\right)\left(\frac{1\ \text{mol CsBr units}}{6.022 \times 10^{23}\ \text{units}}\right)}{\left[(428\ \text{pm})\left(\frac{1\ \text{m}}{10^{12}\ \text{pm}}\right)\left(\frac{10^2\ \text{cm}}{1\ \text{m}}\right)\right]^3} = \underline{4.51\ \text{g cm}^{-3}}$$

---

19. At about 1000°C iron undergoes a transition from the body-centered cubic structure to the face-centered cubic structure. The edge-length of the unit cell increases from 286 pm to 363 pm. Compare the densities of these two forms of iron, and explain the difference.

In the body-centered cubic structure the number of Fe atoms per unit cell is **2**, while in the face-centered cubic structure the number of Fe atoms per unit cell is **4** (see Problem 6). Thus, for the low-temperature (body-centered cubic) form of iron the density is given by:

$$\text{Density} = \frac{\text{mass of unit cell}}{\text{volume of unit cell}} = \frac{\text{mass of 2 Fe atoms}}{\text{volume of unit cell}}$$

$$= \frac{(2\ \text{Fe atoms})\left(\frac{55.85\ \text{g Fe atoms}}{1\ \text{mol Fe atoms}}\right)\left(\frac{1\ \text{mol Fe}}{6.022 \times 10^{23}\ \text{Fe atoms}}\right)}{\left[(286\ \text{pm})\left(\frac{1\ \text{m}}{10^{12}\ \text{pm}}\right)\left(\frac{10^2\ \text{pm}}{1\ \text{m}}\right)\right]^3} = \underline{7.93\ \text{g cm}^{-3}}$$

and for the high-temperature (face-centered cubic) form of iron, the density is given by:

$$\text{Density} = \frac{\text{mass of unit cell}}{\text{volume of unit cell}} = \frac{\text{mass of 4 Fe atoms}}{\text{volume of unit cell}}$$

$$= \frac{(4\text{ Fe atoms})\left(\frac{55.85\text{ g Fe atoms}}{1\text{ mol Fe atoms}}\right)\left(\frac{1\text{ mol Fe}}{6.022\times10^{23}\text{ Fe atoms}}\right)}{\left[(363\text{ pm})\left(\frac{1\text{ m}}{10^{12}\text{ pm}}\right)\left(\frac{10^{2}\text{ pm}}{1\text{ m}}\right)\right]^{3}} = \underline{7.76\text{ g cm}^{-3}}$$

Thus, as anticipated the density of iron decreases from 7.93 g cm^{-3} to 7.76 g cm^{-3} with the increase in temperature.

---

21. Silver has the copper (face-centered cubic) structure and a density 10.5 g·cm^{-3}. How many silver atoms are contained in a cube of silver with an edge length of 1.00 mm?

In the face-centered cubic structure of Ag(s) there are **4** Ag atoms per unit cell, and from the density of 10.5 g cm^{-3} we can calculate the volume of the unit cell, and hence the number of Ag atoms in a cube of volume **(1 mm)^3**:

$$V_{cell} = \frac{\text{mass of unit cell}}{\text{density}} = \frac{(4\text{ Ag atoms})\left(\frac{107.9\text{ g Ag atoms}}{1\text{ mol Ag atoms}}\right)\left(\frac{1\text{ mol Ag atoms}}{6.022\times10^{23}\text{ Ag atoms}}\right)}{\left(\frac{10.50\text{ g Ag atoms}}{1\text{ cm}^3}\right)} = \underline{6.826\times10^{-23}\text{ cm}^3}$$

$$\text{Thus for }(1\text{ mm})^3\text{, number of Ag atoms} = (1\text{ mm})^3\left(\frac{1\text{ cm}}{10\text{ mm}}\right)^3\left(\frac{4\text{ Ag atoms}}{6.826\times10^{23}\text{ cm}^3}\right) = \underline{5.86\times10^{19}\text{ atoms}}$$

---

22. Select the larger ion from each of the following pairs of ions, and justify your choice:
(a) Na^+, F^- (b) Na^+, K^+ (c) F^-, Cl^- (d) Na^+, Mg^{2+} (e) S^{2-}, Cl^-

---

The sizes of ions are determined by (1) their core charges, and (2) the number of filled electron shells. For *isoelectronic* ions, the core charge is the sole determining factor.

(a) Na^+ and F^- Na^+ and F^- are *isoelectronic*; both have the electron configuration as Ne, $1s^22s^22p^6$. However, they have different core charges: Na^+ +9 and F^- +7, so F^- is the larger ion. **F^- > Na^+** (135 pm versus 102 pm).

(b) Na^+ and K^+ Na^+ and K^+ are both singly charged cations of Group I elements and have the same core charge of +9, but K^+ has one more filled shell than Na^+ and is thus the larger ion. **K^+ > Na^+** (138 pm versus 102 pm).

(c) F^- and Cl^- F^- and Cl^- are both singly charged anions of Group VII elements and have the same core charge of +7, but Cl^- has one more filled shell than F^- and is thus the larger ion. **Cl^- > F^-** (181 pm versus 135 pm).

(d) Na^+ and Mg^{2+} Na^+ and Mg^{2+} are *isoelectronic* and both have the same electron configuration as neon, $1s^22s^22p^6$. However Na^+ has a core charge of +9 and Mg^{2+} has a core charge of +10, so Na^+ is the larger ion. **Na^+ > Mg^{2+}** (102 pm versus 72 pm).

(e) S^{2-} and Cl^- S^{2-} and Cl^- are *isoelectronic* with the same electron configuration as argon, $1s^22s^22p^63s^23p^6$. But S^{2-} has core charge +6 and Cl^- has core charge +7, so S^{2-} is the larger ion. **S^{2-} > Cl^-** (185 pm versus 180 pm).

25.* Mercury(II) sulfide has the zinc sulfide structure. The shortest distance between the center of an Hg^{2+} ion and the center of an S^{2-} ion is 253 pm. Calculate the density of mercury(II) sulfide.

The Hg^{2+} ions in $Hg^{2+}S^{2-}$(s) are situated at the lattice points and the S^{2-} ions are situated one-quarter of the distance along a face-diagonal **b**, so **¼b = 253 pm**, **b = 1012 pm**. To determine the density, we first have to calculate the cell-edge **a**, using **b = $\sqrt{3}$a**, so **a = 584 pm** (see Problem 13), from which we can calculate the volume of the cubic cell. The unit cell contains **four** HgS formula units, hence its mass can also be calculated:

$$\text{Density} = \frac{\text{mass of 4 HgS formula units}}{\text{volume of unit cell}}$$

$$\text{Density} = \frac{(4 \text{ HgS units})(\frac{232.7 \text{ g HgS}}{1 \text{ mol HgS units}})(\frac{1 \text{ mol HgS units}}{6.022 \times 10^{23} \text{ HgS units}})}{[(584 \text{ pm})(\frac{1 \text{ m}}{10^{12} \text{ pm}})(\frac{10^2 \text{ cm}}{1 \text{ m}})]^3} = \underline{7.76 \text{ g cm}^{-3}}$$

---

28. (a) Why does the vapor pressure of a liquid increase with increasing temperature? (b) At 20 °C, the vapor pressure of benzene is 75.0 mm Hg and that of toluene is 50.0 mm Hg. Which of these hydrocarbons is expected to have the higher normal boiling point? (c) The normal boiling point of diethyl ether, $(C_2H_5)_2O$, methanol, CH_3OH, and propanone (acetone), $(CH_3)_2CO$, are, respectively, 34.5, 64.5, and 56.1°C. Which of these liquids will have the highest vapor pressure at 25°C, and which the lowest?

(a) The average kinetic energy of the molecules in the liquid increases with increase in temperature, so that a greater proportion of them have sufficient energy to escape from the liquid into the vapor phase. Thus, vapor pressure increases with increase in temperature.

(b) A liquid boils when its vapor pressure is equal to the external pressure and **normal boiling point** is the boiling point when the external pressure is **1 atmosphere.** Since benzene has a higher vapor pressure at 20°C than does toluene, and this difference is expected to be maintained as the temperature is raised, the vapor pressure of benzene will achieve 1 atmosphere at a lower temperature than that of toluene. Thus **toluene** is expected to have the higher normal boiling point.

(c) Assuming that the relative differences in the vapor pressures of these liquids maintains the same order as the temperature is lowered, the liquid with the lowest normal boiling point, **diethyl ether**, is expected to have the highest vapor pressure, and the liquid with the **highest** normal boiling point, **methanol**, is expeced to have the **lowest** vapor pressure, at 25°C.

---

30. Explain what effect (if any) each of the following has on the vapor pressure of a liquid: (a) the surface area of the liquid (b) the volume of the container (c) temperature (d) intermolecular forces (e) the volume of the liquid

---

(a) For a given temperature, the vapor pressure is determined only by the number of molecules with sufficient energy to leave the liquid, and at equilibrium the numbers that leave and return will be constant per unit surface area. Thus, the amount of surface area has no effect on the vapor pressure, although it does affect the rate of evaporation.

(b) The volume of liquid is unchanged although its surface area will change depending on the shape of the container, however, for the reasons in part (a) the vapor pressure remains unchanged.

(c) The vapor pressure of a liquid increases with increasing temperature because more of its molecules have sufficient energy to escape from the liquid into the vapor phase, and thus they exert a greater pressure.

(d) The stronger the intermolecular forces the less easily molecules can escape into the vapor phase, and thus the smaller the vapor pressure at a given temperature.

(e) The volume of the liquid does not affect its vapor pressure for the same reasons given in part (a).

---

33. What types of interactions (bonds and/or intermolecular forces) must be overcome to melt each of the following solids? (a) BaO (b) diamond (c) I_2 (d) P_4O_{10} (e) copper (f) graphite

---

(a) **Barium oxide,** BaO(s) is composed of Ba^{2+} and O^{2-} ions and forms a high melting **ionic network solid in which** ionic bonds **have to be broken for it to melt.**

(b) Diamond **is a very high melting solid consisting of C atom joined in three dimensions in a** covalent network solid in which each C atom is bonded by covalent bonds to 4 other C atoms, some of which have to be broken for it to melt.

(c) **Iodine** is a **covalent solid** composed of I_2 molecules, For it to melt, only weak **London (dispersion) intermolecular forces** have to be overcome. In fact on heating it is readily changed directly from solid to gas (sublimes).

(d) **Phosphorus(V) oxide** is a **covalent solid** consisting of covalent P_4O_{10} cage molecules bonded by **London** (dispersion) **forces**, some of which have to be overcome for it to melt.

(e) **Copper** is a **metallic network solid** in which Cu atoms are bonded by strong metallic bonds, some of which have to be broken for it to melt.

(f) **Graphite** is a high melting *two-dimensional* **covalent network solid** consisting of C atoms arranged in planes that are attracted to each other by weak **London** forces. For it to melt, some of the strong carbon-carbon covalent bonds have to be broken.

---

34. Both NaCl and MgO are ionic solids, yet MgO(s) melts at a temperature about 2000 K higher than NaCl(s). Explain why.

For these ionic solids to melt, some ionic bonds have to be broken. The large difference in the melting points of NaCl(s) and MgO(s) must be due to a large difference in the strengths of the ionic bonds in the two substances. NaCl(s) is composed of Na^+ and Cl^- ions, whereas MgO(s) is composed of Mg^{2+} and O^{2-} ions. Thus, one reason for MgO to have a higher melting point than NaCl is attributable to the fact that the doubly charged Mg^{2+} and O^{2-} ions will attract each other by electrostatic attraction more strongly than the singly charged Na^+ and Cl^- ions. Electrostatic force is proportional to the magnitude of the charges involved but also inversely proportional to their distance apart (Coulomb's law: $\mathbf{F \propto (Q_1Q_2/r^2)}$), so a second contributing factor is the distance apart of the ions in these solids. Na^+ and Mg_{2+} are *isoelectronic* but Mg^{2+} (with a core charge of +10) is considerably smaller than Na^+ (with a core charge of +9). Similarly, O^{2-} is much smaller than Cl^-, because although these ions are not *isoelectronic*, O^{2-} is formed from oxygen in Group VI and Period 2, and has a core charge of +6, while Cl^-, formed from chlorine in Group VII and Period 3, with a core charge of +7, has one more shell of electrons. The ionic radii are Na^+, 102 pm; Mg^{2+}, 72 pm, Cl^-, 181 pm, and O^{2-}, 140 pm, so the Na-Cl distance in NaCl(s) is (102 + 181) = 283 pm, while the Mg-O distance in MgO(s) is significantly less at (72 + 140) = 212 pm. It is important to note that NaCl and MgO have the same type of lattice; different structures would also affect the distance between ions, and thus the mp.

---

36. Which molecule in each of the following pairs has a dipole moment? (a) $BeCl_2$, OCl_2 (b) PF_3, BF_3 (c) PF_5, ClF_5

For any molecule, this type of problem is solved by first writing its Lewis structure and then categorizing it in terms of the AX_nE_m nomenclature. As we proved in Problem 35, molecules with AX_2, AX_3, AX_4, AX_5, and AX_6 geometries have zero dipole moments, while molecules with AX_2E, AX_2E_2, AX_3E, and AX_5E geometries have a dipole moment.

	Molecule	Lewis Structure	Type	μ	Molecule	Lewis Structure	Type	μ
(a)	$BeCl_2$	:Cl—Be—Cl:	AX_2	No	OCl_2	:Cl—O—Cl:	AX_2E_2	Yes
(b)	PF_3	:F—P—F: with :F: below P	AX_3E	Yes	BF_3	:F—B—F: with :F: below B	AX_3	No

(c)	PF_5	:F: :F—P—F: :F: :F:	AX_5	No	ClF_5	:F: :F—Cl—F: :F: :F:	AX_5E	Yes

38. Which substance of each of the following pairs would be expected to have the higher boiling point? Justify your choice: (a) ClF, BrF (b) BrCl, Cl_2 (c) KBr, BrCl (d) Na, Br_2

(a) **ClF and BrF** Although they are both gases under normal conditions **BrF** should have the higher boiling point. The electronegativity difference between Br and F is greater than that between Cl and F, so Br-F is more polar than Cl-F, and Br is a larger atom than F and therefore more polarizable than F. Thus, both dipole-dipole forces and London forces should be stronger in BrF than in ClF. **BrF has the higher boiling point.**

(b) **BrCl and Cl_2** BrCl has a small dipole moment and Cl_2 does not, and Br is a larger atom than Cl and is thus more polarizable. Thus, dipole-dipole forces and relatively stronger London forces act between BrCl molecules in BrCl(ℓ), while only weaker London forces act between Cl_2 molecules in Cl_2(g). **BrCl has the higher boiling point.**

(c) **KBr and BrCl** Here there are significant differences because liquid (molten) KBr contains K^+ and Br^- ions that attract each other by strong inter-ionic electrostatic forces, while BrCl molecules are polar covalent and BrCl molecules in BrCl(ℓ) attract each other only by weak London forces and even weaker dipole-dipole attractions. **KBr has the higher boiling point.**

(d) **Na and Br_2** Sodium is a metal and molten sodium contains Na atoms that attract each other by strong metallic bonds, although each Na atom contributes only one electron each to the bonding. Br_2 contains nonpolar covalent Br_2 molecules and Br atoms are large and therefore have a quite large polarizability, so although the only attractive forces are London forces they are relatively large and bromine is a liquid under normal conditions. Nevertheless metallic bonds are much stronger than any London forces and **sodium has the higher boiling point.** (Bromine boils at 58°C while sodium boils at 880°C.)

---

40. What types of intermolecular force predominate in each of the following? (a) I_2(s) (b) CaO(s) (c) CO_2(g) (d) $CHCl_3$(ℓ) (e) HF(ℓ)

(a) **Iodine** consists of nonpolar covalent I_2 molecules that in the solid attract each other only by London forces.

(b) **Calcium oxide** consists of Ca^{2+} and O^{2-} ions that attract each other in the solid by electrostatic ion-ion interactions (ionic bonding).

(c) **Carbon dioxide gas** consists of nonpolar linear CO_2 molecules that attract each other only by very weak London forces, especially because the molecules in a gas are far apart.

(d) Liquid **chloromethane** consists of tetrahedral polar $CHCl_3$ molecules (because not all the bond dipoles cancel each other when one of the atoms in an AX_3Y molecule is dissimilar. $CHCl_3$ molecules attract each other by London and dipole-dipole forces. The London forces predominate as shown by the fact that $CHCl_3$(ℓ) has a lower bp (62°C) than CCl_4(ℓ)(77°C), for example.

(e) Liquid **hydrogen fluoride** consists of polar $^{\delta+}H—F^{\delta-}$ molecules which attract each other by London and dipole-dipole forces and, most significantly, because fluorine is a small highly electronegative atom, form strong intermolecular hydrogen bonds.

---

42. What is the strongest intermolecular attraction (or bond) that must be broken when each of the following substances melts? (a) benzene (b) ethanol (c) ethane (d) barium oxide (e) chlorine (f) hydrogen chloride

---

(a) **Benzene** consists of C_6H_6 molecules and is a liquid under ordinary conditions. To melt benzene, all that is required is to separate the nonpolar molecules of solid sufficiently for the molecules to be able to flow. Thus the only intermolecular attractions that have to be overcome are weak London forces.

(b) **Ethanol** consists of C_2H_5OH molecules and is liquid under ordinary conditions. The dominant intermolecular attractions both in the solid and liquid are hydrogen bonds between the -OH groups of ethanol molecules, some of which must be broken for ethanol to melt.

(c) **Ethane** consists of nonpolar covalent C_2H_6 and is a gas under ordinary conditions. To melt ethane only weak London forces between C_2H_6 molecules have to be overcome.

(d) **Barium oxide** is a solid consisting of a regular array of Ba^{2+} and O^{2-} ions under ordinary conditions. To melt BaO(s) sufficient has to be supplied to overcome some of the strong electrostatic forces (ionic bonds) between Ba^{2+} and O^{2-} ions.

(e) **Chlorine** consists of nonpolar Cl_2 molecules and is a gas under ordinary conditions. To melt $Cl_2(s)$, only relatively weak London forces between Cl_2 molecules have to be overcome.

(f) **Hydrogen chloride** consists of polar HCl molecules and is a gas under ordinary conditions. The H-Cl bonds are insufficiently polar to form hydrogen bonds, so the forces to be overcome when HCl(s) melts are dipole-dipole and London attractions between HCl molecules.

---

44. Nonpolar tetrachloromethane boils at 76.5 °C, whereas polar trichloromethane (chloroform) boils at 61.7 °C. Explain why the boiling point of polar chloroform is lower than that of nonpolar tetrachloromethane.

Tetrachloromethane, $CCl_4(\ell)$, consists of nonpolar tetrahedral CCl_4 molecules and its physical properties are largely determined by the four highly polarizable Cl atoms per molecule. $CHCl_3$ molecules in **trichloromethane** (chloroform), $CHCl_3(\ell)$, are also tetrahedral but, in contrast to nonpolar CCl_4 molecules, are polar with a small dipole moment. Nevertheless, $CHCl_3$ boils at a lower temperature (61.7°C) than CCl_4 (76.5°C) because its molecules contain only three Cl atoms and a H atom that is much less polarizable than a Cl atom. CCl_4 has a higher polarizability than $CHCl_3$ and, as is the case with many other substances, it is the consequent London forces that dominate in determining physical properties such as boiling point.

45.* Suppose that 5.00 L of air is saturated with water vapor at 25 °C and then completely dried by bubbling through sulfuric acid. If this process increases the mass of the sulfuric acid by 0.115 g, what is the vapor pressure of water at 25°C?

The increase in the mass of the sulfuric acid must be the mass of water that saturates 5.00 L of air at 25°C, which expressed as moles is:

$$0.115 \text{ g } H_2O = (0.115 \text{ g } H_2O)\left(\frac{1 \text{ mol } H_2O}{1.02 \text{ g } H_2O}\right) = \underline{6.382 \times 10^{-3} \text{ mol } H_2O}$$

The vapor pressure of water is the pressure exerted by this amount of gaseous water in a volume of 5.00 L at 25°C:

$$p_{H_2O} = \frac{nRT}{V} = \frac{(6.382 \times 10^{-3} \text{ mol})(0.0821 \text{ atm L mol}^{-1} \text{ K}^{-1})(298 \text{ K})}{5.00 \text{ L}} \; \underline{0.0312 \text{ atm}}$$

Expressed as mm Hg: $0.0312 \text{ atm} = (0.0312 \text{ atm})\left(\frac{760 \text{ mm}}{1 \text{ atm}}\right) = \underline{23.7 \text{ mm Hg}}$

Water and the Hydrogen Bond

47. Which of the following substances are associated by hydrogen bonding in the liquid state? Explain your answers: (a) ammonia (b) sodium chloride (c) hydrogen fluoride (d) hydrogen (e) methane (f) lithium hydride (g) methanol (h) acetic acid, $CH_3CO_2H(\ell)$

For a substance to form hydrogen bonds it must have highly polar X-H covalent bonds and X must have at least one lone pair of electrons (see Problem 46).

(a) **Ammonia** consists of $:NH_3$ molecules with polar $^{\delta-}$N—H$^{\delta+}$ bonds and a lone pair on the N atom. Nitrogen is a small atom of high electronegativity from Period 2, and thus ammonia forms hydrogen bonds in liquid ammonia.

(b) **Sodium chloride** contains no hydrogen and forms an ionic melt, $NaCl(\ell)$, in which there is no possibility of hydrogen bond formation.

(c) The H—F molecules in liquid **hydrogen fluoride** satisfy the criteria for hydrogen bond formation; they form zig-zag chains of HF molecules connected by hydrogen bonds, --H—F:---H—F:--, in $HF(\ell)$. Hydrogen fluoride is the only hydrogen halide that is a liquid at temperatures close to ordinary conditions.

(d) **Hydrogen,** H_2, has no lone pairs, is nonpolar, and cannot form hydrogen bonds.

(e) **Methane** consists of covalent CH_4 molecules. They have no unshared electron pairs, so are incapable of forming hydrogen bonds.

(f) **Lithium hydride,** LiH, consists of Li^+ and $:H^-$ ions and cannot form hydrogen bonds.

(g) **Methanol** consists of H_3C-OH molecules with highly polar $^{\delta-}$O-H$^{\delta+}$ bonds, where the O atoms have two highly localized lone pairs. It forms hydrogen bonds between O—H groups in the liquid; the C—H bonds are insufficiently polar to participate in hydrogen bonding.

(h) **Acetic acid,** H_3C-C(O)OH, has a highly polar $^{\delta-}$O-H$^{\delta+}$ bond and two O atoms with highly localized unshared electron pairs. It forms strong hydrogen bonds.

---

49. Water has a greater density at 0°C than does ice, whereas liquid bromine is less dense than solid bromine at its melting point. Account for these differences.

The smaller density of ice at its freezing point compared to that of water at the same temperature is a consequence of the different extents to which water molecules form hydrogen bonds in the solid and liquid. Ice has a structure in which each water molecule is hydrogen bonded to four others - two via the H atoms of H_2O, and two utilizing each of the lone pairs on the O atom. This gives ice an open, cage-like structure. When ice melts, some of the hydrogen bonds between water molecules are broken and the H_2O molecules that break away slip into some of the cavities of the remaining cage-like structure. This collapse of the ice-structure **increases** the density because in liquid water at the freezing point there are more molecules packed into a given volume than there are in ice.

In contrast, there are only weak intermolecular (London) forces between Br_2 molecules. These have their maximum strength in solid bromine where they hold the Br_2 molecules in a regular array. When solid bromine melts, the intermolecular forces decrease and the molecules are farther apart, allowing Br_2 molecules to slip past each other, so the bromine becomes a liquid. Thus, in bromine (and the majority of substances), the molecules are farther apart in the liquid than in the solid; there is a smaller number of molecules in a given volume than in the solid, and the solid has a **higher** density than the liquid.

Solutions

51. Make qualitative predictions about each of the following solubilities, and explain your answers: (a) HCl(g) in water and in pentane, $C_5H_{12}(\ell)$; (b) water in liquid HF and in gasoline; (c) chloroform (trichloromethane), $CHCl_3(\ell)$, in water and in carbon tetrachloride (tetrachloromethane), $CCl_4(\ell)$; (d) naphthalene, $C_{10}H_8(s)$, in water and in benzene, $C_6H_6(\ell)$; (e) $N_2(g)$ and hydrogen cyanide, HCN(g), in water; (f) benzene, $C_6H_6(\ell)$, in toluene (methylbenzene), $C_7H_8(\ell)$, and in water

---

Substances dissolving in each other increases entropy but for two substances to mix, the enthalpy of solution must not be too unfavorable, which can be discussed qualitatively in terms of changes in the intermolecular solute-solute, solute-solvent, and solvent-solvent interactions that is summarized in the aphorism "like dissolves like".

(a) **HCl gas** is very soluble in water but insoluble in pentane. This is because HCl is a polar molecule, $^{\delta+}$H-Cl$^{\delta-}$, that readily interacts with polar water molecules; indeed, it interacts to the extent of transferring protons quantitatively between HCl and water to give a solution of the strong acid H_3O^+ Cl^-. Hydrogen bonds between water molecules are replaced by hydrogen bonds between HCl and water, so the process is energetically favorable. In contrast, polar HCl(g) is insoluble in nonpolar pentane because these two substances cannot form any strong intermolecular interactions. Therefore **ΔH** is **positive.**

(b) **Water** is very soluble in liquid HF but insoluble in pentane. It can form strong hydrogen bonds with polar HF, so the solution process involves replacement of hydrogen bonds in the solvent HF by solute-solvent hydrogen bonds, which is energetically favorable. On the other hand, nonpolar pentane is insoluble because these two substances cannot form any strong intermolecular interactions. Therefore **ΔH** is **positive.**

(c) **Chloroform** is insoluble in water but very soluble in tetrachloromethane. Its slightly polar molecules cannot hydrogen bond with water, so the energy of interaction is too small to disrupt the hydrogen bonds between the water molecules in liquid water. In tetrachloromethane, the solution process involves the replacement of London forces between CCl_4 molecules by similar solute-solvent interactions, so is favorable.

(d) **Naphthalene** and benzene are both nonpolar aromatic hydrocarbons and therefore naphthalene is expected to be soluble in benzene. It is however insoluble in polar water.

(e) **Nitrogen** has only a slight solubility in water while **hydrogen cyanide** is soluble. Nitrogen is negligibly soluble in water because its small nonpolar :N≡N: molecules have a low polarizability and cannot interact sufficiently strongly to break the hydrogen bonds between water molecules. In contrast, hydrogen cyanide consists of H—C≡N: molecules with significantly polar $^{\delta+}$H—C^{-} bonds that interact strongly with polar water molecules and form hydrogen bonds with them.

(f) **Benzene** is miscible with **toluene** but insoluble in **water**. Benzene, $C_6H_6(\ell)$, and toluene, $C_6H_5{\cdot}CH_3(\ell)$, methylbenzene, are both aromatic hydrocarbons. Thus for benzene and toluene solute-solute, solute-solvent, and solvent-solvent intermolecular interactions involve only London forces and the entropy of mixing dominates. In contrast, benzene is insoluble in water because nonpolar benzene molecules cannot mix with polar, hydrogen bonded water. Here the positive enthalpy of mixing dominates.

General Problems

53. In terms of the radii of the ions involved, explain why the structure of sodium chloride is different from that of cesium chloride.

In an **ionic crystal** the ions in the infinite lattice are arranged so that cations are surrounded by as many anions as possible, and anions are surrounded by as many cations as possible. The sodium ion, Na^+, is a relatively small ion (radius 102 pm) and the maximum number of Cl^- ions that can surround each Na^+ ion in NaCl(s) is **six**. Since there are exactly the same number of Na^+ and Cl^- ions overall, each Cl^- ion also surrounds itself with **six** Na^+ ions. In each case the arrangements are octahedral, which accounts for the **cubic face-centered structure** of **sodium chloride**.

In contrast, in **cesium chloride** the cesium ion, Cs^+ (radius 170 pm) is much larger than the Na^+ ion and can surround itself with **eight** Cl^- ions at the corners of a cube. The Cl^- ions are also surrounded by a cubic arrangement of **eight** Cs^+ ions, which gives CsCl(s) the **body-centered cubic structure.**

---

55. Explain why: (a) water has a boiling point some 200°C higher than is expected in comparison with the boiling points of H_2S, H_2Se, and H_2Te; (b) sulfuric acid, $(HO)_2SO_2$, is a liquid whereas F_2SO_2 is a gas; (c) water has its maximum density at 3.98 °C; (d) methanol, $CH_3OH(\ell)$, is miscible with water whereas hexane, $C_6H_{14}(\ell)$, is immiscible; (e) sodium oxide is readily soluble in water whereas magnesium oxide is insoluble.

(a) All the molecules of the Group VI hydrides are covalent molecules of the AX_2E_2 angular type with dipole moments. H_2S, H_2Se, and H_2Te are normally gases and their boiling points increase systematically from H_2S to H_2Te, consistent with the expected increase in the magnitude of the London forces as the size and polarizability of the central atom increases. Water would also be a gas if the only intermolecular forces were London forces, but among the Group VI hydrides, only H_2O molecules are capable of forming hydrogen bonds, which gives it its anomalous boiling point of 100°C.

(b) **Sulfuric acid**, $(HO)_2SO_2$, and **sulfuryl fluoride**, F_2SO_2, molecules are both AX_4 tetrahedal. The only important intermolecular forces between F_2SO_2 molecules are weak London forces and weak dipole-dipole forces, whereas sulfuric acid molecules, with polar $^{\delta-}O—H^{\delta+}$ bonds can form much stronger hydrogen bonds.

(c) Ice is less dense than water at the melting point of 0°C because of its hydrogen bonded cage-like structure. When ice melts, some hydrogen bonds are broken, the cage structure partially collapses, and the density increases. Above 0°C, the structure continues to collapse as more hydrogen bonds are broken. Water has its maximum density at 3.98°C because this is the temperature above which the expected density increase due to increasing collapse of the water structure begins to be outweighed by the expansion of water due to increased thermal vibrations of its molecules. As the temperature is further increased the density decreases.

(d) **Methanol**, $CH_3OH(\ell)$, is a hydrogen bonded substance, as is water. It is soluble in water in all proportions (miscible) because hydrogen bonds between CH_3OH molecules of methanol, and between H_2O molecules of liquid water, are readily replaced by hydrogen bonds between CH_3OH and H_2O molecules. **Hexane** molecules, C_6H_{14}, can interact with themselves or with water only through London forces. These are too weak to disrupt the hydrogen bonded structure of water, so hexane is insoluble in water.

(e) **Sodium oxide**, $Na_2O(s)$, and **magnesium oxide**, MgO(s), are both ionic solids. For an ionic solid to dissolve in water, its ions have to be separated. Comparing the sizes of the ions in these oxides and their charges, the Na^+ ions in $(Na^+)_2O^{2-}(s)$ has a smaller charge and is considerably larger than the Mg^{2+} ions in $Mg^{2+}O^{2-}(s)$, so the ionic bonds in Na_2O are considerably weaker than those in MgO. The difficulty of separating the Mg^{2+} and O^{2-} ions of MgO is apparently so great that MgO(s) is insoluble while $Na_2O(s)$, with weaker ionic bonds is soluble and dissolves in water to give a solution of NaOH(aq).

---

57. Explain the differences in the boiling points of each of the following pairs of substances:

	Substance	bp (°C)	Substance	bp (°C)
(a)	$(CH_3)O$	35	C_2H_5OH	79
(b)	HF	20	HCl	−85
(c)	CCl_4	76	LiCl	1360

(a) The higher boiling point of ethanol compared to dimethyl ether is due to hydrogen bonds between ethanol molecules, whereas there are only weak London forces between ether molecules.

(b) The higher boiling point of hydrogen fluoride is due to its hydrogen bonded structure, which is not present in liquid hydrogen chloride.

(c) Lithium chloride is composed of Li^+ and Cl^- ions, whereas tetrachloromethane contains nonpolar covalent CCl_4 molecules. To boil LiCl strong ionic bonds have to be broken, while to boil CCl_4 only weak London forces have to be overcome.

---

59. (a) Explain why real gases deviate from ideal behaviour especially at high pressures and low temperatures. To what factors can this be ascribed? (b) From among each of the following pairs, which gas would you expect to show the greater deviation from ideal gas behaviour, and why? (i) bromine and fluorine (ii) carbon monoxide and nitrogen (iii) hydrogen chloride, HCl, and hydrogen fluoride, HF

(a) An ideal gas is defined as composed of molecules of zero volume between which no intermolecular forces act. In real gases, neither assumption is exactly true. Molecules do occupy some space and intermolecular attractions may be small but not negligible. As pressure is increased a gas is compressed. Its molecules are pushed closer together and there is a commensurate increase in the magnitude of the intermolecular attractions. The observed pressure becomes detectably less than the ideal pressure because the molecules now have a tendency to stick to each other, which slows them down. This decreases the pressure P to less than the ideal pressure. The affect will increase in importance at low temperatures where the average kinetic energies of the gas molecules are relatively small. Also at high pressure, the actual volume of the molecules themselves becomes significant compared with the measured volume of gas. The actual space in which molecules are free to move is less than the measured volume of gas, V. Thus, $P_{ideal}V_{ideal} = nRT$ has to be modified to an equation like $(P+a)(V-b) = nRT$. Where a and b depend on the particular gas considered, to take into account these effects.

(b) (i) **Bromine** is expected to be more nonideal than **fluorine** on account of its larger molecules and its stronger London forces due to the greater polarizability of Br_2 compared to F_2.

(ii) **Carbon monoxide** and **nitrogen** are expected to have molecules of comparable size (CO versus NN) and polarizability. The only significant difference is that $^{\delta-}$:C≡O:$^{\delta+}$ is polar while :N≡N: is nonpolar. CO might be expected to be more nonideal than N_2 due to dipole-dipole interactions between its molecules in addition to London forces.

(iii) **Hydrogen chloride and hydrogen fluoride** are both polar molecules, $^{\delta+}$H—X:$^{\delta-}$, but because of the greater difference in electronegativity between H and F than between H and Cl, and the small size of fluorine, HF alone can form hydrogen bonds of significant strength. Thus HF(g) is expected to be more nonideal than HCl(g).

60.* Elemental boron melts at 2300°C and is almost as hard as diamond. It is a poor electrical conductor as both a solid and as a liquid. Would you expect (a) boron to be a molecular solid, an ionic solid, or a covalent network solid? (b) Would you expect solid boron to be soluble in water?

(a) Boron, like carbon is a nonmetal. The high melting point is indicative of a network solid and since only one type of nonmetal atom is involved it must be a **covalent network solid,** as evidenced by the observations that it is a poor electrical conductor as both a solid and as a liquid.

(b) No, because the intermolecular attractions between water and solid boron would be weak London forces, insufficient to break either strong boron-boron bonds in the solute or hydrogen bonds in the solvent.

CHAPTER 12

Acid-Base Equilibria and the pH Scale

1. What are the molar concentrations of the ions in each of the following aqueous solutions? (a) 10^{-5}-M HNO_3 (b) 0.0023-M HCl (c) 0.113-M $HClO_4$ (d) 0.034-M HBr (e) 10^{-3}-M NaOH (f) 0.145-M $Ba(OH)_2$

All of the solutions are aqueous solutions of simple strong acids or bases, so each is fully ionized:

(a) **10^{-5} M H_3O^+(aq), 10^{-5} M NO_3^-(aq)** (b) **0.0023 M H_3O^+(aq), 0.0023 M Cl^-(aq)**

(c) **0.113 M H_3O^+(aq), 0.113 M ClO_4^-(aq)** (d) **0.034 M H_3O^+(aq), 0.034 M Br^-(aq)**

(e) **10^{-3} M Na^+(aq), 10^{-3} M OH^-** (f) **0.145 M Ba^{2+}(aq), 0.290 M OH^-(aq)**

---

3. What are the pH values of the solutions in Problem 1?

We use the expressions: **pH = -log [H_3O^+], pOH = -log [OH^-],** and **pH + pOH = 14.00** (25°C)

(a) **pH** = -log 10^{-5} = -(-5.00) = **5.00** (b) **pH** = -log 0.0023 = -(-2.64) = **2.64**

(c) **pH** = -log 0.113 = -(-0.95) = **0.95** (d) **pH** = -log 0.034 = -(-1.47) = **1.47**

(e) pOH = -log 10^{-3} = -(-3.00) = 3.00, **pH** = 14.00 - 3.00 = **11.00**
(f) pOH = -log 0.290 = -(-0.54) = 0.54, **pH** = 14.00 - pOH = 14.00 - 0.54 = **13.46.**

Note: Remember that in solving problems of the following kind we follow the following steps:
1. Write the balanced equation for the reaction in aqueous solution; 2. Write the expression for the equilibrium constant; 3. Write expressions for the initial concentrations, and for the concentrations at equilibrium, and 4. Substitute the equilibrium concentrations in the equilibrium constant expression, and solve for any unknowns.

5. What are (a) the H_3O^+ ion concentration, (b) the percent ionization, and (c) the pH of a 0.010-M solution of hydrocyanic acid, HCN(aq)?

(a)

	HCN(aq)	+ $H_2O(\ell)$ →	H_3O^+(aq)	+ CN^-(aq)	
initially:	0.010	-	0	0	mol L^{-1}
at equilibrium:	0.010-x	-	x	x	mol L^{-1}

$$K_a = \frac{[H_3O^+][CN^-]}{[HCN]} = \frac{x^2}{0.010 - x} = 4.9 \times 10^{-10} \text{ mol L}^{-1}$$

Assuming x << 0.010, $x^2 = 4.9 \times 10^{-12}$; $x = 2.2 \times 10^{-6}$; $[H_3O^+]$ = $\underline{2.2 \times 10^{-6} \text{ mol L}^{-1}}$

and the assumption, x << 0.010 was justified.

(b) $$\text{Percentage ionization} = \frac{2.2 \times 10^{-6} \text{ mol L}^{-1}}{0.010 \text{ mol L}^{-1}} \times 100\% = \underline{0.022\%} \text{ (which is very small)}$$

(c) **pH** = -log [H_3O^+] = -log (2.2 x 10^{-6}) = -(-5.66) = **5.66.**

---

7. Of a 0.0010-M solution of HNO_3(aq) and a 0.200-M solution of acetic acid, which has the lower pH?

We need to calculate [H_3O^+], and hence the pH, of each solution:

Nitric acid, $HNO_3(aq)$ is a **strong acid**, thus $[H_3O^+] = 0.0010$ mol L^{-1}; pH = -log 0.0010 = -(3.00) = 3.00, and for the solution of the weak acid acetic acid (CH_3CO_2H = HA):

$$HA(aq) + H_2O(\ell) \rightarrow H_3O^+(aq) + A^-(aq)$$

	HA	H_2O	H_3O^+	A^-	
initially:	0.200	-	0	0	mol L^{-1}
at equilibrium:	0.200-x	-	x	x	mol L-1

$$K_a = \frac{[H_3O^+][A^-]}{[HA]} = \frac{x^2}{0.200 - x} = 1.8 \times 10^{-5} \text{ mol L}^{-1}$$

Assuming x << 0.200, $x^2 = 3.6 \times 10^{-6}$; $x = 1.9 \times 10^{-3}$; $[H_3O^+] = 1.9 \times 10^{-3}$ mol L^{-1}

and the assumption, x << 0.200 was justified.

Hence, **pH** = -log (1.9×10^{-3}) = -(-2.72) = **2.72**

Of the two solutions, **0.200 M acetic acid has the lower pH** (is slightly more acidic)

---

9. What are (a) the OH^- ion concentration, (b) the percent ionization, and (c) the pH of a 0.080-M solution of aniline, $C_6H_5NH_2$.

$$C_6H_5NH_2(aq) + H_2O(\ell) \rightarrow C_6H_5NH_3^+(aq) + OH^-(aq)$$

	$C_6H_5NH_2$	H_2O	$C_6H_5NH_3^+$	OH^-	
initially:	0.08	-	0	0	mol L^{-1}
at equilibrium:	0.08-x	-	x	x	mol L^{-1}

$$K_b = \frac{[C_6H_5NH_3^+][OH^-]}{[C_6H_5NH_2]} = \frac{x^2}{0.080 - x} = 4.3 \times 10^{-10} \text{ mol L}^{-1}$$

Assuming x << 0.080, $x^2 = 3.44 \times 10^{-11}$; $x = 5.9 \times 10^{-6}$; $[OH^-] = 5.9 \times 10^{-6}$ mol L^{-1}

and the assumption, x << 0.08 was justified.

(b) $$\%\text{ ionization} = \frac{5.9 \times 10^{-6} \text{ mol L}^{-1}}{0.08 \text{ mol L}^{-1}} \times 100\% = 0.007\%$$ (x << 0.08 justified)

(c) pOH = -log $[OH^-]$ = -log (5.9×10^{-6}) = -(-5.23) = 5.23; **pH** = 14.00 - pOH = 14.00-5.23 = **8.77**

---

11. Your muscles may ache after strenuous exercise because lactic acid, $pK_a = 3.08$, forms faster than it is metabolized to CO_2 and water. What is the pH of the fluid in muscle when the lactic acid concentration reaches 1.0×10^{-3} mol·L^{-1}?

$pK_a = 3.08$; $K_a = 10^{-3.08} = 8.3 \times 10^{-4}$ mol L^{-1}, and for lactic acid

$$HA(aq) + H_2O(\ell) \rightarrow H_3O^+(aq) + A^-(aq)$$

	HA	H_2O	H_3O^+	A^-	
initially:	1.0×10^{-3}	-	0	0	mol L^{-1}
at equilibrium:	(1.0×10^{-3} -x)	-	x	x	mol L^{-1}

so we must solve the quadratic equation: $x^2 + (8.3 \times 10^{-4})x - (8.3 \times 10^{-7}) = 0$; **x = 5.9 x 10^{-4}**

$$K_a = \frac{[H_3O^+][A^-]}{[HA]} = \frac{x^2}{(1x10^{-3}) - x} = 8.3 \text{ x } 10^{-4} \text{ mol L}^{-1}$$

Assuming $x << 1x10^{-3}$, $x^2 = 8.3 \text{ x } 10^{-7}$; $\underline{x = 9.1 \text{ x } 10^{-4}} \approx 10^{-3}$

Hence, $[H_3O^+] = 5.9x10^{-4}$ mol L^{-1}; **pH** = -log $[H_3O^+]$ = -log (5.9 x 10^{-4}) = -(-3.23) = **3.23**

---

13. Citric acid has a pK_a of 3.10. What is the pH of a 0.10-M solution of citric acid?

This is a straightforward calculation of $[H_3O^+]$, and then pH, like Problems 5 and 11:

pK_a = 3.10; $K_a = 10^{-3.10}$ = 7.9 x 10^{-4} mol L^{-1}, and for citric acid

	HA(aq) +	$H_2O(\ell)$ →	$H_3O^+(aq)$ +	$A^-(aq)$	
initially:	0.10	-	0	0	mol L^{-1}
at equilibrium:	0.10-x	-	x	x	mol L^{-1}

$$K_a = \frac{[H_3O^+][A^-]}{[HA]} = \frac{x^2}{0.10 - x} = 7.9 \text{ x } 10^{-4} \text{ mol L}^{-1}$$

Assuming $x << 1x10^{-3}$, $x^2 = 7.9 \text{ x } 10^{-5}$; $\underline{x = 8.9 \text{ x } 10^{-3}} \approx 10\%$ of x

so the quadratic equation must be solved: $x^2 + (7.9 \text{ x } 10^{-4})x - (7.9 \text{ x } 10^{-5}) = 0$; x = **8.5 x 10^{-3}**

Hence, $[H_3O^+] = 8.5x10^{-3}$ mol L^{-1}; **pH** = -log $[H_3O^+]$ = -log (8.5x10^{-3}) = -(-2.07) = **2.07**

Acid-Base Properties of Anions, Cations, and Salts

15. Give the formulas and names of the conjugate acids of each of the following bases: (a) NH_3 (b) F^- (c) OH^- (d) H_2O (e) H^-

---

Transfer of a proton to a base gives its *conjugate acid*:

(a) $NH_3 + H^+ \rightarrow$ **NH_4^+ ammonium ion**; (b) $F^- + H^+ \rightarrow$ **HF hydrogen fluoride** (hydrofluoric acid in water)
(c) $OH^- + H^+ \rightarrow$ **H_2O water**; (d) $H_2O + H^+ \rightarrow$ **H_3O^+ hydronium ion**; (e) $H^- + H^+ \rightarrow$ **H_2 hydrogen.**

---

16. Give the formulas and names of the conjugate bases of each of the following acids: (a) HF (b) HNO_3 (c) $HClO_4$ (d) H_2O (e) H_3O^+

Loss of a proton from an acid gives its *conjugate base*:

(a) $HF \rightarrow H^+ +$ **F^- fluoride ion**;
(b) $HNO_3 \rightarrow H^+ +$ **NO_3^- nitrate ion**;
(c) $HClO_4 \rightarrow H^+ +$ **ClO_4^- perchlorate ion**;
(d) $H_2O \rightarrow H^+ +$ **OH^- hydroxide ion**;
(e) $H_3O^+ \rightarrow H^+ +$ **H_2O water.**

---

19. For each of the following acids, name the conjugate base, give its formula, and calculate its pK_b at 25°C: (a) H_2CO_3, (b) H_2S, (c) HNO_2, (d) H_3PO_4

Note: K_a and K_b for a conjugate acid-base pair are related by: **$K_a.K_b = K_w = 1.00 \text{ x } 10^{-14}$ (25°C)**

(a) HCO_3^-, <u>hydrogencarbonate ion</u>: $K_b = \frac{K_w}{K_a} = \frac{1.00 \times 10^{-14}\ mol^2\ L^{-2}}{3.5 \times 10^{-7}\ mol\ L^{-1}} = \underline{2.9 \times 10^{-8}\ mol\ L^{-1}}$

(b) HS^-, <u>hydrogensulfide ion</u>: $K_b = \frac{K_w}{K_a} = \frac{1.00 \times 10^{-14}\ mol^2\ L^{-2}}{9.1 \times 10^{-8}\ mol\ L^{-1}} = \underline{1.1 \times 10^{-7}\ mol\ L^{-1}}$

(c) NO_2^-, <u>nitrite ion</u>: $K_b = \frac{K_w}{K_a} = \frac{1.00 \times 10^{-14}\ mol^2\ L^{-2}}{4.5 \times 10^{-4}\ mol\ L^{-1}} = \underline{2.2 \times 10^{-11}\ mol\ L^{-1}}$

(d) $H_2PO_4^{2-}$, <u>dihydrogenphosphate ion</u>: $K_b = \frac{K_w}{K_a} = \frac{1.00 \times 10^{-14}\ mol^2\ L^{-2}}{7.5 \times 10^{-3}\ mol\ L^{-1}} = \underline{1.3 \times 10^{-12}\ mol\ L^{-1}}$

---

20. What is the pH of a 0.050-M aqueous solution of KF?

KF(aq) is fully dissociated into K^+(aq) and F^-(aq); K^+(aq) has no acid or basic properties, but F^-(aq) is the anion of a *weak acid*. Thus, it behaves as a **weak base** and is partially converted to its conjugate acid, HF(aq):

	$F^-(aq)$ +	$H_2O(\ell)$ ⇄	HF(aq) +	$OH^-(aq)$	
initially:	0.050	-	0	0	mol L^{-1}
at equilibrium:	0.050-x	-	x	x	mol L^{-1}

and the appropriate ionization constant, $K_b(F^-)$, is calculated from K_a(HF). Thus:

$$K_b(F^-) = \frac{[HF][OH^-]}{[F^-]} = \frac{x^2}{0.050-x} = \frac{K_w}{K_a(HF)} = \frac{10^{-14}\ mol^2\ L^{-2}}{3.5 \times 10^{-4}\ mol\ L^{-1}} = 2.9x10^{-11}\ mol\ L^{-1}$$

Assuming $x << 0.050$, $x^2 = 1.45 \times 10^{-12}$; $x = 1.20 \times 10^{-6}$; $[OH^-] = \underline{1.20 \times 10^{-6}\ mol\ L^{-1}}$

$$pH = 14.00 - pOH = 14.00 - [-\log (1.20x10^{-6})] = 14.00 - [-(-5.92)] = \underline{8.08}$$

---

22. (a) What are the equilibrium concentrations of NH_4^+, NH_3, OH^-, and H_3O^+, in a 0.020-M solution of NH_4Cl(aq)? (b) What is the pH of the solution in (a)?

NH_4Cl(aq) is fully ionized, $NH_4Cl(aq) \rightarrow NH_4^+(aq) + Cl^-(aq)$, and NH_4^+(aq) behaves as a weak acid, while Cl^-, the conjugate base of the strong acid HCl(aq), has no acid or base properties:

	$NH_4^+(aq)$ +	$H_2O(\ell)$ ⇄	$H_3O^+(aq)$ +	$NH_3(aq)$	
initially:	0.020	-	0	0	mol L^{-1}
at equilibrium:	0.020-x	-	x	x	mol L^{-1}

and the appropriate ionization constant, $K_a(NH_4^+)$, is calculated from $K_b(NH_3)$. Thus:

$$K_a(NH_4^+) = \frac{[H_3O^+][NH_3]}{[NH_4^+]} = \frac{x^2}{0.020-x} = \frac{K_w}{K_b(NH_3)} = \frac{10^{-14}\ mol^2\ L^{-2}}{1.8 \times 10^{-5}\ mol\ L^{-1}} = 5.6 \times 10^{-10}\ mol\ L^{-1}$$

For $x << 0.020$, $x^2 = 1.12 \times 10^{-11}$; $x = 3.35 \times 10^{-6}$; $[H_3O^+] = \underline{3.35 \times 10^{-6}\ mol\ L^{-1}}$

$$pH = -\log (3.35 \times 10^{-6}) = \underline{5.47}$$

Thus, we have: $[NH_4^+] = 0.020\text{-x} = \mathbf{0.020\ mol\ L^{-1}}$; $[NH_3] = [H_3O^+] = \mathbf{3.4 \times 10^{-6}\ mol\ L^{-1}}$, and

$$[OH^-] = \frac{K_w}{[H_3O^+]} = \frac{10^{-14}\ mol^2\ L^{-2}}{3.4 \times 10^{-6}} = \underline{2.9 \times 10^{-9}\ mol\ L^{-1}}$$

---

24. Suggest a suitable acid-base reaction by which each of the following salts could be prepared. Would you expect a 0.10-M solution of each salt to be acidic, basic, or neutral? Explain your answers: (a) ammonium nitrate; (b) ammonium chloride; (c) calcium sulfate; (d) potassium acetate; (e) aluminum chloride; (f) sodium iodide.

(a) $NH_3(aq) + HNO_3(aq) \rightarrow NH_4NO_3(aq)$

and the **ammonium nitrate** will be fully dissociated into $NH_4^+(aq)$ ions and $NO_3^-(aq)$ ions in solution. NO_3^- is the conjugate base of the strong acid HNO_3 and will have no acidic or basic properties in water; NH_4^+ is the conjugate acid of the weak base $NH_3(aq)$ and will behave as a <u>weak acid</u>, The solution is **acidic** due to the reaction:

$$NH_4^+(aq) + H_2O(\ell) \rightleftarrows NH_3(aq) + H_3O^+(aq)$$

(b) $NH_3(aq) + HCl(aq) \rightarrow NH_4Cl(aq)$

and the ammonium choride will be fully dissociated into $NH_4^+(aq)$ ions and $Cl^-(aq)$ ions in solution. Cl^- is the conjugate base of the strong acid HCl and will have no acidic or basic properties in water; NH_4^+ is the conjugate acid of the weak base $NH_3(aq)$ and will behave as a <u>weak acid</u>. The solution is **acidic** due to the reaction:

$$NH_4^+(aq) + H_2O(\ell) \rightleftarrows NH_3(aq) + H_3O^+(aq)$$

(c) $Ca(OH)_2(s) + H_2SO_4(aq) \rightarrow CaSO_4(aq)$

and the calcium sulfate will be fully dissociated into $Ca^{2+}(aq)$ ions and $SO_4^{2-}(aq)$ ions in solution. SO_4^{2-} is the conjugate base of the moderately strong acid HSO_4^- and will be a <u>very weak base</u>, while Ca^{2+} has no acidic or basic properties in water. The solution will be weakly **basic** due to the reaction:

$$SO_4^{2-}(aq) + H_2O(\ell) \rightleftarrows HSO_4^-(aq) + OH^-(aq)$$

(d) $CH_3CO_2H(aq) + KOH(aq) \rightarrow CH_3CO_2K(aq) + H_2O(\ell)$

and potassium acetate will be fully dissociated into $K^+(aq)$ ions and $CH_3CO_2^-(aq)$ ions in solution. $CH_3CO_2^-$ is the conjugate base of the weak acid CH_3CO_2H and will behave as a <u>weak base</u> in water, while K^+ has no acid or base properties. The solution is **basic** due to the reaction:

$$CH_3CO_2^-(aq) + H_2O(\ell) \rightleftarrows CH_3CO_2H(aq) + OH^-(aq)$$

(e) $Al(OH)_3(s) + 3HCl(aq) \rightarrow AlCl_3(aq) + 3H_2O(\ell)$

and the aluminum chloride will be fully dissociated into $Al(H_2O)_6^{3+}(aq)$ ions and $Cl^-(aq)$ ions in solution. Cl^- is the conjugate base of the strong acid HCl and will have no acidic or basic properties in water, but strongly hydrated $Al(H_2O)_6^{3+}$ behaves as a <u>weak acid</u>. The solution will be weakly **acidic** due to the reaction:

$$Al(H_2O)_6^{3+}(aq) + H_2O(\ell) \rightleftarrows Al(H_2O)_5(OH)^{2+}(aq) + H_3O^+(aq)$$

(f) $NaOH(aq) + HI(aq) \rightarrow NaI(aq) + H_2O(\ell)$

and the sodium iodide formed will be fully dissociated into $Na^+(aq)$ ions and $I^-(aq)$ ions. Neither Na^+ nor I^- have acid or base properties; Na^+ is too weakly hydrated to be acidic and I^- is the conjugate base of the strong acid HI(aq), and is therefore too weakly basic to behave as a base in water. The solution will be **neutral.**

---

25. A 30.0-g sample of phosphoric acid, H_3PO_4, is dissolved in water to give 500 mL of solution. (a) What volume of 0.200-M NaOH(aq) is needed to react completely with the H_3PO_4 to form $Na_3PO_4(aq)$ quantitatively, and (b) will

the resulting solution be acidic, basic, or neutral?

(a) The balanced equation for the reaction is:

$$H_3PO_4(aq) + 3NaOH(aq) \rightarrow Na_3PO_4(aq) + 3H_2O(\ell).$$

First, we calculate the moles of phosphoric acid:

$$\text{moles of } H_3PO_4 = (30.0 \text{ g } H_3PO_4)\left(\frac{1 \text{ mol } H_3PO_4}{97.99 \text{ g } H_3PO_4}\right) = \underline{0.3062 \text{ mol}}$$

and now we can calculate the moles of NaOH required for complete reaction, and then the volume of 0.200-M NaOH(aq) containing this number of moles:

$$\text{moles NaOH required} = (0.3062 \text{ mol } H_3PO_4)\left(\frac{3 \text{ mol NaOH}}{1 \text{ mol } H_3PO_4}\right) = \underline{0.9186 \text{ mol}}$$

$$\text{Thus, volume of 0.200 M NaOH(aq)} = (0.9186 \text{ mol NaOH})\left(\frac{1 \text{ L}}{0.200 \text{ mol NaOH}}\right) = \underline{4.59 \text{ L}}$$

(b) The final solution contains Na_3PO_4, the salt of a weak acid and a strong base, and will be fully dissociated into Na^+(aq) and PO_4^{3-}(aq) ions. Na^+ has no acid or base properties; PO_4^{3-} is the conjugate base of the weak acid HPO_4^{2-} and behaves as a weak base. Thus, the final solution is **basic**, due to the reaction:

$$PO_4^{3-}(aq) + H_2O(\ell) \rightleftarrows HPO_4^{2-}(aq) + OH^-(aq)$$

---

27. What is the pH of a 0.100-M solution of (a) $AlCl_3$(aq) and (b) sodium hydrogensulfate, $NaHSO_4$(aq)?

(a) $AlCl_3$(aq) contains Cl^-(aq) ions, with no acid or base properties, and $Al(H_2O)_6^{3+}$(aq) ions that are weakly acidic (K_a = 7.2 x 10^{-6} mol L^{-1}), so we have:

	$Al(H_2O)_6^{3+}(aq)$ +	$H_2O(\ell)$ →	$H_3O^+(aq)$ +	$Al(H_2O)_5(OH)^{2+}(aq)$	
initially:	0.10	-	0	0	mol L^{-1}
at equilibrium:	0.15-x	-	x	x	mol L^{-1}

$$K_a = \frac{[H_3O^+][Al(H_2O)_5(OH)^{2+}]}{[Al(H_2O)_6^{3+}]} = \frac{x^2}{0.10 - x} = 7.2 \times 10^{-6} \text{ mol L}^{-1}$$

Assuming x << 0.10, x^2 = 7.2 x 10^{-7}; x = 8.49 x 10^{-4} ; $[H_3O^+]$ = $\underline{8.5 \times 10^{-4} \text{ mol L}^{-1}}$; pH = -log (8.5 x 10^{-4}) = $\underline{3.07}$

(b) Sodium hydrogen sulfate, $NaHSO_4$(aq), is fully dissociated into Na^+(aq) and HSO_4^-(aq) ions in aqueous solution. Na^+ has no acid or base properties, but we have to decide if HSO_4^-, the conjugate base of H_2SO_4 and the conjugate acid of SO_4^{2-}, behaves as a weak base or as a weak acid. Since H_2SO_4 is a strong acid, HSO_4^- will be too weak a base to be protonated in water, whereas HSO_4^- is a weak acid (K_a = 1.2 x 10^{-2} mol L^{-1}), so we have:

	$HSO_4^-(aq)$ +	$H_2O(\ell)$ →	$H_3O^+(aq)$ +	$SO_4^{2-}(aq)$	
initially:	0.10	-	0	0	mol L^{-1}
at equilibrium:	0.10-x	-	x	x	mol L^{-1}

$$K_a(HSO_4^-) = \frac{[H_3O^+][SO_4^{2-}]}{[HSO_4^-]} = \frac{x^2}{0.10 - x} = 1.2 \times 10^{-2} \text{ mol L}^{-1}$$

Assuming x << 0.10, x^2 = 1.2 x 10^{-3}; x = 0.034 (≈ 30% initial $[HSO_4^-]$)

so we have to solve the quadratic equation, x^2 + (1.2x10^{-2})x - (1.2x10^{-3}) = 0, which gives x = $\underline{0.0292 \text{ mol L}^{-1}}$.

Thus, **pH** = -log (0.0292) = **1.53.**

Buffer Solutions

Note: In problems involving **buffer solutions,** it is best not to try to remember equations such as the **Henderson-Hasselbalch equation** but to start with the K_a expression for the weak acid, or the K_b expression for its conjugate base. Remember that in a **buffer solution** at equilibrium, the amounts of weak acid and base are the *stoichiometric amounts.*

28. What are the pH values of (a) a buffer solution containing 0.30-M ammonia and 0.25-M ammonium chloride, and (b) a buffer solution made from equal volumes of 0.10-M HCN(aq) and 0.10-M NaCN(aq)?

(a) This **buffer solution** contains the *weak base* NH_3(aq) and its *conjugate acid* NH_4^+(aq), and we can use the given concentrations directly. Substituting in the expression for $K_b(NH_3)$ gives:

$$K_b(NH_3) = 1.8 \times 10^{-5} \text{ mol L}^{-1} = \frac{[NH_4^+][OH^-]}{[NH_3]} = [OH^-]\frac{[NH_4^+]}{[NH_3]} = [OH^-](\frac{0.25 \text{ mol L}^{-1}}{0.30 \text{ mol L}^{-1}})$$

$$[OH^-] = \frac{0.30}{0.25}(1.8 \times 10^{-5} \text{ mol L}^{-1}) = \underline{2.16 \times 10^{-5} \text{ mol L}^{-1}}$$

$$pH = 14.00 - pOH = 14.00 - \log(2.16 \times 10^{-5}) = 14.00 - 4.67 = \underline{9.33}$$

Equally well, we could have started with $K_a(NH_4^+)$:

$$K_a(NH_4^+) = \frac{K_w}{K_b(NH_3)} = \frac{10^{-14} \text{ mol}^2 \text{ L}^{-2}}{1.8 \times 10^{-5} \text{ mol L}^{-1}} = 5.6 \times 10^{-10} \text{ mol L}^{-1}$$

$$= \frac{[H_3O^+][NH_3]}{[NH_4^+]} = [H_3O^+]\frac{[NH_3]}{[NH_4^+]} = [H_3O^+]\frac{0.30 \text{ mol L}^{-1}}{0.25 \text{ mol L}^{-1}}$$

$$[H_3O^+] = \frac{0.25}{0.30}(5.6 \times 10^{-10} \text{ mol L}^{-1}) = \underline{4.67 \times 10^{-10} \text{ mol L}^{-1}}; \quad pH = -\log(4.7 \times 10^{-10}) = \underline{9.33}$$

(b) This **buffer solution** contains equal amounts of the *weak acid* HCN(aq) and its *conjugate base* CN^-, so

$$K_a(HCN) = 4.9 \times 10^{-10} \text{ mol L}^{-1} = \frac{[H_3O^+][CN^-]}{[HCN]} = [H_3O^+]\frac{[CN^-]}{[HCN]} = [H_3O^+]\frac{x \text{ mol L}^{-1}}{x \text{ mol L}^{-1}}$$

$$[H_3O^+] = \frac{1}{1}(4.9 \times 10^{-10} \text{ mol L}^{-1}) = \underline{4.9 \times 10^{-10} \text{ mol L}^{-1}}; \quad pH = -\log(4.9 \times 10^{-10}) = \underline{9.31}$$

---

30. What are the pH values of solutions prepared by mixing: (a) 25.0 mL of 0.100 M-NaOH(aq) and 50.0 mL of 0.100-M acetic acid; (b) 15.0 mL of 0.0100-M NH_3(aq) and 25.0 mL of 0.0100-M NH_4Cl(aq)?

(a) NaOH(aq) and CH_3CO_2H(aq) react to give $CH_3CO_2^-Na^+$(aq), so we have first to calculate the extent of this reaction, using the initial amounts of NaOH and CH_3CO_2H, and we can use moles, rather than mol L^{-1}, throughout:

initial moles NaOH = (25.0 mL)(0.100 mol L^{-1}) = <u>2.50 mmol</u>

initial moles CH_3CO_2H = (50.0 mL)(0.100 mol L^{-1}) = <u>5.00 mmol</u>

and we can now calculate the stoichiometric amounts of reactants and products in the final solution:

	$NaOH(aq)$ +	$CH_3CO_2H(aq)$ →	$CH_3CO_2Na(aq)$ +	$H_2O(\ell)$	
initially:	2.50	5.00	0	-	mmol
finally:	0	2.50	2.50	-	mmol

Thus, the final solution contains <u>equal</u> amounts of the weak acid CH_3CO_2H and its conjugate base $CH_3CO_2^-$, so

$[H_3O^+] = K_a$; <u>or</u> $pH = pK_a = \log(1.8 \times 10^{-5}) = \mathbf{4.74}$.

(b) This is a NH_4^+—NH_3 buffer solution and, since the solutions have the same concentration:

$$K_a(NH_4^+) = [H_3O^+]\frac{[NH_3]}{[NH_4^+]} = 5.6 \times 10^{-10} \text{ mol L}^{-1} = [H_3O^+](\frac{15.0 \text{ mL}}{25.0 \text{ mL}})$$

$$[H_3O^+] = \frac{25.0}{15.0}(5.6 \times 10^{-10} \text{ mol L}^{-1}) = 9.33\text{x}10^{-10} \text{ mol L}^{-1} ; \quad pH = -\log(9.33 \times 10^{-10}) = \underline{9.03}$$

32.* What is the change in pH when 1.00 mL of 1.00 M NaOH(aq) is added to 100 mL of a solution containing 0.18-M $NH_3(aq)$ and 0.10-M $NH_4Cl(aq)$?

First, we calculate the pH of the original buffer solution, using the amounts of NH_3 and NH_4^+ it contains:

$$K_a(NH_4^+) = [H_3O^+]\frac{[NH_3]}{[NH_4^+]} = 5.6 \times 10^{-10} \text{ mol L}^{-1} = [H_3O^+](\frac{0.18 \text{ mol L}^{-1}}{0.10 \text{ mol L}^{-1}})$$

$$[H_3O^+] = 3.11 \times 10^{-10} \text{ mol L}^{-1} ; \quad pH = -\log(3.11 \times 10^{-10}) = \underline{9.51}$$

Then we consider the affect of adding 1.00 mL of 1.00 M NaOH(aq) to the solution, which reacts as follows:

$$NH_4^+(aq) + OH^-(aq) \rightarrow NH_3(aq) + H_2O(\ell)$$

Initially: the solution contains: $(100 \text{ mL})(0.180 \text{ mol L}^{-1})$ = <u>18.0 mol</u> NH_3

and $(100 \text{ mL})(0.100 \text{ mol L}^{-1})$ = <u>10.0 mmol</u> NH_{4+}

and moles of OH^- *added*: = $(1.00 \text{ mL})(1.00 \text{ mol L}^{-1})$ = <u>1.00 mmol</u>

Thus, we have:

	$NH_4^+(aq)$ +	$OH^-(aq)$ →	$NH_3(aq)$ +	$H_2O(\ell)$	
initially:	10.0	1.00	18.0	-	mmol
finally:	9.0	0	19.0	-	mmol

and we can calculate the pH of the <u>new</u> buffer solution:

$$K_a(NH_4^+) = [H_3O^+]\frac{[NH_3]}{[NH_4^+]} = 5.6 \times 10^{-10} \text{ mol L}^{-1} = [H_3O^+](\frac{19.0 \text{ mmol}}{9.0 \text{ mmol}})$$

$$[H_3O^+] = 2.65 \times 10^{-10} \text{ mol L}^{-1}; \; pH = -\log(2.65 \times 10^{-10}) = \underline{9.58}$$

34*. (a) What is the pH of a buffer solution prepared by mixing 25.0 mL of a 0.020-M solution of aniline, $C_6H_5NH_2$, with 10.0 mL of a 0.030-M solution of anilinium chloride, $C_6H_5NH_3^+Cl^-$? (b) What is the pH of the solution in (a) after 1.00 mL of 0.040-M $HNO_3(aq)$ is added? (c) What will the pH be if 2.00 mL of 0.030-M KOH(aq) is added to the solution in (a) rather than nitric acid?

(a) The pH of this buffer solution depends on $K_a(C_6H_5NH_3^+)$ and the mole ratio $C_6H_5NH_2/C_6H_5NH_3^+$

$$\text{mol } C_6H_5NH_2 = (25.0 \text{ mL})(0.0200 \text{ mol L}^{-1}) = \underline{0.500 \text{ mmol}}$$
$$\text{mol } C_6H_5NH_3^+ = (10.0 \text{ mL})(0.0300 \text{ mol L}^{-1}) = \underline{0.300 \text{ mmol}}$$

$$K_a(C_6H_5NH_3^+) = \frac{K_w}{K_b(C_6H_5NH_2)} = \frac{10^{-14} \text{ mol}^2 \text{ L}^{-2}}{4.3 \times 10^{-10} \text{ mol L}^{-1}} = [H_3O^+](\frac{\text{mol base}}{\text{mol acid}}) = \underline{2.3 \times 10^{-5} \text{ mol L}^{-1}}$$

$$[H_3O^+] = (\frac{0.300}{0.500})(2.3 \times 10^{-5} \text{ mol L}^{-1}) = \underline{1.38 \times 10^{-5} \text{ mol L}^{-1}} \; ; \quad pH = -\log(1.38\times10^{-5}) = \underline{4.86}$$

(b) Addition of the strong acid $HNO_3(aq)$ to the buffer solution converts some of the aniline to anilinium salt.

$$\text{moles of added } HNO_3 = (1.00 \text{ mL})(0.040 \text{ mol L}^{-1}) = \underline{0.040 \text{ mmol}}$$

	$C_6H_5NH_2(aq)$ +	$HNO_3(aq)$ →	$C_6H_5NH_3^+(aq)$ +	$NO_3^-(aq)$	
initially:	0.500	0.040	0.300	0	mmol
finally:	0.460	0	0.340	0.040	mmol

$$K_a = [H_3O^+](\frac{\text{base}}{\text{acid}}) = [H_3O^+](\frac{0.460}{0.340}) = 2.3 \times 10^{-5} \text{ mol L}^{-1} \; ; \quad [H_3O^+] = \underline{1.70 \times 10^{-5} \text{ mol L}^{-1}} \; ; \quad \underline{pH = 4.77}$$

(c) Addition of the strong base KOH(aq) to the solution converts some of the anilinium salt to aniline.

$$\text{moles of added KOH} = (2.00 \text{ mL})(0.030 \text{ mol L}^{-1}) = \underline{0.060 \text{ mmol}}$$

	$C_6H_5NH_3^+(aq)$ +	$OH^-(aq)$ →	$C_6H_5NH_2(aq)$ +	$H_2O(\ell)$	
initially:	0.300	0.060	0.500	0	mmol
finally:	0.240	0	0.560	0.060	mmol

$$K_a = [H_3O^+](\frac{\text{base}}{\text{acid}}) = [H_3O^+](\frac{0.560}{0.240}) = 2.3 \times 10^{-5} \text{ mol L}^{-1} \; ; \quad [H_3O^+] = \underline{9.86 \times 10^{-6} \text{ mol L}^{-1}} \; ; \quad \underline{pH = 5.01}$$

The Measurement of pH

35. (a) In using an indicator, why is it important to add the smallest amount possible to the solution being investigated? (b) For methyl orange indicator (pK_a = 4.2), at what pH values is $[In^-]/[HIn]$ = 5:1, 1:1, and 1:5?

(a) An indicator is an acid or a base. Unless a minute amount is added, the reaction of the indicator with the system changes its pH by more than an insignificant amount.

(b) For any indicator, we have the equilibrium: $HIn(aq) + H_2O(\ell) \rightleftarrows H_3O^+(aq) + In^-(aq)$, and

$$K_a(HIn) = [H_3O^+]\frac{[In^-]}{[HIn]} \; ; \; pK_a = pH - \log\frac{[In^-]}{[HIn]} \; ; \; pH = pK_a + \log\frac{[In^-]}{[HIn]}$$

(i) $pH = 4.2 + \log\frac{5}{1} = 4.2 + 0.7 = \underline{4.9}$; (ii) $pH = 4.2 + \log\frac{1}{1} = 4.2 + 0 = \underline{4.2}$ (iii) $pH = 4.2 + \log\frac{1}{5} = 4.2 - 0.7 = \underline{3.5}$

---

36. (a) Estimate the pH of a colorless aqueous solution that turns yellow when methyl red is added to it and yellow when bromothymol blue is added to it; (b) What color would you expect a solution containing methyl red to be when the pH is 5.0?

(a). **Methyl red** (pK_a 5.2) is **yellow** in its **base** In^- form and **red** in its **acid** HIn form, and is effective in the pH range 4.2 to 6.2; thus, the color will be a definite *yellow* at a pH of 6.2, or a greater pH. **Bromothymol blue (pK_a 7.1) is blue in its basic In^- form and yellow in its acid HIn form, and is effective in the pH range 6.0 to 7.8; thus, the color will be a definite *yellow* at a pH of 6.0, or a lower pH. We can conclude that the pH of the solution must be in the range:**

$$\mathbf{6.0 < pH < 6.2}$$

(b) For pH = 5.0, from the expression for pK_{HIn}:

$$K_a = [H_3O^+]\frac{[In^-]}{[HIn]} \; ; \quad pK_a = pH - \log\frac{[In^-]}{[HIn]} \; ; \quad 5.20 = 5.0 - \log\frac{[In^-]}{[HIn]} \; ; \quad \log\frac{[In^-]}{[HIn]} = -0.20 \; ; \quad \frac{[In^-]}{[HIn]} = \underline{0.63}$$

The ratio $[In^-]$: [HIn] is approximately 2:3, so the color will be roughly 2 parts *yellow* and 3 parts *red*, or orange.

---

38. When each of the following indicators is added to a 0.10-M solution of a weak acid the colors are as follows: methyl violet - violet; methyl orange - yellow; methyl red - orange; bromothymol blue - yellow. What is the approximate K_a of the acid?

(a) **Methyl violet** is in its violet In^- form, therefore pH > 3.0

(b) **Methyl orange** is in its yellow In^- form, therefore pH > 4.4

(c) **Methy red** is red in its HIn form and yellow in its In^- form. The fact that the solution is orange indicates that [HIn] ≈ $[In^-]$, and means that pH ≈ pK_a(HIn) = 5.2.

(d) **Bromothymol blue** is in its yellow HIn form, therefore pH < 6.0.

Thus, we can conclude that the pH of the 0.10 M solution of the weak acid HA, is close to 5.2,

<u>or</u> $[H_3O^+] = 10^{-5.2}$ mol L^{-1} = **6.3×10^{-6} mol L^{-1}**

$$HA(aq) + H_2O(\ell) \rightleftarrows H_3O^+(aq) + A^-(aq)$$

	HA(aq)	$H_2O(\ell)$	$H_3O^+(aq)$	$A^-(aq)$	
initially:	0.10	-	0	0	mol L^{-1}
at equilibrium:	0.10	-	6.3×10^{-6}	6.3×10^{-6}	mol L^{-1}

$$K_a(HA) = \frac{[H_3O^+][A^-]}{[HA]} = \frac{(6.3 \times 10^{-6}\ \text{mol L}^{-1})^2}{0.10\ \text{mol L}^{-1}} = \underline{4.0 \times 10^{-10}\ \text{mol L}^{-1}}$$

---

40. What color would each of the following indicators give in a 0.10-M solution of aqueous ammonium chloride using the indicators: (a) methyl red (b) methyl orange (c) bromothymol blue.

We first calculate the pH of 0.10-M NH_4Cl(aq), using the method of Problem 23, which gives **pH = 5.13.**

(a) <u>Methyl red</u> has pK_a = 5.2: $\log\frac{[In^-]}{[HIn]} = pH - pK_a = 5.1 - 5.2 = \underline{-0.1}$; $\frac{[In^-]}{[HIn]} = \underline{0.8}$

(b) <u>Methyl orange</u> has pK_a = 4.2: $\log\frac{[In^-]}{[HIn]} = pH - pK_a = 5.1 - 4.2 = \underline{0.9}$; $\frac{[In^-]}{[HIn]} = 7.9$

(c) <u>Bromothymol blue</u> has pK_a = 7.1: $\log\frac{[In^-]}{[HIn]} = pH - pK_a = 5.1 - 7.1 = \underline{-2.0}$; $\frac{[In^-]}{[HIn]} = 0.01$

In (a), the expected color is approximately 50% yellow and 50% red, so the solution is **orange.**
In (b), the In^- color will dominate, so the solution is **yellow.**
In (c), the HIn color will dominate, so the solution is **yellow.**

Gas-Phase Equilibria

Remember in writing equilibrium constant expressions that products appear in the numerator (top) of the expression and reactants in the denominator (bottom).

42. Write the equilibrium constant expressions K_c and K_p for each of the following reactions:
(a) $2SO_2(g) + O_2(g) \rightleftarrows 2SO_3(g)$; (b) $SO_2(g) + \frac{1}{2}O_2(g) \rightleftarrows SO_3(g)$;
(c) $P_4(g) + 5O_2(g) \rightleftarrows P_4O_{10}(g)$; (d) $PCl_5(g) \rightleftarrows PCl_3(g) + Cl_2(g)$

These are all examples of homogeneous equilibria:

(a) $K_c = \dfrac{[SO_3]^2}{[SO_2]^2[O_2]}$ $K_p = \dfrac{p_{SO_3}^2}{p_{SO_2}^2 \cdot p_{O_2}}$ (b) $K_c = \dfrac{[SO_3]}{[SO_2][O_2]^{1/2}}$ $K_p = \dfrac{p_{SO_3}}{p_{SO_2} \cdot p_{O_2}^{\frac{1}{2}}}$

(c) $K_c = \dfrac{[P_4O_{10}]}{[P_4][O_2]^5}$ $K_p = \dfrac{p_{P_4O_{10}}}{p_{P_4} \cdot p_{O_2}^5}$ (d) $K_c = \dfrac{[PCl_3][Cl_2]}{[PCl_5]}$ $K_p = \dfrac{p_{PCl_3} \cdot p_{Cl_2}}{p_{PCl_5}}$

---

44. At 425°C, $K_c = 300\ mol^{-2}\cdot L^2$ for the reaction in which methanol, $CH_3OH(g)$, is synthesized from hydrogen and carbon monoxide: $2H_2(g) + CO(g) \rightleftarrows CH_3OH(g)$. (a) When the concentrations of H_2, CO, and CH_3OH are each $0.10\ mol\cdot L^{-1}$, is the system at equilibrium at 425°C? (b) If the reaction is not at equilibrium, will the concentration of CH_3OH be greater than or less than $0.10\ mol\cdot L^{-1}$ when equilibrium is established?

(a) We first write the reaction quotient expression, Q, (with the same form as the equilibrium constant expression). When the system is at equilibrium, $Q = K_c$.

$$Q = \frac{[CH_3OH]}{[H_2]^2[CO]} = \frac{0.10\ mol\ L^{-1}}{(0.10\ mol\ L^{-1})^2(0.10\ mol\ L^{-1})} = \underline{1.0 \times 10^2\ mol^{-2}\ L^2}$$

$Q < K_c = 300\ mol^{-2}\ L^2$, so the system is ***not at equilibrium.***

(b) To attain equilibrium Q must *increase,* so the value of the numerator in the expression must increase and the value of the denominator must decrease; **$[CH_3OH]$ must increase,** so its concentration will be **greater** than 0.10 mol L^{-1} at equilibrium.

---

45. For the reaction $H_2(g) + CO_2(g) \rightleftarrows H_2O(g) + CO(g)$ at 600 K, the following concentrations, in moles per liter, were found at equilibrium: $[H_2] = 0.600$, $[CO_2] = 0.459$, $[H_2O] = 0.500$, $[CO] = 0.425$. Calculate the values of (a) K_c and (b) K_p?

(a) $$K_c = \frac{[H_2O][CO]}{[H_2][CO_2]} = \frac{(0.500\ mol\ L^{-1})(0.425\ mol\ L^{-1})}{(0.600\ mol\ L^{-1})(0.459\ mol\ L^{-1})} = \underline{0.772}$$

(b) For this reaction, K_c has *no units*, and $\mathbf{K_p = K_c}$.

---

47. At 1000 K, iodine molecules dissociate into iodine atoms, with $K_c = 3.76 \times 10^{-5}$ mol·L^{-1} for the equilibrium, $I_2(g) \rightleftarrows 2I(g)$. At this temperature what are (a) the equilibrium concentrations of $I_2(g)$ and I(g) after 1.00 mol of I_2 has initially been introduced into a 2.00-L flask, and (b) the percentage dissociation of I_2?

	$I_2(g)$	$\rightleftarrows$ $2I(g)$	
initially:	0.500	0	mol L^{-1}
at equilibrium:	0.500-x	0.200-x	mol L^{-1}

For the system at equilibrium:

$$K_c = \frac{[I]^2}{[I_2]} = \frac{4x^2}{0.500-x} = 3.76 \times 10^{-5} \text{ mol L}^{-1}$$

Since K_c is small, we can assume x << 0.500 , giving: $4x^2 = 1.88 \times 10^{-5}$; $\underline{x = 2.17 \times 10^{-3}}$

and the equilibrium concentrations are: $[I_2] = (0.500\text{-x}) =$ **0.498 mol L^{-1}**, and $[I] = 2x =$ **0.00434 mol L^{-1}**.

(We should now substitute these values in the K_c expression, to check the accuracy of our answer.)

$$\text{(b)} \quad \%\ \text{dissociation } I_2 = \frac{x \text{ mol L}^{-1}}{0.500 \text{ mol L}^{-1}} = \left(\frac{0.00217 \text{ mol L}^{-1}}{0.500 \text{ mol L}^{-1}}\right) \times 100\% = \underline{0.434\%}$$

---

49. The equilibrium constant K_c at 698 K is 54.4 for the reaction, $H_2(g) + I_2(g) \rightleftarrows 2HI(g)$. A reaction mixture contains 0.10 mol·L^{-1} of both H_2 and I_2 and 1.00 mol·L^{-1} of HI. (a) Has the reaction reached equilibrium? (b) If not, calculate the concentrations of H_2, I_2, and HI at equilibrium. (c) What would be the concentrations at equilibrium if we started with just 1.2 mol·L^{-1} of HI alone?

	$H_2(g)$ +	$I_2(g)$ $\rightleftarrows$	$2HI(g)$	
initially:	0.10	0.10	1.0	mol L^{-1}
at equilibrium:	0.10 + x	0.10 + x	1.0 - 2x	mol L^{-1}

Now we can calculate the value of the *reaction quotient*, Q, and compare it to the value of the equilibium constant, K_c:

$$\text{(a)} \qquad Q = \frac{[HI]^2}{[H_2][I_2]} = \frac{(1.00 \text{ mol L}^{-1})^2}{(0.10 \text{ mol L}^{-1})(0.10 \text{ mol L}^{-1})} = \underline{100}$$

$Q > K_c = 54.4$, so the mixture is *not at equilibrium.*

(b) $Q > K_c$, so to achieve equilibrium some HI must be converted to H_2 and I_2:

$$K_c = \frac{(1.0-2x)^2}{(0.10+x)^2} = 54.4 \ ; \quad \frac{1.0-2x}{0.10+x} = 7.38$$

$$1.0-2x = 0.74+1.0x \ ; \quad 9.38x = 0.26 \ ; \quad \underline{x = 0.028}$$

At equilibrium the concentrations are: $[H_2] = [I_2] =$ **0.13 mol L^{-1}**; $[HI] =$ **0.94 mol L^{-1}**.

(c)

	$H_2(g)$ +	$I_2(g)$ $\rightleftarrows$	$I_2(g)$	
initially:	0	0	1.2	mol L^{-1}
at equilibrium:	x	x	1.2-2x	mol L^{-1}

$$K_c = \frac{[HI]^2}{[H_2][I_2]} = \frac{(1.2-x)^2}{x^2} = 54.4 \ ; \quad \frac{1.2-x}{x} = 7.38 \ ; \quad 9.38x = 1.2$$

Thus, x = **0.128**, and the equilibrium concentrations are: $[H_2]$ = $[I_2]$ = **0.13 mol L^{-1}**; [HI] = **0.94 mol L^{-1}**: as they were in part (b). Indeed, the calculation was not really necessary because throughout we have the same system, *for which there is only one equilibrium position*. In other words, the overall composition of the system is the same in both cases; hence the same equilibrium concentrations are achieved at the same temperature.

---

51. A 5.00-L flask containing 1 mol of HI(g) is heated to 800°C. What is the percent dissociation of the HI, given that value of the equilibrium constant K_c is 6.34 x 10^{-4} at 800°C for the reaction 2HI(g) ⇄ H_2(g) + I_2(g).

The initial concentration of HI(g) is 1.00 mol/5.00 L = **0.200 mol L^{-1}**.

	2HI(g) ⇄	H_2(g) +	I_2(g)	
initially:	0.200	0	0	mol L^{-1}
at equilibrium:	0.200-x	x	x	mol L^{-1}

$$K_c = \frac{[H_2][I_2]}{[HI]^2} = \frac{x^2}{(0.200-x)^2} = 6.34\text{x}10^{-4} \ ; \quad \frac{x}{0.200-x} = 2.52\text{x}10^{-2}$$

$$x = 5.04 \text{ x}10^{-3} - (5.04\text{x}10^{-2})x \ ; \quad x = \frac{5.04\text{x}10^{-3}}{1.050} = \underline{4.80\text{x}10^{-3}}$$

and the equilibrium concentrations are: [HI] = **0.190 mol L^{-1}**; $[H_2]$ = $[I_2]$ = **4.80x10^{-3} mol L^{-1}**.

$$\% \text{ dissociation HI} = \left(\frac{9.60\text{x}10^{-3} \text{ mol L}^{-1}}{0.200 \text{ mol L}^{-1}}\right) \text{x } 100\% = \underline{4.80\%}$$

---

53. The equilibrium constant K_p = 3.2 x 10^2 atm$^{-½}$ at 425°C for the reaction, SO_2(g) + ½O_2(g) ⇄ SO_3(g). At 525°C the value of K_p decreases to 33 atm$^{-½}$. Is the reaction as written exothermic or endothermic?

K_p can only decrease with increased temperature if there is a shift in favor of more reactants, i.e., from products to reactants. Thus, the *reverse* reaction must be *endothermic*, so the reaction as written (the forward reaction) must be **exothermic**.

Note: Remember that ΔH°, the standard reaction enthalpy is calculated from standard reaction enthalpies of formation, $\Delta H_f°$, and the number of moles of reactants and products in the balanced equation:

$$\Delta H° = \sum[n_p(\Delta H_f°)_p] - \sum[n_r(\Delta H_f°)_r]$$

Endothermic reactions are favored by an *increase* in temperatures, and **exothermic reactions** by a *decrease* in temperature. If reaction *decreases* the number of moles of gas in the system, *increased pressure* **favors formation of products**, and if reaction *increases* the number of moles of gas in the system, *decreased pressure* favors the **formation of products** at equilibrium.

55. For the equilibrium, C_2H_4(g) + H_2(g) ⇄ C_2H_6(g), ΔH = -137 kJ. How will each of the following changes affect the equilibrium concentration of C_2H_6(g)? (a) doubling the volume of the reaction vessel; (b) increasing the temperature at constant pressure; (c) adding more H_2 to the reaction vessel; (d) increasing the pressure by adding helium to the reaction vessel at constant volume.

This problem is similar to Problem 54, except that you are told that it is an *exothermic reaction*:

(a) Doubling the volume of the reaction vessel decreases the pressure and, thus, favors the *reverse* reaction because it is this reaction which tends to increase the pressure and thus oppose the change. The consequence is that the yield of $C_2H_6(g)$ **decreases.**

(b) Increasing the temperature will favor the *endothermic* reverse reaction and **decrease** the amount of $C_2H_6(g)$.

(c) When $H_2(g)$ is added, $Q < K$, and the position of equilibrium has to be restored by forming **more** $C_2H_6(g)$.

(d) The partial pressures of reactants or products are unaffected; addition of He(g) will have **no effect.**

Heterogeneous Equilibria

Note: We express equilibrium constant expressions in concentrations **mol L^{-1}** for **K_c**, and in **partial pressures** for **K_p**. but for **heterogeneous equilibria** we omit any component that is a pure liquid or solid.

56. Write the expressions for the equilibrium constants K_c and K_p for each of the following reactions:
(a) $H_2O(g) \rightleftarrows H_2(g) + \frac{1}{2}O_2(g)$; (b) $H_2O(\ell) \rightleftarrows H_2(g) + \frac{1}{2}O_2(g)$; (c) $2H_2O(\ell) \rightleftarrows 2H_2(g) + O_2(g)$;
(d) $H_2(g) + O_2(g) \rightleftarrows H_2O_2(\ell)$; (e) $2HgO(s) \rightleftarrows 2Hg(\ell) + O_2(g)$.

(a) $K_p = \dfrac{p_{H_2} \cdot p_{O_2}^{1/2}}{p_{H_2O}}$; $K_c = \dfrac{[H_2][O_2]^{1/2}}{[H_2O]}$ (b) $K_p = p_{H_2} \cdot p_{O_2}^{1/2}$; $K_c = [H_2][O_2]^{1/2}$

(c) $K_p = p_{H_2}^2 \cdot p_{O_2}$; $K_c = [H_2]^2[O_2]$ (d) $K_p = \dfrac{1}{p_{H_2} \cdot p_{O_2}}$; $K_c = \dfrac{1}{[H_2][O_2]}$ (e) $K_p = p_{O_2}$; $K_c = [O_2]$

---

58. In which direction will the equilibrium, $C(s) + 2H_2(g) \rightleftarrows CH_4(g)$, $\Delta H° = -75$ kJ, shift in response to each of the following changes in conditions? (a) increasing the temperature; (b) increasing the volume of the reaction vessel; (c) increasing the partial pressure of hydrogen; (d) adding more carbon.

(a) The reaction as written is exothermic; increasing the temperature favors the endothermic **reverse reaction.**

(b) Increasing the volume of the reaction vessel decreases the pressure, which favors the **reverse reaction,** because it is this reaction which best counters the decrease in pressure.

(c) Increasing the partial pressure of $H_2(g)$ favors the **forward reaction**; it gives a decreased value for Q which can only be increased again by increasing the concentration of product.

(d) Increasing the amount of carbon has **no effect,** because it does not appear in the equilibrium constant expression for this *heterogeneous equilibrium* and its concentration is constant and unaffected by its amount.

Gibbs Free Energy and the Equilibrium Constant

Remember that $\Delta G°$, the standard reaction free energy is calculated from standard free energies of formation, $\Delta G_f°$, and the number of moles of reactants and products in the balanced equation, using:

$$\Delta G° = \sum [n_p(\Delta G_f°)_p] - \sum [n_r(\Delta G_f°)_r]$$

To calculate K_p from ΔG, we use the formula $K_p = e^{-\Delta G/RT}$, (with $R = 8.314$ J K^{-1} mol^{-1}).

---

60. (a) From the data in Table 9.4, calculate the standard reaction Gibbs free energy for the reaction: $2SO_2(g) + O_2(g) \rightarrow 2SO_3(g)$; (b) What is the value of K_p at 298 K? (c) Is this reaction spontaneous at 298 K?

(a) For the reaction as written:

$$\Delta G° = [2\Delta G_f°(SO_3,g)] - [2\Delta G_f°(SO_2,g) + \Delta G_f°(SO_2,g)] = [2(-371.1)] - [2(-300.1) + (0)] = \mathbf{-142.0\ kJ}$$

(b)

$$-\frac{\Delta G°}{RT} = \frac{-(-142.0 \text{ kJ mol}^{-1})(\frac{10^3 \text{ J}}{1 \text{ kJ}})}{(8.314 \text{ J K}^{-1} \text{ mol}^{-1})(298 \text{ K})} = \underline{57.3} \ ; \ K_p = e^{57.3} = \underline{7.7 \times 10^{24} \text{ atm}^{-1}}$$

(c) The reaction is **spontaneous**, as indicated by the large negative value of $\Delta G°$, and confirmed by the very large value of K_p.

---

62. Consider the reaction, $CH_4(g) + 2O_2(g) \rightarrow CO_2(g) + 2H_2O(g)$: (a) Using data from Table 9.4, decide if this reaction is spontaneous at 298 K; (b) What is the value of K_p at 298 K? (c) Explain the fact that a mixture of methane and oxygen shows a negligible extent of reaction at room temperature, even after a very long time.

(a)

$$\Delta G° = [\Delta G_f°(CO_2,g) + 2\Delta G_f°(H_2O,g)] - [\Delta G_f°(CH_4,g) + 2\Delta G_f°(O_2,g)]$$
$$= [(-394.4) + (2(-228.6)] - [(-50.8) + 2(0)] = \mathbf{-800.8\ kJ}$$

$\Delta G°$ is <u>negative</u>, so the reaction is **spontaneous.**

(b) At 25°C (298 K), we can write:

$$\frac{-\Delta G}{RT} = -\frac{(-800.8 \text{ kJ})(\frac{10^3 \text{ J}}{1 \text{ kJ}})}{(8.314 \text{ J K}^{-1})(298 \text{ K})} = \underline{323} \ ; \ K_p = e^{323} = \underline{10^{140} \text{ atm}}$$

(c) Although the reaction is spontaneous, it is very slow at room temperature and proceeds to equilibrium negligibly slowly. Nevertheless, when initiated by a spark methane ignites and burns readily.

64*. For the reaction, $2C(s,\text{graphite}) + H_2(g) \rightarrow C_2H_2(g)$, the standard reaction Gibbs free energy is 209 kJ. (a) Is this reaction a practical route for the synthesis of ethyne (acetylene), $C_2H_2(g)$, at room temperature?
(b) Would the reaction be expected to be spontaneous at high temperature? (c) Assuming that $\Delta H°$ and $\Delta S°$ are independent of temperature, calculate the equilibrium constant K_p at 1200 K.

(a) Since $\Delta G° > 0$, the reaction is **not spontaneous**; the value of the equilibrium constant will be small and only a very small amount of $C_2H_2(g)$ will be present in the equilibrium mixture at room temperature. This **is not** a practical route at room temperature for the synthesis of acetylene.

(b) The reaction will be spontaneous at high temperature if ΔG becomes negative, which we can assess by first calculating $\Delta H°$ and $\Delta S°$ for the reaction and then determining what happens to $\Delta G = \Delta H° - T\Delta S°$ as the temperature increases, assuming that $\Delta H°$ and $\Delta S°$ are temperature independent:

$$\Delta H° = [\Delta H_f°(C_2H_2,g)] - [2\Delta H_f°(C,s,gr) + \Delta H_f°(H_2,g)] = [(228.0)] - [2(0) + (0)] = \mathbf{228.0\ kJ.}$$
$$\Delta S° = [S°(C_2H_2,g)] - [2\Delta S°(C,s,gr) + \Delta S°(H_2,g)] = [(200.8)] - [2(5.8) + (130.6)] = \mathbf{58.6\ J\ K^{-1}.}$$

Thus, $\Delta G = \Delta H^0 - T.\Delta S° = 228.0$ kJ - T(0.0586 kJ), which we can solve to obtain **T** for $\mathbf{\Delta G = 0}$, because *above* this temperature ΔG *becomes negative.*

$$\Delta G = \Delta H^{\circ} - T.\Delta S^{\circ} = (228.0 \text{ kJ}) - (0.0586 \text{ kJ K}^{-1})T = 0$$

$$\text{Thus, } T = \frac{228.0 \text{ kJ}}{0.0586 \text{ kJ K}^{-1}} = \underline{3890 \text{ K}}$$

(c) At 1200 K, ΔG = [228.0 kJ - (1200 K)(0.0586 kJ K^{-1})] = **157.7 kJ**, and at this temperature we can write:

$$\frac{-\Delta G}{RT} = -\frac{(-157.7 \text{ kJ})(\frac{10^3 \text{ J}}{1 \text{ kJ}})}{(8.314 \text{ J K}^{-1})(1200 \text{ K})} = \underline{-15.8} \; ; \; K_p = e^{-15.8} = \underline{1.4 \times 10^{-7}}$$

General Problems

66. From the data that follows: (a) Calculate the pH of pure water at the temperatures given, and (b) deduce whether the autoionization of water is exothermic or endothermic.

t (°C)	0	10	30	50	100
$K_w \times 10^{14}$ (mol² L⁻²)	0.115	0.293	1.47	5.48	51.3

(a) $K_w = [H_3O^+][OH^-]$, and in pure water $[H_3O^+] = [OH^-]$, so $[H_3O^+]^2 = K_w$ **for pure water.**
and taking negative logarithms of each side: $-2\log [H_3O^+] = -\log K_w$, or **pH = ½p$K_w$**.

t (°C)	0	10	30	50	100
$K_w \times 10^{14}$ ($mol^2 L^{-2}$)	0.115	0.293	1.47	5.48	51.3
pK_w	14.94	14.53	13.83	13.26	12.29
pH = ½pK_w	**7.47**	**7.27**	**6.92**	**6.63**	**6.15**

(b) According to Le Châtelier's principle, the decrease in K_w with increasing temperature means that the forward reaction evolves heat. Thus, the self-ionization of water is an **exothermic reaction.**

---

68. Which of the following are buffer solutions? Explain. (a) 25 mL 0.10-M HNO_3(aq) and 25 mL 0.10-M $NaNO_3$(aq); (b) 25 mL 0.10-M HNO_2(aq) and 25 mL 0.10-M $NaNO_2$(aq); (c) 25 mL 0.10-M acetic acid and 25 mL 0.15-M KOH(aq); (d) 25 mL 0.10-M acetic acid and 25 mL 0.05-M KOH(aq).

A **buffer solution** contains a *weak acid* and *its conjugate base* in approximately equal amounts.

(a) Nitric acid, HNO_3(aq), is a *strong acid*, so this solution is a solution containing H_3O^+(aq), NO_3^-(aq), and Na^+(aq), and is **not** a buffer solution.

(b) Nitrous acid, HNO_2(aq), is a *weak acid*, so this solution contains a **weak acid** and a **salt of the weak acid** and is a **buffer solution.**

(c) The *weak acid* acetic acid, $CH_3CO_2H(aq)$, reacts with KOH(aq) to give the *salt* potassium acetate, $CH_3CO_2K(aq)$, but is a buffer solution only if the final solution contains both $CH_3CO_2H(aq)$ and $CH_3CO_2K(aq)$. We have initially:

$$25 \text{ mL } 0.10\text{-M acetic acid} = (25 \text{ mL})(0.10 \text{ mol L}^{-1}) = \underline{2.5 \text{ mmol}}\ CH_3CO_2H$$

$$25 \text{ mL } 0.15\text{-M KOH(aq)} = (25 \text{ mL})(0.15 \text{ mol L}^{-1}) = \underline{3.8 \text{ mmol}}\ KOH$$

	$CH_3CO_2H(aq)$ +	$KOH(aq) \rightarrow$	$CH_3CO_2K(aq)$ +	$H_2O(\ell)$	
initially:	2.5	3.8	0	-	mmol
finally:	0	1.3	2.5	-	mmol

Thus, the final solution contains potassium acetate and unreacted strong base KOH(aq), the *salt of a weak acid plus a strong base*, which is **not** a buffer solution.

(d) As in part (c) we have to determine the constituents of the final solution. Initially, we have:

$$25 \text{ mL } 0.10\text{-M acetic acid} = (25 \text{ mL})(0.10 \text{ mol L}^{-1}) = \underline{2.5 \text{ mmol}}\ CH_3CO_2H$$

$$25 \text{ mL } 0.05\text{-M KOH(aq)} = (25 \text{ mL})(0.05 \text{ mol L}^{-1}) = \underline{1.3 \text{ mmol}}\ KOH$$

	$CH_3CO_2H(aq)$ +	$KOH(aq) \rightarrow$	$CH_3CO_2K(aq)$ +	$H_2O(\ell)$	
initially:	2.5	1.3	0	-	mmol
finally:	1.2	0	1.3	-	mmol

Thus, the final solution contains acetic acid and potassium acetate, *a weak acid and the salt of the weak acid*, which is a **buffer solution.**

70.* From Table 12.5, select suitable indicators for detecting the "equivalence" point in each of the following titrations: (a) 0.10-M NaOH(aq) with 0.10 M-HCl(aq); (b) 0.10-M acetic acid with 0.10-M NaOH(aq); (c) 0.10-M $NH_3(aq)$ with 0.10-M HCl(aq).

In each of these titrations, **moles of acid = moles of base** at the equivalence point, and the solutions are then 0.05-M solution of the respective salts, for which we can then calculate their corresponding pH values and select the appropriate indicators:

(a) The solution at the equivalence point is 0.05-M NaCl(aq), fully dissociated into $Na^+(aq)$ and $Cl^-(aq)$ ions. Neither Na^+ nor Cl^- (the conjugate base of the strong acid HCl(aq)) have acid or base properties in water. Thus, this solution is neutral with **pH 7.00**. A suitable indicator has $pK_a = pH \pm 1 = 7 \pm 1$. Bromothymol blue, $pK_a = 7.1$ would be ideal.

(b) The solution at the equivalence point is 0.05-M $CH_3CO_2Na(aq)$, fully dissociated into $Na^+(aq)$ and acetate, $CH_3CO_2^-(aq)$, ions. Na^+ has no acid or base properties, whereas $CH_3CO_2^-$ (the conjugate base of the weak acid CH_3CO_2H) is a weak base in water and for this solution **pH > 7**. Calculation of the pH by the usual methods gives **pH = 8.72**. So, a suitable indicator has $pK_a = pH \pm 1 = 8.7 \pm 1$. **Thymol blue**, $pK_a = 8.2$, would be suitable.

(c) The solution at the equivalence point is 0.05-M $NH_4Cl(aq)$, fully dissociated into $NH_4^+(aq)$ and $Cl^-(aq)$ ions. Cl^- (the conjugate base of the strong acid HCl(aq)) has no basic properties in water, whereas NH_4^+ (the conjugate acid of the weak base NH_3) is a weak acid in water and for this solution **pH < 7**. Calculation of the pH by the usual methods gives **pH = 5.28**. So, a suitable indicator has $pK_a = pH \pm 1 = 5.3 \pm 1$. **Methyl red**, $pK_a = 5.2$, would be very suitable.

CHAPTER 13

Electrochemical Cells

Standard Reduction Potentials

1. Using Table 13.1, explain why (a) metals such as magnesium, aluminum, zinc, iron, and nickel react readily with dilute sulfuric acid, whereas (b) copper and silver react only with hot concentrated sulfuric acid.

(a) Mg, Al, Zn, Fe, and Ni all have standard oxidation potentials:

$$Mg^{2+}(aq) + 2e^- \rightarrow Mg(s), \quad E_{red}^\circ = -2.36\ V$$
$$Mg(s) \rightarrow Mg^{2+}(aq) + 2e^-, \quad E_{ox}^\circ = +2.36\ V$$

$$Al^{3+}(aq) + 3e^- \rightarrow Al(s), \quad E_{red}^\circ = -1.66\ V$$
$$Al(s) \rightarrow Al^{3+}(aq) + 2e^-, \quad E_{ox}^\circ = +1.66\ V$$

$$Zn^{2+}(aq) + 2e^- \rightarrow Zn(s), \quad E_{red}^\circ = -0.76\ V$$
$$Zn(s) \rightarrow Zn^{2+}(aq) + 2e^-, \quad E_o^\circ = +0.76\ V$$

$$Fe^{2+}(aq) + 2e^- \rightarrow Fe(s), \quad E_{red}^\circ = -0.44\ V$$
$$Fe(s) \rightarrow Fe^{2+}(aq) + 2e^-, \quad E_{ox}^\circ = +0.44\ V$$

$$Ni^{2+}(aq) + 2e^- \rightarrow Ni(s), \quad E_{red}^\circ = -0.25\ V$$
$$Ni(s) \rightarrow Ni^{2+}(aq) + 2e^-, \quad E_{ox}^\circ = +0.25\ V$$

that are more positive than that the reduction potential of $H_3O^+(aq)$:

$$2H_3O^+(aq) + 2e^- \rightarrow H_2(g) + 2H_2O(\ell), \quad E_{red}^\circ = 0.00\ V$$

so all these metals are oxidized by $H_3O^+(aq)$ ions in aqueous solution because combination of their oxidation half-reactions with the reduction half-reaction for $H_3O^+(aq)$ gives *overall* a **positive** cell potential.

In contrast, the standard oxidation potential of copper and silver:

$$Cu^{2+}(aq) + 2e^- \rightarrow Cu(s), \quad E_{red}^\circ = +0.34\ V$$
$$Cu(s) \rightarrow Cu^{2+}(aq) + 2e^-, \quad E_{ox}^\circ = -0.34\ V$$

$$Ag^+(aq) + e^- \rightarrow Ag(s), \quad E_{red}^\circ = +0.80\ V$$
$$Ag(s) \rightarrow Ag^+(aq) + e^-, \quad E_{ox}^\circ = -0.80\ V$$

are both more negative than the reduction potential of $H_3O^+(aq)$, so these metals are not oxidized by $H_3O^+(aq)$ ions in aqueous solution, because combination of their oxidation half-reactions with the reduction half-reaction for $H_3O^+(aq)$ gives a **negative** cell potential, and it is the *reverse reaction* that is spontaneous.

Redox potentials are not valid for concentrated aqueous solution, but since both Cu and Ag are oxidized by concentrated sulfuric acid, which contains H_2SO_4 molecules which are a stronger oxidizing agent than H_3O^+ ions, we can conclude that the reduction potential, E_{red} for the half-reaction

$$H_2SO_4(\ell) + 2e^- + 2H^+(aq) \rightarrow SO_2(aq) + 2H_2O(\ell)$$

is **at least +0.80 V**, since combination of this reduction potential with the *oxidation potentials* of either copper or silver must give a **positive** cell potential.

---

2. Using Table 13.1, explain why copper dissolves in dilute nitric acid but not in dilute hydrochloric acid.

In dilute HCl(aq), the oxidizing agent could be $H_3O^+(aq)$ or $Cl^-(aq)$ ions, while in dilute nitric acid it could be $H_3O^+(aq)$ or $NO_3^-(aq)$. From Table 13.1, the E°_{ox} value (= $-E^\circ_{red}$) for the oxidation half-reaction of copper is:

$$Cu \rightarrow Cu^{2+} + 2e^-, \quad E^\circ_{ox} = -0.34\ V \quad .... (1)$$

which we can combine with any of the reduction half-reactions

$$2H_3O^+ + 2e^- \rightarrow H_2(g) + 2H_2O, \quad E^\circ_{red} = 0.00\ V \quad .... (2)$$
$$NO_3^- + 2H^+ + e^- \rightarrow NO_2(g) + H_2O, \quad E^\circ_{red} = +0.80\ V \quad .... (3)$$
$$NO_3^- + 4H^+ + 3e^- \rightarrow NO(g) + 2H_2O, \quad E^\circ_{red} = +0.97\ V \quad .... (4)$$

For (1) and (2): $E°_{cell} = E°_{red} + E°_{ox} = -0.34$ V, which is **negative,** so no reaction occurs.

For (1) and (3): $E°_{cell} = E°_{red} + E°_{ox} = +0.46$ V, and for (1) and (4), $E°_{cell} = E°_{red} + E°_{ox} = +0.63$ V, so both of these reactions are spontaneous under standard conditions.

In conclusion: Copper does not react with HCl(aq), but with HNO_3(aq), the possible reactions are:

$$Cu + 2HNO_3 \rightarrow Cu^{2+} + 2NO_2(g) + H_2O(\ell)$$

$$3Cu + 2HNO_3 + 6H^+ \rightarrow 3Cu^{2+} + 2NO(g) + 4H_2O(\ell)$$

---

4. Place 1.00-M solutions of the metal ions Ca^{2+}, Fe^{3+}, Al^{3+}, Cu^{2+}, Ni^{2+}, and Na^+ in order of increasing oxidizing strength.

Oxidizing strength (the ease of electron gain) increases as the standard reduction potential, $E°_{red}$, increases, so that the order of oxidizing strengths is the order in which the $E°_{red}$ values for the ions in question becomes more positive:

$$Ca^{2+} + 2e^- \rightarrow Ca(s) \qquad E°_{red} = -2.87 \text{ V}$$

$$Na^+ + e^- \rightarrow Na(s) \qquad E°_{red} = -2.71 \text{ V}$$

$$Al^{3+} + 3e^- \rightarrow Al(s) \qquad E°_{red} = -1.66 \text{ V}$$

$$Ni^{2+} + 2e^- \rightarrow Ni(s) \qquad E°_{red} = -0.25 \text{ V}$$

$$Cu^{2+} + 2e^- \rightarrow Cu(s) \qquad E°_{red} = +0.34 \text{ V}$$

$$Fe^{3+} + e^- \rightarrow Fe^{2+} \qquad E°_{red} = +0.77 \text{ V}$$

i.e., $Ca^{2+} < Na^+ < Al^{3+} < Ni^{2+} < Cu^{2+} < Fe^{3+}$

---

6. Under standard conditions, which of the following will be oxidized by dichromate ion, $Cr_2O_7^{2-}$, in acidic aqueous solution? (a) F^-(aq) (b) Cl^-(aq) (c) Br^-(aq) (d) I^-(aq) (e) Fe^{2+}(aq)

In each case we calculate $E°_{cell}$ for the reaction by adding together $E°_{red}$ for the half-reaction for the reduction of $Cr_2O_7^{2-}$(aq) to Cr^{3+}(aq), and $E°_{ox}$ for the half-reaction in which the species in question is oxidized in aqueous solution:

(a)

$$Cr_2O_7^{2-} + 6e^- + 14H^+ \rightarrow 2Cr^{3+} + 7H_2O \qquad E°_{red} = +1.33 \text{ V}$$

$$[2F^- \rightarrow F_2(g) + 2e^-] \qquad E°_{ox} = -E°_{red} = -2.87 \text{ V}$$

$$Cr_2O_7^{2-} + 6F^- + 14H^+ \rightarrow 2Cr^{3+} + 3F_2 + 7H_2O \qquad E°_{cell} = -1.54 \text{ V}$$

$E°_{cell}$ is **negative**, so we conclude that the reaction as written is **not spontaneous** (although the reverse reaction would be), so that fluoride ion is **not** oxidized by dichromate ion in aqueous solution under standard conditions.

(b)

$$Cr_2O_7^{2-} + 6e^- + 14H^+ \rightarrow 2Cr^{3+} + 7H_2O \qquad E°_{red} = +1.33 \text{ V}$$

$$[2Cl^- \rightarrow Cl_2(g) + 2e^-] \qquad E°_{ox} = -E°_{red} = -1.36 \text{ V}$$

$$Cr_2O_7^{2-} + 6Cl^- + 14H^+ \rightarrow 2Cr^{3+} + 3Cl_2 + 7H_2O \qquad E°_{cell} = -0.03 \text{ V}$$

$E°_{cell}$ is **negative**, so we conclude that the reaction as written is *not spontaneous* (although the reverse reaction would be), so that chloride ion is **not** oxidized by dichromate ion in aqueous solution under standard conditions.

(c)

$$Cr_2O_7^{2-} + 6e^- + 14H^+ \rightarrow 2Cr^{3+} + 7H_2O \qquad E°_{red} = +1.33 \text{ V}$$

$$3[2Br^- \rightarrow Br_2(g) + 2e^-] \qquad E°_{ox} = -E°_{red} = -1.09 \text{ V}$$

$$Cr_2O_7^{2-} + 6Br^- + 14H^+ \rightarrow 2Cr^{3+} + 3Br_2 + 7H_2O \qquad E°_{cell} = +0.24 \text{ V}$$

$E°_{cell}$ is **positive**, so we conclude that the reaction as written is *spontaneous*; bromide ion is oxidized to Br_2 by dichromate ion in aqueous solution under standard conditions.

(d)

$$Cr_2O_7^{2-} + 6e^- + 14H^+ \rightarrow 2Cr^{3+} + 7H_2O \qquad E^\circ_{red} = +1.33\ V$$
$$3[2I^- \rightarrow I_2(g) + 2e^-] \qquad E^\circ_{ox} = -E^\circ_{red} = -0.54\ V$$
$$Cr_2O_7^{2-} + 6Br^- + 14H^+ \rightarrow 2Cr^{3+} + 3Br_2 + 7H_2O \qquad E^\circ_{cell} = +0.79\ V$$

E°_{cell} is **positive**, so we conclude that the reaction as written is *spontaneous*; iodide ion is oxidized to I_2 by dichromate ion in aqueous solution under standard conditions.

(e)

$$Cr_2O_7^{2-} + 6e^- + 14H^+ \rightarrow 2Cr^{3+} + 7H_2O \qquad E^\circ_{red} = +1.33\ V$$
$$6[Fe^{2+} \rightarrow Fe^{3+} + e^-] \qquad E^\circ_{ox} = -E^\circ_{red} = -0.77\ V$$
$$Cr_2O_7^{2-} + 6Fe^{2+} + 14H^+ \rightarrow 2Cr^{3+} + 6Fe^{3+} + 7H_2O \qquad E^\circ_{cell} = +0.56\ V$$

E°_{cell} is **positive**, so we conclude that the reaction as written is *spontaneous*; Fe^{2+} ion is oxidized to I_2 by dichromate ion in aqueous solution under standard conditions.

Alternatively, we could have obtained the answers more easily simply by considering the relative positions of each half-reaction written as **a reduction** in the table of standard reduction potentials. Any half-reaction has a greater tendency to proceed to the **left** than any reaction **below** it in the table, In this case, only half-reactions with E°_{red} values **above** that for the reduction of $Cr_2O_7^{2-}(aq)$ to $Cr^{3+}(aq)$ will be spontaneous under standard conditions:

In summary: $Br^-(aq)$, $I^-(aq)$, and $Fe^{2+}(aq)$ are **oxidized** by $Cr_2O_7^{2-}(aq)$ in acidic solution.

---

7. Use standard half-cell reduction potentials to predict which of the following reactions will occur at standard conditions in acidic aqueous solution. For each predicted reaction, give the balanced equation:
(a) $H_2O_2(aq) + Cu^{2+}(aq) \rightarrow Cu(s) + O_2(g)$; (b) $Ag^+(aq) + Fe^{2+}(aq) \rightarrow Ag(s) + Fe^{3+}(aq)$;
(c) $I^-(aq) + NO_3^-(aq) \rightarrow I_2(s) + NO(g)$

We use the method described in the the answer to Problem 6, with the relevant E°_{red} values from Table 13.1:

(a)

$$H_2O_2 \rightarrow O_2 + 2H^+ + 2e^- \qquad E^\circ_{ox} = -0.68\ V$$
$$Cu^{2+} + 2e^- \rightarrow Cu(s) \qquad E^\circ_{red} = +0.34\ V$$
$$H_2O_2 + Cu^{2+} \rightarrow Cu(s) + O_2 + 2H^+ \qquad E^\circ_{cell} = -0.34\ V$$

E°_{cell} is negative; the reaction as written is not expected to occur under standard conditions.

(b)

$$Fe^{2+} \rightarrow Fe^{3+} + e^- \qquad E^\circ_{ox} = -0.77\ V$$
$$Ag^+ + e^- \rightarrow Ag(s) \qquad E^\circ_{red} = +0.80\ V$$
$$Ag^+ + Fe^{2+} \rightarrow Ag(s) + Fe^{3+} \qquad E^\circ_{cell} = +0.03\ V$$

E°_{cell} is positive; the reaction as written is expected to be spontaneous under standard conditions.

(c)

$$3[2I^- \rightarrow I_2(s) + 2e^-] \qquad E^\circ_{ox} = -0.54\ V$$
$$2[NO_3^- + 3e^- + 4H^+ \rightarrow NO + 2H_2O] \qquad E^\circ_{red} = +0.97\ V$$
$$6I^- + 2NO_3^- \rightarrow 3I_2(s) + 2NO(g) + 4H_2O \qquad E^\circ_{cell} = +0.43\ V$$

E°_{cell} is positive; the reaction as written is expected to be spontaneous under standard conditions.

---

9. Predict which of each of the following reactions occurs under standard conditions. For those that do take place, complete and balance the equations: (a) $Mn^{2+}(aq) + Cr_2O_7^{2-}(aq) \rightarrow MnO_4^-(aq) + Cr^{3+}(aq)$ (acidic solution); (b) $O_2(g) + Br^-(aq) \rightarrow Br_2(aq)$ (acidic solution); (c) $Br_2(aq) + Cl^-(aq) \rightarrow Br^-(aq) + Cl_2(aq)$; (d) $I^-(aq) + Cl_2(aq) \rightarrow I_2(aq) + Cl^-(aq)$.

(a) In Table 13.1, the half-reaction for reduction of MnO_4^- to Mn^{2+} lies below the half-reaction for the reduction of $Cr_2O_7^{2-}$, thus **no reaction occurs** under standard conditions.

(b) The product of the reduction of O_2 is not specified, but it must be water, H_2O. From Table 13.1, the half-reaction for the reduction of $O_2(g)$ to H_2O lies below the half-reaction for the reduction of Br_2 to Br^-, thus $O_2(g)$ oxidizes $Br^-(aq)$ to $Br_2(aq)$, and is reduced to $H_2O(\ell)$, under standard conditions.

$O_2(g) + 4e^- + 4H^+ \rightarrow 2H_2O$	$E°_{red} = +1.23$ V
$2[2Br^- \rightarrow Br_2 + 2e^-]$	$E°_{ox} = -1.09$ V
$O_2(g) + 4Br^- + 4H^+ \rightarrow 2H_2O + 2Br_2$	$E°_{cell} = +0.14$ V

(c) In Table 13.1, the half-reaction for reduction of Br_2 to Br^- lies below the half-reaction for the reduction of Cl_2 to Cl^- thus no reaction occurs under standard conditions.

(d) From Table 13.1, the half-reaction for the reduction of Cl_2 to Cl^- lies below the half-reaction for the reduction of I_2 to I^-, thus $Cl_2(aq)$ oxidizes $I^-(aq)$ to $I_2(aq)$, and is reduced to $Cl^-(aq)$, under standard conditions.

$Cl_2 + 2e^- \rightarrow 2Cl^-$	$E°_{red} = +1.36$ V
$2I^- \rightarrow I_2 + 2e^-$	$E°_{ox} = -0.54$ V
$Cl_2 + 2I^- \rightarrow 2Cl^- + I_2$	$E°_{cell} = +0.82$ V

---

10. Predict which of the following metals should be oxidized by $O_2(g)$ at a pressure of 1 atm in 1 M acidic aqueous solution at room temperature: (a) Ag (b) Cu (c) Ca (d) Zn (e) Al

---

From Table 13.1, the relevant **standard oxidation** potentials **are:**

(a)	$Ag(s) \rightarrow Ag^+ + e^-$;	$E°_{ox} = -E°_{red} = -0.80$ V
(b)	$Cu(s) \rightarrow Cu^{2+} + 2e^-$;	$E°_{ox} = -E°_{red} = -0.34$ V
(c)	$Ca(s) \rightarrow Ca^{2+} + 2e^-$;	$E°_{ox} = -E°_{red} = +2.87$ V
(d)	$Zn(s) \rightarrow Zn^{2+} + 2e^-$;	$E°_{ox} = -E°_{red} = +0.76$ V
(e)	$Al(s) \rightarrow Al^{3+} + 3e^-$;	$E°_{ox} = -E°_{red} = +1.66$ V

From Table 13.1, $E°_{red}$ for the half-reaction, $O_2 + 4e^- + 4H^+ \rightarrow 2H_2O$, has the value +1.23 V, so any metal with $E°_{ox} > -1.23$ **V** will be oxidized by oxygen in acidic solution under standard conditions. Inspection of the above values shows that **all of these metals will be oxidized under these conditions.**

---

13. (a) Illustrate the construction of the electrochemical cell $Zn(s)|Zn^{2+}(aq)\|Ni^{2+}(aq)|Ni(s)$, showing the direction of electron flow in the external circuit. (b) What is the standard cell voltage, $E°_{cell}$?

(a) See Problem 12. Electrons move through the external circuit from the zinc anode to the Ni anode.

(b) The half-cell reactions, the overall cell-reaction, and the standard cell voltage are:

<u>cathode:</u>	$Ni^{2+} + 2e^- \rightarrow Ni(s)$	$E°_{red} = +0.25$ V
<u>anode:</u>	$Zn(s) \rightarrow Zn^{2+} + 2e^-$	$E°_{ox} = +0.76$ V
	$Zn(s) + 2Ag^+ \rightarrow 2Ag(s) + Zn^{2+}$	**$E°_{cell} = +1.01$ V**

---

15. Consider the reaction $Fe(s) + 2H^+ \rightarrow Fe^{2+}(aq) + H_2(g)$: (a) Draw an electrochemical cell in which this reaction takes place; (b) What are the charge carriers in the wire that connnects the two electrodes; (c) At which electrode does reduction occur? Is this the anode or the cathode? (d) Give the equation for the reaction that occurs at the cathode

---

(a) The cell is **$Fe(s)|Fe^{2+}(aq)\|H^+(aq)|H_2(g)$, Pt(s)**, in which iron is the anode in the anode compartment, and hydrogen gas bubbling over a platinum wire is the cathode in the cathode compartment. The cell diagram should show the two compartments joined by a "salt bridge", with electrons moving in the external circuit (a wire joining the iron and platinum electrodes) from the iron anode to the platinum cathode.

(b) Electrons carry the charge in the external circuit (the wire joining the iron and platinum electrodes).

(c) **Reduction** occurs at the **Pt cathode**, where H^+ ions are reduced to $H_2(g)$, (and **oxidation** occurs at the **Fe anode**, where Fe(s) is oxidized to $Fe^{2+}(aq)$ ions).

(d) The reaction at the **cathode** is $\mathbf{2H^+(aq) + 2e^- \rightarrow H_2(g)}$.

---

16. Draw the experimental arrangements for each of the standard electrochemical cells for which the overall reactions are given as follows. In each case, place the anode compartment on the left and indicate the direction of electron flow in the external circuit:

(a) $Zn(s) + Br_2(aq) \rightarrow Zn^{2+}(aq) + 2Br^-(aq)$; (b) $Pb(s) + 2Ag^+(aq) \rightarrow Pb^{2+}(aq) + 2Ag(s)$

In **each cell** the electrode at which **oxidation** occurs (the **anode**) is placed on the **left**, and the electrode where **reduction** occurs (the **cathode**) goes on the **right**. In each **half-cell**, the contact between the electrode and the solution in its half-cell is denoted by |, and the **salt bridge** joining the cells is denoted by ‖. **The electrons in the exernal circuit should be shown flowing from the *anode* to the *cathode*.**

(a) Zn(s) is oxidized at the anode ($Zn(s) \rightarrow Zn^{2+}(aq) + 2e^-$), and in all probability forms this electrode, and $Br^-(aq)$ is reduced at the cathode ($Br_2(aq) + 2e^- \rightarrow 2Br^-(aq)$), where an inert electrode in contact with the bromine is required, which is probably a platinum wire. Thus, the cell diagram is:

$$\mathbf{Zn(s)|Zn^{2+}(aq)\|Br^-(aq)|Br_2(aq),Pt(s)}$$

(b) Pb(s) functions as the anode where Pb(s) is oxidized ($Pb(s) \rightarrow Pb^{2+}(aq) + 2e^-$), and in all probability the cathode is Ag(s), where $Ag^+(aq)$ is reduced ($Ag^+(aq) + e^- \rightarrow Ag(s)$). Thus, the cell diagram is:

$$\mathbf{Pb(s)|Pb^{2+}(aq)\|Ag^+(aq)|Ag(s)}$$

---

18. What are the standard reaction Gibb's free energies for each of the reactions in Problem 14?

(a) $Al(s)|Al^{3+}(aq)\|Cu^{2+}(aq)|Cu(s)$ (b) $Pb(s)|Pb^{2+}(aq)\|Ag^+(aq)|Ag(s)$ (c) $Ag(s)|Ag^+(aq)\|Cl^-(aq)|Cl_2(g),Pt(s)$

Note: Recollect that standard reaction Gibb's free energy and the standard cell potential are related by:

$$\mathbf{\Delta G^\circ = -nFE^\circ_{cell}}$$

where **n** is the number of electrons transferred in the reaction, **F** is the Faraday constant (96 485 C · mol^{-1}), and **E°_{cell} is the standard cell potential. Substituting 1 F = 96.485 kJ · V^{-1} · mol-1**, with E°_{cell} in *volts*, gives

$$\mathbf{\Delta G^\circ = -96.485(nE^\circ_{cell})\ kJ}$$

(a) For $\mathbf{2Al(s) + 3Cu^{2+}(aq) \rightarrow 2Al^{3+}(aq) + 3Cu(s)}$, E°_{cell} = **2.00** V, and **n** = **6**, since six electrons are transferred in the reaction as written ($2Al \rightarrow 2Al^{3+} + 6e^-$). Thus:

$$\Delta G^\circ = 6(-96.485\ kJ \cdot V^{-1})(2.00\ V) = \mathbf{-1.16 \times 10^3\ kJ}$$

(b) For $\mathbf{2Ag^+(aq) + Pb(s) \rightarrow 2Ag(s) + Pb^{2+}(aq)}$, E°_{cell} = **0.93** V, and **n** = **2**, since two electrons are transferred in the reaction as written ($Pb(s) \rightarrow Pb^{2+}(aq) + 2e^-$). Thus:

$$\Delta G^\circ = 2(-96.485\ kJ \cdot V^{-1})(0.93\ V) = \mathbf{-179\ kJ}$$

(c) For $\mathbf{2Ag(s) + Cl_2(g) \rightarrow 2Ag^+(aq) + 2Cl^-(aq)}$, E°_{cell} = **0.56** V, and **n** = **2**, since two electrons are transferred in the reaction as written ($2Ag(s) \rightarrow 2Ag^+(aq) + 2e^-$), Thus:

$$\Delta G^\circ = 2(-96.485\ kJ \cdot V^{-1})(0.56\ V) = \mathbf{-108\ kJ}$$

---

20. Separate beakers contain a piece of iron immersed in 1.00 M $FeSO_4$(aq), and a piece of copper immersed in 1.00 M $CuSO_4$(aq); the solutions are then connected by a salt bridge, and the metals are connected by a conducting wire: (a) Which metal dissolves, and which metal increases in mass? (b) What will be the initial voltage between the metals? (c) As the reaction proceeds, which of the two solutions increases in concentration, and which decreases? (d) Does the initial voltage increase, or decrease, with time?

Since $E°_{red}$ for $Cu^{2+} + 2e^- \rightarrow Cu(s)$ lies below $E°_{red}$ for $Fe^{2+} + 2e^- \rightarrow Fe(s)$ in Table 13.1, Cu^{2+}(aq) will be reduced and Fe(s) will be oxidized.

<u>cathode:</u>	$Cu^{2+} + 2e^- \rightarrow Cu(s)$	$E°_{red} = +0.34$ V
<u>anode:</u>	$Fe(s) \rightarrow Fe^{2+} + 2e^-$	$E°_{ox} = +0.44$ V
	$Cu^{2+} + Fe(s) \rightarrow Cu(s) + Fe^{2+}$	$E°_{cell} = \underline{+0.78 \text{ V}}$

The reaction proceeds *spontaneously* as written.

(a) Iron dissolves and the copper increases in mass.

(b) The initial voltage is $E°_{cell}$ = **+0.78 V**.

(c) Since iron dissolves to form Fe^{2+}(aq), the concentration of the Fe^{2+}(aq) solution **increases**, and the concentration of the Cu^{2+}(aq) solution **decreases**.

(d) Initially, $E°_{cell}$ = **+0.78 V**. As equilibrium is approached, the voltage **decreases** and at equilibrium $\mathbf{E_{cell} = 0}$.

---

21. Consider a room temperature electrochemical cell composed of Cu(s)|Cu^{2+}(aq) and Ag(s)|Ag^+(aq) half-cells. What is the cell voltage when: (a) all the ions have concentrations of 1 M; (b) [Cu^{2+}] is 2.0 M and [Ag^+] is 0.05 M

(a) Since $E°_{red}$ for $Ag^+ + e^- \rightarrow Ag(s)$ lies below $E°_{red}$ for $Cu^{2+} + 2e^- \rightarrow Cu(s)$ in Table 13.1, Ag^+(aq) will be reduced and Cu(s) will be oxidized:

<u>cathode:</u>	$2[Ag^+ + e^- \rightarrow Ag(s)]$	$E°_{red} = +0.80$ V
<u>anode:</u>	$Cu(s) \rightarrow Cu^{2+} + 2e^-$	$E°_{ox} = -0.34$ V
	$2Ag^+ + Cu(s) \rightarrow 2Ag(s) + Cu^{2+}$	$E°_{cell} = \underline{+0.46 \text{ V}}$

The reaction will proceed spontaneously as written and since the concentrations are initially 1 M, the initial voltage will be the standard cell voltage of +0.46 V.

(b) The cell reaction $2Ag^+(aq) + Cu(s) \rightarrow 2Ag(s) + Cu^{2+}(aq)$ is a **heterogeneous reactions**, so neither [Ag(s)] nor [Cu(s)] appear in the **reaction quotient expression, Q. and the number of moles of electrons transferred in the reaction is** n = 2. Using the *Nernst equation*, the voltage, E, is given by:

$$E = E° - \frac{RT}{nF}\ln Q = E° - \frac{0.0257}{2}\ln\frac{[Cu^{2+}]}{[Ag^+]^2} = 0.46\text{ V} - \frac{0.0257}{2}\ln\frac{2.0}{(0.05)^2} = (0.46 - 0.09)\text{ V} = \underline{0.37\text{ V}}$$

---

22. The voltage of the electrochemical cell Zn(s)|Zn^{2+}(aq,x M)‖Cr^{3+}(aq,0.001 M)|Cr(s) is 0.00 V at 25 °C. What is the Zn^{2+}(aq) concentration, x mol L^{-1}?

The cell diagram tells us that Zn(s) is oxidized and Cr^{3+} is reduced, so for this cell:

<u>cathode:</u>	$2[Cr^{3+} + 2e^- \rightarrow Cr(s)]$	$E°_{red} = -0.74$ V
<u>anode:</u>	$3[Zn(s) \rightarrow Zn^{2+} + 2e^-]$	$E°_{ox} = +0.76$ V
	$2Cr^{3+} + 3Zn(s) \rightarrow 2Cr(s) + 3Zn^{2+}$	$E°_{cell} = \underline{+0.02 \text{ V}}$

The reaction proceeds spontaneously as written and $E°_{cell}$ = **+0.02 V**.

The cell reaction $2Cr^{3+}(aq) + 3Zn(s) \rightarrow 2Cr(s) + 3Zn^{2+}(aq)$ is **heterogeneous**, so that neither [Zn(s)] nor [Cr(s)] appear in the reaction quotient expression, Q. and the number of moles of electrons transferred in the reaction is n = 6. Using the Nernst equation, for a voltage of 0.00 V we have:

$$E = E° - \frac{RT}{nF} \ln Q = E° - \frac{0.0257}{6} \ln \frac{[Zn^{2+}]^3}{[Cr^{3+}]^2} = 0.00 = 0.02\ V - \frac{0.0257}{6} \ln \frac{x^3}{(0.001)^2}$$

$$\text{Whence:} \quad \ln \frac{x^3}{(0.001)^2} = 4.67; \quad \frac{x^3}{(0.001)^2} = 1.07 \times 10^2\ ; \quad x^3 = 1.07 \times 10^{-4}\ ; \quad \underline{x = 0.047}$$

When $E_{cell} = 0$, the reaction has achieved equilibrium and $[Zn^{2+}]$ = **0.047 M.**

---

24. For the electrochemical cell $Pb(s)|Pb^{2+}(aq)\|Cu^{2+}(aq)|Cu(s)$ at 25°C, calculate the cell voltage for each of the following ion concentrations: (a) $[Pb^{2+}$, 1.0 M; $[Cu^{2+}]$, 1.0 M; (b) $[Pb^{2+}$, 1.0 M; $[Cu^{2+}]$, 1.0×10^{-5} M; (c) $[Pb^{2+}$, 1.0×10^{-3} M; $[Cu^{2+}]$, 1.0×10^{-2} M; (d) $[Pb^{2+}$, 6.0×10^{-5} M; $[Cu^{2+}]$, 2.0×10^{-2} M;

(a) The voltage will be $E°_{cell}$, since all reactants are in their standard states:

cathode: $Cu^{2+} + 2e^- \rightarrow Cu(s)$	$E°_{red} = +0.34$ V
anode: $Pb(s) \rightarrow Pb^{2+} + 2e^-$	$E°_{ox} = +0.47$ V
$Cu^{2+} + Pb(s) \rightarrow Cu(s) + Pb^{2+}$	$E°_{cell} = \underline{+0.47\ V}$

(b) The cell reaction $Cu^{2+}(aq) + Pb(s) \rightarrow Cu(s) + Pb^{2+}(aq)$ is **heterogeneous**, so that neither [Cu(s)] nor [Pb(s)] appear in the **reaction quotient expression, Q; the number of electrons transferred in the reaction** is n = 2. So for this calculation, and in subsequent parts, we use the Nernst equation in the form:

$$E = E° - \frac{RT}{nF} \ln Q = E° - \frac{0.0257}{2} \ln \frac{[Pb^{2+}]}{[Cu^{2+}]}$$

$$E = 0.47\ V - \frac{0.0257}{2} \ln \frac{1.0}{(1.0 \times 10^{-5})} = 0.47 - \frac{0.0257}{2} \ln (1.0 \times 10^5) = (0.47 - 0.18)\ V = \underline{0.29\ V}$$

$$\text{(c)} \quad E = 0.47\ V - \frac{0.0257}{2} \ln \frac{1.0 \times 10^{-3}}{1.0 \times 10^{-2}} = 0.47 - \frac{0.0257}{2} \ln 0.10 = [0.47 - (-0.03)]V = \underline{0.50\ V}$$

$$\text{(d)} \quad E = 0.47\ V - \frac{0.0257}{2} \ln \frac{6.0 \times 10^{-5}}{2.0 \times 10^{-2}} = 0.47 - \frac{0.0257}{2} \ln (0.0030) = [0.47 - (-0.07)]V = \underline{0.54\ V}$$

Batteries

28. A possible source of power for a heart pacemaker is to implant a zinc electrode and a platinum electrode directly into the body. When inserted into the oxygen-containing body fluid, these electrodes form a "biogalvanic" cell in which zinc if oxidized and oxygen is reduced: (a) Write equations for the anode and cathode reactions. (b) Estimate the standard voltage of such a cell. (c) If a current of 40 μA is drawn from the cell, how often will a zinc electrode of mass 5.0 g have to be replaced?

(a) Anode: $Zn(s) \rightarrow Zn^{2+}(aq)) + 2e^-$; cathode: $O_2(g) + 4H^+(aq) + 4e^- \rightarrow 2H_2O(\ell)$

(b) From the data in Table 13.1:

Oxidation	$2[Zn(s) \rightarrow Zn^{2+}(aq)) + 2e^-]$	$E°_{ox}$	= +0.76 V (anode)
Reduction	$O_2(g) + 4H^+(aq) + 4e^- \rightarrow 2H_2O(\ell)$	E_{ored}	= +1.23 V (cathode)
Overall	$2Zn(s) + O_2(g) + 4H^+(aq) \rightarrow 2Zn^{2+}(aq) + 2H_2O(\ell)$	$E°_{cell}$	= **1.99 V**

(c) In the limit, the cell loses its effectiveness when all of the zinc anode has been oxidized to $Zn^{2+}(aq)$, so we need to calculate the coulombs of electricity to achieve this, and, hence, assuming that the current remains close to 40 μA, the time it takes for 5.0 g of Zn to dissolve:

$$\text{Coulombs generated} = (5.0\text{ g Zn})\left(\frac{1\text{ mol Zn}}{65.38\text{ g Zn}}\right)\left(\frac{2\text{ mol e}^-}{1\text{ mol Zn}}\right)\left(\frac{96\ 500\text{ C}}{1\text{ mol e}^-}\right) = \underline{1.48 \times 10^4\text{ C}}$$

$$\text{Time taken} = \left(\frac{1.48 \times 10^4\text{ C}}{40\ \mu\text{A}}\right)\left(\frac{1\text{ A s}}{1\text{ C}}\right)\left(\frac{10^6\mu\text{A}}{1\text{ A}}\right)\left(\frac{1\text{ hr}}{3600\text{ s}}\right)\left(\frac{1\text{ day}}{24\text{ hr}}\right)\left(\frac{1\text{ year}}{365\text{ day}}\right) = \underline{12\text{ year}}$$

---

29. (a) Write the two half-reactions for a fuel cell in which ethane, $C_2H_6(g)$, is oxidized in acid solution by $O_2(g)$ to $CO_2(g)$ and water. (b) How many liters of ethane at STP would be needed to generate a current of 0.500 A for 6.00 h? (c) How many liters of oxygen at STP would be consumed at the cathode during this time?

(a) Carbon, oxidation number -3 in $C_2H_6(g)$, is oxidized to carbon oxidation number +4 in $CO_2(aq)$, and $O_2(g)$ is **reduced** to $H_2O(\ell)$, so the **oxidation** and **reduction** half-reactions are:

oxidation $C_2H_6(g) + 4H_2O(\ell) \rightarrow 2CO_2(aq) + 14H^+(aq) + 14e^-$

reduction $O_2(g) + 4H^+(aq) + 4e^- \rightarrow 2H_2O(\ell)$

and adding **twice** the first equation to **seven** times the second equation gives the ovell balanced equation:

$$2C_2H_6(g) + 7O_2(g) \rightarrow 4CO_2(aq) + 6H_2O(\ell)$$

which is a process in which 14 mol e^- are generated per mol of $C_2H_6(g)$ used.

(b) We first calculate the moles of electrons generated and then the mol $C_2H_6(g)$ consumed:

$$\text{Mol } C_2H_6 = (6.00\text{ hr})\left(\frac{3600\text{ s}}{1\text{ hr}}\right)(0.50\text{ A})\left(\frac{1\text{ C}}{1\text{ A s}}\right)\left(\frac{1\text{ mol e}^-}{96\ 500\text{ C}}\right)\left(\frac{1\text{ mol }C_2H_6}{14\text{ mol e}^-}\right) = \underline{8.0\text{x}10^{-3}\text{ mol}}$$

$$V_{C_2H_6} = \frac{nRT}{P} = \frac{(8.0\text{x}10^{-3}\text{ mol})(0.0821\text{ mol L mol}^{-1}\text{ K}^{-1})(273.1\text{ K})}{1\text{ atm}} = = \underline{0.18\text{ L}}$$

$$V_{O_2} = (0.18\text{ L }C_2H_6)\left(\frac{7\text{ mol }O_2}{2\text{ mol }C_2H_6}\right) = \underline{0.63\text{ L}}$$

0.18 L C_2H_6 would be consumed at the ***anode*** and **0.63 L O_2** at the ***cathode*** at STP.

Corrosion

30. (a) What sort of reactions cause metals to corrode? (b) Which of iron and aluminum is expected to corrode more easily in the presence of oxygen and water, and how do the oxides of these metals differ in their subsequent behaviour?

(a) Corrosion reactions are oxidation reactions in which metals are eaten away and eventually destroyed, especially in the natural environment, for example, by moist air or sea water.

(b) The standard cell potentials for the reactions in which the metals are oxidized and water is reduced give a qualitative measure of the relative ease of corrosion:

For iron:

Oxidation $2[Fe(s) \rightarrow Fe^{2+}(aq)) + 2e^-]$ $E°_{ox} = +0.44$ V

Reduction $O_2(g) + 4H^+(aq) + 4e^- \rightarrow 2H_2O(\ell)$ $E_{ored} = +1.23$ V

$2Fe(s) + O_2(g) + 4H^+(aq) \rightarrow 2Fe^{2+}(aq) + 2H_2O(aq)$ $E°_{cell} = +1.67$ V

For **aluminum:**

Oxidation $4[Al(s) \rightarrow Al^{3+}(aq)) + 3e^-]$ $E°_{ox} = +1.66$ V

Reduction $3[O_2(g) + 4H^+(aq) + 4e^- \rightarrow 2H_2O(\ell)]$ $E_{ored} = +1.23$ V

$4Al(s) + 3O_2(g) + 12H^+(aq) \rightarrow 4Al^{3+}(aq) + 6H_2O(\ell))$ $E°_{cell} = +2.89$ V

Thus, aluminum is much more easily oxidized, and more easily corroded than iron. Subsequently, the reaction with $O_2(g)$ from the air gives an oxide. $Fe^{2+}(aq)$ ions formed from iron are further oxidized to $Fe^{3+}(aq)$ and form hydrated $Fe_2O_3(s)$, while aluminum forms $Al_2O_3(s)$. These oxides have quite different properties. Whereas $Al_2O_3(s)$ binds strongly to Al and forms a tough protective coating over the surface, protecting it and preventing further corrosion, hydrated $Fe_2O_3(s)$ is porous, easily cracks, and flakes off relatively, exposing iron to further corrosion.

---

32. Write equations for the half-reactions and overall cell reactions to show how metals such as (a) magnesium, and (b) zinc protect iron by cathodic protection.

In each case, the iron is protected, because when joined to (a) magnesium, or (b) zinc, the iron and the other metal form the electrodes of an electrochemical cell. In these cells, iron becomes the **cathode** and the other metal the **anode,** and is thus preferentially oxidized - as expected from the relative values of the standard oxidation potentials of these metals.

(a) **Magnesium**

anode: $2[Mg(s) \rightarrow Mg^{2+}(aq) + 2e^-]$ oxidation

cathode: $O_2(g) + 4e^- + 4H^+(aq) \rightarrow 2H_2O(\ell)$ reduction

$2Mg(s) + O_2(g) + 4H^+(aq) \rightarrow 2Mg^{2+}(aq) + 2H_2O(\ell)$

(a) **Zinc**

anode: $2[Zn(s) \rightarrow Zn^{2+}(aq) + 2e^-]$ oxidation

cathode: $O_2(g) + 4e^- + 4H^+(aq) \rightarrow 2H_2O(\ell)$ reduction

$2Zn(s) + O^2(g) + 4H^+(aq) \rightarrow 2Zn^{2+}(aq) + 2H_2O(\ell)$

Electrolysis

35. For inert electrodes, write equations for the electrode reactions that occur during the electrolysis of: (a) molten aluminum chloride, and (b) a dilute aqueous solution of aluminum chloride. (c) Explain why the products of electrolysis in (a) are different from those in (b).

(a) $AlCl_3(s)$ is an ionic substance consisting of Al^{3+} and Cl^- ions. On electrolysis of molten $AlCl_3$, Al^{3+} ions are reduced at the *cathode*, and Cl^- ions are oxidized at the *anode*:

cathode: $2[Al^{3+} + 3e^- \rightarrow Al(\ell)]$ reduction

anode: $3[2Cl^- \rightarrow Cl_2(g) + 2e^-]$ oxidation

$2AlCl_3(\ell) \rightarrow 2Al(\ell) + 3Cl_2(g)$ overall

(b) $AlCl_3(aq)$ contains $Al^{3+}(aq)$ and $Cl^-(aq)$ ions. On electrolysis of concentrated $AlCl_3(aq)$, water is more readily reduced than $Al^{3+}(aq)$ ions at the *cathode*, but at the *anode* Cl^- ions are oxidized at a faster rate than water due to *overvoltage*, even though water is more easily oxidized than Cl^-.:

cathode:	$3[H_2O(\ell) + 2e^- \rightarrow H_2(g) + 2OH^-]$	reduction
anode:	$3[2Cl^- \rightarrow Cl_2(g) + 2e^-]$	oxidation
	$2AlCl_3(\ell) \rightarrow 2Al(OH)_3(s) + 3H_2(g) + 3Cl_2(g)$	overall

In molten $AlCl_3$ the only species that can be oxidized is Cl^- and the only species that can be reduced is Al^{3+}. In aqueous solution using **inert** electrodes, $Al^{3+}(aq)$ or the solvent, $H_2O(\ell)$, could be reduced, and $Cl^-(aq)$ or the solvent $H_2O(\ell)$ could be oxidized. Which occurs depends on the relative reduction (or oxidation) potentials and other factors such as **overvoltage**. For $Al^{3+}(aq)$, $E°_{red}$ is -1.66 V, and for $H_2O(\ell)$, $E°_{red}$ is -0.83 V, so water is *more easily reduced* than $Al^{3+}(aq)$; for $Cl^-(aq)$, $E°_{ox}$ is -1.36 V, and for $H_2O(\ell)$, $E°_{ox}$ is -1.23 V, but, because of *overvoltage*, $Cl^-(aq)$ is oxidized rather than $H_2O(\ell)$.

---

36. (a) Draw a cell for the electrolysis of HBr(aq) with platinum electrodes. (b) Label the anode and the cathode. (c) Write equations for the electrode reactions and for the overall cell reaction. (d) Indicate the direction of electron flow in the external circuit and the directions in which the ions move in the solution.

The electrodes are inert, as is the solvent water (as can be seen from the relevant oxidation/reduction potentials below.) $H_3O^+(aq)$ ions from the HBr(aq) are reduced at the cathode, and $Br^-(aq)$ ions from the HBr(aq) are oxidized at the anode:

cathode:	$2H^+ + 2e^- \rightarrow H_2(g)$	reduction
anode:	$2Br^- \rightarrow Br_2(\ell) + 2e^-$	oxidation
	$2HBr(aq) \rightarrow H_2(g)) + Br_2(\ell)$	overall

In the external circuit, electrons from the current source move from *anode to cathode*, in the solution, H_3O^+ ions move to the *cathode*, where they are *reduced*, and $Br^-(aq)$ ions move to the *anode*, where they are *oxidized*.

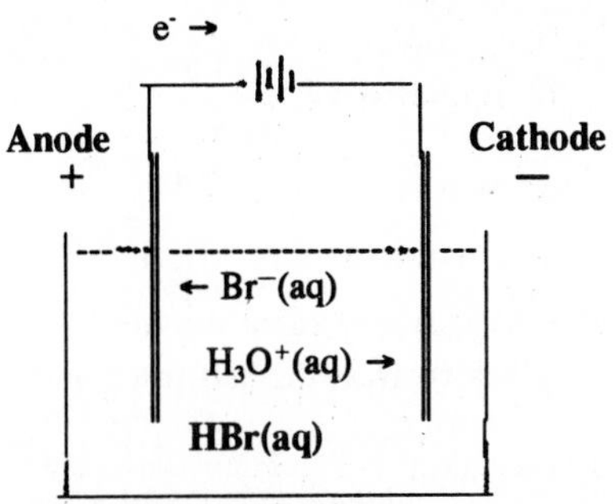

---

38. How many coulombs of electricity are required for each of the following reductions? (a) 1.00 mol of $Cu^{2+}(aq)$ to Cu(s); (b) 1.00 mol of $Fe^{3+}(aq)$ to $Fe^{2+}(aq)$; (c) 1.00 mol of $MnO_4^-(aq)$ to $Mn^{2+}(aq)$; (d) 1.00 mol of $Cr_2O_7^{2-}(aq)$ to $Cr^{3+}(aq)$

In each case we write the half-equation for the reduction reaction, and deduce moles of electrons (number of Faradays) from moles of reactant or product, and finally convert moles of electrons to coulombs, using the identity

$$1\ F = 96\ 500\ C$$

(a)

$$Cu^{2+}(aq) + 2e^- \rightarrow Cu(s)$$

$$\text{Coulombs} = (1.00\ \text{mol}\ Cu^{2+})\left(\frac{2\ \text{mol}\ e^-}{1\ \text{mol}\ Cu^{2+}}\right)\left(\frac{1\ F}{1\ \text{mol}\ e^-}\right)\left(\frac{96\ 500\ C}{1\ F}\right) = \underline{1.93 \times 10^5\ C}$$

(b)

$$Fe^{3+}(aq) + e^- \rightarrow Fe^{2+}(aq)$$

$$\text{Coulombs} = (1.00 \text{ mol } Fe^{3+})(\frac{1 \text{ mol } e^-}{1 \text{ mol } Fe^{3+}})(\frac{1 \text{ F}}{1 \text{ mol } e^-})(\frac{96\ 500 \text{ C}}{1 \text{ F}}) = \underline{9.65 \times 10^4 \text{ C}}$$

(c) $MnO_4^-(aq) + 5e^- + 8H^+(aq) \rightarrow Mn^{2+}(aq) + 4H_2O(\ell)$

$$\text{Coulombs} = (1.00 \text{ mol } MnO_4^-)(\frac{5 \text{ mol } e^-}{1 \text{ mol } MnO_4^-})(\frac{1 \text{ F}}{1 \text{ mol } e^-})(\frac{96\ 500 \text{ C}}{1 \text{ F}}) = \underline{4.83 \times 10^5 \text{ C}}$$

(d) $Cr_2O_7^{2-}(aq) + 6e^- + 14H^+(aq) \rightarrow 2Cr^{3+}(aq) + 7H_2O(\ell)$

$$(\text{Coulombs} = 1.00 \text{ mol } Cr_2O_7^{2-})(\frac{6 \text{ mol } e^-}{1 \text{ mol } Cr_2O_7^{2-}})(\frac{1 \text{ F}}{1 \text{ mol } e^-})(\frac{96\ 500 \text{ C}}{1 \text{ F}}) = \underline{5.79 \times 10^5 \text{ C}}$$

---

40. How many coulombs of electricity are required to produce each of the following? (a) 50.0 mL of $O_2(g)$ at STP, from $Na_2SO_4(aq)$; (b) 50.0 kg of Al(s) from molten Al_2O_3; (c) 20.0 g of calcium from molten calcium chloride; (d) 5.00 g of silver from aqueous silver nitrate.

This problem is similar to Problems 38 and 39, except that for each we calculate the moles of product first and then use the balanced half-equation for the oxidation or reduction reaction, as appropriate, to calculate the amount of current (coulombs of electricity) required.

(a) We first calculate the moles of $O_2(g)$ produced:

$$\text{(a)} \quad n_{O_2} = \frac{PV}{RT} = \frac{(1 \text{ atm})(50.0 \text{ mL})(\frac{1 \text{ L}}{10^3 \text{ mL}})}{(0.0821 \text{ atm L mol}^{-1} \text{ K}^{-1})(273.1 \text{ K})} = \underline{2.23 \times 10^{-3} \text{ mol}}$$

Although in $Na_2SO_4(aq)$ the $Na^+(aq)$ and $SO_4^{2-}(aq)$ ions carry the current, it is the water that is oxidized at the anode and reduced at the cathode, and $O_2(g)$ results from the half-reaction: $2H_2O(\ell) \rightarrow O_2(g) + 4e^- + 4H^+(aq)$.

$$\text{Coulombs} = (2.23 \times 10^{-3} \text{ mol } O_2)(\frac{4 \text{ mol } e^-}{1 \text{ mol } O_2})(\frac{1 \text{ F}}{1 \text{ mol } e^-})(\frac{96\ 500 \text{ C}}{1 \text{ F}}) = \underline{861 \text{ C}}$$

(b) Al^{3+} ions in ionic $Al_2O_3(\ell)$ are reduced to Al(s): $Al^{3+} + 3e^- \rightarrow Al(s)$

$$\text{Coulombs} = (50.0 \text{ kg Al})(\frac{10^3 \text{ g}}{1 \text{ kg}})(\frac{1 \text{ mol Al}}{26.98 \text{ g Al}})(\frac{3 \text{ mol } e^-}{1 \text{ mol Al}})(\frac{1 \text{ F}}{1 \text{ mol } e^-})(\frac{96\ 500 \text{ C}}{1 \text{ F}}) = \underline{5.37 \times 10^8 \text{ C}}$$

(c) Ca^{2+} ions in ionic $CaCl_2(\square)$ are reduced to Ca(s): $Ca^{2+} + 2e^- \rightarrow Ca(s)$

$$\text{Coulombs} = (20.0 \text{ g Ca})(\frac{1 \text{ mol Ca}}{40.08 \text{ g Ca}})(\frac{2 \text{ mol } e^-}{1 \text{ mol Ca}})(\frac{1 \text{ F}}{1 \text{ mol } e^-})(\frac{96\ 500 \text{ C}}{1 \text{ F}}) = \underline{9.63 \times 10^4 \text{ C}}$$

(d) $Ag^+(aq)$ ions from ionic $AgNO_3(aq)$ are reduced to Ag(s): $Ag^+(aq) + e^- \rightarrow Ag(s)$

$$\text{Coulombs} = (5.00 \text{ g Ag})(\frac{1 \text{ mol Ag}}{107.9 \text{ g Ag}})(\frac{1 \text{ mol } e^-}{1 \text{ mol Ag}})(\frac{1 \text{ F}}{1 \text{ mol } e^-})(\frac{96\ 500 \text{ C}}{1 \text{ F}}) = \underline{4.47 \times 10^3 \text{ C}}$$

---

42. What masses of sodium hydroxide and chlorine are produced by the electrolysis of aqueous sodium chloride for 3.00 h using a current of 0.200 A?

Water is more easily reduced at the cathode than $Na^+(aq)$ and, for **concentrated** NaCl(aq), $Cl^-(aq)$ is more easily oxidized at the anode than $H_2O(\ell)$. Thus:

cathode: $2H_2O(\ell) + 2e^- \rightarrow 2OH^-(aq) + H_2(g)$; **anode:** $2Cl^-(aq) \rightarrow Cl_2(g) + 2e^-$

$$\text{moles of electrons} = (3.00\ \text{hr})\left(\frac{3600\ \text{s}}{1\ \text{hr}}\right)(2.00\ \text{A})\left(\frac{1\ \text{C}}{1\ \text{A s}}\right)\left(\frac{1\ \text{mol e}^-}{96\ 500\ \text{C}}\right) = \underline{0.0224\ \text{mol e}^-}$$

$$\text{mass of NaOH} = (0.0224\ \text{mol e}^-)\left(\frac{2\ \text{mol NaOH}}{2\ \text{mol e}^-}\right)\left(\frac{40.00\ \text{g NaOH}}{1\ \text{mol NaOH}}\right) = \underline{0.896\ \text{g}}$$

$$\text{mass of } Cl_2 = (0.0224\ \text{mol e}^-)\left(\frac{1\ \text{mol } Cl_2}{2\ \text{mol e}^-}\right)\left(\frac{70.90\ \text{g } Cl_2}{1\ \text{mol } Cl_2}\right) = \underline{0.794\ \text{g}}$$

---

44. How long will it take to deposit 16.0 g of silver from an $AgNO_3(aq)$ solution, using a current of 6.00 A?

$Ag^+(aq)$ is reduced to Ag(s), according to the half-equation: $Ag^+(aq) + e^- \rightarrow Ag(s)$

$$\text{Coulombs} = (16.0\ \text{g Ag})\left(\frac{1\ \text{mol Ag}}{107.9\ \text{g Ag}}\right)\left(\frac{1\ \text{mol e}^-}{1\ \text{mol Ag}}\right)\left(\frac{96\ 500\ \text{C}}{1\ \text{mol e}^-}\right) = \underline{1.431 \times 10^4\ \text{C}}$$

$$\text{Time} = (1.431 \times 10^4\ \text{C})\left(\frac{1\ \text{A s}}{1\ \text{C}}\right)\left(\frac{1}{6.00\ \text{A}}\right)\left(\frac{1\ \text{min}}{60\ \text{s}}\right) = \underline{39.7\ \text{min}}$$

---

46. How long would a current of 1.50 A have to be passed through a solution containing chromium(III) sulfate for a steel object of surface area 0.10 m^2 to be plated with chromium (density 7.1 g cm^{-3}) to a thickness of 0.10-mm?

Chromium(III) sulfate solution, $Cr_2(SO_4)_3(aq)$ contains $Cr^{3+}(aq)$ ions, which are reduced in the half-reaction:

$$Cr^{3+}(aq) + 3e^- \rightarrow Cr(s)$$

$$\text{Volume of Cr} = (0.10\ \text{m}^2)\left(\frac{10^2\ \text{cm}}{1\ \text{m}}\right)^2(0.10\ \text{mm})\left(\frac{1\ \text{cm}}{10\ \text{mm}}\right) = \underline{10.0\ \text{cm}^3}$$

$$\text{moles Cr} = (10.0\ \text{cm}^3)\left(\frac{7.1\ \text{g Cr}}{1\ \text{cm}^3}\right)\left(\frac{1\ \text{mol Cr}}{52.00\ \text{g Cr}}\right) = \underline{1.37\ \text{mol Cr}}$$

$$\text{Time} = (1.37\ \text{mol Cr})\left(\frac{3\ \text{mol e}^-}{1\ \text{mol Cr}}\right)\left(\frac{96\ 500\ \text{C}}{1\ \text{mol e}^-}\right)\left(\frac{1\ \text{A s}}{1\ \text{C}}\right)\left(\frac{1}{1.50\ \text{A}}\right)\left(\frac{1\ \text{hr}}{3600\ \text{s}}\right)\left(\frac{1\ \text{day}}{24\ \text{hr}}\right) = \underline{3.1\ \text{days}}$$

---

48. When a current of 0.500 A was passed through a solution containing an unknown $M^{2+}(aq)$ ion for exactly 2 h, 1.98 g of the metal M were deposited. (a) What is the atomic mass of M; (b) What is the metal?

In this case the reduction half-reaction is $M^{2+}(aq) + 2e^- \rightarrow M(s)$, and we can calculate the moles of metal produced:

$$\text{moles of M} = (0.500 \text{ A})(2 \text{ hr})(\frac{3600 \text{ s}}{1 \text{ hr}})(\frac{1 \text{ C}}{1 \text{ A s}})(\frac{1 \text{ mol M}}{2 \text{ mol e}^-})(\frac{1 \text{ mol e}^-}{96\ 500 \text{ C}}) = \underline{1.865 \times 10^{-2} \text{ mol M}}$$

$$\text{Thus:} \quad \text{Molar mass of M} = \frac{1.98 \text{ g}}{1.865 \times 10^{-2} \text{ mol}} = \underline{106 \text{ g mol}^{-1}}$$

Reference to the period table identifies the metal as the *transition metal* **palladium** (*atomic mass* 106.4 g mol^{-1}) in Period 5, which like Ni above it in Period 4 should form an M^{2+} ion.

---

50. Two electrolysis cells are connected so that the same current is passed through each. In the first, Fe^{3+}(aq) is reduced to Fe(s), and in the second, Cu^{2+}(aq) is reduced to Cu(s). After passage of the current for 30.00 min, 1.030 g of iron were deposited in the first cell. How many grams of copper were deposited in the second cell?

The two reduction half-reactions are: $Fe^{3+}(aq) + 3e^- \rightarrow Fe(s)$, and $Cu^{2+}(aq) + 2e^- \rightarrow Cu(s)$, and the time taken irrelevant because we do not need to know how many coulombs were passed through the cell;. For the *same* amount of current passed:

$$\text{mass of Cu} = (1.030 \text{ g Fe})(\frac{1 \text{ mol Fe}}{55.85 \text{ g Fe}})(\frac{3 \text{ mol e}^-}{1 \text{ mol Fe}})(\frac{1 \text{ mol Cu}}{2 \text{ mol e}^-})(\frac{63.55 \text{ g Cu}}{1 \text{ mol Cu}}) = \underline{1.758 \text{ g Cu}}$$

---

52. The same electric current was passed for 1.00 h through each of three electrolytic cells fitted with platinum electrodes and containing $CuSO_4$(aq), $AgNO_3$(aq), and H_2SO_4(aq), respectively. During this time, 0.106 g of copper was deposited at the cathode in the first cell. Calculate: (a) The average current, in milliamps; (b) the mass of silver deposited at the cathode in the second cell, and (c) the total volume of dry gas, measured at 20°C and a pressure of 750 mm Hg, liberated in the third cell.

The electrodes are inert, and Cu^{2+}(aq) from $CuSO_4$(aq) in the first cell, and Ag^+(aq) from $AgNO_3$(aq) in the second cell, are both more easily reduced than water, so:

(a) At the cathode in the first cell, where $Cu^{2+}(aq) + 2e^- \rightarrow Cu(s)$:

$$\text{Moles of electrons used} = (0.106 \text{ g Cu})(\frac{1 \text{ mol Cu}}{63.55 \text{ g Cu}})(\frac{2 \text{ mol e}^-}{1 \text{ mol Cu}}) = \underline{3.336 \times 10^{-3} \text{ mol}}$$

$$\text{Current} = (\frac{3.336 \times 10^{-3} \text{ mol e}^-}{1.00 \text{ hr}})(\frac{96\ 500 \text{ C}}{1 \text{ mol e}^-})(\frac{1 \text{ A s}}{1 \text{ C}})(\frac{1 \text{ hr}}{3600 \text{ s}})(\frac{10^3 \text{ mA}}{1 \text{ A}}) = \underline{89.4 \text{ mA}}$$

(b) At the cathode in the second cell: $Ag^+(aq) + e^- \rightarrow Ag(s)$

$$\text{Mass of Ag} = (3.336 \times 10^{-3} \text{ mol e}^-)(\frac{1 \text{ mol Ag}}{1 \text{ mol Ag}})(\frac{107.9 \text{ g Ag}}{1 \text{ mol Ag}}) = \underline{0.360 \text{ g}}$$

In the third cell, the solution contains H_3O^+(aq) and SO_4^{2-}(aq) from the ionization of the strong acid H_2SO_4(aq), and H_3O^+(aq) is reduced at the cathode. However, water, rather than SO_4^{2-}(aq), is oxidized at the anode:

$$2H_3O^+(aq) + 2e^- \rightarrow H_2(g) + 2H_2O(\ell) \quad \text{and} \quad 2H_2O(\ell) \rightarrow O2(g) + 4H\!+\!(aq) + 4e^-$$

$$\text{moles of } H_2 = (3.336 \times 10^{-3} \text{ mol } e^-)(\frac{1 \text{ mol } H_2}{2 \text{ mol } e^-}) = \underline{1.668 \times 10^{-3} \text{ mol } H_2}$$

$$\text{moles of } O_2 = (3.336 \times 10^{-3} \text{ mol } e^-)(\frac{1 \text{ mol } O_2}{4 \text{ mol } e^-}) = \underline{0.834 \times 10^{-3} \text{ mol } O_2}$$

The total pressure, and hence the volume, is due to the total moles of H_2 and O_2 gas = (1.668 + 0.834) x 10^{-3} mol = 2.502 x10^{-3} mol, the pressure is 750 mm Hg, and the temperature is 20°C (293.1 K). Thus:

$$V_{gas} = \frac{nRT}{P} = \frac{(2.502 \times 10^{-3} \text{ mol})(0.0821 \text{ atm L mol}^{-1} \text{ K}^{-1})(293.1 \text{ K})}{(750 \text{ mm Hg})(\frac{1 \text{ atm}}{760 \text{ mm Hg}})} = 0.0625 \text{ L}; \quad (\underline{62.5 \text{ mL}})$$

General Problems

53. (a) How long would a current of 1.50 A have to be passed through a 5.00-M $Cr_2(SO_4)_2(aq)$ solution to coat a metal object of surface area 1.00 m^2 with a 0.0100-mm layer of chromium (density 7.21 g cm^{-3})? (b) Why do you suppose the coating on chromium-plated steel is generally very thin?

(a) Chromium(III) sulfate solution, $Cr_2(SO_4)_3(aq)$ contains $Cr^{3+}(aq)$ ions, which are reduced in the half-reaction:

$Cr^{3+}(aq) + 3e^- \rightarrow Cr(s)$ (Note: The concentration of the solution is presumably only of practical significance).

$$\text{Volume of Cr} = (1.00 \text{ m}^2)(\frac{10^2 \text{ cm}}{1 \text{ m}})^2(0.10 \text{ mm})(\frac{1 \text{ cm}}{10 \text{ mm}}) = \underline{1.00 \times 10^2 \text{ cm}^3}$$

$$\text{moles Cr} = (1.00 \times 10^2 \text{ cm}^3)(\frac{7.1 \text{ g Cr}}{1 \text{ cm}^3})(\frac{1 \text{ mol Cr}}{52.00 \text{ g Cr}}) = \underline{13.7 \text{ mol Cr}}$$

$$\text{Time taken} = (13.7 \text{ mol Cr})(\frac{3 \text{ mol } e^-}{1 \text{ mol Cr}})(\frac{96\,500 \text{ C}}{1 \text{ mol } e^-})(\frac{1 \text{ A s}}{1 \text{ C}})(\frac{1}{1.50 \text{ A}})(\frac{1 \text{ hr}}{3600 \text{ s}})(\frac{1 \text{ day}}{24 \text{ hr}}) = \underline{31 \text{ days}}$$

(b) The fact that the time taken to electroplate the steel, the current used, and amount of chromium consumed are all large, makes this an expensive process.

---

55. Calculate the equilibrium constant for the reaction in which $I^-(aq)$ is oxidized by $Fe^{3+}(aq)$ to give $I_2(s)$ and $Fe_{2+}(aq)$ under standard conditions.

The reaction and the standard cell potential are:

reduction	$2[Fe^{3+} + e^- \rightarrow Fe^{2+}]$	$E^\circ_{red} = +0.77$ V
oxidation	$2I^- \rightarrow I_2(s) + 2e^-$	$E^\circ_{ox} = -0.54$ V
overall	$2Fe^{3+} + 2I^- \rightleftarrows 2Fe^{2+} + I_2(s)$	**$E^\circ_{cell} = +0.20$ V**

and we can now use the standard cell potential and the Nernst equation to calculate the equilibrium constant, because at equilibrium:

$$E_{cell} = 0, \quad Q = [Fe^{2+}]^2/([Fe^{3+}]^2[I^-]^2) = Keq, \text{ and } n = 2:$$

$$E = E^\circ - \frac{RT}{nF} \ln Q = 0 = 0.20 \text{ V} - \frac{0.0257}{2} \ln K_{eq}; \quad \ln K_{eq} = \frac{2(0.20)}{0.0257} = \underline{15.6}; \quad K_{eq} = e^{15.6} = \underline{6.0 \times 10^6 \text{ mol}^{-2} L^2}$$

57.* Predict qualitatively the affect of adding each of the following on the cell voltage of the cell:

$$Zn(s)|Zn^{2+}(aq,\ 1\ M)\|\ Cu^{2+}(aq,\ 1\ M)|Cu(s)$$

(a) 2-M $Zn^{2+}(aq)$ to the $Zn(s)|Zn^{2+}(aq)$ half-cell; (b) Zn(s) to the $Zn(s)|Zn^{2+}(aq)$ half-cell; (c) a drop or two of dilute NaOH(aq) to the $Zn(s)|Zn^{2+}(aq)$ half-cell; (d) concentrated $NH_3(aq)$ to the $Cu(s)|Cu^{2+}(aq)$ half-cell.

In this **standard cell,** Zn(s) is oxidized to $Zn^{2+}(aq)$ at the anode, $Cu^{2+}(aq)$ is reduced to Cu(s) at the cathode, and this cell reaction will *spontaneously* go to equilibrium:) $Zn(s) + Cu^{2+}(aq) \rightleftarrows Zn^{2+}(aq) + Cu(s)$ in a process in which $Cu^{2+}(aq)$ is replaced to some extent by $Zn^{2+}(aq)$. The problem is concerned with the way in which changes in concentrations of *reactants* or *products* affect the initial voltage. Thus, we should examine the **Nernst equation:**

$$E_{cell} = E^{\circ}_{cell} - \frac{0.0257\ V}{n} \ln Q$$

$$\text{with } n = 2, \quad Q = \frac{[Zn^{2+}]}{[Cu^{2+}]} = 1, \quad E_{cell} = E^{\circ}_{cell}, \text{ initially.}$$

(a) **Adding 2-M $Zn^{2+}(aq)$** increases $[Zn^{2+}]$ and Q is greater than 1; **ln Q** is **positive** and $E_{cell} < E^{\circ}_{cell}$; **the initial voltage** *decreases.*

(b) **Adding Zn(s)** has *no affect*, because this is a heterogeneous equilibrium and [Zn(s)] remains the same, whatever the amount, and does not affect the value of Q.

(c) **Added NaOH(aq)** reacts with $Zn^{2+}(aq)$ to give a precipitate of $Zn(OH)_2(s)$, which decreases $[Zn^{2+}]$, and results in an *increase* in the value of Q. **ln Q** becomes **negative** and $E_{cell} > E^{\circ}_{cell}$; **the initial voltage** *increases.*

(d) **Addition of a concentrated $NH_3(aq)$** *decreases* $[Cu^{2+}(aq)]$ because the $Cu(NH_3)_4^{2+}(aq)$ complex ion is formed:

$$Cu^{2+}(aq) + 4NH_3(aq) \rightarrow Cu(NH_3)_4^{2+}(aq)$$

Thus, the value of Q is *greater than* than 1; **ln Q** is **positive** and and $E_{cell} < E^{\circ}cell$; **the initial voltage** *decreases.* (Because there is less tendency for $Cu^{2+}(aq)$ to be reduced.

CHAPTER 14

Functional Groups

1. Write the structural formulas of and name each of the following noncyclic compounds containing three carbon atoms: (a) a primary alcohol (b) a secondary alcohol (c) an aldehyde (d) a ketone (e) a carboxylic acid (f) an ether (g) an amide (h) a primary amine

---

(a) $CH_3CH_2CH_2OH$	(b) CH_3CHCH_3 with OH on central C	(c) $CH_3CH_2C{=}O$ with H on C	(d) CH_3CCH_3 with =O on central C
1-propanol	**2-propanol**	**propanal**	**propanone**
(e) $CH_3CH_2\text{-}C{=}O$ with OH on C	(f) $CH_3\text{-}O\text{-}CH_3$	(g) $CH_3CH_2\text{-}C{=}O$ with NH_2 on C	(h) $CH_3CH_2CH_2NH_2$
propanoic acid	**methoxyethane**	**propanamide**	**propylamine**

---

3. Name the functional groups in each of the following:

(a) $HO\text{-}C(=O)\text{-}CH_2\text{-}CH_2\text{-}CH(NH_2)\text{-}C(=O)\text{-}O^-$ Na^+

(a)
Monosodium glutamate (MSG)
(a flavor enhancer in cookery)

(b) benzene ring bearing H_3CO, HO– and –C(=O)–H

(b)
Vanallin
(used as vanilla flavoring)

(a) **$-CO_2H$, carboxylic acid; $-CO_2^-$, carboxylate; $-NH_2$, amino-.**

(b) **$-OCH_3$, methoxy-; $-OH$, phenoxy-** (hydroxyl); **$-C(O)H$ aldehyde.**

---

5. Name each of the compounds:

(a) $CH_3\text{-}C(=O)\text{-}H$ (b) $CH_3CH_2\text{-}C(=O)\text{-}CH_3$ (c) $CH_3CH_2\text{-}C(=O)\text{-}OH$ (d) $(CH_3)_2C(H)\text{-}C(=O)\text{-}NH_2$ (e) $(CH_3)_3CNH_2$ (f) $CH_3CH_2CH(OH)CH_2CH_3$

(a) **ethanal** (acetaldehyde); (b) **2-butanone** (ethylmethyl ketone); (c) **propanoic acid;**
(d) **2-methylpropanamide;** (e) **2,2-dimethylpropanamine;** (f) **3-pentanol**

---

7. Identify the functional groups in each of the following: (a) the thyroid gland hormone thyroxine; (b) ascorbic acid (vitamin C); (c) aspartame (Nutra-Sweet)

(a) HO–(diiodophenyl)–O–(diiodophenyl)–CH_2–C(H)(NH_2)–C(=O)–OH, with I, I, I, I on the rings

(b) CH_2OH–HO–CH– attached to a five-membered lactone ring (O, C=O) bearing HO and OH on the C=C

$$\begin{array}{c} \quad\ \ O \qquad\qquad\quad\ O \\ \quad\ \ \| \qquad\qquad\quad\ \| \\ H_2N-CH-C-NH-CH-C-OCH_3 \\ | \qquad\qquad\quad | \qquad \\ CH_3 \qquad\quad CH_2 \quad \\ \qquad\qquad\quad | \qquad \\ \qquad\qquad\quad C_6H_5 \end{array}$$

(c)

(a) **–I**, iodo; **HO–**, phenoxy- (hydroxyl); **–NH_2**, primary amino-; **–CO(OH)**, carboxylic acid.

(b) **HO–C(H)-CH_2-**, secondary alcohol; **–CH_2OH**, primary alcohol;
<u>RING</u>: cyclic ester (**–O–** (in ring) and **C=O**); **-OH**, alcohol; **C=C** (in ring), alkene.

(c) **H_2N–**, primary amino-; **–C(O)N(H)–**, amido-; **–C(O)OCH_3**, ester.

<u>Haloalkanes (Alkyl Halides)</u>

9. What reaction or reactions and other reactants would you use to synthesize each of the following from ethyl bromide? (a) ethyl iodide (b) ethyl acetate (c) diethyl ether (d) acetic acid

---

(a) **Ethyl iodide** would result from the reaction of ethyl bromide with excess sodium iodide:

$$C_2H_5Br + NaI \rightarrow C_2H_5I + NaBr$$

(b) Hydrolysis of ethyl bromide with NaOH gives **ethanol**, C_2H_5OH, a sample of which could be oxidized to **acetic acid**, $CH_3C(O)OH$, (ethanoic acid), with dichromate ion:

$$C_2H_5Br + NaOH \rightarrow C_2H_5OH + NaBr; \qquad C_2H_5OH + 2[O] \rightarrow CH_3C(O)OH + H_2O$$

The ethanol and acetic acid could then be reacted to give the *ester* **ethyl acetate**:

$$C_2H_5OH + CH_3CO_2H \rightarrow C_2H_5O\text{-}C(O)CH_3 + H_2O$$

(c) Hydrolysis of ethyl bromide with NaOH gives **ethanol**, C_2H_5OH, which when reacted with excess NaOH gives **sodium ethoxide**, $C_2H_5O^-Na^+$. Reaction of sodium ethoxide with ethyl bromide then gives **diethyl ether**, $(C_2H_5)_2O$, (ethoxyethane), and sodium bromide:

$$C_2H_5Br + NaOH \rightarrow C_2H_5OH + NaBr; \qquad C_2H_5OH + NaOH \rightarrow C_2H_5O^-Na^+ + H_2O$$

$$C_2H_5O^- Na^+ + C_2H_5Br \rightarrow (C_2H_5)_2O + NaBr$$

(d) Hydrolysis of ethyl bromide with NaOH gives **ethanol**, C_2H_5OH, a sample of which could be oxidized (via **ethanal**, $C_2H_5C(H)O$, to **acetic** (ethanoic) **acid**, $CH_3C(O)OH$:

$$C_2H_5Br + NaOH \rightarrow C_2H_5OH + NaBr$$

$$C_2H_5OH + [O] \rightarrow CH_3C(O)H + H_2O; \qquad CH_3C(O)H + [O] \rightarrow CH_3C(O)OH + H_2O$$

<u>Alcohols and Ethers</u>

11. Give the systematic names of

(a) [benzene ring with CH_2CH_3 substituent]-OH (b) $(CH_3)_2CH\text{-}OCH_3$ (c) $(CH_3CH_2CH_2)_2O$ (d) $HS\text{-}CH_2\text{-}CH(NH_2)\text{-}C(=O)\text{-}OH$ (e) $(CH_3)_2C(H)\text{-}SH$

(a) **2-ethylphenol** (2-ethylhydroxybenzene); (b) **2-methoxypropane**, (methyl isopropyl ether);
(c) **dipropyl ether**, (propoxypropane); (d) **2-amino-3-thiol-propanoic acid;** (e) **propane-2-thiol.**

---

13. Classify each of the following alcohols as a primary, secondary, or tertiary alcohol and give the systematic name: (a) $(CH_3)_2C(OH)CH_2CH_3$ (b) $CH_3CH(OH)CH_2CH_3$ (c) $(CH_3)_3CCH_2OH$ (d) $(CH_3)_2CHOH$ (e) $(CH_3CH_2)_2CH(OH)$

Classification of alcohols as primary, secondary, or tertiary depends on the number of carbon atoms attached to the carbon atom to which the —OH group is attached: *primary* **one**; *secondary* **two**; *tertiary* **three**

CH_3–C(H_3C)(OH)–CH_2–CH_3	H_3C–CH(OH)–CH_2–CH_3	H_3C–C(CH_3)$_2$–CH_2OH	H_3C–CH(CH_3)–OH	H_3C–CH(OH)–CH(H)–CH_2–CH_3
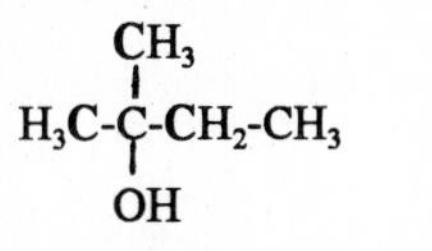 **tertiary**	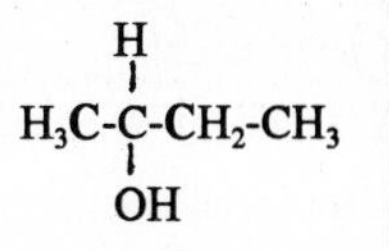 **secondary**	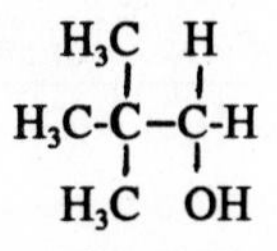 **primary**	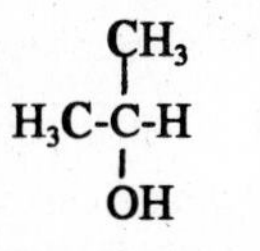 **secondary**	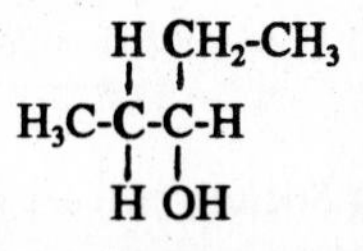 **secondary**
2-methyl-2-butanol	**2-butanol**	**2,2-dimethyl-1-propanol**	**2-propanol**	**3-pentanol**

---

15. What alkenes might be formed by dehydration of each of the following? (a) $CH_3CH_2CH(OH)CH(CH_3)_2$ (b) $(CH_3CH_2)_3COH$ (c) $C_6H_5CH(OH)CH_2CH_3$

In each case we remove the elements of water (H + OH) to form a C=C bond, so it is useful first to write the structural formulas (in some cases there is more than one possible product):

(a) H_3C–CH_2–CH(OH)–CH(CH_3)–CH_3 → (H_3C)(H)C=C(H)(CH(CH_3)$_2$) and (H)(H_3C–CH_2)C=C(CH_3)$_2$

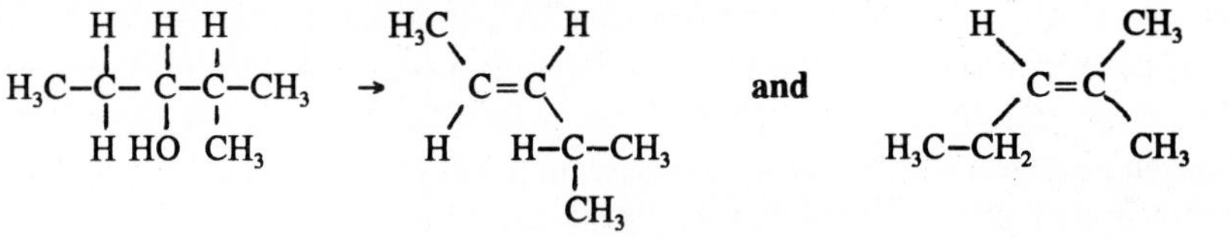

trans- (and cis-) 4-methyl-2-pentene **2-methyl-2-pentene**

(b) H_3C–CH_2–C(OH)(CH_2–CH_3)$_2$ → (H)(H_3C)C=C(CH_2–CH_3)$_2$ **2-ethyl-2-pentene**

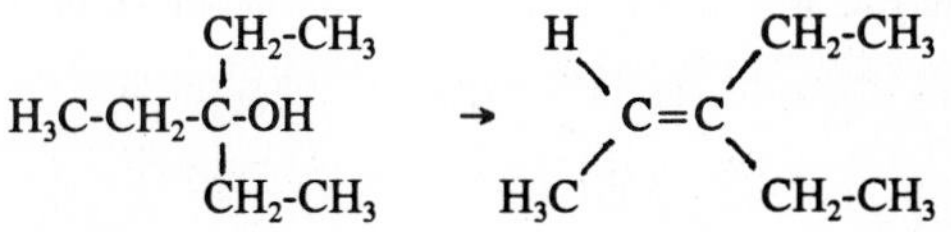

(c) C_6H_5–CH(OH)–CH_2–CH_3 → C_6H_5–CH=C(CH_3)$_2$ **2-methyl-1-phenyl-1-propene**

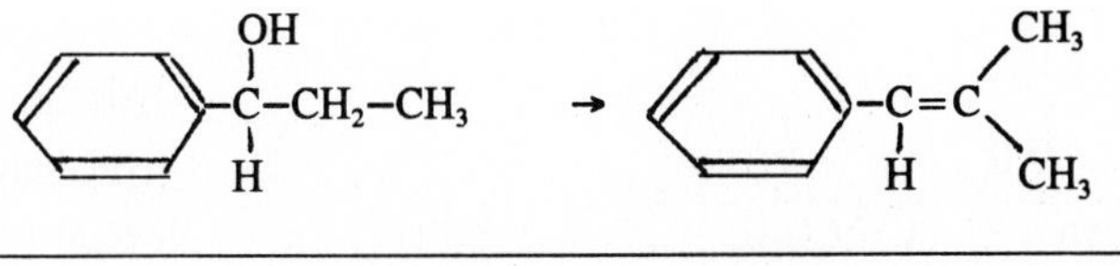

---

17. What reactions occur when phenol is dissolved in (a) HCl(aq), and (b) NaOH(aq)?

Phenol behaves in (a) as a weak base and is protonated on the **—OH** group, and in (b) as a weak acid, losing a proton to give the **phenoxide** ion:

(a) $C_6H_5OH(aq) + HCl(aq) \rightleftarrows C_6H_5OH_2^+(aq) + Cl^-(aq)$

(b) $C_6H_5OH(aq) + NaOH(aq) \rightleftarrows C_6H_5O^-(aq) + Na^+(aq) + H_2O(\ell)$

---

18. Complete and balance each of the following equations, and name the organic product:
(a) $C_2H_5OH(g) + O_2(g) \rightarrow$ (using a Cu catalyst at 600 °C) (b) $CH_3(CH_2)_2OH(\ell) + NaOH(s) \rightarrow$
(c) $CH_3OH(\ell) + Na(s) \rightarrow$ (d) $C_2H_5OH(\ell) + H_2SO_4(conc) \rightarrow$
(e) $CH_3CH{=}CH_2(g) + H_2O(g) \rightarrow$ (using a catalyst and high T and P)

(a) In this catalyzed reaction, **ethanol** (a primary alcohol) is *oxidized* to **ethanal** (an aldehyde):

$$2C_2H_5OH + O_2(g) \rightarrow 2CH_3CHO(g) + 2H_2O(g)$$

(b) **Propanol** (a primary alcohol) is a *weak acid* and form **sodium propoxide:**

$$CH_3(CH_2)_2OH(\ell) + NaOH(s) \rightarrow CH_3(CH_2)_2O^-Na^+(s) + H_2O(\ell)$$

(c) Sodium *reduces* **methanol** (an alcohol) to hydrogen gas and **sodium methoxide:**

$$2CH_3OH(\ell) + 2Na(s) \rightarrow 2CH_3O^-Na^+ + H_2(g)$$

(d) Concentrated sulfuric acid *dehydrates* **ethanol** (an alcohol), removing the elements of water, to form **ethene:**

$$C_2H_5OH(\ell) + H_2SO_4(conc) \rightarrow H_2C{=}CH_2(g) + H_3O^+ + HSO_4^-$$

(e) **Propene** (an alkene) *adds water* as H and OH across the double bond, giving two alcohols - the primary alcohol **1-propanol,** $CH_3CH_2CH_2OH$, and the secondary alcohol **2-propanol,** $CH_3CH(OH)CH_3$.

$$CH_3CH{=}CH_2(g) + H_2O(g) \rightarrow CH_3CH_2CH_2OH(g) \text{ and } CH_3CH(OH)CH_3$$

---

20. (a) Explain why the structural isomers dimethyl ether (bp -23 °C) and ethanol (bp 78.3 °C) differ significantly in their boiling points? (b) Describe a simple chemical test that could distinguish ethanol from dimethyl ether?

(a) The boiling point of ethanol, CH_2CH_2OH, (78°C) is significantly higher than that of dimethyl ether, $H_3C\text{-}O\text{-}CH_3$, (-23°C), because the former has a polar O-H bond (see Problem 19) that can participate in intermolecular hydrogen bonds.

(b) A simple chemical test that would distinguish ethanol from diethyl ether is its reaction with sodium metal to give $H_2(g)$, and towards which dimethyl ether is inert (indeed sodium is often stored in dry ether):

$$2C_2H_5OH + 2Na \rightarrow 2C_2H_5O^-Na^+ + H_2$$

An alternative test would be addition of ethanol to an acidified aqueous solution of potassium dichromate, which oxidizes the primary alcohol ethanol to acetaldehyde and then to acetic acid, during which **orange** $Cr_2O_7^{2-}(aq)$ is reduced to **green** $Cr^{3+}(aq)$, whereas the ether is again unreactive.

$Cr_2O_7^{2-} + 6e^- + 14H^+ \rightarrow 2Cr^{3+} + 7H_2O$	reduction
$3[CH_3CH_2OH \rightarrow CH_3CHO + 2e^- + 2H^+]$	oxidation
$3CH_3CH_2OH + Cr_2O_7^{2-} + 8H^+ \rightarrow 3CH_3C(H)O + 2Cr^{3+} + 7H_2O$	overall

and

$Cr_2O_7^{2-} + 6e^- + 14H^+ \rightarrow 2Cr^{3+} + 7H_2O$	reduction
$3[CH_3CHO + H_2O \rightarrow CH_3CO_2H + 2e^- + 2H^+]$	oxidation
$3CH_3CHO + Cr_2O_7^{2-} + 8H^+ \rightarrow 3CH_3CO_2H + 2Cr^{3+} + 3H_2O$	overall

---

22. Name the products of heating 2-butanol with (a) sulfuric acid, (b) hydrobromic acid, and (c) oxalic acid.

(a) **2-butanol** (a secondary alcohol) is dehydrated by **sulfuric acid** in a reaction where the -OH group and a H atom on an adjacent C atom are removed as water, to give an alkene. Thus, two alkenes are formed:

$$H_3C\text{-}CH_2\text{-}CH(OH)\text{-}CH_3 \rightarrow H_3CCH_2CH{=}CH_2 \text{ and } H_3C\text{-}CH{=}CH\text{-}CH_3 + H_2O$$

2-butanol → **1-butene** and **2-butene** (*cis-* and *trans-*)

(b) **Hydrobromic acid** first protonates **2-butanol**, and **bromide ion** then replaces a water molecule to give **2-bromobutane**, $CH_3CH_2CH(Br)CH_3$, for the overall nucleophilic substitution reaction:

$$CH_3\text{-}CH_2\text{-}CH(OH)\text{-}CH_3 + HBr \rightarrow CH_3\text{-}CH_2\text{-}CH(Br)\text{-}CH_3 + H_2O$$

(c) **Oxalic acid,** $(CO_2H)_2$, is a diprotic carboxylic acid and can in principle form two *esters*, **1-methylpropyl oxalate** and **di(1-methylpropyl) oxalate:**

$$HOOC\text{-}COOH + H_3C\text{-}CH(OH)\text{-}CH_2\text{-}CH_3 \rightarrow HOOC\text{-}C(=O)\text{-}O\text{-}CH(CH_3)(CH_2CH_3) + H_2O$$

$$HOOC\text{-}COOH + 2\,H_3C\text{-}CH(OH)\text{-}CH_2\text{-}CH_3 \rightarrow (CH_3CH_2)(CH_3)HC\text{-}O\text{-}C(=O)\text{-}C(=O)\text{-}O\text{-}CH(CH_3)(CH_2CH_3) + 2H_2O$$

24.* Suggest ways you could prepare each of the following: (a) 2-propanol from 1-propanol; (b) ethylene glycol from ethanol; (c) oxalic acid, $(CO_2H)_2$, from ethylene glycol.

(a) **2-propanol** is $H_3C\text{-}CH(OH)\text{-}CH_3$ with the alcohol group -OH on carbon-2, whereas **1-propanol** is $CH_3CH_2CH_2OH$, with the alcohol group -OH on carbon-1. To convert 1-propanol to 2-propanol, the first step is to **dehydrate** it with concentrated H_2SO_4 to give the alkene **1-propene:**

$$H_3C\text{-}CH_2\text{-}CH_2\text{-}OH \rightarrow H_3C\text{-}CH{=}CH_2$$

Reaction of this with a hydrogen halide, such as HBr(g), forms both **bromopropane** and **2-bromopropane.** Hydrolysis of the the latter, *2-bromopropane*, with NaOH(aq) gives the required product, **2-propanol.**

$$H_3C\text{-}CH{=}CH_2 + HBr \rightarrow H_3C\text{-}CH_2\text{-}CH_2Br \textbf{ and } H_3C\text{-}CH(Br)\text{-}CH_3$$

$$H_3C\text{-}CH(Br)\text{-}CH_3 + OH^- \rightarrow H_3C\text{-}CH(OH)\text{-}CH_3 + Br^-$$

(b) *Dehydration* of **ethanol,** CH_3CH_2OH, with concentrated sulfuric acid gives **ethene,** which undergoes an *addition reaction* with bromine, Br_2, to give **1,2-dibromoethane,** $BrCH_2\text{-}CH_2Br$, which could be *hydrolyzed* with NaOH(aq) to give the required product **ethylene glycol,** $HO\text{-}CH_2CH_2\text{-}OH$:

$$CH_3CH_2OH + H_2SO_4 \rightarrow H_2C{=}CH_2 + H_3O^+ + HSO_4^-$$

$$H_2C{=}CH_2 + Br_2 \rightarrow BrCH_2\text{-}CH_2Br$$

$$BrCH_2\text{-}CH_2Br + 2OH^- \rightarrow HO\text{-}CH_2\text{-}CH_2\text{-}OH$$

(c) Ethylene glycol, 1,2-ethanediol, is a *primary* alcohol, so its *oxidation* with, for example, acidified $K_2Cr_2O_7(aq)$ gives the *dicarboxylic acid*, **oxalic acid** (ethanedioic acid):

$$HOCH_2\text{-}CH_2OH + 2[O] \rightarrow HO\text{-}C(O)\text{-}C(O)\text{-}OH + 2H_2O$$

Thiols and Disulfides

27. The α-amino acid cysteine, $HS\text{-}CH_2\text{-}CH(NH_2)\text{-}CO_2H$, forms a disulfide when oxidized. What is the structure of this disulfide?

Disulfides are readily formed from thiols by their oxidation; a reaction in which an oxygen atom combines with the hydrogen atoms of the **S–H** bonds of two thiol molecules to form an **S–S bond**, Thus for **cysteine**:

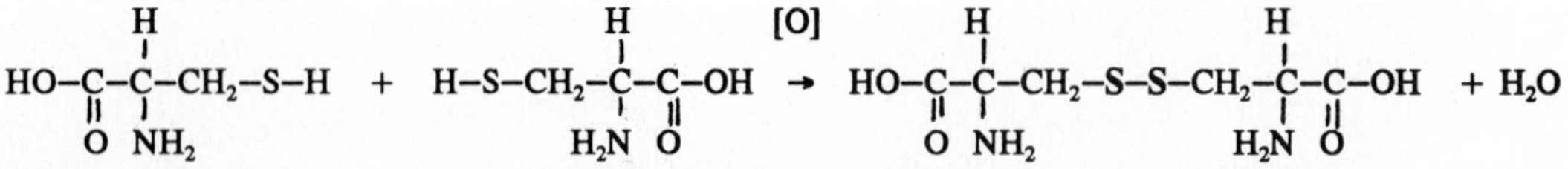

```
        H                     H         [O]        H                      H
        |                     |                    |                      |
HO–C–C–CH2–S–H  +  H–S–CH2–C–C–OH  →  HO–C–C–CH2–S–S–CH2–C–C–OH  + H2O
   ||  |                   |  ||          ||  |                  |  ||
   O  NH2                 H2N O           O  NH2                H2N O
```

Aldehydes and Ketones

28. Draw structural formulas for each of the following aldehydes and ketones: (a) butanal; (b) 2-pentanone; (c) 3-methyl-2-butanone; (d) 3,3-dimethylhexanal; (e) acetone; (f) formaldehyde.

The name gives us the number of C atoms in the longest carbon chain. The ending **-al** designates **an aldehyde,** and the ending **-one** designates a **ketone.** For aldehydes, the -C(H)O group always comes at the end of the carbon chain, so its C atom is always labeled carbon-1. For ketones, the position of the C atom of the C=O group is given by the name. For example, (a) **Butanal** is an **aldehyde** with a four carbon atom chain, and no other substituents, and (b) **2-pentanone** is a **ketone** with a five carbon atom chain, no other substituents, and the carbonyl group, C=O, at position-2.

Name	Type	Structural Formula
(a) butanal	aldehyde	$CH_3CH_2CH_2$–C=O with H on the C (–C(H)=O)
(b) 2-pentanone	ketone	$CH_3CH_2CH_2$–C–CH_3 with =O on the C (C=O)
(c) 3-methyl-2-butanone	ketone	CH_3–CH–C–CH_3 H_3C on CH; O (double bond) on C
(d) 3,3-dimethylhexanal	aldehyde	$CH_3CH_2CH_2$–C(CH_3)(CH_3)–CH_2–C(H)=O
(e) acetone (propanone)	ketone	H_3C–C–CH_3 with =O on the C (C=O)

(f) formaldehyde (methanal)	aldehyde	H–C(H)=O

---

30. The following aldehydes and ketones may be prepared by the oxidation of a suitable alcohol. In each case name the alcohol and draw its structural formula: (a) ethanal; (b) propanone; (c) 2-methylpropanal; (d) 2-pentanone; (e) cyclopentanone.

Aldehydes result in the first stage of the oxidation of **primary alcohols**, and **ketones** from the oxidation of **secondary alcohols**. Here, (a) and (c) are aldehydes that result from partial oxidation of the corresponding primary alcohols, and (b), (d), and (e) are ketones and result from oxidation of the corresponding secondary alcohols:

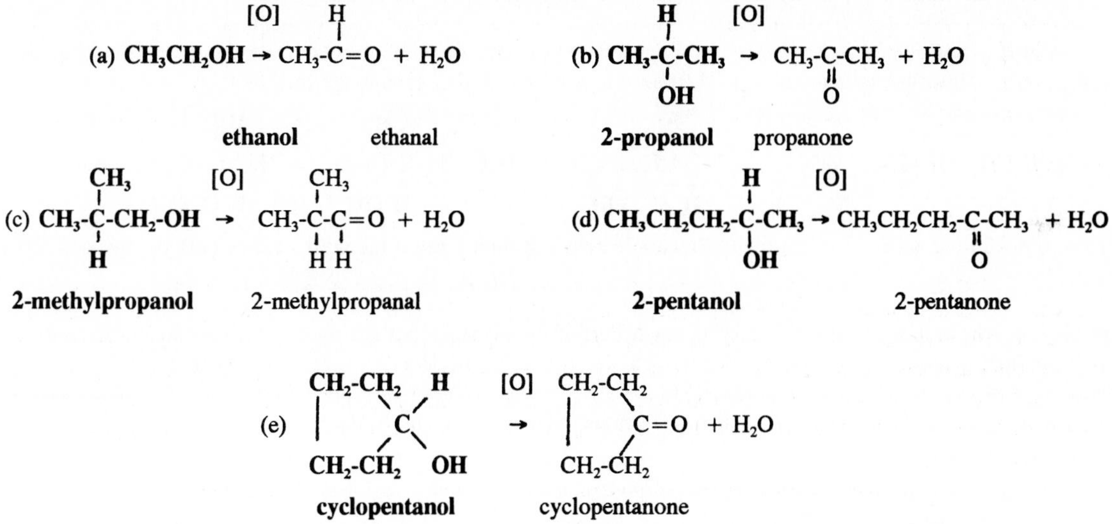

(a) **CH_3CH_2OH** $\xrightarrow{[O]}$ CH_3-C(H)=O + H_2O

ethanol ethanal

(b) **CH_3-C(H)(OH)-CH_3** $\xrightarrow{[O]}$ CH_3-C(=O)-CH_3 + H_2O

2-propanol propanone

(c) **CH_3-C(H)(CH_3)-CH_2-OH** $\xrightarrow{[O]}$ CH_3-C(H)(CH_3)-C(H)=O + H_2O

2-methylpropanol 2-methylpropanal

(d) **$CH_3CH_2CH_2$-C(H)(OH)-CH_3** $\xrightarrow{[O]}$ $CH_3CH_2CH_2$-C(=O)-CH_3 + H_2O

2-pentanol 2-pentanone

(e) (CH_2-CH_2)(CH_2-CH_2)C(H)(OH) $\xrightarrow{[O]}$ (CH_2-CH_2)(CH_2-CH_2)C=O + H_2O

cyclopentanol cyclopentanone

---

32. How are each of the following made industrially? (a) formaldehyde; (b) acetaldehyde; (c) acetone; (d) 2-butanone.

Industrially, aldehydes and ketones are made conveniently by the *dehydrogenation* of alcohols at around 500 to 600°C, using a catalyst, such as copper or silver. There is no loss of carbon atoms. **Primary alcohols** give the corresponding **aldehydes** and **secondary alcohols** give the corresponding ketones. Thus:

(a) **Formaldehyde,** $H_2C=O$, *methanal,* is made from **methanol:** $CH_3OH \rightarrow H_2C=O + H_2$

(b) **Acetaldehyde,** H_3C-C(H)O, *ethanal,* comes from **ethanol:** $C_2H_5OH \rightarrow CH_3CHO + H_2$

(c) **Acetone,** $(CH_3)_2C=O$, *propanone,* comes from **2-propanol:** $CH_3CH(OH)CH_3 \rightarrow (CH_3)_2C=O + H_2$

(d) **2-Butanone,** $CH_3CH_2C(O)CH_3$, comes from **2-butanol:** $CH_3CH_2CH(OH)CH_3 \rightarrow CH_3CH_2C(O)CH_3 + H_2$

---

34. A compound was found by analysis and molar mass determination to have the molecular formula $C_5H_{12}O$. Oxidation converts it to a compound with the molecular formula $C_5H_{10}O$, which gives the characteristic reactions of a ketone. Suggest two or more structural formulas for the original compound.

Straight chain ketones with molecular formula $C_5H_{10}O$ and thus *five* C atoms can have the carbonyl group, C=O, only at positions 2 or 3 and, thus, one of the following structural formulas:

CH_3-C(O)-CH_2-CH_2-CH_3 **or** CH_3-CH_2-C(O)-CH_2-CH_3
2-pentanone **3-pentanone**

which must have resulted from the oxidation of the corresponding *secondary alcohols*:

CH_3-CH(OH)-CH_2-CH_2-CH_3 and CH_3-CH_2-CH(OH)-CH_2-CH_3
2-pentanol **3-pentanol**

A *third* possibility is that the ketone is **3-methyl-2-butanone,** $CH_3C(O)CH(CH_3)_2$, resulting from the oxidation of **3-methyl-2-butanol,** $CH_3CH(OH)CH(CH_3)_2$, because this is *isomeric* with 2-pentanone.

Carboxylic Acids and Esters

35. Draw and name the four carboxylic acids with the molecular formula $C_5H_{10}O_2$.

The **-CO_2H** group comes at the end of the longest carbon-chain, so the molecular formula can be rewritten as $C_4H_{10}CO_2H$. Thus, we have to consider the possible *isomeric* **-C_4H_{10}** groups, giving:

H_3C-CH_2-CH_2-CH_2-C(=O)-OH	H_3C-CH(CH_3)-CH_2-C(=O)-OH	H_3C-CH_2-CH(CH_3)-C(=O)-OH	H_3C-C(CH_3)(H_3C)-C(=O)-OH
pentanoic acid	**3-methylbutanoic acid**	**2-methylbutanoic acid**	**2,2-dimethylpropanoic acid**

---

37. Name the carboxylic acids formed by oxidation of the alcohols and aldehydes: (a) 3-ethyl-1-hexanol; (b) 4,4-dimethylpentanal; (c) 4-methylbenzaldehyde; (d) 2,3,3-trimethylbutanol; (e) heptanal.

In each case, the resulting carboxylic acid has the same number of C atoms as the *primary alcohol* or *aldehyde* from which it is formed:

(a) **3-ethylhexanoic acid** (b) **4,4-dimethylpentanoic acid** (c) **4-methylbenzoic acid**
(d) **2,3,3-trimethylbutanoic acid** (e) **heptanoic acid**

---

38. Draw the structural formulas of three different esters with the molecular formula $C_5H_{10}O_2$ and name them.

An *ester*, **RO-C(O)R'** contains a R-O group and a R'-C=O group:

(i) **R** could be **-C_4H_9** and **R'** could be **H**, and there are *four* isomeric C_4H_9- groups:

-$CH_2CH_2CH_2CH_3$	**-CH(CH_3)CH_2CH_3**	**-CH_2CH(CH_3)$_2$** and	**-C(CH_3)$_3$**
1-butyl	**2-butyl**	**1-methylpropyl**	**2-methylpropyl**

(ii) **R** and **R'** could be **-CH_3** or -C_3H_7, where there are two possibilities for -C_3H_7:

-CH_2-CH_2-CH_3 or **-CH(CH_3)$_2$**
1-propyl **2-propyl**

(iii) Both R and R' could be **-C_2H_5, ethyl.**

Thus, you can select any *three* from a total of ***nine*** possible isomers:

1. H_3C-CH_2-CH_2-CH_2-O-C(=O)-H	2. H_3C-CH_2-CH(CH_3)-O-C(=O)-H	3. (H_3C)$_2$CH-CH_2-O-C(=O)-H	4. (H_3C)$_3$C-O-C(=O)-H
1-butyl methanoate	2-butyl methanoate	1-methylpropyl methanoate	2-methylpropyl methanoate

5. $H_3C\text{-}O\text{-}C(=O)\text{-}CH_2\text{-}CH_2\text{-}CH_3$ methyl butanoate

6. $H_3C\text{-}O\text{-}C(=O)\text{-}CH(CH_3)\text{-}CH_3$ methyl 2-methylpropanoate

7. $H_3C\text{-}CH_2\text{-}CH_2\text{-}O\text{-}C(=O)\text{-}CH_3$ 1-propyl ethanoate

8. $H_3C\text{-}C(=O)\text{-}O\text{-}CH(CH_3)\text{-}CH_3$ 2-propyl ethanoate

9. $H_3C\text{-}CH_2\text{-}O\text{-}C(=O)\text{-}CH_2\text{-}CH_3$ ethyl propanoate

---

40. Name the ester that could be obtained from each of the following reactions, and draw its structure: (a) methanoic acid and methanol (b) butanoic acid and ethanol (c) ethanoic acid and butanol (d) propionic acid and 2-propanol (e) 1 mol of phosphoric acid and 1 mol of ethanol.

Elimination of a molecule of H_2O between an alcohol HO-R and an acid R'C(O)OH gives an ester, RO-C(O)R':

(a) $H_3C\text{-}O\text{-}C(=O)\text{-}H$ methyl methanoate

(b) $H_3C\text{-}CH_2\text{-}O\text{-}C(=O)\text{-}CH_2\text{-}CH_2\text{-}CH_3$ ethyl butanoate

(c) $H_3C\text{-}CH_2\text{-}CH_2\text{-}CH_2\text{-}O\text{-}C(=O)\text{-}CH_3$ butyl ethanoate

(d) $H_3C\text{-}CH(CH_3)\text{-}O\text{-}C(=O)\text{-}CH_2\text{-}CH_3$ 2-propyl propanoate

(e) $H_3C\text{-}CH_2\text{-}O\text{-}P(=O)(OH)\text{-}OH$ ethyl dihydrogenphosphate

---

42. Write equations to describe how ethyl ethanoate could be made from ethene as the starting material.

Ethyl ethanoate is the ester resulting from the reaction of ethanol and ethanoic (acetic) acid. **Ethanol** may be synthesized by addition of H_2O across the double bond of ethene, a reaction that is catalyzed by acid:

$$H_2C{=}CH_2 + H\text{–}OH \rightarrow CH_3CH_2OH$$

Ethanol is oxidized by acidified dichromate to **ethanoic acid**

$$3CH_3CH_2OH + 2Cr_2O_7^{2-} + 16H_3O^+ \rightarrow 3CH_3CO_2H + 4Cr^{3+} + 27H_2O$$

and reaction of this ethanol with the ethanoic acid prepared above gives **ethyl ethanoate**:

$$CH_3CH_2OH + HO\text{-}C(=O)\text{-}CH_3 \rightarrow CH_3CH_2\text{-}O\text{-}C(=O)\text{-}CH_3 + H_2O$$

ethanol ethanoic acid **ethyl ethanoate**

Amines, Amides, and Amino Acids

44. Draw structures for and name: (a) primary, (b) secondary, and (c) tertiary amines with the molecular formula C_3H_9N.

A **primary amine**, RNH_2, has *one* N-C bond, a **secondary amine**, RR'NH, has *two*, and a **tertiary amine**, RR'R"N, has *three*:

(a) The alkyl group R in a *primary amine* $C_3H_7NH_2$ has the formula $-C_3H_7$, for which there are *two* structural possibilities:

CH_3-CH_2-CH_2- 1-propyl or $(CH_3)_2C(H)$- 2-propyl

Thus, there are *two* possible *primary amines* with the formula C_3H_9N:

$$CH_3\text{-}CH_2\text{-}CH_2\text{-}NH_2 \qquad CH_3\text{-}\underset{\displaystyle CH_3}{\overset{\displaystyle H}{C}}\text{-}NH_2$$

1-propylamine **2-propylamine**

(b) The alkyl groups in a *secondary amine* **RR'NH** with the formula C_3H_8NH, have to be $-CH_3$ and $-C_2H_5$:

$$CH_3\text{-}CH_2\text{-}\underset{\displaystyle H}{N}\text{-}CH_3$$

ethylmethylamine

(c) The alkyl groups in a *tertiary amine* **RR'R"N** with the formula C_3H_9N all have to be $-CH_3$ groups:

$$H_3C\text{-}\underset{\displaystyle CH_3}{N}\text{-}CH_3$$

trimethylamine

---

47. (a) Draw the structural formulas of three different amides with the molecular formula C_4H_9NO, and name them. (b) From what carboxylic acids and amines could each of the amides in part (a) be prepared?

(a) (i) The general formula R–C(O)NH_2, is one possibility, and in this case **R** must have the formula C_3H_7-, which is a **propyl** *or* an **isopropyl** group. (ii) Another possibility is **R–C(O)NH(R')**, where **R** and **R'** would have to add up to C_3H_8, so one could be $-CH_3$ and the other $-C_2H_5$. (iii) A third possibility is **R–C(O)NR'R"**, where R, R', and R" would have to add up to C_3H_9, so all would have to be $-CH_3$ groups. Thus, the **three** structures can be chosen from among:

$$H_3C\text{-}CH_2\text{-}CH_2\text{–}\underset{\displaystyle O}{\underset{\|}{C}}\text{–}NH_2 \qquad H_3C\text{-}\underset{\displaystyle H}{\overset{\displaystyle CH_3}{C}}\text{-}\underset{\displaystyle O}{\underset{\|}{C}}\text{–}NH_2 \qquad H_3C\text{–}\underset{\displaystyle O}{\underset{\|}{C}}\text{–}\underset{\displaystyle H}{N}\text{–}CH_2\text{-}CH_3 \qquad H_3C\text{-}CH_2\text{–}\underset{\displaystyle O}{\underset{\|}{C}}\text{–}\underset{\displaystyle H}{N}\text{–}CH_3$$

(i) butanamide **2-methylpropanamide** **(ii) N-ethylethanamide** **N-methylpropanamide**

$$H_3C\text{–}\underset{\displaystyle O}{\underset{\|}{C}}\text{–}\underset{\displaystyle CH_3}{N}\text{–}CH_3$$

(iii) N,N-dimethylethanamide

(b) An **amide** is formed by reacting the appropriate **amine**, **RR'NH**, with the appropriate **carboxylic acid**, **R"CO_2H**, to give the salt, **R"CO_2^- RR'NH_2^+**, which is then decomposed to the amide and water by heating:

$$R''CO_2H + RR'NH_2 \rightarrow [R''CO_2^- \; RR'NH_2^+] \rightarrow R''C(O)NRR' + H_2O$$

(i)

$$H_3CCH_2CH_2CO_2H + NH_3 \rightarrow H_3CCH_2CH_2C(O)NH_2 + H_2O$$

butanoic acid **ammonia** butanamide

$$(H_3C)_2C(H)CO_2H + NH_3 \rightarrow H_3CCH_2CH_2C(O)NH_2 + H_2O$$

2-methylpropanoic acid **ammonia** 2-methylpropanamide

(ii) $H_3CCO_2H + C_2H_5NH_2 \rightarrow H_3CC(O)N(H)C_2H_5 + H_2O$
ethanoic acid ethylamine N-ethylethanamide

$H_3CCH_2CO_2H + CH_3NH_2 \rightarrow H_3CCH_22C(O)N(H)CH_3 + H_2O$
propanoic acid methylamine N-methylpropanamide

(iii) $H_3CCO_2H + (CH_3)_2NH \rightarrow H_3CC(O)N(CH_3)_2 + H_2O$
ethanoic acid dimethylamine N,N-dimethylethanamide

---

48. (a) Arrange the following in the order of increasing basicity: ammonia (pK_b 4.74), methylamine (pK_b 3.41), dimethylamine (pK_b 3.27), trimethylamine (pK_b 4.19), and aniline (pK_b 9.39). (b) What is the pH of an 0.100-M solution of (i) the strongest, and (ii) the weakest of these bases?

(a) $pK_b = -\log K_b$, so the larger the value of K_b, the weaker the base, and in order of increasing basicity:

aniline < ammonia < trimethylamine < methylamine < dimethylamine

(b) The *weakest* of the organic bases is **aniline** ($K_b = 4.07 \times 10^{-10}$ M) and the *strongest* is **dimethylamine** ($K_b = 5.37 \times 10^{-4}$ mol L^{-1}). In each case the equilibrium position (Chapter 12) is given by:

In each case, we have the equilibrium:

	B(aq)	+ $H_2O(\ell)$ ⇄	BH^+(aq)	+ OH^-(aq)	
initially	0.100	-	0	0	mol L^{-1}
at equilibrium	0.100-x	-	x	x	mol L^{-1}

$$K_b = \frac{[BH^+][OH^-]}{[B]} = \frac{x^2}{0.100 - x} \text{ mol L}^{-1} = \frac{x^2}{0.100} \text{ mol L}^{-1} \quad \text{(provided } x << 0.100)$$

(i) Aniline

$$\frac{x^2}{0.100} = 4.07 \times 10^{-10} \text{ mol L}^{-1}\ ; \quad [OH^-] = x = \sqrt{4.07 \times 10^{-11}} = \underline{6.38 \times 10^{-6} \text{ mol L}^{-1}}$$

$$pOH = -\log(6.38 \times 10^{-6}) = 5.20\ ; \quad \underline{pH} = 14.00 - pOH = \underline{8.80}$$

(ii) Dimethylamine:

$$\frac{x^2}{0.100} = 5.37 \times 10^{-4} \text{ mol L}^{-1}\ ; \quad [OH^-] = x = \sqrt{5.37 \times 10^{-5}} = \underline{7.33 \times 10^{-3} \text{ mol L}^{-1}}$$

$$pOH = -\log(7.33 \times 10^{-3}) = 2.14\ ; \quad \underline{pH} = 14.00 - pOH = \underline{11.86}$$

---

50. (a) What do the terms "chiral" and "achiral" mean? (b) Use the terms "chiral" or "achiral" to describe a shoe, a house key, a baseball, a nail, and a screw.

(a) **Chiral** is a term used to describe molecules (or objects) that have *nonsuperimposable* mirror images of each other and are therefore *isomers*; they have no plane of symmetry and are described as *right-hand* and *left-hand* isomers. **Achiral** describes molecules (or objects) that have *superimposable* mirror images and are identified through having a *plane of symmetry*, such that the plane of symmetry divides a molecule into two-halves which are reflections of each other in the symmetry plane.

(b) **Chiral objects** have no plane of symmetry, whereas **achiral objects** have a plane of symmetry. The easiest way to distinguish chiral and achiral objects is to look for this plane, or to see if the mirror images are superimposable (*chiral*) or nonsuperimposable (*achiral*).

A **shoe** is "right-foot" or "left-foot" and is *chiral*; a yale **house key** (the most familiar type) is *achiral*; by examining the disposition of the seam on a **baseball** it can be seen to be *chiral*; a **nail** is *achiral*; and a **screw** is "right-hand" or "left-hand" and is *chiral*.

---

51. Categorize as chiral or achiral: (a) 2-butanol, (b) 2-chloro-1-propanol, (c) 2-bromo-2-chlorobutane, (d) 2-aminoethanoic acid, and (e) 2-aminopropanoic acid.

A **chiral** organic molecule has at least *one* AXX'X"X"' tetrahedral carbon atom to which *four different* atoms or groups are bonded. Thus, we draw the structural formulas and examine each for this property.

(a) **2-butanol**

```
          CH3
H3C-CH2-C-OH
          H
```

The C atom indicated is bonded to four different groups. **2-butanol** is **chiral.**

(b) **2-chloro-1-propanol**

```
          Cl
H3C-CH2-C-OH
          H
```

The C atom indicated is bonded to four different groups. **2-chloro-1-propanol** is **chiral.**

(c) **2-bromo-2-chlorobutane**

```
          Br
H3C-CH2-C-CH3
          Cl
```

The C atom indicated is bonded to four different groups. **2-bromo-2-chlorobutane** is **chiral.**

(d) **2-aminoethanoic acid**

```
     H
H2N-C-C=O
     H OH
```

The C atom indicated is the only tetrahedrally substituted C atom but is *not* bonded to four different groups. **2-aminoethanoic acid** is **achiral.**

(e) **2-aminopropanoic acid**

```
     NH2
H3C-C-C=O
     H OH
```

The C atom indicated is bonded to four different groups. **2-aminopropanoic acid** is **chiral.**

Determining the Structure of Organic Compounds

53. Suggest structures for each of the following: (a) A compound of molecular formula C_4H_8O with three peaks in its nmr spectrum of relative intensity 3:3:2, and a strong stretching frequency in its infrared spectrum at 1720 cm^{-1}. (b) A compound of molecular formula $C_4H_7ClO_2$ with four peaks in its nmr spectrum of relative intensity 3:2:1:1, and an infrared spectrum with a broad absorption band in the range 2500-3000 cm^{-1} and an intense band at 1715 cm^{-1}.

(a) A strong stretching frequency at **1720** $\mathbf{cm^{-1}}$ is indicative of a carbonyl group (C=O), and the NMR spectrum suggests **two** *non identical* $\mathbf{CH_3}$ **groups** and **one** $\mathbf{CH_2}$ **group**, which is consistent with the structure of the ketone:

```
H3C-C-CH2-CH3
    ||
    O
```

2-butanone

(b) The intense band in the IR spectrum at **1715** $\mathbf{cm^{-1}}$ is characteristic of a carbonyl group, C=O, and the broad band in the region **2500-3000** $\mathbf{cm^{-1}}$ is indicative of an O-H group, accounting for the two O atoms in the formula. Thus, the NMR spectrum must have a peak due to O—H, which leaves the other intensities to be assigned to **one** $\mathbf{CH_3}$ **group, one** $\mathbf{CH_2}$ **group**, and **a CH group**, which is what is expected for the structures:

```
H3C-CH2-CH-C-OH        H3C-CH-CH2-C-OH        Cl-CH2-CH-C-OH
        |  ||              |      ||                 |  ||
        Cl O               Cl     O                  CH3 O
```

2-chlorobutanoic acid | **3-chlorobutanoic acid** | **3-chloro-2-methylpropionic acid**

---

55. The peaks in the mass spectrum of the parent ion of a chloroalkane correspond to masses of 78 u and 80 u. There are two important isotopes of chlorine, ^{35}Cl and ^{37}Cl. (a) What are the possible structures of the compound? (b) How could they be distinguished by proton NMR spectroscopy?

(a) Since the two isotopes ^{35}Cl and ^{37}Cl differ in mass by **2u**, and the masses of the parent ion differ by the *same mass*, one must contain **one** ^{35}Cl atom and the other must contain **one** ^{37}Cl atom.

Subtracting the mass of one Cl atom of the appropriate isotope from the mass of each parent ion gives as *residual mass* of (78 - 35)u, <u>or</u> (80 - 37)u, = **43 u.**

43u must be the mass of the alkyl group of an alkyl halide of formula $C_nH_{2n-3}Cl$, and since the H atoms contribute a small fraction of the total mass, **n must be 3.**

The **chloroalkane** in question is thus C_3H_7Cl, with the possible structures:

$H_3C{-}CH_2{-}CH_2{-}Cl$ or $H_3C{-}\underset{\displaystyle Cl}{\underset{|}{C}}H{-}CH_3$

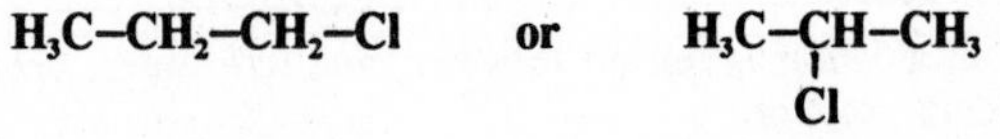

1-chloropropane **2-chloropropane**

(b) **1-Chloropropane** has **one** CH_3 **group**, and **two** CH_2 **groups**, *all* of which have different chemical environments. Thus, **three** principal peaks, with relative intensities **3:2:2** would be observed in its NMR spectrum.

2-Chloropropane has **two** *identical* CH_3 **groups**, and a chemically different **C-H group**. Thus, its NMR spectrum would consist of **two principal peaks**, with relative intensities **6:1.**

56.* Suggest a likely structure for each of the compounds with the following molecular formulas and numbers of peaks in their proton NMR spectra, with the indicated relative intensities: (a) C_2H_6O with a single peak; (b) C_2H_6O with three peaks of intensity 3:2:1; (c) $C_4H_{10}O$, with three peaks of intensity 6:3:1; (d) $C_5H_{10}O_2$, with two peaks of intensity 3:2.

(a) Observation of **one** peak only in the NMR spectrum of C_2H_6O implies that all **six H** atoms are in the same chemical environment, which can only be true if they constitute the H atoms of two -CH_3 groups, which must be attached to the single O atom. Thus, the compound is **dimethyl ether** (methoxy methane) $H_3C{-}O{-}CH_3$.

(b) The **3:2:1** ratio of intensities of the peaks due to the **six** H atoms of C_2H_6O means that they constitute chemically different groups containing 1, 2, and 3 hydrogen atoms, respectively. This is what is expected if the compound is **ethanol,** $H_3C{-}CH_2{-}OH$.

(c) The spectral intensities of **6:3:1** indicate that the **ten** H atoms of $C_4H_{10}O$ are present as **two** identical CH_3 groups, **one** CH_3 group that is different from the other two, and **one** CH or OH group. One possible structure that can be discarded on this basis is $(CH_3)_3C$-OH, *tertiary-butanol* (2-methyl-2-propanol), for which all three CH_3 groups are chemically similar, and for which only <u>two</u> nmr peaks of relative intensities 9:1 are expected. Another possibility is $(CH_3)_2(H)C$-OCH_3, **2-methoxypropane**, which would give the observed NMR spectrum:

$$H_3C{-}\overset{\displaystyle CH_3}{\overset{|}{\underset{\displaystyle CH_3}{\underset{|}{C}}}}{-}OH \qquad H{-}\overset{\displaystyle CH_3}{\overset{|}{\underset{\displaystyle CH_3}{\underset{|}{C}}}}{-}O{-}CH_3$$

t-butanol (9:1) **2-methoxypropane (6:3:1)**

(d) The spectral intensities of 3:2 indicate that the **ten** H atoms in $C_5H_{10}O_2$ are present as *two* identical CH_3 groups and *two* identical CH_2 groups, leaving just one more C atom and the **two O** atoms to be accommodated in chemically similar environments (otherwise they would make the CH_3 and CH_2 groups nonidentical). The additional C atom cannot be bonded to any H atoms, so the only likely possibility is that it is bonded to *two* C atoms and *two* O atoms. This suggests that most likely structure is that of **1,1-dimethoxycyclopropane.**

$$\begin{array}{c} H_3C \quad CH_3 \\ O \quad O \\ C \\ H_2C - CH_2 \end{array}$$

1,1-dimethoxycyclopropane

General Problems

59. Classify each of the following as an acid or a base in water, or both, or as having neither acidic nor basic properties, and explain your choice: (a) ethanol, (b) ethylamine, (c) pentane, (d) bromoethane, (e) phenol, (f) ethyl acetate, (g) methyl dihydrogenphosphate, (h) aminobenzene (aniline), (i) ethanamide.

(a) **Ethanol**, C_2H_5OH, like all alcohols is related to water, and like water, behaves both as a very weak acid and as a very weak base in *aqueous* solution:

$$C_2H_5OH + NaOH \rightleftarrows C_2H_5O^- + Na^+ + H_2O \quad \textbf{ACID}; \quad C_2H_5OH + H_3O^+ \rightleftarrows C_2H_5OH_2^+ + H_2O \quad \textbf{BASE}$$

(b) Ethylamine, $C_2H_5NH_2$, is related to $:NH_3$, ammonia; like ammonia it behaves as a weak base in water:

$$C_2H_5NH_2 + H_2O \rightleftarrows C_2H_5NH_3^+ + OH^- \quad \textbf{BASE}$$

(c) **Pentane**, C_5H_{12}, like all saturated hydrocarbons has *neither* acidic *nor* basic properties in water.

(d) **Bromoethane**, C_2H_5Br, a substituted alkane, is *neither* an acid *nor* a base in water.

(e) **Phenol**, C_6H_5OH, like all *alcohols* is a weak acid and a very weak base in water:

$$C_6H_5OH + H_2O \rightleftarrows H_3O^+ + C_6H_5O^- \quad \textbf{ACID}; \quad C_6H_5OH + H_2O \rightleftarrows C_6H_5OH_2^+ + OH^- \quad \textbf{BASE}$$

(f) **Ethylacetate**, $C_2H_5CO_2CH_3$, contains a carbonyl, C=O, group and is a weak base in water:

$$CH_3{-}CH_2{-}O{-}\underset{\underset{\ddot{O}:}{\|}}{C}{-}CH_3 + H_3O^+ \rightleftarrows H_3C{-}CH_2{-}O{-}\underset{\underset{+\ddot{O}-H}{\|}}{C}{-}CH_3 + H_2O \quad \textbf{BASE}$$

(g) **Methyl dihydrogenphosphate**, $CH_3O\text{-}PO(OH)_2$, like phosphoric acid, is a weak acid in water:

$$CH_3{-}O{-}\underset{\underset{O}{\|}}{\overset{\overset{OH}{|}}{P}}{-}O{-}H + H_2O \rightleftarrows CH_3{-}O{-}\underset{\underset{O}{\|}}{\overset{\overset{OH}{|}}{P}}{-}O^- + H_3O^+ \quad \textbf{ACID}$$

(h) **Aminobenzene**, $C_6H_5NH_2$, aniline, like ammonia and methylamine is a weak base in water:

$$C_6H_5NH_2 + H_2O \rightleftarrows C_6H_5NH_3^+ + OH^- \quad \textbf{BASE}$$

(i) **Ethanamide**, $CH_3C(O)NH_2$, contains a carbonyl, C=O, group and is a weak base in water:

$$CH_3C(O)NH_2 + H_2O \rightleftarrows CH_3{-}\underset{\underset{+O-H}{\|}}{C}{-}NH_2 + OH^- \quad \textbf{BASE}$$

---

60. Give reaction schemes (one or more steps) for the preparation of each of the following. For each, name the starting organic reactant and give the reaction conditions: (a) ethyl bromide from an alkane; (b) ethyl bromide from an alkene; (c) ethanal from an alcohol; (d) diethyl ether from an alcohol; (e) propanoic acid from an alcohol.

(a) **Ethyl bromide**, C_2H_5Br, results from the photochemical chain reaction between bromine, Br_2, and **ethane**, C_2H_6, in the gas phase: Overall

$$C_2H_6(g) + Br_2(g) \rightarrow C_2H_5Br(g) + HBr(g)$$

(b) **Ethyl bromide**, C_2H_5Br, results from the addition of HBr to the double bond of **ethene**, $H_2C{=}CH_2$, in the gas phase:

$$C_2H_4(g) + HBr(g) \rightarrow C_2H_5Br(g)$$

(c) An aldehyde is the product of either the *dehydrogenation*, or the *oxidation*, of the corresponding *primary* alcohol. **Ethanal**, CH_3CHO, acetaldehyde, results from *either* the dehydrogenation in the gas phase, using a catalyst at high temperature, *or* the oxidation, using an oxidizing agent such as $K_2Cr_2O_7$ in acidic solution, of **ethanol**, C_2H_5OH:

$$H_3C{-}CH_2{-}OH \rightarrow H_3C{-}\underset{\displaystyle H}{\underset{|}{C}}{=}O + H_2 \quad \textbf{dehydrogenation}$$

$$H_3C{-}CH_2{-}OH + [O] \rightarrow H_3C{-}\underset{\displaystyle H}{\underset{|}{C}}{=}O + H_2O \quad \textbf{oxidation}$$

(d) An ether results from reaction between an alkoxide (such as sodium alkoxide) and a **haloalkane** (such as a bromoalkane). **Diethyl ether** is formed by the reaction between **sodium ethoxide**, $C_2H_5O^-\ Na^+$, and **bromoethane**, C_2H_5Br, ethyl bromide. Sodium ethoxide results from the reaction of **ethanol** with NaOH(aq), *or* reaction directly with sodium:

$$C_2H_5OH + NaOH \rightarrow C_2H_5O^-\ Na^+ + H_2O \quad \textbf{or} \quad 2C_2H_5OH + 2Na \rightarrow 2C_2H_5O^-\ Na^+ + H_2(g)$$

and ethyl bromide from the reaction of ethanol with HBr:

$$C_2H_5OH + HBr \rightarrow C_2H_5Br + H_2O$$

and then reaction of the **sodium ethoxide** with the **ethyl bromide** gives **diethyl ether** (ethylethoxide):

$$C_2H_5O^-\ Na^+ + C_2H_5Br \rightarrow (C_2H_5)_2O + NaBr$$

(e) A carboxylic acid results from the oxidation (via the aldehyde) of the primary alcohol with the same number of carbon atoms. **Propanoic acid** results from the oxidation of **1-propanol**, C_3H_7OH:

$$CH_3CH_2{-}CH_2OH + 2[O] \rightarrow CH_3CH_2{-}\underset{\displaystyle OH}{\underset{|}{C}}{=}O + H_2O$$

---

62. In terms of suitable balanced equations, describe the manufacture of each of the following: (a) synthesis gas; (b) methanol from synthesis gas; (c) methanal from methanol; (d) methylamine from methanol

---

(a) $$CH_4(g) + H_2O(g) \xrightarrow[\text{Ni catalyst}]{700\text{-}800^\circ C} CO(g) + 3H_2(g)$$

(b) $$CO(g) + 2H_2(g) \xrightarrow[Al_2O_3(s)\text{ catalyst}]{250^\circ C\text{ at }50\text{-}100\text{ atm}} CH_3OH(g)$$

(c) $$CH_3OH(g) \xrightarrow[\text{Cu catalyst}]{500\text{-}600^\circ C} H_2C{=}O(g) + H_2(g)$$

(d) $$CH_3OH(g) + NH_3(g) \xrightarrow[Al_2O_3(s)\text{ catalyst}]{400^\circ C} CH_3NH_2(g) + H_2O(g)$$

---

64. Each of the compounds with the following structural formulas may be synthesized from an alkene, or an alkyne, and another reactant. In each case give the name and structure of the alkene or alkyne, and name the other reactant: (a) $CH_3CH(OH)CH_3$ (b) $CH_3CBr_2CBr_2CH_3$ (c) $CH_3CH{=}CHCH_3$ (d) $CH_3C(Br){=}CH_2$

(a) An **alcohol** results from catalytic addition of **water** across the double bond of an **alkene**, in this case **propene**:

$$CH_3\text{-}\underset{\displaystyle H}{\underset{|}{C}}{=}CH_2 + H_2O \rightarrow CH_3\text{-}\underset{\displaystyle OH}{\underset{|}{C}H}\text{-}CH_3$$

propene **2-propanol**

(b) **2,2,3,3-tetrabromobutane** results from reaction of **2-butyne** with **bromine:**

$$H_3C\text{-}C\equiv C\text{-}CH_3 + 2Br_2 \rightarrow H_3C\text{-}CBr_2\text{-}CBr_2\text{-}CH_3$$

2-butyne **2,2,3,3-tetrabromobutane**

(c) **2-butene** results from the addition of **hydrogen** to **2-butyne:**

$$H_3C\text{-}C\equiv C\text{-}CH_3 + H_2 \rightarrow H_3C\text{-}\underset{H}{\underset{|}{C}}=\underset{H}{\underset{|}{C}}\text{-}CH_3$$

2-butyne **2-butene (cis & trans)**

(d) **2-bromopropene** results from the addition of hydrogen bromide to **propyne** (1-bromopropene would also form):

$$H_3C\text{-}C\equiv C\text{-}H + HBr \rightarrow H_3C\text{-}\underset{Br}{\underset{|}{C}} = \underset{H}{\underset{|}{C}}\text{-}H$$

propyne **2-bromopropene**

66.* A mass of 1.00 g of a primary alcohol was completely oxidized to a carboxylic acid, which required 83.3 mL of 0.20-M NaOH(aq) to reach the equivalence point in a titration. (a) Calculate the molar mass of the alcohol, deduce its molecular formula, write its structural formula, and give its systematic name. (b) Name and give the structure of another alcohol that is isomeric with the alcohol in (a), and discuss the possible products of its oxidation with acidic $K_2Cr_2O_7$(aq).

(a) A *primary alcohol* must have the general formula **RCH_2OH**, and the reactions described are:

(i) Oxidation of the alcohol to a carboxylic acid: $R\text{-}CH_2\text{-}OH + \rightarrow R\text{-}CO_2H + H_2O$

(ii) The neutralization reaction: $R\text{-}CO_2H + NaOH \rightarrow Na^+ + RCO_2^- + H_2O$

The initial moles of alcohol, and hence its *molar mass*, can be calculated from the titration results:

$$\text{mol } RCH_2OH = (83.3 \text{ mL})\left(\frac{1 \text{ L}}{10^3 \text{ mL}}\right)\left(\frac{0.200 \text{ mol NaOH}}{1 \text{ L}}\right)\left(\frac{1 \text{ mol } RCO_2H}{1 \text{ mol NaOH}}\right)\left(\frac{1 \text{ mol } RCH_2OH}{1 \text{ mol } RCO_2H}\right) = \underline{1.666 \times 10^{-2} \text{mol}}$$

$$\text{molar mass of } RCH_2OH = \frac{1.00 \text{ g } RCH_2OH}{1.666 \times 10^{-2} \text{ mol } RCH_2OH} = \underline{60.0 \text{ g mol}^{-1}}$$

Rewriting the general formula for a *primary alcohol* ass $C_nH_{2n+1}OH$, and subtracting the OH mass of 17.01 g mol^{-1} from the experimental molar mass of 60.0 g mol^{-1}, gives 43.0 g mol^{-1} for the C_nH_{2n+1} entity. Hence, **n = 3**, and the alcohol is **C_3H_7OH, 1-propanol**, (molar mass 60.09 g mol^{-1}):

$$CH_3\text{–}CH_2\text{–}CH_2\text{–}OH$$

1-propanol

(b) Isomeric with 1-propanol is the *secondary alcohol* **2-propanol:**

$$H_3C\text{-}\overset{H}{\overset{|}{\underset{OH}{\underset{|}{C}}}}\text{-}CH_3$$

2-propanol

In general, *primary* alcohols are oxidized by $K_2Cr_2O_7$(aq) to *aldehydes* and then to *carboxylic acids*, whereas *secondary* alcohols are oxidized only to *ketones*. Thus, the product of oxidation by $Cr_2O_7^{2-}$(aq) of **2-propanol** is the *ketone* **2-propanone** :

$$H_3C\text{-}\underset{O}{\underset{\|}{C}}\text{-}CH_3$$

2-propanone

68.* A mass of 256 mg of a compound containing carbon, hydrogen, and oxygen was burned in excess oxygen to give 512 mg CO_2 and 209 mg H_2O. At 100°C, 156 mg of the compound occupied a volume of 93.3 mL at a pressure of 882 torr. (a) Suggest a structure for the compound. (b) Identify its functional group. (c) Suggest tests that would confirm the presence of this functional group.

(a) As in Problem 67, we first use the analytical data to calculate the *empirical formula*, and then use the gas data to determine the *molar mass*. However, we have first to calculate the moles of oxygen and moles of hydrogen in the sample, and their masses, and then we can obtain the mass and moles of oxygen in the sample, by difference, from the mass of the sample of the compound:

$$\text{moles of C} = (0.512\ \text{g C})\left(\frac{1\ \text{mol } CO_2}{44.01\ \text{g } CO_2}\right)\left(\frac{1\ \text{mol C}}{1\ \text{mol } CO_2}\right) = \underline{1.16 \times 10^{-2}\ \text{mol}}$$

$$\text{moles of H} = (0.209\ \text{g } H_2O)\left(\frac{1\ \text{mol } H_2O}{18.02\ \text{g } H_2O}\right)\left(\frac{2\ \text{mol H}}{1\ \text{mol } H_2O}\right) = \underline{2.32 \times 10^{-2}\ \text{mol}}$$

Thus, the sample contains: $(1.16 \times 10^{-2}\ \text{mol C})\left(\frac{12.01\ \text{g C}}{1\ \text{mol C}}\right) = \underline{0.139\ \text{g C}}$

$$(2.32 \times 10^{-2}\ \text{mol H})\left(\frac{1.008\ \text{g H}}{1\ \text{mol H}}\right) = \underline{0.0234\ \text{g H}}$$

By difference, the 256 mg sample contains : $(0.256 - 0.139 - 0.023)\ \text{g} = \underline{0.094\ \text{g O}}$

Thus, sample contains: $(0.094\ \text{g O})\left(\frac{1\ \text{mol O}}{16.00\ \text{g O}}\right) = \underline{5.9 \times 10^{-3}\ \text{mol O}}$

<u>Mole (atom) ratio</u> C : H : O = 1.16 : 2.32 : 0.59 = <u>2.0 : 4.0 : 1.0</u>

Therefore, the **empirical formula** is C_2H_4O (formula mass **44.05 u**), and we can now calculate the **molar mass:**

$$\text{moles of gas} = \frac{PV}{RT} = \frac{(882\ \text{torr})\left(\frac{1\ \text{atm}}{760\ \text{torr}}\right)(93.3\ \text{mL})\left(\frac{1\ \text{L}}{10^3\ \text{mL}}\right)}{(0.0821\ \text{atm L mol}^{-1}\ \text{K}^{-1})(373.1\ \text{k})} = \underline{3.535 \times 10^{-3}\ \text{mol}}$$

$$\text{molar mass} = \frac{0.156\ \text{g compound}}{3.535 \times 10^{-3}\ \text{mol compound}} = \underline{44.1\ \text{g mol}^{-1}}$$

Thus, the molecular mass is the same as the empirical formula mass, so the **molecular formula is C_2H_4O.**

There are two known compounds with the molecular formula C_2H_4O: **ethanal** and **ethylene oxide** with the structures:

```
   H                H  H
   |                |  |
 H-C-C=O          H-C--C-H
   | |               \/
   H H               O
 ethanal        ethylene oxide
```

(c) **Ethanal** is readily oxidized to **ethanoic acid.** Familiar tests include the *"silver mirror test,"* in which $Ag(NH_3)_2^+(aq)$ is *reduced* to **metallic silver,** and *reduction* of **orange** $Cr_2O_7^{2-}(aq)$ to **green** $Cr^{3+}(aq)$, or **purple** $MnO_4^-(aq)$ to **colorless** $Mn^{2+}(aq)$, in *aqueous acidic solution.*

Ethylene oxide would be unaffected by the above reagents (no tests for it were given in the text).

69.* An organic compound A contains 52.1% C, 13.1% H, and 34.8% O by mass. At 100°C and 1.00 atm, 0.230 g of A occupies a volume of 153 mL. (a) Find the empirical and molecular formulas of A, and draw the possible structural formulas. (b) If the compound reacts with sodium to give hydrogen and a sodium salt, write the balanced equation for the reaction and name A and its salt. (c) What volume of hydrogen at 25°C and a pressure of 730 torr would result from the reaction of 0.250 g of A with excess sodium?

(a) As in Problems 67 and 68, we first calculate the **empirical formula**, the **molar mass**, and the **molecular formula**:

	C	H	O
mass in 100 g sample	52.1 g	13.1 g	34.8 g
moles in 100 g sample	$\frac{52.1\ g}{12.01\ g\ mol^{-1}}$	$\frac{13.1\ g}{1.008\ g\ mol^{-1}}$	$\frac{34.8\ g}{16.00\ g\ mol^{-1}}$
	= 4.34 mol	**= 13.0 mol**	**= 2.18 mol**
Ratio of moles (atoms)	$\frac{4.34}{2.18}$	$\frac{13.0}{2.18}$	$\frac{2.18}{2.18}$
	1.99	**5.96**	**1.00**

Giving the **empirical formula** C_2H_6O (formula mass **46.07** u). From the gas data:

$$\text{moles of sample} = \frac{PV}{RT} = \frac{(1.00\ atm)(153\ mL)(\frac{1\ L}{10^3\ mL})}{(0.0821\ atm\ L\ mol^{-1}\ K^{-1})(373.1\ K)} = \underline{4.995 \times 10^{-3}\ mol}$$

$$\text{molar mass} = \frac{0.230\ g}{4.995 \times 10^{-3}\ mol} = \underline{46.0\ g\ mol^{-1}}$$

Thus, the **molecular formula** is the same as the empirical formula, i.e., C_2H_6O.

Two isomers have this molecular formula:

$CH_3{-}CH_2{-}OH$ and $H_3C{-}O{-}CH_3$

ethanol **diethyl ether**

(b) Since the substance reacts with sodium with the evolution of hydrogen it must be **ethanol**:

$$2C_2H_5OH + 2Na \rightarrow 2C_2H_5O^- \ Na^+ + H_2(g)$$

sodium ethoxide hydrogen

(c) We first calculate the moles of ethanol, then the moles of hydrogen evolved, and the volume of this hydrogen:

$$\text{moles of } H_2 = (0.250\ g\ ethanol)(\frac{1\ mol\ ethanol}{46.07\ g\ ethanol})(\frac{1\ mol\ H_2}{2\ mol\ ethanol}) = \underline{2.713 \times 10^{-3}\ mol}$$

$$\text{Volume of } H_2 = \frac{nRT}{P} = \frac{(2.713 \times 10^{-3}\ mol)(0.0821\ atm\ L\ mol^{-1}\ K^{-1})(298\ K)}{(730\ torr)(\frac{1\ atm}{760\ torr})}$$

$$= \underline{6.91 \times 10^{-2}\ L}\ ;\ (69.1\ mL)$$

71.* (a) Of three isomers with the molecular formula C_4H_8, two have strong infrared frequencies close to 2200 cm^{-1}, and the third has a strong infrared frequency at 1630 cm^{-1}. Suggest their likely structures, and name them.
(b) Given a sample known to be one of the isomers in (a), could it be unambiguously identified from a combination of data from infrared and proton NMR spectroscopic data?

(a) In these *hydrocarbons* the only possible functional groups are C=C, expected to have a strong infrared frequency in the range **1600 to 1700 cm^{-1}**, and C≡C, expected to have a strong infrared frequency in the range **2100 to 2300 cm^{-1}**. Thus, *two* of the isomers contain C≡C bonds and the *third* contains at least one C=C bond.

There are two **acetylenes** with the molecular formula C_4H_6:

$H_3C–C≡C–CH_3$ (**2-butyne**) and **$H–C≡C–CH_2–CH_3$** (**1-butyne**)

so these are the two isomers with strong infrared frequencies close to 2200 cm^{-1}.

The third isomer, with *one* or *more* C=C bonds, could be a *cyclic alkene*, for which there are *three* possibilities, namely **cyclobutene**, **1-methylcyclopropene**, and **2-methylcyclopropene**, or it could be the *diene* **1,3-butadiene**:

```
H2C – CH2          H3C—C—H              H—C—H              H  H
 |     |               / \                 / \               |  |
H–C = C–H          H–C = C–H          H3C–C = C–H      H2C=C—C=CH2
cyclobutene    1-methylcyclopropene  2-methylcyclopropene  1,3-butadiene
```

(b) The *infrared spectrum* alone will distinguish the **butenes** from the **butynes**, and the anticipated *proton NMR spectra* are shown below:

Compound	Number of NMR Peaks	Intensities
1-Butyne	three	1:2:3
2-Butyne	one	-
Cyclobutene	two	2:1
1-Methylcyclopropene	three	1:2:3
2-Methylcyclopropene	three	1:2:3
1,3-Butadiene	two	2:1

For the *alkynes*, **2-butyne** has a **single** peak in its NMR spectrum, whereas **1-butyne** has **three peaks** with relative intensities **1:2:3**. So for these the combination of *infrared* and *NMR* data unambiguously distinguishes one from the other.

For the **third isomer**, the situation is less straightforward: **Cyclobutene** has **two** NMR peaks with relative intensities **2:1**,but so does **1,3-butadiene.** Similarly, **1-methylcyclopropene** and **2-methylcyclopropene** both have NMR spectra with *three peaks* of intensity 1:2:3. Thus, it would be possible from the combination of *infrared* and *NMR* data only to distinguish whether the compound was one of the butenes or one of the cyclopropenes. [In fact, it is possible to distinguish between these pairs of isomers with the same number of NMR peaks, by making use of the *fine structure* in their NMR spectra (the splitting of the main peaks), the nature of which depends on the numbers of different hydrogen nuclei in identical groups that are in close proximity.]

In conclusion, the combination of *infrared* and *NMR data* will unambiguously identify the sample if it is one of the butyne isomers, but if the infrared spectrum shows the sample to contain a C=C bond, the NMR spectrum cannot so unambiguously distinguish the butenes from the cyclopropenes, unless we make use of the fine structure.

CHAPTER 15

Reaction Rate

1. For the reaction $C_2H_6(g)+2H_2O(g) \rightarrow 2CO(g)+5H_2(g)$, the initial rate of formation of CO(g) in a particular experiment was 1.00 mL s^{-1}. What was the initial rate expressed in terms of (a) the formation of $H_2(g)$, and (b) the disappearance of $C_2H_6(g)$?

(a) From the balanced equation, 5 mol H_2 are formed for every 2 mol of CO formed, i.e., H_2 is formed at 2.5 times the rate that CO is formed:

$$+\frac{\Delta[H_2]}{\Delta t} = +\frac{5}{2}\frac{\Delta[CO]}{\Delta t} = \frac{5}{2}(1.00\ \text{mL s}^{-1}) = \underline{2.50\ \text{mL s}^{-1}}$$

(b) Similarly, from the balanced equation, 1 mol C_2H_6 is consumed for every 2 mol CO formed, i.e., C_2H_6 disappears at one-half the rate that CO is formed:

$$-\frac{\Delta[C_2H_6]}{\Delta t} = +\frac{1}{2}\frac{\Delta[CO]}{\Delta t} = \frac{1}{2}(1.00\ \text{mL s}^{-1}) = \underline{0.500\ \text{mL s}^{-1}}$$

---

3. For each of the following reactions, express the rate in terms of the change in concentration of the reactant with time, and relate this to the rate of formation of each of the products:

(a) $2HI(g) \rightarrow H_2(g) + I_2(g)$; (b) $2N_2O(g) \rightarrow 2N_2(g) + O_2(g)$; (c) $2N_2O_5(g) \rightarrow 4NO_2(g) + O_2(g)$

(a) $-\frac{\Delta[HI]}{\Delta t} = +2\frac{\Delta[H_2]}{\Delta t} = +2\frac{\Delta[I_2]}{\Delta t}$; (b) $-\frac{\Delta[N_2O]}{\Delta t} = +\frac{\Delta[N_2]}{\Delta t} = +2\frac{\Delta[O_2]}{\Delta t}$; (c) $-\frac{\Delta[N_2O_5]}{\Delta t} = +\frac{1}{2}\frac{\Delta[NO_2]}{\Delta t} = +2\frac{\Delta[O_2]}{\Delta t}$

---

6. The reaction $CH_3OH(aq)+HCl(aq) \rightarrow CH_3Cl(aq) + H_2O(\ell)$ was followed by measuring the change in hydronium ion concentration with time, to give the results in the following table. Calculate the average reaction rate for each time interval.

Time (min)	$[H_3O^+]$ (mol L^{-1})
0	1.85
80	1.66
159	1.53
314	1.31
628	1.02

Since HCl(aq) is fully dissociated to $H_3O^+(aq)$ and $Cl^-(aq)$, the reaction rate can be measured by the rate at which H_3O^+ is used up:

t (min)	$[H_3O^+]$ (mol L^{-1})	Δt (min)	$\Delta[H_3O^+]$ (mol L^{-1})	Average Rate (mol L^{-1} min^{-1})
0	1.85			
80	1.66	80	-0.19	2.4×10^{-3}
159	1.53	79	-0.13	1.6×10^{-3}
314	1.31	155	-0.22	1.4×10^{-3}
628	1.02	314	-0.29	0.92×10^{-3}

Note that as $[H_3O^+]$ decreases with increasing time, the rate also decreases.

Effect of Reaction Concentration on Reaction Rate

9. The rate constant for a first-order reaction at a certain temperature is $3.7 \times 10^{-2}\ s^{-1}$. For an initial concentration of 0.040 mol L^{-1}, what is the initial rate, (a) in moles per liter per second, and (b) in moles per liter per hour?

For a first order reaction: **Rate = k[A]**

(a) Rate = k[A] = $(3.7\times10^{2}\ s^{-1})(0.040\ mol\ L^{-1})$ = **1.5×10^{-3} mol L^{-1} s^{-1}.**

(b) **Rate = $(1.5\times10^{-3}\ mol\ L^{-1}\ s^{-1})(\frac{3600\ s}{1\ h})$ = 54 mol L^{-1} h^{-1}**

---

11. For the reaction $2NO(g)+H_2(g) \rightarrow N_2O(g)+H_2O(g)$, the following initial rate data were obtained at a given temperature. What are (a) the rate law for the reaction, and (b) the value of the rate constant?

Exp	[NO] (mol L^{-1})	[H_2] (mol L^{-1})	Initial Rate (mol L^{-1} min^{-1})
1	0.150	0.800	0.500
2	0.075	0.800	0.125
3	0.150	0.400	0.250

Rate = $k[NO]^x[H_2]^y$, and from the given data:

$0.500 = k[0.150]^x[0.800]^y$ .... (1)

$0.125 = k[0.075]^x[0.800]^y$ .... (2)

$0.250 = k[0.150]^x[0.400]^y$ .... (3)

Dividing equation (1) by equation (2), $2.00^x = 4.00$; **x = 2**, and

Dividing equation (1) by equation (3), $1.00^y = 2.00$; **y = 1**

Thus, the rate law is: **Rate = $k[NO]^2[H_2]$**

The value of the rate constant is obtained from the data from any of the experiments. For example, for EXP 1:

$$k = \frac{\text{rate}}{[NO]^2[H_2]} = \frac{0.500\ mol\ L^{-1}\ min^{-1}}{(0.150\ mol\ L^{-1})^2(0.800\ mol\ L^{-1})} = \underline{27.8\ mol^{-2}\ L^{2}\ min^{-1}}$$

---

13. One of the major irritants in smog is formaldehyde (methanal), $CH_2O(g)$, formed by the reaction between ethene and ozone, $O_3(g)$, in the atmosphere: $2C_2H_4(g) + 2O_3(g) \rightarrow 4CH_2O(g) + O_2(g)$. From the following initial rate data deduce the rate equation for this reaction:

Exp	[O_3] (M)	[C_2H_4] (M)	Initial Rate (M s^{-1})
1.	0.5×10^{-7}	1.0×10^{-8}	1.0×10^{-12}
2.	1.5×10^{-7}	1.0×10^{-8}	3.0×10^{-12}
3.	1.0×10^{-7}	2.0×10^{-8}	4.0×10^{-12}

The rate equation has the form: **Rate = $k[C_2H_4]^x[O_3]^y$**, and from the given data:

$1.0 \times 10^{-12} = [1.0 \times 10^{-8}]^x[0.5 \times 10^{-7}]^y$ .... (1)

$3.0 \times 10^{-12} = [1.0 \times 10^{-8}]^x[1.5 \times 10^{-7}]^y$ .... (2)

$4.0 \times 10^{-12} = [2.0 \times 10^{-8}]^x[1.0 \times 10^{-7}]^y$ .... (3)

Dividing equation (2) by equation (1): $(3.00)^y = 3.0$; **y = 1**

Dividing equation (3) by equation (1): $(2.0)^x(2.0) = 4.0$; $x = 1$

Thus the rate equation is: **Rate = $k[C_2H_4][O_3]$**, *first order* in C_2H_4, *first order* in O_3, and *second order overall.*

---

15. The decomposition of gaseous hydrogen peroxide to O_2(g) and H_2O(g) is a first order reaction. Experimentally, at a given temperature, the initial concentration of H_2O_2 decreased to half in 17.0 min. (a) What is the rate constant of the reaction? (b) What fraction of the initial H_2O_2 remains after (i) 51.0 min, and (ii) ten half-lives?

(a) From the integrated rate law for a first order reaction: $k_1t_{1/2} = \ln 2 = \mathbf{0.693}$, and for $t_{1/2} = 17.0$ min

$$k_1 = \frac{0.693}{17.0 \text{ min}} = \underline{4.08 \times 10^{-2} \text{ min}^{-1}}$$

(b) 51.0 min equals *three* half-lives; after 51.0 min $[H_2O_2]$ will have been reduced to $(\frac{1}{2})^3$ = **1/8th** of its initial value.
(c) After ten-half lives, the fraction of the initial $[H_2O_2]$ remaining is $(\frac{1}{2})^{10}$, or **1/1024th** of its initial value.

---

17. The rate of decomposition of H_2O_2 in aqueous solution at a particular temperature was measured by withdrawing samples of the reaction mixture after given times of reaction. These were immediately titrated with acidified $KMnO_4$(aq) to determine the amount of H_2O_2 remaining, which gave the following results:

t (min)	0	10	20
mL $KMnO_4$(aq)	22.8	13.8	8.3

Without plotting the data: (a) show that the reaction is first order in H_2O_2, (b) find the value of the rate constant, (c) calculate the half-life of the reaction. (The balanced equation for the reaction is not required.)

(a) The volume of $KMnO_4$(aq), **V mL**, used to titrate the H_2O_2 remaining at time t is proportional to $[H_2O_2]$. Thus the data should satisfy the integrated first order rate equation, in the form: $\ln V_t = \ln V_o - k_1t$, where V_o is the initial volume of $KMnO_4$(aq), and V_t is the volume after time t. And we can write for the initial volume V_o and each of the ensuing volumes V_t:

$$\ln 13.8 = \ln 22.8 - k(10 \text{ min}); \quad k = \frac{\ln(\frac{22.8}{13.8})}{10 \text{ min}} = \underline{0.050 \text{ min}^{-1}}$$

$$\ln 8.3 = \ln 22.8 - k(20 \text{ min}); \quad k = \frac{\ln(\frac{22.8}{8.3})}{20 \text{ min}} = \underline{0.050 \text{ min}^{-1}}$$

Showing that each pair fits the same straight line, because the slope is the same for each pair of data:

$$\mathbf{k = 0.050 \text{ min}^{-1}}$$

(b) For a first-order rate law, $kt_{1/2} = \ln 2 = 0.693$:

$$(0.050 \text{ min}^{-1})t_{\frac{1}{2}} = 0.693\,; \qquad t_{\frac{1}{2}} = \frac{0.693}{0.050 \text{ min}^{-1}} = \underline{13.9 \text{ min}}$$

---

19. Annual production of the insecticide DDT amounted to about 7.5×10^7 kg in the 1960s. In 1972, DDT was banned for general use in the USA. At ordinary temperatures, the half-life of DDT in soil is about 10 years. How long will it take for 1000 kg of DDT sprayed on the ground in 1965 to decrease to 1.0 g?

If the number of half-lives for the decrease to 1.0 g DDT to occur is x, then: $(\frac{1}{2})^x = 10^{-6}$

$$\text{Fraction of DDT remaining} = \left(\frac{1.0 \text{ g DDT}}{1000 \text{ kg DDT}}\right)\left(\frac{1 \text{ kg}}{10^3 \text{ g}}\right) = 10^{-6}$$

x(log 0.5) = log(10^{-6}) ; -0.3010x = -6; x = **20**; i.e., Time taken = 20(10 years) = **200 years.**

---

22. For the reaction between chlorine and carbon monoxide to form the poisonous war gas phosgene: $Cl_2(g) + CO(g) \rightarrow COCl_2(g)$, the following mechanism has been proposed:

$Cl_2 \rightleftarrows 2Cl$	(Fast equilibrium)	(1)
$Cl + CO \rightleftarrows COCl$	(Fast equilibrium)	(2)
$COCl + Cl_2 \rightarrow COCl_2 + Cl$	(Slow)	(3)

(a) In terms of the reactants CO and Cl_2, what is the expected rate law? (b) What name is given to molecules such as COCl and Cl?

(a) The reaction rate is determined by the rate of the *slowest step* in the series of steps that constitute the *reaction mechanism*, in this case **step 3**, so the expected rate law is:

$$\textbf{Rate} = \mathbf{k_3[COCl][Cl_2]}$$

However, this involves the concentration of the *intermediate* COCl, and cannot be tested experimentally without first expressing it entirely in terms of reactants:

For step 2: $K_2 = \frac{[COCl]}{[Cl][CO}$; i.e., $[COCl] = K_2[Cl][CO]$

For step 1: $K_1 = \frac{[Cl]^2}{[Cl_2]}$; i.e., $[Cl] = K_1^{\frac{1}{2}}[Cl_2]^{\frac{1}{2}}$

and substituting into the above rate law gives:

$$\text{Rate} = k_3[CO][Cl] = k_3K_2K_1^{\frac{1}{2}}[Cl]^{\frac{1}{2}}[CO][Cl_2] = \underline{k[Cl_2]^{\frac{3}{2}}[CO]}, \text{ which is the expected rate law.}$$

(b) **CO and COCl are reaction intermediates** (short-lived reactive species formed in one step of a reaction mechanism and used up in a following step).

---

23. A proposed mechanism for the first-order decomposition of $N_2O_5(g)$ consists of the steps:

$N_2O_5 \rightarrow NO_2 + NO_3$	(Slow)	..... (1)
$NO_3 \rightarrow NO + O_2$	(Fast)	..... (2)
$NO + N_2O_5 \rightarrow NO_2 + N_2O_5$	(Fast)	..... (3)
$N_2O_4 \rightleftarrows 2NO_2$	(Fast equilibrium)	..... (4)

(a) What is the molecularity of each step of this reaction? (b) Is the proposed mechanism consistent with an observed first-order rate law?

(a) **Steps 1, 2, and 4** are **unimolecular**, and **step 3** is **bimolecular**.

(b) From the slowest step, *step 1*: **Rate** = $\mathbf{k[N_2O_5]}$

25*. For the reaction $2NO(g) + O_2(g) \rightarrow 2NO_2(g)$, the rate law is: rate = $k[NO]^2[O_2]$. (a) Explain why this reaction is unlikely to occur by a one-step in termolecular process. (b) Devise two multi-step mechanisms for this reaction that are consistent with the observed rate law and do not involve the simultaneous collision of three molecules.

(a) Because the simultaneous collision of *three molecules*, even ignoring the need for a suitable steric orientation, is a very unlikely event.

(b) Each of the following mechanisms is consistent with the observed rate law:

(i) $NO + NO \rightleftarrows N_2O_2$ (Fast equilibrium) - For which: $K_1 = [N_2O_2]/[NO]^2$

$N_2O_2 + O_2 \rightarrow 2NO_2$ (Slow)

Thus: $\text{Rate} = k_2[N_2O_2][O_2] = k_2K_1[NO]^2[O_2] = k[NO]^2[O_2]$

(ii) $NO + O_2 \rightleftarrows NO_3$ (Fast equilibrium) - For which: $K_1 = [NO_3]/[NO][O_2]$

$NO_3 + NO \rightarrow 2NO_2$ (Slow)

Thus: $\text{Rate} = k_2[NO_3][NO] = k_2K_1[NO]^2[O_2] = k[NO]^2[O_2]$

Activation Energy and the Effect of Temperature on Reaction Rate:

28. Rate constants at several different temperatures for the reaction, $2HI(g) \rightarrow H_2(g)+I_2(g)$, are given in the following table:

T (°C)	k (mol^{-1} L s^{-1})
302	1.18×10^{-6}
356	3.33×10^{-5}
374	8.96×10^{-5}
410	5.53×10^{-4}
427	1.21×10^{-3}

(a) Plot ln k versus 1/T. (b) Obtain the value for the activation energy of the reaction, E_a, from the slope of the plot. (b) Calculate the value of the rate constant at 400°C.

The Arrhenius Equation, $k = Ae^{\frac{-E_a}{RT}}$ gives: $\ln k = \ln A - \frac{E_a}{RT}$

(a) Your plot of ln k versus 1/T should be a straight line, *slope* $-E_a/R = -2.24 \times 10^4$ K, from which:

(b) $$E_a = (8.314 \text{ J K}^{-1} \text{ mol}^{-1})(\frac{1 \text{ kJ}}{10^3 \text{ J}})(2.24\times10^4 \text{ K}) = \underline{186 \text{ kJ mol}^{-1}}$$

(c) From the Arrhenius equation for temperatures T_1 and T_2:

$$k_1 = Ae^{\frac{-E_a}{RT_1}} ; \quad k_2 = Ae^{\frac{-E_a}{RT_2}} ; \quad \frac{k_2}{k_1} = e^{\frac{E_a}{R}(\frac{1}{T_1}-\frac{1}{T_2})}$$

Selecting, for example, $k_1 = 1.18\times10^{-6}$ mol L^{-1} s^{-1} for $T_1 = 302$°C (575 K), and for k_2 $T_2 = 400$°C (673 K):

$$\ln\frac{k_2}{k_1} = \frac{E_a}{R}(\frac{1}{T_2}-\frac{1}{T_1}) = \frac{(186 \text{ kJ mol}^{-1})(\frac{10^3 \text{ J}}{1 \text{ kJ}})}{8.314 \text{ J K}^{-1} \text{ mol}^{-1}}(\frac{1}{575}-\frac{1}{673}) \text{ K}^{-1} = 5.67$$

Thus, $$\frac{k_2}{1.18 \times 10^{-6} \text{ mol}^{-1} \text{ L s}^{-1}} = e^{5.67} = 2.90 \times 10^2 ; \quad k_2 = \underline{3.42 \times 10^{-4} \text{ mol}^{-1} \text{ L s}^{-1}}$$

---

30. The activation energy of a reaction is 100 kJ mol^{-1}. To what temperature must the reaction mixture be raised for its rate constant to have exactly twice the value it has at 27 °C?

Using the Arrhenius equation given in Problem 28, with T_1 = 27°C (300 K), E_a = 100 kJ mol^{-1}, and k_2/k_1 = 2.

$$\frac{k_2}{k_1} = e^{\frac{E_a}{R}(\frac{1}{T_1}-\frac{1}{T_2})} = 2 \; ; \qquad \ln\frac{k_2}{k_1} = \frac{(100 \text{ kJ mol}^{-1})(\frac{10^3 \text{ J}}{1 \text{ kg}})}{8.314 \text{ J K}^{-1} \text{ mol}^{-1}}(\frac{1}{300 \text{ K}} - \frac{1}{T_2}) = 0.693$$

$$\frac{1}{T_2} = \frac{39.40}{(1.202 \times 10^4 \text{ K})} = 3.278 \times 10^{-3} \text{ K}^{-1}; \qquad T_2 = \underline{305 \text{ K}} \quad (32°\text{C})$$

---

33. For the reaction, $CO(g) + NO_2(g) \rightarrow CO_2(g) + NO(g)$, the rate constant at 425°C has the value 1.3 mol^{-1} L s^{-1}, and at 525°C the value is 23 mol^{-1} L s^{-1}. Find (a) the activation energy of the reaction, and (b) the rate constant at 298°C.

(a)
$$\ln\frac{k_2}{k_1} = \frac{E_a}{R}(\frac{T_2 - T_1}{T_1T_2}) = \ln(\frac{23 \text{ mol}^{-1} \text{ L s}^{-1}}{1.3 \text{ mol}^{-1} \text{ L s}^{-1}}) = \ln 17.7 = (\frac{E_a}{8.314 \text{ J K}^{-1} \text{ mol}^{-1}})(\frac{798 - 698}{798 \times 698}) \text{ K}^{-1}$$

$$(2.159 \times 10^{-5} \text{ J}^{-1} \text{ mol}) E_a = 2.87 \; ; \quad E_a = (1.33 \times 10^5 \text{ J mol}^{-1})(\frac{1 \text{ kJ}}{10^3 \text{ J mol}^{-1}}) = \underline{133 \text{ kJ mol}^{-1}}$$

(b)
$$\ln\frac{k_2}{k_1} = \frac{E_a}{R}(\frac{T_2 - T_1}{T_1T_2}) = \frac{(133 \text{ kJ mol}^{-1})(\frac{10^3 \text{ J}}{1 \text{ kJ}})}{8.314 \text{ J K}^{-1} \text{ mol}^{-1}}(\frac{571 - 698}{571 \times 698}) \text{ K}^{-1} = -5.10$$

$$\frac{k_2}{k_1} = 6.10 \times 10^{-3} \; ; \quad k_2 = (6.10 \times 10^{-3})(1.3 \text{ mol}^{-1} \text{ L s}^{-1}) = \underline{7.9 \times 10^{-3} \text{ mol}^{-1} \text{ L s}^{-1}}$$

---

35. What is the activation energy for the first-order reaction $C_2H_5Cl(g) \rightarrow C_2H_4(g) + HCl(g)$ if the value of the rate constant is 3.5 x 10^{-8} s^{-1} at 600 K, and 1.6 x 10^{-6} s^{-1} at 650 K?

$$\ln\frac{k_2}{k_1} = \frac{E_a}{R}(\frac{T_2 - T_1}{T_1T_2}) = \ln(\frac{1.6 \times 10^{-6} \text{ s}^{-1}}{3.5 \times 10^{-8} \text{ s}^{-1}}) = \ln 45.7 = (\frac{E_a}{8.314 \text{ J K}^{-1} \text{ mol}^{-1}})(\frac{650 - 600}{600 \times 650}) \text{ K}^{-1} = 7.03$$

$$(1.542 \times 10^{-5} \text{ J}^{-1} \text{ mol}) E_a = 7.03 \; ; \qquad E_a = (4.56 \times 10^5 \text{ J mol}^{-1})(\frac{1 \text{ kJ}}{10^3 \text{J}}) = \underline{456 \text{ kJ mol}^{-1}}$$

Catalysis

36. State the effect that a catalyst has on each of the following: (a) The rate of a reaction; (b) The activation energy of a reaction; (c) The enthalpy change of a reaction; (d) The temperature at which a reaction has a given rate; (e) The equilibrium position of a reaction

(a) A **catalyst** increases (speeds up) the rate of a reaction.

(b) A **catalyst** usually lowers the *activation energy*, E_a, of a reaction by providing an alternative mechanism (pathway); in heterogeneous catalysis, it may also improve the *steric factor* by bringing reactant molecules together in more favorable orientations for the reaction to take place.

(c) Since it does not appear in the balanced equation for the reaction, a **catalyst** does not affect the reaction enthalpy.

(d) A **catalyst** decreases the temperature at which a reaction will proceed at a given rate, because at a given temperature it increases the proportion of collisions between molecules with sufficient energy to react.

(e) A **catalyst** does not affect the position of equilibrium (since it does not affect the value of the Gibbs reaction free energy, ΔG), even though it speeds up the rate at which equilibrium is achieved.

---

38. For a reaction at a given temperature, does a catalyst affect (a) the value of the rate constant, and (b) the equilibrium position?

(a) At a given temperature, T, a principal role of a catalyst is to lower the value of the activation energy, E_a, for a reaction, so that less energy is required to form the *activated complex* (transition state) that goes on to form products, which increases the value of the rate of the reaction, and thus the value of the rate constant **k** (which is equal to the *reaction rate* when all the concentrations of reactants in the rate equation are 1 mol L^{-1}). In other words, by lowering E_a, at a given temperature, a greater proportion of collisions have sufficient energy to lead to reaction, which increases the value of the rate constant.

(b) Nevertheless, a catalyst is neither a reactant nor a product and does not appear in the equilibrium constant expression, K_{eq}, so the value of the *equilibrium constant* at a given temperature is unaffected. A catalyst provides an alternative mechanism for the reaction but K_{eq} is independent of the mechanism, and the equilibrium concentrations of reactants and products remain the same at a given temperature. A catalyst simply increases the speed at which the state of equilibrium at a given temperature is attained.

---

40. A platinum catalyst reduces the activation energy for the hydrogenation of ethene from 180 kJ mol^{-1} to 80 kJ mol^{-1}. At what temperature will the rate of the catalyzed reaction be the same as that of the uncatalyzed reaction at 1000°C?

$$\text{Using } k = Ae^{-\frac{E_a}{RT}} : \quad \frac{\text{rate catalyzed reaction}}{\text{rate uncatalyzed reaction}} = \frac{k_{cat}}{k_{uncat}} = 1 = \frac{Ae^{-\frac{80\text{ kJ mol}^{-1}}{RT}}}{Ae^{-\frac{180\text{ kJ mol}^{-1}}{R(1073\text{ K})}}}$$

$$\ln 1 = 0 = \frac{80\text{ kJ mol}^{-1}}{RT} - \frac{180\text{ kJ mol}^{-1}}{R(1073\text{ K})}; \quad \text{Whence: } \frac{80}{T} = \frac{180}{1073\text{ K}}; \quad \underline{T = 477\text{ K}} \quad (204°\text{C})$$

43*. The reaction of acetone with iodine in aqueous solution occurs according to the equation: $CH_3COCH_3(aq) + I_2(aq) \rightarrow CH_3COCH_2I(aq) + H^+(aq) + I^-(aq)$. In the presence of a weak base B, the rate law is rate = $k[CH_3COCH_3][B]$. (a) Suggest a three-step mechanism that is consistent with this rate law. (b) What is the function of the weak base B?

Since the base B does not appear in the balanced equation for the reaction, it must be a *catalyst*, and for the catalyzed mechanism to be consistent with the observed rate law, the rate determining step (the slowest step) must involve bimolecular collisions between $(CH_3)_2CO$ molecules and molecules of the base, and the catalyst must finally be regenerated, suggesting:

```
               H                      H
               |                      |
Step 1:  B: + H-C-C-CH3  →  BH+ + ⁻:C-C-CH3          slow
               | ||                   | ||
               H O                    H O
```

Step 2:	$I_2 + {}^-:CH_2\text{-}\underset{\underset{O}{\parallel}}{C}\text{-}CH_3 \rightarrow ICH_2\text{-}\underset{\underset{O}{\parallel}}{C}\text{-}CH_3 + I^-$	fast
Step 3:	$BH^+ \rightleftarrows B: + H^+$	fast equilibrium

Chain Reactions

44. (a) Describe each step in the following mechanism for the reaction of bromine with methane as chain initiation, chain propagation, or chain termination:

$$Br_2 \rightarrow 2Br \quad (1)$$
$$Br + CH_4 \rightarrow CH_3 + HBr \quad (2)$$
$$CH_3 + Br_2 \rightarrow CH_3Br + Br \quad (3)$$
$$2Br \rightarrow Br_2 \quad (4)$$

(b) What is the maximum wavelength of light that could catalyze this reaction?

(a) **Step 1** is *chain initiation*, because it initiates the chain by producing free radical bromine atom intermediates; **steps 2 and 3** are *chain propagation* steps in which bromine atoms first generate CH_3 radicals (step 2), which in turn generate more bromine atoms (step 3), and these reactions continue until the Br free radicals (atoms) are removed in **step 4**, the *chain termination* step.

(b) From Table 6.4, the bond dissociation energy of $Br_2(g)$ is 190 kJ mol^{-1}, which is the minimum energy required to dissociate $Br_2(g)$ in to Br(g) atoms. Thus, from Planck's relationship **E = hν = hc/λ**:

$$\lambda = \frac{hc}{E} = \frac{(6.63 \times 10^{-34}\text{ J s})(3.00 \times 10^{8}\text{ m s}^{-1})(\frac{1\text{ kJ}}{10^3\text{ J}})}{(190\text{ kJ mol}^{-1})(\frac{1\text{ mol Br}_2}{6.022 \times 10^{23}\text{ Br}_2\text{ molecules}})} = (6.30 \times 10^{-7}\text{ m})(\frac{10^9\text{ nm}}{1\text{ m}}) = \underline{630\text{ nm}}$$

---

45. The following three-step mechanism has been proposed for the gas-phase reaction in which tetrachloromethane is formed from chlorine and trichloromethane (chloroform): $CHCl_3(g) + Cl_2(g) \rightarrow CCl_4(g) + HCl(g)$.

$$Cl_2 \rightleftarrows 2Cl \quad \text{(Fast equilibrium)} \quad (1)$$
$$Cl + CHCl_3 \rightarrow CCl_3 + HCl \quad (2)$$
$$Cl + CCl_3 \rightarrow CCl_4 \quad (3)$$

The observed rate law is rate = $k[CHCl_3][Cl_2]^{½}$. Is the proposed mechanism consistent with the observed rate law when step 2 is much slower than step 3, or when step 3 is much slower than step 2?

(i) For **step 2** as the rate-determining (slow) step:

$$\textbf{Rate} = k[Cl][CH_3Cl] = kK_1^{½}[Cl_2]^{½}[CH_3Cl] = \mathbf{k'[Cl_2]^{½}[CH_3Cl]}$$

which is *consistent* with the observed rate law.

(ii) For **step 3** as the rate-determining (slow) step:

$$\textbf{Rate} = k[Cl][CCl_3] = kK_1K_2[Cl_2][CH_3Cl] = \mathbf{k'[Cl_2]^{½}[CHCl_3][HCl]^{-1}}$$

which is *inconsistent* with the observed rate law.

General Problems

47. In a reaction of ozone with ammonia at 500 K, according to equation $5O_3(g) + 6NH_3(g) \rightarrow 6NO(g) + 9H_2O(g)$,

the rate of increase in the pressure of NO(g) in a given time interval was 1095 mm Hg s^{-1}. For the same time interval, what was (a) the rate of disappearance of ozone in moles per liter per second, and (b) the rate of disappearance of ammonia?

It is useful to first express the rate of increase in pressure of NO in **mol L^{-1} s^{-1}**. From the ideal gas equation:

$$PV = nRT\ ; \qquad mol\ L^{-1} = \frac{n}{V} = \frac{P}{RT}$$

$$\frac{d[NO]}{dt} = \frac{1}{RT}\left(\frac{dP_{NO}}{dt}\right) = \frac{(1095\ mm\ Hg)\left(\frac{1\ atm}{760\ mm\ Hg}\right)}{(0.082\ atm\ L\ mol^{-1}\ K^{-1})(500\ K)} = \underline{3.51 \times 10^{-2}\ mol\ L^{-1}\ s^{-1}}$$

and from the balanced equation:

$$+\frac{d[NO]}{dt} = -\frac{d[NH_3]}{dt} = -\frac{6}{5}\frac{d[O_3]}{dt}$$

$$\text{Rate of disappearance of } O_3 = -\frac{d[O_3]}{dt} = \frac{5}{6}\frac{d[NO]}{dt} = \frac{5}{6}(3.51 \times 10^{-2}\ mol\ L^{-1}\ s^{-1}) = \underline{2.92 \times 10^{-2}\ mol\ L^{-1}\ s^{-1}}$$

$$\text{Rate of disappearance of } NH_3 = -\frac{d[NH_3]}{dt} = \frac{d[NO]}{dt} = \underline{2.51 \times 10^{-2}\ mol\ L^{-1}\ s^{-1}}$$

---

48. For the reaction $C_2H_4(g) + H_2(g) \rightarrow C_2H_6(g)$, $\Delta H° = -137$ kJ, explain why low temperature, high pressure, and a catalyst are beneficial to obtaining a high yield of ethane.

According to *Le Châtelier's principle*: (i) low temperature favors a high equilibrium concentration of products, in this case of $C_2H_6(g)$, when the forward reaction is exothermic because a decrease in temperature is partially countered by adding more heat to the system; (ii) high pressure favors the forward reaction when there are less moles of gaseous products than of gaseous reactants, because decrease in the number of moles tends to oppose the concentration increase due to increased pressure; and (iii) a catalyst will speed up the rate at which equilibrium is achieved, even though it cannot change the equilibrium concentration of ethane.

---

51. The reaction of an alkyl halide with hydroxide ion, $RX + OH^- \rightarrow ROH + X^-$, occurs by one of two mechanisms depending on the nature of the alkyl group R. Describe (a) the two possible mechanisms, and (b) how the actual mechanism could be established for a particular alkyl halide.

(a) Depending on the nature of R in RX, the reaction is by *either* a *single-step* bimolecular mechanism, such as:

$$H{-}\ddot{\underset{\cdot\cdot}{O}}{:}^- + H_3C{-}X \rightarrow \left[\begin{array}{c} H \\ | \\ HO{---}C{---}X \\ / \quad \backslash \\ H \qquad H \end{array}\right]^- \rightarrow HO{-}CH_3 + {:}\ddot{\underset{\cdot\cdot}{X}}{:}^-$$

with the *second order* rate law: **Rate = k_2[RX][OH^-]**

or by a *two-step* mechanism in which the rate-determining (slowest) step is unimolecular, for instance:

Step 1 $(CH_3)_3C{-}X \rightarrow (CH_3)_3C^+ + X^-$ SLOW

Step 2 $(CH_3)_3C^+ + OH^- \rightarrow (CH_3)_3C{-}OH$ FAST

with the *first-order* rate law: **Rate = k_1[RX]**

(b) The simplest way to distinguish which of the two mechanisms a particular alkyl halide follows is to first assume that the reaction is *first order* and use the integrated rate equation for a first order rate law:

$$\ln[RX]_t = \ln[RX]_o - k_1 t$$

where $[RX]_o$ is the initial concentration of the alkyl halide, RX, and $[RX]_t$ is the concentration after time t. When the reaction is first order, a plot of $\ln[RX]_t$ versus time t should be a straight line with a slope $-k_1$. If such a plot gives a straight line we can assume the reaction is first order. If not, then it is not first order but is second order.

---

53. At high temperature, chloroethane (ethyl chloride) decomposes in a first order reaction to ethene and hydrogen chloride gas. In an experiment at 800 K, an initial concentration of chloroethane of 0.0200 mol L^{-1} decreased to 0.0033 mol L^{-1} in 340 s. What are (a) the balanced equation for the reaction, (b) the value of the rate constant, and (c) the half-life of the reaction?

(a) $$C_2H_5Cl(g) \rightarrow H_2C{=}CH_2(g) + HCl(g)$$

(b) From the integrated rate law for a first order reaction: $\ln[A]_t = \ln[A]_o - k_1 t$ and the data gives:

$$\ln(0.0033) = \ln(0.0200) - k_1(340\ s)$$

$$-5.714 - (-3.912) = -k_1(340\ s); \quad k_1 = 5.3 \times 10^{-3}\ s^{-1}$$

(c) From the integrated rate law, at the half-life $t_{1/2}$:

$$\ln \tfrac{1}{2}[A]_o = \ln[A]_o - k_1 t_{1/2};\ k_1 t_{1/2} = \ln 2 = 0.693, \text{ and for } k_1 = 5.3\times10^{-3}\ s^{-1},\ t_{1/2} = 131\ s$$

54.* In acidic aqueous solution, hydrogen peroxide oxidizes bromide ions to bromine. Experimentally, when the initial concentration of H_2O_2 was doubled, and $[Br^-]$ and the pH were held constant, the rate doubled. The rate also doubled when $[Br^-]$ was doubled and $[H_2O_2]$ and the pH were held constant. When the pH alone was changed from 1.00 to 0.400, the rate increased four-fold. (a) Write the balanced equation for the reaction; (b) Find the rate law; (c) If under given conditions the rate of disappearance of Br^- is 7.2×10^{-3} mol L^{-1} s^{-1}, what is the rate of disappearance of H_2O_2, and the rate of appearance of Br_2? (d) What is the effect on the rate of increasing the pH? (e) If the initial solution is diluted with water so that its volume is doubled, what would be the effect on the initial reaction rate?

(a)

$2Br^- \rightarrow Br_2 + 2e^-$	oxidation
$H_2O_2 + 2e^- + 2H^+ \rightarrow 2H_2O$	reduction
$2Br^- + H_2O_2 + 2H^+ \rightarrow Br_2 + 2H_2O$	overall

(b) The rate law has the form: **Rate** $= k[H_2O_2]^x[Br^-]^y[H^+]^z$, and since the rate *doubles* on doubling either $[H_2O_2]$ or $[Br^-]$, keeping the other and $[H^+]$ constant, $x = y = 1$. Since on changing $[H^+]$ *from pH 1.00 to 0.400* (which is by a factor of 4, from 0.10 mol L^{-1} to 0.40 mol L^{-1}), the rate also increases by a factor of *four*, **z is also 1**, and the rate law is:

$$\text{Rate} = k[H_2O_2][Br^-][H^+] \quad \text{(which could also be written as: Rate} = k[H_2O_2][HBr])$$

(c) From the balanced equation, the *initial* rate of disappearance of Br^- is *twice* the initial rate of disappearance of H_2O_2, and *twice* the initial rate of formation of Br_2, or:

$$-\frac{\Delta[Br^-]}{\Delta t} = \frac{\Delta[H^+]}{\Delta t} = -2\frac{\Delta[H_2O_2]}{\Delta t} = +2\frac{\Delta[Br_2]}{\Delta t} = 7.2 \times 10^{-3}\ mol\ L^{-1}\ s^{-1}$$

$$-\frac{\Delta[H_2O_2]}{\Delta t} = \frac{\Delta[Br_2]}{\Delta t} = \underline{3.6 \times 10^{-3}\ mol\ L^{-1}\ s^{-1}}$$

(d) Changing the pH affects the rate of the reaction but not the value of the rate constant at a particular temperature.

15.54

(e) Doubling the volume decreases each concentration by a factor of 2 (by one-half), so that the overall change in the initial rate is by a factor of (½)(½)(½)= one-eighth.

CHAPTER 16

Oxides and Oxoacids of Nitrogen

2. Name the compounds with each of the following formulas. (a) $NaNO_3$(s) (b) KNO_2(s) (c) N_2O_4(s) (d) NO_2(g) (e) NO(g) (f) N_2O(g) (g) HNO_2(aq) (h) HNO_3(ℓ)

(a) **sodium nitrate** (b) **potassium nitrite** (c) **dinitrogen tetraoxide** (d) **nitrogen dioxide** (e) **nitrogen monoxide** (nitric oxide) (f) **dinitrogen monoxide** (nitrous oxide) (g) **nitrous acid** (h) **nitric acid**

---

4. Draw Lewis structures, including resonance structures where appropriate, for each of the following. Find the bond order, and hence arrange them in order of expected decreasing NO bond length: (a) N_2O (b) NO (c) NO_2^- (d) NO_3^- (e) NO^+

(a) $^-$:N=N$^+$=Ö: ↔ :N≡N$^+$–Ö:$^-$ (b) $^{½-}$:N=O:$^{½+}$ ↔ :N=Ö: ↔ $^-$:N=O:$^+$ (c) :Ö=N–Ö:$^-$ ↔ $^-$:Ö-N=Ö:

(d) Ö:‖N$^+$(–:Ö:$^-$)(–:Ö:$^-$) ↔ :Ö:$^-$|N$^+$(–:Ö:$^-$)(=Ö:) ↔ :Ö:$^-$|N$^+$(=Ö)(–:Ö:$^-$) (e) :N≡O:$^+$

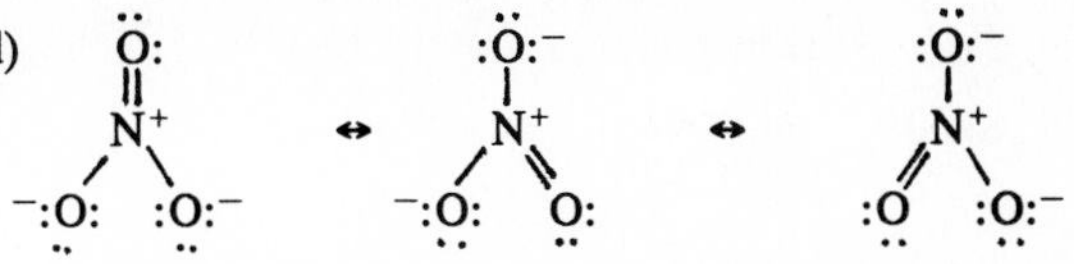

Molecule or Ion	NO Bond Order
(a) N_2O	½(2 + 1) = 1½
(b) NO	2 to 2½
(c) NO_2^-	½(2 + 1) = 1½
(d) NO_3^-	⅓(2 + 1 + 1) = 1⅓
(e) NO^+	3

Bond length *decreases* with *increasing* bond order, so the **NO bond lengths** are expected to decrease in the sequence:

$$NO_3^- > N_2O = NO_2^- > NO > NO^+$$

---

6. Using the VSEPR model, categorize each of the following (where nitrogen is the central atom) in terms of the AE_nX_m nomenclature, and predict the approximate geometry of each: (a) NO_2^-, (b) NO_2, (c) NO_2^+

Molecule or Ion	Lewis Structure	AX_nE_m Type	Geometry
(a) NO_2^-	:Ö=N–Ö:$^-$	AX_2E	angular
(b) NO_2	$^{½-}$:Ö–N$^{½+}$=Ö:	AX_2E	angular
(c) NO_2^+	:Ö=N$^+$=Ö:	AX_2	linear

---

8. Write a balanced equation for each of the following reactions. State whether each is an acid-base or an oxidation-reduction reaction. For the acid-base reactions, indicate which reactant is the acid and which is the base; for the redox reactions state which element is oxidized and which is reduced: (a) the photochemical decomposition of nitric acid to $NO_2(g)$, $O_2(g)$, and $H_2O(\ell)$; (b) the decomposition upon heating of lead(II) nitrate to give PbO(s), $NO_2(g)$, and $O_2(g)$; (c) the reaction of nitrous acid with water.

(a) N (oxidation number +5) in HNO_3 is reduced to N(+4) in NO_2, and the oxygen must come from the oxidation of HNO_3, in which O(-2) in HNO_3 is oxidized to O(0) in O_2:

$$2[HNO_3 + e^- + H^+ \rightarrow NO_2 + H_2O] \quad \text{reduction}$$

$$2HNO_3 \rightarrow 2NO_2 + O_2 + 2e^- + 2H^+ \quad \text{oxidation}$$

$$\mathbf{4HNO_3 \rightarrow 4NO_2 + O_2 + 2H_2O} \quad \textbf{oxidation-reduction}$$

N in nitric acid is reduced, and O in nitric acid is oxidized

(b) Pb^{2+} occurs both in reactants and products, so its oxidation state remains Pb(+2). But N(+5) in NO_3^- is reduced to N(+4) in NO_2, and O(-2) in NO_3^- is oxidized to O(0) in O_2. Since the reaction is not in aqueous solution, where we would use H^+ ions or OH^- ions to balance any charges, here we will use O^{2-} ions:

$$2[NO_3^- + e^- \rightarrow NO_2 + O^{2-}] \quad \text{reduction}$$

$$2NO_3^- \rightarrow 2NO_2 + O_2 + 2e^- \quad \text{oxidation}$$

$$\mathbf{4NO_3^- \rightarrow 4NO_2 + O_2 + 2O^{2-}}$$

and we can now add $2Pb^{2+}$ to both sides of the equation, to give:

$$\mathbf{2Pb(NO_3)_2 \rightarrow 2PbO + 4NO_2 + O_2} \quad \textbf{oxidation-reduction}$$

N in nitrate ion is oxidized and O in nitrate ion is reduced

(c) Nitrous acid is a weak acid and ionizes to give hydronium and nitrite ions in aqueous solution:

$$HNO_2(aq) + H_2O(l) \rightleftarrows H_3O^+(aq) + NO_2^-(aq) \quad \text{acid-base}$$

However, nitrous acid is not very stable in water and slowly reacts to form nitric oxide and nitric acid. Thus N(+3) in HNO_2 is oxidized to N(+5) in HNO_3, and at the same time is also reduced to N(+2) in NO:

$$2[HNO_2 + e^- + H^+ \rightarrow NO + H_2O] \quad \text{reduction}$$

$$HNO_2 + H_2O \rightarrow HNO_3 + 2e^- + 2H^+ \quad \text{oxidation}$$

$$\mathbf{3HNO_2 \rightarrow HNO_3 + 2NO + H_2O} \quad \textbf{oxidation-reduction}$$

N in nitrous acid is *both* oxidized and reduced

---

10. A useful method for the quantitative determination of nitrate ion in aqueous solution is to determine the amount of nitrogen monoxide gas formed by the reaction, $2NO_3^-(aq) + 3H_2SO_4(aq) + 3Hg(\ell) \rightarrow 3HgSO_4(s) + 2H_2O(\ell) + 2OH^-(aq) + 2NO(g)$. If 3.26 L of NO(g) at STP was obtained from 2.00 L of a solution containing nitrate ion, what was the concentration of nitrate ion in the solution?

From the gas data:

$$\text{moles of NO(g)} = \frac{PV}{RT} = \frac{(1\ \text{atm})(3.26\ \text{L})}{(0.0821\ \text{atm L mol}^{-1}\ \text{K}^{-1})} = \underline{0.1454\ \text{mol}}$$

$$\text{moles of } NO_3^- \text{ ion} = (0.1454\ \text{mol NO})\left(\frac{2\ \text{mol } NO_3^-}{2\ \text{mol NO}}\right) = \underline{0.1454\ \text{mol}}$$

$$[NO_3^-] = \frac{0.1454\ \text{mol}}{2.00\ \text{L}} = \underline{0.0727\ \text{mol L}^{-1}}$$

---

11. (a) Calculate the standard enthalpy change for the reaction $4NH_3(g) + 5O_2(g) \rightarrow 4NO(g) + 6H_2O(g)$. (Use data from Table 9.1 to calculate $\Delta H°$.) (b) Under what conditions of temperature and pressure would the equilibrium concentration of NO(g) be maximized?

(a) We use the relationship between $\Delta H°$ for a reaction and the $\Delta H_f°$ values of the products and reactants:

$$\Delta H° = \Sigma[n_p(\Delta H_f°)_p] - \Sigma[n_r(\Delta H_f°)_r]$$

$$\Delta H° = [(4 \text{ mol})\{\Delta H_f°(NO,g)\} + (6 \text{ mol})\{\Delta H_f°(H_2O,g)\}] - [(4 \text{ mol})\{\Delta H_f°(NH_3,g)\} + (5 \text{ mol})\{\Delta H_f°(O_2,g)\}]$$

$$= [4(90.3) + 6(-241.8)] - [4(-46.2) + 5(0)] = \mathbf{-904.8\ kJ}$$

(b) The reaction is **exothermic**, so will be favored by **low temperature**. Since 9 mol of gases react to give 10 mol of gases, **low pressure** will favor the forward reaction (Le Châtelier's principle).

---

12. For the room temperature reaction between nitrogen monoxide and oxygen, $2NO(g) + O_2(g) \rightarrow 2NO_2(g)$, the following initial rate data were obtained:

[NO] (mol L^{-1})	[O_2] (mol L^{-1})	Initial Rate (mol L^{-1} s^{-1})
0.010	0.020	0.014
0.010	0.010	0.007
0.020	0.040	0.114
0.040	0.020	0.227

Determine (a) the order of the reaction with respect to NO and O_2, and (b) the value of the rate constant.

(a) The rate equation will have the form, **rate** = $[NO]^x[O_2]^y$, and from the results of the first two experiments, the rate is halved when [NO] is kept constant and [O_2] is halved, hence rate $\propto$ [O_2], or **y = 1**. Similarly, from the first and last experiment, the rate increases by a factor of **0.227/0.014** = **16.2** when [O_2] is kept constant and [NO] is increased by a **factor of 4**. Thus, **rate** $\propto$ $[NO]^2$, or **x = 2**, and combining these results gives:

$$\textbf{rate} = k[NO]^2[O_2]$$

Second order in NO and **first order** in O_2

(b) From the rate law, **k = rate/**$[NO]^2[O_2]$, and from the data given:

[NO] (mol L^{-1})	[O_2] (mol L^{-1})	rate (mol L^{-1} s^{-1})	k (mol^{-2} L^2 s^{-1})
0.010	0.020	0.014	7.0×10^3
0.010	0.010	0.007	7.0×10^3
0.020	0.040	0.014	7.1×10^3
0.040	0.020	0.227	7.1×10^3
		Average	$\mathbf{7.1 \times 10^3}$

Ozone and Peroxides

14. Write the formulas of each of the following compounds: (a) ozone (b) sodium oxide (c) sodium peroxide (d) barium oxide (e) barium peroxide (f) hydrogen peroxide

---

(a) O_3 (b) Na_2O (c) Na_2O_2 (d) BaO (e) BaO_2 (f) H_2O_2

---

16. (a) Draw Lewis resonance structures for the ozone molecule, O_3, and calculate the bond order. (b) Why is ozone not given the Lewis structure

```
   ..
   O:
  / \
:O—O:
 ..  ..
```

(a) There are **two** possible Lewis (*resonance*) structures:

$$:\ddot{O}=\dot{O}^{+}-\ddot{O}:^{-} \leftrightarrow {}^{-}:\ddot{O}-\dot{O}^{+}=\ddot{O}:$$

oxygen-oxygen bond order = ½(2 + 1) = 1½

(b) The structure given,

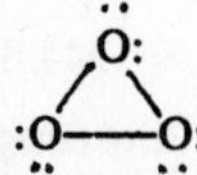

would seem to be a possible structure, but would imply that O_3 has the shape of an ***equilateral triangle*** with all the O–O distances the same, three single bonds and all OOO bond angles equal to 60°. The actual structure of the O_3 molecule is **angular** with <OOO = **117°**, consistent with the Lewis structures we drew in part (a), which imply an **AX_2E type** molecule based on a trigonal planar arrangement of two bonds and an unshared electron pair on the central oxygen atom. Thus, the expected bond angle should be approximately 120°, close to that found.

---

18. (a) Describe how an aqueous solution of hydrogen peroxide can be prepared in the laboratory from BaO_2(s); (b) Explain why H_2O_2(aq) sometimes behaves as an oxidizing agent and sometimes as a reducing agent; (c) Write balanced equations for each of the following reactions of H_2O_2(aq) in acidic solution: (i) oxidation of sulfur dioxide to sulfate ion; (ii) reduction of ozone to water; (iii) oxidation of iodide ion to iodine; (iv) oxidation of nitrite ion to nitrate ion.

(a) When barium is heated in oxygen it reacts to form **barium peroxide**, BaO_2(s), rather than the oxide BaO(s):

$$Ba(s) + O_2(g) \rightarrow BaO_2(s)$$

Reaction of barium peroxide with H_2SO_4(aq) then gives H_2O_2(aq) and a precipitate of insoluble $BaSO_4$(s), which can be removed from the solution by filtration:

$$BaO_2(s) + H_2SO_4(aq) \rightarrow BaSO_4(s) + H_2O_2(aq)$$

(b) H_2O_2 contains oxygen in the **-I** oxidation state that can *either* be oxidized to give O_2(g), with oxygen in the **0** oxidation state *or* reduced to water, H_2O, with oxygen in the **-II** oxidation state.

$H_2O_2 \rightarrow O_2(g) + 2e^- + 2H^+$ **oxidation;** $H_2O_2 + 2e^- + 2H^+ \rightarrow 2H_2O$ **reduction**

(c) (i)

$SO_2 + 2H_2O \rightarrow SO_4^{2-} + 2e^- + 4H^+$ oxidation

$H_2O_2 + 2e^- + 2H^+ \rightarrow 2H_2O$ reduction

$SO_2 + H_2O_2 \rightarrow SO_4^{2-} + 2H^+$

(ii)

$H_2O_2 \rightarrow O_2(g) + 2e^- + 2H^+$ oxidation

$O_3 + 2e^- + 2H^+ \rightarrow O_2(g) + H_2O$ reduction

$H_2O_2 + O_3 \rightarrow 2O_2(g) + H_2O$

(iii)

$3I^- \rightarrow I_3^- + 2e^-$ oxidation

$H_2O_2 + 2e^- + 2H^+ \rightarrow 2H_2O$ reduction

$3I^- + H_2O_2 + 2H^+ \rightarrow I_3^- + 2H_2O$

(iv)

$NO_2^- + H_2O \rightarrow NO_3^- + 2e^- + 2H^+$ oxidation

$H_2O_2 + 2e^- + 2H^+ \rightarrow 2H_2O$ reduction

$NO_2^- + H_2O_2 \rightarrow NO_3^- + H_2O$

---

19. The standard reduction potential for the half-reaction, $O_3(g) + 2H^+(aq) + 2e^- \rightarrow O_2(g) + H_2O(\ell)$, is 2.07 V. On the basis of standard reduction potentials from Table 13.1 decide which of the following oxidations: (a) H_2O to H_2O_2; (b) Mn^{2+} to MnO_4^-; (c) Cr^{3+} to $Cr_2O_7^{2-}$; (d) Co^{2+} to Co^{3+}; (e) NO(g, 1 atm) to NO_3^- could be achieved under standard conditions in acidic solution using (i) ozone; (ii) oxygen.

In each case, we use standard potentials to calculate the standard cell potentials, $E_{cell}°$, for the reaction in question when the calculated $E_{cell}° > 0$, a reaction is spontaneous.

For oxidation with $O_2(g)$: $O_2(g) + 4e^- + 4H^+ \rightarrow 2H_2O$, $E°_{red} = 1.23$ V

For oxidation with $O_3(g)$: $O_3(g) + 2e^- + 2H^+ \rightarrow O_2(g) + H_2O$, $E°_{red} = 2.07$ V

(a) (i)

$2[2H_2O \rightarrow H_2O_2 + 2e^- + 2H^+]$ $E°_{ox} = -E°_{red} = -1.78$ V

$O_2(g) + 4e^- + 4H^+ \rightarrow 2H_2O$ $E°_{red} = +1.23$ V

$2H_2O + O_2 \rightarrow 2H_2O_2$ $E°_{cell} = -0.55$ V **not spontaneous**

(ii)

$2H_2O \rightarrow H_2O_2 + 2e^- + 2H^+$ $E°_{ox} = -E°_{red} = -1.78$ V

$O_3 + 2e^- + 2H^+ \rightarrow O_2 + H_2O$ $E°_{red} = +2.07$ V

$H_2O + O_3 \rightarrow H_2O_2 + O_2$ $E°_{cell} = +0.29$ V **spontaneous**

(b) (i)

$4[Mn^{2+} + 4H_2O \rightarrow MnO_4^- + 5e^- + 8H^+]$ $E°_{ox} = -E°_{red} = -1.49$ V

$5[O_2 + 4e^- + 4H^+ \rightarrow 2H_2O]$ $E°_{red} = +1.23$ V

$4Mn^{2+} + 5O_2 + 6H_2O \rightarrow 4MnO_4^- + 12H^+$ $E°_{cell} = -0.26$ V **not spontaneous**

(ii)

$2[Mn^{2+} + 4H_2O \rightarrow MnO_4^- + 5e^- + 8H^+]$ $E°_{ox} = -E°_{red} = -1.49$ V

$5[O_3 + 2e^- + 2H^+ \rightarrow O_2 + H_2O]$ $E°_{red} = +2.07$ V

$2Mn^{2+} + 5O_3 + 3H_2O \rightarrow 2MnO_4^- + 5O_2 + 6H^+$ $E°_{cell} = +0.58$ V **spontaneous**

(c) (i)

$2[2Cr^{3+} + 7H_2O \rightarrow Cr_2O_7^{2-} + 6e^- + 14H^+]$ $E°_{ox} = -E°_{red} = -1.33$ V

$3[O_2 + 4e^- + 4H^+ \rightarrow 2H_2O]$ $E°_{red} = +1.23$ V

$4Cr^{3+} + 3O_2 + 8H_2O \rightarrow 2Cr_2O_7^{2-} + 16H^+$ $E°_{cell} = -0.10$ V **not spontaneous**

(ii)

$2Cr^{3+} + 7H_2O \rightarrow Cr_2O_7^{2-} + 6e^- + 14H^+$ $E°_{ox} = -E°_{red} = -1.33$ V

$3[O_3 + 2e^- + 2H^+ \rightarrow O_2 + H_2O]$ $E°_{red} = +2.07$ V

$2Cr^{3+} + 3O_3 + 4H_2O \rightarrow 2Cr_2O_7^{2-} + 3O_2 + 8H^+$ $E°_{cell} = +0.74$ V **spontaneous**

(d) (i)

$4[NO + 2H_2O \rightarrow NO_3^- + 3e^- + 4H^+]$ $E°_{ox} = -E°_{red} = -0.97$ V

$3[O_2(g) + 4e^- + 4H^+ \rightarrow 2H_2O]$ $E°_{red} = +1.23$ V

$4NO + 3O_2 + 2H_2O \rightarrow 4NO_3^- + 4H^+$ $E°_{cell} = 0.26$ V **spontaneous**

(ii)

$2[NO + 2H_2O \rightarrow NO_3^- + 3e^- + 4H^+]$ $E°_{ox} = -E°_{red} = -0.97$ V

$3[O_3 + 2e^- + 2H^+ \rightarrow O_2 + H_2O]$ $E°_{red} = +2.07$ V

$2NO + 3O_3 + H_2O \rightarrow 2NO_3^- + 3O_2 + 2H^+$ $E°_{cell} = +1.10$ V **spontaneous**

Summary:

(a) **Only** the **oxidation of NO(g)** is achieved using $O_2(g)$

(b) **All** the **oxidations** are achieved using $O_3(g)$

Note: The same conclusions could have been reached by considering the positions of the respective half-reactions in the table of standard reduction potentials: (a) All the reductions are above $E°_{red}$ for $O_3(g) + 2e^- + 2H^+ \rightarrow O_2(g) + H_2O$, in Table 13.1, so all the respective oxidations by $O_3(g)$ occur under standard conditions.(b)But only $NO_3^- + 3e^- + 4H^+ \rightarrow NO + 2H_2O$ comes above $O_2(g) + 4e^- + 4H^+ \rightarrow 2H_2O$ in the table; so among these oxidations, only the oxidation of NO(g) to $NO_3^-(aq)$ is achieved by $O_2(g)$ under standard conditions.

Pollution of the Stratosphere

21. (a) Why is the spectrum of solar radiation different at the earth's surface than it is outside the atmosphere? (b) Account for the fact, that, of the major constituents of the atmosphere, nitrogen and argon remain unchanged in the stratosphere, whereas oxygen decomposes although it is stable in the troposphere.

(a) In passing through the stratosphere, part of the solar radiation is absorbed by molecules in the atmosphere, notably radiation in the ultra-violet spectrum with $\lambda < 240$ nm, which is absorbed by $O_2(g)$ and dissociates it into O atoms:

$$O_2(g) \rightarrow 2O(g)$$

O atoms so formed react with more O_2 to form ozone, O_3, which absorbs radiation with $\lambda < 320$ nm in dissociating to reform O_2 and O atoms:

$$O_3(g) \rightarrow O_2(g) + O(g)$$

and $O_3(g)$ also absorbs some of the radiation with λ up to 350 nm, which raises its molecule to an excited electronic state:

$$O_3 \rightarrow O_3^*$$

Thus, almost all of the radiation with $\lambda < 240$ nm is absorbed by O_2 molecules, and ozone absorbs most of the radiation in the range from 240 to 350 nm. The result is that most of the low wavelength electromagnetic spectrum does not reach the earth's surface.

(b) **Argon gas** consists as Ar atoms which when they interact with radiation from the sun can form only excited states. On the other hand, **nitrogen** and **oxygen** consist of N_2 and O_2 molecules, respectively, and could, in principle, both be dissociated into atoms by photons from the sun, if radiation of the appropriate frequencies reaches the stratosphere with significant intensity. We can calculate the maximum wavelengths that *might* cause this to happen from the respective bond energies:

For N_2, $BE(N{\equiv}N) = 941 \text{ kJ mol}^{-1}$; $\lambda_{max} = \dfrac{1.20 \times 10^5 \text{ nm kJ mol}^{-1}}{941 \text{ kJ mol}^{-1}} = \underline{128 \text{ nm}}$

For O_2 , $BE(O{=}O) = 498 \text{ kJ mol}^{-1}$; $\lambda_{max} = \dfrac{1.20 \times 10^5 \text{ nm kJ mol}^{-1}}{498 \text{ kJ}^{-1}} = \underline{241 \text{ nm}}$

Reference to Figure 16.2 shows that light with $\lambda < 128$ pm that reaches the stratosphere is of negligible intensity, insufficient to dissociate any appreciable number of N_2 molecules, whereas between 200 nm and 240 nm there are sufficient photons to dissociate O_2 molecules into O atoms. On the other hand, all of the radiation with $\lambda < 240$ pm has already been absorbed (filtered out) by the time light from the sun reaches the troposphere, so close to earth O_2 molecules remain undissociated.

---

23. Use standard enthalpies of formation from Table 9.1 to find the reaction enthalpies for both steps in the conversion of $O_3(g)$ to $O_2(g)$ catalyzed by NO(g): (a) $NO(g) + O_3(g) \rightarrow NO_2(g) + O_2(g)$; (b) $NO_2(g) + O_3(g) \rightarrow 2O_2(g) + NO(g)$.

(a) $\Delta H° = [\Delta H_f°(NO_2,g) + \Delta H_f°(O_2,g)] - [\Delta H_f°(NO,g) + \Delta H_f°(O_3,g)] = [(33.2) + (0)] - [(90.3) + (142.7)]$

$$= \mathbf{-199.8 \text{ kJ}}$$

(b) $\Delta H° = [\Delta H_f°(NO,g) + \Delta H_f°(O_2,g)] - [\Delta H_f°(NO_2,g) + \Delta H_f°(O,g)] = [(+90.3) + (0)] - [(+33.2) + (+249)]$

$$= \mathbf{-192 \text{ kJ}}$$

---

25. For the reaction: $O(g) + O_3(g) \rightarrow 2O_2(g)$, the rate constant is 5.56×10^6 mol^{-1} L s^{-1} at 300 K and 1.38×10^5 mol^{-1} L s^{-1} at 200 K. (a) What is the activation energy for this reaction? (b) What is the rate constant for the reaction at -55°C, the temperature of the stratosphere?

(a) We calculate the activation energy from the Arrhenius equation:

$$k = Ae^{-\frac{E_a}{RT}} ; \quad \text{For temperature } T_2 > T_1 : \quad \ln\frac{k_2}{k_1} = \frac{E_a}{R}\left(\frac{T_2 - T_1}{T_1 T_2}\right) = \ln\frac{5.56 \times 10^6 \text{ mol}^{-1} \text{ L s}^{-1}}{1.38 \times 10^5 \text{ mol}^{-1} \text{ L s}^{-1}}$$

$$= \ln 40.29 = \underline{3.696} = \frac{E_a}{8.314 \text{ J K}^{-1} \text{ mol}^{-1}}\left(\frac{300 - 200}{200 \times 300}\right) \text{K}^{-1} = 2.004 \times 10^{-4} \, E_a \text{ J mol}^{-1}$$

$$\text{Whence,} \quad E_a = \left(\frac{3.696}{(2.004 \times 10^{-4})\left(\frac{1 \text{ kJ}}{10^3 \text{ J}}\right)}\right) = \underline{18.4 \text{ kJ mol}^{-1}}$$

(b) For -55°C = (273-55) = 218 K

$$\ln\frac{k_3}{k_1} = \ln\frac{k_3}{1.38 \times 10^5 \text{ mol}^{-1} \text{ L s}^{-1}} = \frac{(18.4 \text{ kJ mol}^{-1})\left(\frac{10^3 \text{ J}}{1 \text{ kJ}}\right)}{8.314 \text{ J K}^{-1} \text{ mol}^{-1}}\left(\frac{218 - 200}{200 \times 218}\right) \text{K}^{-1} = 0.9137$$

$$\frac{k_3}{1.38 \times 10^5 \text{ mol}^{-1} \text{ L s}^{-1}} = 2.493 ; \quad k_3 = \underline{3.44 \times 10^5 \text{ mol}^{-1} \text{ L s}^{-1}}$$

---

27. (a) Give the major reasons why the maximum ozone concentration is found at an altitude of about 30 km. (b) Describe the principal reactions that are most probably responsible for the destruction of the ozone layer.

(a) The greater the height above the earth the more ultraviolet light ($\lambda < 320$ nm) is available to form ozone from $O_2(g)$, but the pressure of $O_2(g)$ decreases with increasing height, so the concentration of ozone reaches a maximum at a particular height; this occurs at a height of 25 to 35 km with a steady state concentration of 20-30 ppm, and is called the **ozone layer**.

(b) Ozone is formed naturally by the reaction, $O_2 + O + M \rightarrow O_3 + M^*$, where a third molecule M carries away the excess energy, and is removed by the reactions: $O_3 + h\nu \rightarrow O_2 + O$ (photochemically, $\lambda < 320$ nm), and $O_3 + O \rightarrow 2O_2$. The combination of these reactions leads to a steady state concentration of ozone in the ozone layer. Destruction of the ozone layer is primarily due to reactions *catalyzed* by (1) Cl atoms, and (2) NO molecules. (1) Cl atoms result largely from the photochemical dissociation of the C-Cl bonds in CFCs; for example, $CF_2Cl_2 + h\nu \rightarrow CF_2Cl\cdot + \cdot Cl$, ($\nu < 325$ nm). These Cl atoms catalyze the reaction $O_3 + O \rightarrow 2O_2$, which occurs by the chain mechanism, $O_3 + Cl \rightarrow ClO + O_2$, followed by $ClO + O \rightarrow Cl + O_2$. (2) NO molecules are largely formed by the reaction $N_2O + O \rightarrow NO + O_2$, where the N_2O is formed initially on the earth's surface by the oxidation of ammonia from decomposition of decaying organisms, fertilizers, and animal and human wastes, and eventually diffuses up into the stratosphere. Also, some NO forms in the engines of high-flying aircraft. NO also catalyzes the reaction $O_3 + O \rightarrow 2O_2$, by the two step mechanism: $NO + O_3 \rightarrow NO_2 + O_2$, followed by $NO_2 + O \rightarrow NO + O_2$. There is now good evidence that the combination of (1) and (2) has decreased the steady state O_3 concentration in the ozone layer, not only in polar regions but over all the earth.

---

29. (a) Estimate $\Delta H_f°(ClO,g)$ from the following bond energy data (in kJ·mol^{-1}): Cl_2 242; O_2 498; ClO 205. (b) How is the ClO radical formed in the atmosphere and why is it important in atmospheric chemistry? (c) Use your value of $\Delta H_f°(ClO,g)$ to estimate the standard reaction enthalpies for the reactions: (1) $O_3(g) + Cl(g) \rightarrow ClO(g) + O_2(g)$; (2) $O(g) + ClO(g) \rightarrow O_2(g) + Cl(g)$, involved in the catalytic decomposition of ozone by CFCs.

(a) $\Delta H_f°(ClO,g)$ is ½($\Delta H°$) for the reaction: $O{=}O(g) + Cl{-}Cl(g) \rightarrow 2Cl{-}O(g)$, which is given by:

$\mathbf{\Delta H_f°(ClO,g)} = ½(\Delta H°) = ½[BE(O{=}O) + BE(Cl{-}Cl)] - [2BE(Cl{-}O)] = [498 + 242] - [2(205)]$ kJ = **165 kJ**

(b) ClO is formed by the reaction of Cl atoms with ozone, and a principal source of Cl atoms is from the dissociation of C-Cl bonds in chlorocarbon compounds, such as the CFCs, by light from the sun ($\nu \leq 370$ nm). The reaction in question is $O_3 + O \rightarrow 2O_2$, which is catalyzed by Cl atoms in the two stage chain mechanism:

$O_3 + Cl \rightarrow O_2 + ClO$, *followed by:* $ClO + O \rightarrow Cl + O_2$, which regenerates the Cl atoms.

(c) For the reaction: $O_3(g) + Cl(g) \rightarrow O_2(g) + ClO(g)$

$\mathbf{\Delta H°} = [\Delta H_f°(O_2,g) + \Delta H_f°(ClO,g)] - [\Delta H_f°(O_3,g) + \Delta H_f°(Cl,g)] = [(0) + (165)] - [(143) + (121)]$ kJ = **-99 kJ**

For the reaction: $O(g) + ClO(g) \rightarrow O_2(g) + Cl(g)$

$\mathbf{\Delta H°} = [\Delta H_f°(O_2,g) + \Delta H_f°(Cl,g)] - [\Delta H_f°(O,g) + \Delta H_f°(ClO,g)] = [(0) + (121)] - [(249) + (165)]$ kJ = **-293 kJ**

31.* The conversion of ozone to molecular oxygen in the upper atmosphere, $2O_3(g) \rightarrow 3O_2(g)$, is thought to follow the mechanism:

$O_3 \rightleftarrows O_2 + O$ (Fast equilibrium)

$O + O_3 \rightarrow 2O_2$ (Slow)

(a) What rate law is consistent with this mechanism? (b) Explain why the reaction rate decreases as the concentration of O_2 increases. In other words, why does $[O_2]$ occur in the rate law with a negative order?

(a) The rate law is determined by the second, slow step, for which: **rate = $k[O][O_3]$** and in which O atoms are short-lived reactive intermediates. Thus:

For the first step: $K_{eq} = \dfrac{[O_2][O]}{[O_3]}$; $[O] = K_{eq}\dfrac{[O_3]}{[O_2]}$; Rate $= k[O][O_3] = kK_{eq}\dfrac{[O_3][O_3]}{[O_2]} = \underline{k'\dfrac{[O_3]^2}{[O_2]}}$

(b) The rate is inversely proportional to $[O_2]$, and since O_2 is a product in the first reaction step, any increase in $[O_2]$ will shift the equilibrium involved to the left, thereby decreasing [O], the concentration of the O atoms. Since the rate of the rate determining step is proportional to [O], a decrease in its concentration will decrease the rate.

32.* (a) Calculate ΔG at -55°C and K_p at -55°C for the reaction, $1½O_2(g) \rightleftarrows O_3(g)$, from the data in Table 9.1 and $S°(O_3,g) = 238.8$ J K^{-1} mol^{-1}, $S°(O_2,g) = 205.0$ J K^{-1} mol^{-1}, and $S°(O.g) = 160.9$ J K^{-1} mol^{-1}. Assume that $\Delta H_f°$ and S° do not vary with temperature. (b) If the pressure of $O_2(g)$ is 2.0×10^{-3} atm in the stratosphere, what is the equilibrium pressure of $O_3(g)$? (c) The steady state concentration of $O_3(g)$ is 3.0×10^{15} molecules per liter. Show by calculation whether the system O_3/O_2 is at equilibrium, and comment on the significance of your result.

(a) Assuming that $\Delta H°$ and $\Delta S°$ are temperature independent, we can calculate ΔG at -55°C (218 K) from the equation $\Delta G = \Delta H - T.\Delta S$. From the given data:

$\Delta H° = [\Delta H_f°(O_3,g)] - [1½(\Delta H_f°(O_2,g)] = [+142.7] - [(0)] =$ **+142.7 kJ**

$\Delta S° = [S°(O_3,g] - [1½(S°(O_2,g)] = [(238.8)] - [1½(205.0)]$ J K^{-1} = **-68.7 J K^{-1}**

Thus: $\Delta G = \Delta H° - T\Delta S° = 142.7 \text{ kJ} - (218 \text{ K})(-68.7 \text{ J K}^{-1})(\dfrac{1 \text{ kJ}}{10^3 \text{ J}}) = (142.7 + 15.0) \text{ kJ} = \underline{157.7 \text{ kJ}}$

(b) We can now use the value of ΔG calculated in part (a) to calculate K_p for the reaction from: **$\Delta G = -RT \ln K_p$**

$$\ln K_p = -\frac{\Delta G}{RT} = \frac{-157.7 \text{ kJ mol}^{-1}}{(8.314 \text{ J K}^{-1} \text{ mol}^{-1})(\frac{1 \text{ kJ}}{10^3 \text{ J}})(218 \text{ K})} = \underline{-87.01}\ ; \quad K_p = e^{-87} = \underline{1.63 \times 10^{-38} \text{ atm}^{-\frac{1}{2}}}$$

$$\text{Thus, for } [O_2] = 2 \times 10^{-3} \text{ atm:} \quad K_p = \frac{[O_3]}{[O_2]^{\frac{3}{2}}}; \quad [O_3] = (1.63 \times 10^{-38} \text{ atm}^{-\frac{1}{2}})(2 \times 10^{-3} \text{ atm})^{\frac{3}{2}} = \underline{1 \times 10^{-42} \text{ atm}}$$

(c) From part (b), the pressure of O_3 for the system at equilibrium is 10^{-42} atm, which corresponds to:

$$n = \frac{PV}{RT} = \left(\frac{1 \times 10^{-42} \text{ atm}}{(0.0821 \text{ atm L mol}^{-1} \text{ K}^{-1})(218 \text{ K})}\right)\left(\frac{6.022 \times 10^{23} \text{ molecules}}{1 \text{ mol}}\right) = \underline{3 \times 10^{-20} \text{ molecules L}^{-1}}$$

The actual concentration of O_3 of 3.0×10^{15} molecules L^{-1} is much greater than this equilibrium concentration, so the system is **not at equilibrium**. The significance of the relatively high steady state O_3 concentration lies in the effectiveness of the ozone layer in screening out much of the UV light from the sun.

33.* The mechanism for the reaction of Cl atoms with ozone in the stratosphere (at -55°C) involves the steps: (1) $O_3(g) + Cl(g) \rightarrow ClO(g) + O_2(g)$; (2) $O(g) + ClO(g) \rightarrow O_2(g) + Cl(g)$. (a) Why is the overall reaction described as a chain-reaction, and why are the Cl atoms described as a catalyst? (b) At -55°C the rate constant for step (1) is 5.2×10^9 $mol^{-1}\cdot L\cdot s^{-1}$ and that for step (2) is 2.5×10^{10} $mol^{-1}\cdot L\cdot s^{-1}$. The steady state concentrations are $[O_3] = 5.3 \times 10^{-9}$ $mol\cdot L^{-1}$, $[O] = 8.3 \times 10^{-14}$ $mol\cdot L^{-1}$, $[Cl] = 1.7 \times 10^{-16}$ $mol\cdot L^{-1}$, and $[ClO] = 1.1 \times 10^{-13}$ $mol\cdot L^{-1}$. What is the overall rate of the reaction? (c) For the direct reaction, $O_3(g) + O(g) \rightarrow 2O_2(g)$, the rate constant is 3.5×10^6 $mol^{-1}\cdot L\cdot s^{-1}$. What fraction of the ozone is destroyed under these conditions by the reaction with Cl atoms?

(a) The reaction is a **chain reaction** because it is *initiated* by Cl atoms, which react in step (1) and are then regenerated in step (2), which is a *propagation* step, and then undergo further reaction, so that steps (1) and (2) are continually repeated. Cl atoms are described as a catalyst because they increase the rate of the reaction but, because they are continuously regenerated, they are not used up in the overall reaction, $O_3(g) + O(g) \rightarrow 2O_2(g)$, and do not appear in the overall balanced equation for the reaction.

(b) The **overall rate** is determined by the rate of the slowest step:
For **step 1**:

$$\text{Rate} = k_1[O_3][Cl] = (5.2 \times 10^9 \text{ mol}^{-1} \text{ L s}^{-1})(5.3 \times 10^{-9} \text{ mol L}^{-1})(1.7 \times 10^{-16} \text{ mol L}^{-1}) = \mathbf{4.7 \times 10^{-15} \text{ mol L}^{-1} \text{ s}^{-1}}$$

For step 2:

$$\text{Rate} = k_2[O][ClO] = (2.5 \times 10^{10} \text{ mol}^{-1} \text{ L s}^{-1})(8.3 \times 10^{-14} \text{ mol L}^{-1})(1.1 \times 10^{-13} \text{ mol L}^{-1}) = \mathbf{2.3 \times 10^{-16} \text{ mol L}^{-1} \text{ s}^{-1}}$$

Thus the **overall rate** is: $\mathbf{2.3 \times 10^{-16} \text{ mol L}^{-1} \text{ s}^{-1}}$

(c) For the direct (single step) reaction: rate $= k[O_3][O]$, thus:
For the direct reaction:

$$\text{Rate} = (3.5 \times 10^6 \text{ mol}^{-1} \text{ L s}^{-1})(5.3 \times 10^{-9} \text{ mol L}^{-1})(8.3 \times 10^{-14} \text{ mol L}^{-1}) = \mathbf{1.5 \times 10^{-15} \text{ mol L}^{-1} \text{ s}^{-1}}$$

$$\text{Thus, } \frac{\text{rate catalyzed reaction}}{\text{rate direct reaction}} = \frac{4.7 \times 10^{-15} \text{ mol L}^{-1} \text{ s}^{-1}}{1.5 \times 10^{-15} \text{ mol L}^{-1} \text{ s}^{-1}} = \underline{3.1}$$

Of (31 + 10) = 41 molecules destroyed, 31 are destroyed by reaction with Cl atoms and 10 by the direct reaction.

i.e., a **fraction of 0.76** is destroyed by the reaction with Cl atoms.

Photochemical Smog

36. (a) Explain why the concentrations of oxides of nitrogen are usually reported as NO_x, rather than separately as NO or NO_2. (b) Close to an urban highway the concentration of NO_x is 60 $\mu g \cdot m^{-3}$. Express this as atm, ppm, and mol · L^{-1}, assuming the average molar mass of NO_x is 38 g mol^{-1}, an atmospheric pressure of 1.00 atm, and a temperature of 20°C. (c) Suppose that a city is circular and 25 km across. What is the number of moles of NO_x in the atmosphere to an altitude of 1.0 km if the average NO_x concentration is uniformly 0.04 ppm? (Assume the same conditions as in (b).)

(a) NO(g) readily with O_2(g) to give NO_2(g) (see Demonstration 16.1), so it is not very meaningful or useful to report separately the concentrations involved in air pollution.

(b) Remember that 1 L = 1 dm³, then:

(i) $$60\ \mu g\ m^{-3} = (60\ \mu g\ m^{-3})\left(\frac{1\ g}{10^6\ \mu g}\right)\left(\frac{1\ mol\ NO_x}{38\ g\ NO_x}\right)\left(\frac{1\ m}{10\ dm}\right)^3\left(\frac{1\ dm^3}{1\ L}\right) = \underline{1.58 \times 10^{-9}\ mol\ L^{-1}}$$

(ii) $$60\ \mu g\ m^{-3} = (1.58 \times 10^{-9}\ mol\ L^{-1})(0.0821\ atm\ L\ mol^{-1}\ K^{-1})(273\ K) = \underline{3.54 \times 10^{-8}\ atm}$$

(iii) 1.58×10^{-9} mol NO_x at 1.00 atm, T = 273 K, occupies a volume, $V = \dfrac{nRT}{P}$

$$= \frac{(1.58 \times 10^{-9}\ mol)(0.0821\ atm\ L\ mol^{-1}\ K^{-1})(273\ K)}{1.00\ atm} = 3.54 \times 10^{-8}\ L$$

$$ppm = \left(\frac{3.54 \times 10^{-8}\ L}{1\ L}\right)\left(\frac{10^6\ ppm}{1}\right) = \underline{3.54 \times 10^{-2}\ ppm} \quad (35\ ppb)$$

ANSWER: $\underline{1.6 \times 10^{-9}\ mol\ L^{-1}}$; $\underline{3.5 \times 10^{-8}\ atm}$; $\underline{3.5 \times 10^{-2}\ ppm}$

(c) The air to be considered above the city is a cylinder of radius, **r**, = ½(20 km) = 10 km and height, **h**, 1 km, whose volume is given by $V = \pi r^2 h = \pi(10\ km)^2(1\ km) = 3.14 \times 10^2\ km^3$.

$$\text{Volume of } NO_x = \left(\frac{4 \times 10^{-2}\ L}{10^6\ L\ air}\right)(3.14 \times 10^2\ km^3\ air)\left(\frac{10^3\ m}{1\ km}\right)^3\left(\frac{10\ dm}{1\ m}\right)^3\left(\frac{1\ L}{1\ dm^3}\right) = \underline{3.1 \times 10^6\ L}$$

$$\text{moles of } NO_x = \frac{PV}{RT} = \frac{(1.00\ atm)(3.1 \times 10^6\ L)}{(0.0821\ atm\ L\ mol^{-1}\ K^{-1})(273\ K)} = \underline{1.4 \times 10^5\ mol} \quad (10^5\ mol)$$

---

38. An oil-fired power station consumes 10^6 L of oil daily. The oil has the average composition $C_{15}H_{32}(\ell)$ and a density of 0.80 g cm^{-3}, and the gas emitted from the stacks contains 70 ppm of NO(g). (a) How much NO(g) is emitted per day? (b) Assuming that the stack gases become uniformly mixed to an altitude of 1 km above a circular city 20 km across, what will be the increase in the NO_x concentration per day?

(a) Using the average composition of $C_{15}H_{22}$ for the oil and its complete combustion, the reaction is:

$$C_{15}H_{32}(\ell) + 23O_2(g) \rightarrow 15CO_2(g) + 16H_2O(g)$$

we can use the volume of oil, its density, and the average molar mass of $C_{15}H_{32}$ of **212.4 g mol⁻¹**, to calculate the moles of oil consumed daily, and hence the volume of gases (CO_2(g) and H_2O(g)) produced:

$$10^6 \text{ L oil} = (10^6 \text{ L oil})(\frac{10^3 \text{ mL}}{1 \text{ L}})(\frac{1 \text{ cm}^3}{1 \text{ mL}})(\frac{0.80}{1 \text{ cm}^3})(\frac{1 \text{ mol oil}}{212.4 \text{ g oil}}) = \underline{3.8 \times 10^6 \text{ mol oil}}$$

$$V_{gas} \text{ day}^{-1} = \frac{nRT}{P} = \frac{(3.8 \times 10^6 \text{ mol oil})(\frac{31 \text{ mol gas}}{1 \text{ mol oil}})(0.0821 \text{ atm L mol}^{-1} \text{ K}^{-1})(293 \text{ K})}{1 \text{ atm}} = \underline{2.834 \times 10^9 \text{ L}}$$

$$\text{Thus, volume of NO} = (2.834 \times 10^9 \text{ L gases})(\frac{70 \text{ L NO}_x}{10^6 \text{ L gas}}) = \underline{1.984 \times 10^5 \text{ L}}$$

$$\text{mol NO}_x = \frac{PV}{RT} = \frac{(1.00 \text{ atm})(1.984 \times 10^5 \text{ L NO}_x)}{(0.0821 \text{ atm L mol}^{-1} \text{ K}^{-1})(293 \text{ K})} = \underline{8.3 \times 10^3 \text{ mol NO day}^{-1}}$$

(b) First (see Problem 36), we calculate the volume in liters of the air above the city, which is the volume of a cylinder of radius 10 km and height 1 km, whose volume is **3.14×10^2 km^3**, which we convert to liters (note that any conversion of NO to NO_2 does not affect the moles of NO_x):

$$V_{air} = (3.14 \times 10^2 \text{ km}^3)(\frac{10^3 \text{ m}}{1 \text{ km}})^3(\frac{10 \text{ dm}}{1 \text{ m}})^3(\frac{1 \text{ L}}{1 \text{ dm}^3}) = \underline{3.14 \times 10^{14} \text{ L}}$$

$$[NO_x] = \frac{8.25 \times 10^3 \text{ mol}}{3.14 \times 10^{14} \text{ L}} = \underline{2.63 \times 10^{-11} \text{ mol L}^{-1}}$$

Answers: (a) **8.3×10^3 mol NO**; (b) **2.6×10^{-11} mol L^{-1} NO_x**.

Acid Rain

40. (a) Why does rain normally have a pH of 5.6? (b) Why does $Al^{3+}(aq)$ remain in solution in water at pH < 5, but precipitate as $Al(OH)_3(s)$ at pH 7.4? (c) What is the relation between the answers to (b) and the death of fish in acidified lakes?

(a) Normal rain contains $CO_2(g)$ dissolved from the atmosphere and is therefore a dilute solution of carbonic acid:

$$CO_2(g) + H_2O(\ell) \rightleftarrows H_2CO_3(aq); \quad H_2CO_3(aq) + H_2O(\ell) \rightleftarrows H_3O^+(aq) + HCO_3^-(aq)$$

which is slightly acidic (pH 5.6).

(b) Al remains in solution as soluble $Al(H_2O)_6^{3+}(aq)$, a weak acid, in acidic solutions (pH < 5), but when the solution becomes basic (pH > 7) precipitates as *insoluble* $Al(OH)_3(s)$; in other words, $Al(OH)_3(aq)$ is an amphoteric hydroxide remaining as $Al(OH)_3(s)$ in basic solution, but dissolving as hydrated $Al^{3+}(aq)$ ion in acidic solution.

(c) In lakes having the acidity of normal rain water (pH 5.6), the $Al^{3+}(aq)$ concentration is very low but in lakes polluted by acid rain, with an acidity of pH 5 or less, the concentration of $Al^{3+}(aq)$ is considerably higher due to the leaching out of aluminum from aluminosilicate rocks. When this water enters the gills of fish, it encounters a slightly basic environment of pH 7.4 and precipitates as $Al(OH)_3(s)$, which clogs the gills, leading to death by suffocation.

---

42. Calculate the pH of rainwater if SO_2 is the only acidic gas present and its concentration is 0.12 ppm.

Here the concentration of $SO_2(aq)$ is 0.12 ppm **by mass**, from which we can calculate the concentration in mol L^{-1}. Then we use the ionization constant for the reaction **$SO_2(aq) + 2H_2O(\ell) \rightleftarrows H_3O^+(aq) + HSO_3^-(aq)$**, for which at 25°C, $K_a = 1.2 \times 10^{-2}$ mol L^{-1}, given in Table 12.1, to calculate $[H_3O^+]$ and, hence, the pH:

$$SO_2(aq) + 2H_2O(\ell) \rightleftarrows H_3O^+(aq) + HSO_3^-(aq)$$

$$0.120 \text{ ppm } SO_2 = (0.120 \text{ g } SO_2)\left(\frac{1 \text{ mol } SO_2}{64.07 \text{ g } SO_2}\right)\left(\frac{10^3 \text{ g solution}}{1 \text{ L solution}}\right) = \underline{1.873 \times 10^{-6} \text{ mol L}^{-1}}$$

initially:	1.873×10^{-6}	-	0	0	mol L^{-1}
at equilibrium:	$(1.873 \times 10^{-6}) - x$	-	x	x	mol L^{-1}

$$K_a = \frac{[H_3O^+][HSO_3^-]}{[H_2SO_3]} = \frac{x^2}{(1.873 \times 10^{-6}) - x} = 1.2 \times 10^{-2} \text{ mol L}^{-1}$$

Solving the quadratic equation : $x^2 + (1.2 \times 10^{-2})x - 2.248 \times 10^{-8} = 0$; $x = \dfrac{-1.2 \times 10^{-2} \pm \sqrt{(1.44 \times 10^{-4}) + (9 \times 10^{-8})}}{2}$

$x = 1.85 \times 10^{-6}$; $[H_3O^+] = \underline{1.85 \times 10^{-6} \text{ mol L}^{-1}}$; $pH = -\log(1.85 \times 10^{-6}) = \underline{5.7}$

---

44. A nickel ore has the partial composition Ni 1.4%, Cu 1.3%, Fe 7.2%, and S 9.1% by mass. A plant processes 3.5×10^7 kg of ore per day; 17% of the sulfur is converted to H_2SO_4 and 30% of the sulfur is released to the atmosphere as $SO_2(g)$. Calculate the following: (a) The volume of SO_2 in liters released into the atmosphere per day. (b) The mass of H_2SO_4 in kilograms per day. (c) The mass of SO_2 emitted per kilogram of nickel produced.

(a) We first calculate the moles of $SO_2(g)$ produced per day, and hence the volume of SO_2 (at 20°C and 1 atm):

$$\text{moles of } SO_2 \text{ per day} = (3.5 \times 10^7 \text{ kg ore})\left(\frac{10^3 \text{ g}}{1 \text{ kg}}\right)\left(\frac{9.1 \text{ g S}}{100 \text{ g ore}}\right)\left(\frac{1 \text{ mol S}}{32.06 \text{ g S}}\right)\left(\frac{1 \text{ mol } SO_2}{1 \text{ mol S}}\right)$$

$$= \underline{9.93 \times 10^7 \text{ mol}}$$

$$V_{SO_2} \text{ per day} = \frac{nRT}{P} = \frac{(9.93 \times 10^7 \text{ mol})(0.0821 \text{ atm lmol}^{-1} \text{ K}^{-1})(293 \text{ K})}{1 \text{ atm}} = \underline{2.4 \times 10^7 \text{ L}}$$

(b) We calculate the moles of H_2SO_4 per day, and hence the mass of H_2SO_4:

$$H_2SO_4 \text{ per day} = (3.5 \times 10^7 \text{ kg ore})\left(\frac{9.1 \text{ g S}}{100 \text{ g ore}}\right)\left(\frac{17 \text{ g S}}{100 \text{ g S}}\right)\left(\frac{1 \text{ mol S}}{32.06 \text{ g S}}\right)\left(\frac{1 \text{ mol } H_2SO_4}{1 \text{ mol S}}\right)\left(\frac{98.08 \text{ g } H_2SO_4}{1 \text{ mol } H_2SO_4}\right)$$

$$= \underline{1.82 \times 10^5 \text{ kg}}$$

(c) Assuming complete conversion of the 1.4% Ni in the ore to the metal, 1 kg of ore will give 1.4×10^{-2} kg Ni, and 30% of the 9.1×10^{-2} kg of S per kg of ore will be oxidized to SO_2. Thus:

$$\text{mass of ore giving 1 kg Ni} = (1 \text{ kg ore})\left(\frac{1 \text{ kg Ni}}{1.4 \times 10^{-2} \text{ kg Ni}}\right) = \underline{71.4 \text{ kg ore}}$$

$$71.4 \text{ kg ore gives:} \quad (71.4 \text{ kg ore})\left(\frac{9.1 \text{ g S}}{100 \text{ g ore}}\right)\left(\frac{30 \text{ g S}}{100 \text{ g S}}\right)\left(\frac{1 \text{ mol S}}{32.06 \text{ g S}}\right)\left(\frac{1 \text{ mol } SO_2}{1 \text{ mol S}}\right)\left(\frac{64.06 \text{ g } SO_2}{1 \text{ mol } SO_2}\right) = \underline{3.9 \text{ kg } SO_2}$$

---

45. At 450°C, $K_p = 24 \text{ atm}^{-1/2}$ for the reaction $SO_2(g) + \frac{1}{2}O_2(g) \rightleftarrows SO_3(g)$. At 450°C and initial pressures of 2.0 atm of $SO_2(g)$ and 20 atm of air passed over a catalyst, 97% of the SO_2 is converted to SO_3. Did the reaction reach equilibrium?

Initially, since air contains approximately 20% $O_2(g)$ and 80% $N_2(g)$, the partial pressure of oxygen is 20% of 20 atm = **4.0 atm $O_2(g)$**. 97% of the $SO_2(g)$ is converted to $SO_3(g)$.

Thus, in the final mixture the partial pressure of $SO_2(g)$ is 3% of 2.0 atm = **0.06 atm, and the partial pressure of** SO_3 is 97% of 2.0 atm = **1.94 atm**, and we have:

	$SO_2(g)$	+	$\frac{1}{2}O_2(g)$	$\rightleftarrows$	$SO_3(g)$	
initially:	2.0		4.0		-	atm
finally:	0.06		[4.0 - ½(1.94)]		1.94	atm

$$Q = \frac{p_{SO_3}}{p_{SO_2}\, p_{O_2}^{\frac{1}{2}}} = \frac{1.94 \text{ atm}}{(0.06 \text{ atm})(3.03 \text{ atm})^{\frac{1}{2}}} = \underline{18.6 \text{ atm}^{-\frac{1}{2}}}$$

$Q < K_p$, and therefore the system **did not** reach equilibrium.

General Problems

49. A hot coil of platinum wire is inserted into a flask containing a mixture of ammonia and oxygen. Brown fumes are formed at the surface of the wire. Write balanced equations for the reactions taking place.

We have to consider the oxidation of ammonia, $NH_3(g)$, catalyzed on the surface of a metal, such as platinum. The obvious products are an oxide of nitrogen and water. The brown fumes suggest the formation of $NO_2(g)$, and evokes the **Ostwald process** for the preparation of nitric acid, in which the first step is the oxidation of $NH_3(g)$ to colorless NO(g), using a platinum-rhodium catalyst, which is then oxidized by air to brown $NO_2(g)$:

$$4NH_3(g) + 5O_2(g) \rightarrow 4NO(g) + 6H_2O(g)$$

$$2NO(g) + O_2(g) \rightarrow 2NO_2(g)$$

---

50. In terms of the bond orders in the following molecules and ions, account for the observed trends in the observed bond lengths: (a) O_2, 121 pm (b) O_3, 128 pm (c) O_2^{2-}, 149 pm.

First we have to write the Lewis structures (including resonance structures, if any), which gives us the bond order of each oxygen-oxygen bond:

Species	Valence Electrons	Lewis Structure	Bond Order	Bond Length (pm)
(a) O_2	6 + 6 = **12**	$:\ddot{O}=\ddot{O}:$	**2.00**	121
(b) O_3	3(6) = **18**	$:\ddot{O}=\ddot{O}^{+}-\ddot{\underset{..}{O}}:^{-} \rightleftarrows {}^{-}:\ddot{\underset{..}{O}}-\ddot{O}^{+}=\ddot{O}:$	**1.50**	128
(c) O_2^{2-}	6 + 6 + 2 = **14**	${}^{-}:\ddot{\underset{..}{O}}-\ddot{\underset{..}{O}}:^{-}$	**1.00**	149

As anticipated, **the bond length increases with decreasing bond order.**

---

52. For the reaction, $2N_2O(g) \rightarrow 2N_2(g) + O_2(g)$, the rate constant is 1.1×10^{-3} $mol^{-1}\cdot L\cdot s^{-1}$ at 565°C and 3.8×10^{-3} $mol^{-1}\cdot L\cdot s^{-1}$ at 728 °C. What are: (a) the activation energy of the reaction; (b) the value of the rate constant at 25°C?

(a) We first calculate the activation energy, E_a, using the Arrhenius equation:

$$\ln\frac{k_2}{k_1} = \frac{E_A}{R}\left(\frac{1}{T_1} - \frac{1}{T_2}\right) = \ln\frac{3.8 \times 10^{-3}\ \text{mol}^{-1}\ \text{L s}^{-1}}{1.1 \times 10^{-3}\ \text{mol}^{-1}\ \text{L s}^{-1}} = \ln 3.45 = \underline{1.24}$$

$$\left(\frac{E_A}{(8.31\ \text{J K}^{-1}\ \text{mol}^{-1})(\frac{1\ \text{kJ}}{10^3\ \text{J}})}\right)\left(\frac{(1001 - 838)\ \text{K}}{(838\ \text{K})(1001\ \text{K})}\right) = 1.24\ ; \quad E_A = \underline{53\ \text{kJ mol}^{-1}}$$

(b) $$\ln\frac{3.8 \times 10^{-3}\ \text{mol}^{-1}\ \text{L s}^{-1}}{k_3} = \left(\frac{(53\ \text{kJ mol}^{-1})(\frac{10^3\ \text{J}}{1\ \text{kJ}})}{8.31\ \text{J K}^{-1}\ \text{mol}^{-1}}\right)\left(\frac{(1001 - 298)\text{K}}{(298\ \text{K})(1001\ \text{K})}\right) = \underline{15.0}$$

$$\frac{3.8 \times 10^{-3}\ \text{mol}^{-1}\ \text{L s}^{-1}}{k_3} = e^{15.0} = 3.27 \times 10^6\ ; \quad k_3 = \underline{1.2 \times 10^{-9}\ \text{mol}^{-1}\ \text{L s}^{-1}}$$

---

54. Balance each of the following equations, and in each case classify NO_2 as an oxidizing agent or a reducing agent: (a) $NO_2(g) + I^-(aq) + H_2O(\ell) \rightarrow NO(g) + I_3^-(aq) + OH^-(aq)$; (b) $MnO_4^-(aq) + NO_2(g) + H_2O(\ell) \rightarrow Mn^{2+}(aq) + H_3O^+(aq) + NO_3^-(aq)$; (c) $NO_2(g) + H_2O(\ell) \rightarrow HNO_2(aq) + HNO_3(aq)$.

(a) I (ON -1) in I^- is oxidized to I (ON -⅓) in I_3^-, and N (ON +4) in NO_2 is reduced to N (ON +2) in NO. Thus:

$3I^- \rightarrow I_3^- + 2e^-$	oxidation
$NO_2 + 2e^- + H_2O \rightarrow NO + 2OH^-$	reduction
$3I^- + NO_2 + H_2O \rightarrow I_3^- + NO + 2OH^-$	overall

(b) N (ON +4) in NO_2 is oxidized to N (ON +5) in NO_3^-; Mn (+7) in MnO_4^- is reduced to Mn(+2) in Mn^{2+}:

$5[NO_2 + H_2O \rightarrow NO_3^- + e^- + 2H^+]$	oxidation
$MnO_4^- + 5e^- + 8H^+ \rightarrow Mn^{2+} + 4H_2O$	reduction
$MnO_4^- + 5NO_2 + H_2O \rightarrow Mn^{2+} + 5NO_3^- + 2H^+$	overall

(c) Here N (ON +4) in NO_2 is oxidized to N (ON +5) in HNO_3 *and* reduced to N (ON) + 3 in HNO_2:

$NO_2 + H_2O \rightarrow HNO_3 + e^- + H^+$	oxidation
$NO_2 + e^- + H^+ \rightarrow HNO_2$	reduction
$2NO_2 + H_2O \rightarrow HNO_3 + HNO_2$	overall

In (a) NO_2 is the **oxidizing agent**, whereas in (b) and (c) it is the **reducing agent.**

---

57. A gas X that is an oxide of nitrogen supports the combustion of a variety of substances. X may be prepared by gently heating ammonium nitrate. When 0.1020 g of white phosphorus was burned completely in X, another gas Y and 0.2337 g of an oxide of phosphorus were obtained. Starting with a given volume of X, the volume of Y produced was identical at a given temperature and pressure, but Y was found not to support combustion. The relative densities of X and Y were in the ratio 1.571 to 1.000. (a) Identify X and Y. (b) Draw their Lewis structures. (c) Write balanced equations for the reactions described.

(a) An oxide of nitrogen that is formed by gently heating ammonium nitrate is **dinitrogen monoxide,** $N_2O(g)$.

$$NH_4NO_3(s) \rightarrow N_2O(g) + 2H_2O(g)$$

Assuming that X is $N_2O(g)$, the reaction with white phosphorus, $P_4(s)$, is **either**

$$P_4(s) + 6N_2O(g) \rightarrow P_4O_6(s) + 6N_2(g) \quad \text{or} \quad P_4(s) + 10N_2O(g) \rightarrow P_4O_{10}(s) + 10N_2(g)$$

Both reactions are consistent with the observation that the volume of the product gas is equal to the initial volume of $N_2O(g)$ at the same temperature and pressure (Avogadro's law), and that the product gas does not support combustion, so **Y must be $N_2(g)$.**

From the initial mass of $P_4(s)$ of 0.1020 g, we can calculate the anticipated mass of each oxide (P_4O_6 and P_4O_{10}):

$$\text{(1)}\quad \text{Expected mass of } P_4O_6 = (0.1020\ g\ P_4)\left(\frac{1\ mol\ P_4}{123.9\ g\ P_4}\right)\left(\frac{219.9\ g\ P_4O_6}{1\ mol\ P_4O_6}\right) = \underline{0.1810\ g}$$

$$\text{(2)}\quad \text{Expected mass of } P_4O_{10} = (0.1020\ g\ P_4)\left(\frac{1\ mol\ P_4}{1239\ g\ P_4}\right)\left(\frac{283.9\ g\ P_4O_{10}}{1\ mol\ P_4O_{10}}\right) = \underline{0.2337\ g}$$

The calculation in (2) is consistent with $P_4O_{10}(s)$ as the oxide of phosphorus formed. Since equal volumes of gases at the same temperature and pressure contain equal number of moles of gas, the ratio of their densities is the ratio of their molar masses. If indeed the gases are N_2O and N_2, we expect:

$$\frac{\text{density } N_2O}{\text{density } N_2} = \frac{44.02\ g\ mol^{-1}}{28.02\ g\ mol^{-1}} = \underline{1.571}\ ,\ \underline{\text{as observed}}$$

(b) Lewis structures: $\underline{N_2O}$ $\quad ^{-}{:}\ddot{N}{=}N^{+}{=}\ddot{O}{:}$ $\quad \underline{N_2}$ $\quad {:}N{\equiv}N{:}$

(c)

$$NH_4NO_3(s) \rightarrow N_2O(g) + 2H_2O(g)$$

$$P_4(s) + 10N_2O(g) \rightarrow P_4O_{10}(s) + 10N_2(g)$$

CHAPTER 17

1. Draw the structures of each of the following α-amino acids. Which are dicarboxylic acids? **(a) alanine (b) glycine (c) asparagine (d) glutamic acid**

The α-amino acids are given in Table 17.1:

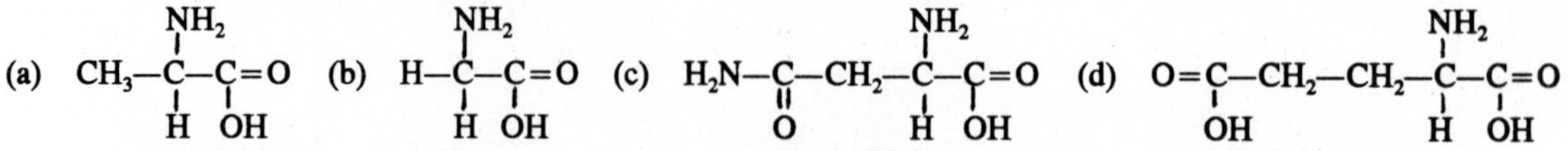

```
         NH2              NH2                      NH2                          NH2
         |                |                        |                            |
(a) CH3—C—C=O   (b) H—C—C=O   (c) H2N—C—CH2—C—C=O   (d) O=C—CH2—CH2—C—C=O
         |  |             |  |             ||       |  |         |              |  |
         H  OH            H  OH            O        H  OH        OH             H  OH
```

alanine **glycine** **aspargine** **glutamic acid**

Of these amino acids, only **glutamic acid** is a **dicarboxylic acid (two $-CO_2H$ groups)**

---

3. An amino acid isolated from animal tissue was believed to be glycine (aminoethanoic acid), $NH_2CH_2CO_2H$. When 0.500 g of the amino acid was completely converted to ammonia, which was then absorbed into 50.0 mL of 0.0500-M HCl(aq), the excess HCl(aq) required 30.57 mL of 0.0600-M NaOH(aq) for neutralization. (a) How many moles of ammonia were neutralized by the NaOH(aq)? (b) How many grams of nitrogen were in the 0.500-g sample of the amino acid? (c) Is the mass percentage of nitrogen in the sample that expected for glycine?

(a) The moles of ammonia neutralized is given by the difference between the total moles of HCl(aq) used the excess HCl(aq) neutralized by the NaOH(aq):

Moles of HCl added = (50.0 ml)(0.0500 mol L^{-1}) = $\underline{2.50\ \text{mmol}}$

Excess HCl(aq) = mol NaOH titrated = (30.57 mL)(0.0600 mol L^{-1}) = $\underline{1.83\ \text{mmol}}$

Mol HCl(aq) neutralized by NH_3(aq) from amino acid = (2.50 - 1.83) mmol = **0.67 mmol**

(b) Moles of nitrogen in sample = mol NH_3(aq) formed = mol NH_3(aq) neutralized by NH_3(aq) = **0.67 mmol**

$$\text{Sample contains:}\quad (0.67\ \text{mmol N})\left(\frac{1\ \text{mol N}}{10^3\ \text{mmol}}\right)\left(\frac{14.01\ \text{g N}}{1\ \text{mol N}}\right) = \underline{9.4 \times 10^{-3}\ \text{g N}}$$

(c) $$\text{Glycine has theoretical mass\% N} = \left(\frac{14.01\ \text{g N}}{1\ \text{mol glycine}}\right)\left(\frac{1\ \text{mol glycine}}{75.07\ \text{g}}\right) \times 100\% = \underline{18.66\ \text{mass \% N}}$$

$$\text{Experiment mass\% N} = (9.4 \times 10^{-3}\ \text{g N})\left(\frac{100\%}{0.0500\ \text{g}}\right) = \underline{19\ \text{mass \%}}$$

Thus, the experimental mass % N in the sample is close to that expected for **glycine.**

---

5. Draw the structures of (a) a dipeptide formed from the α-amino acids histidine and cysteine, and (b) a tripeptide formed from the α-amino acids alanine, glycine, and histidine. (c) Express these peptides in terms of their three-letter symbols.

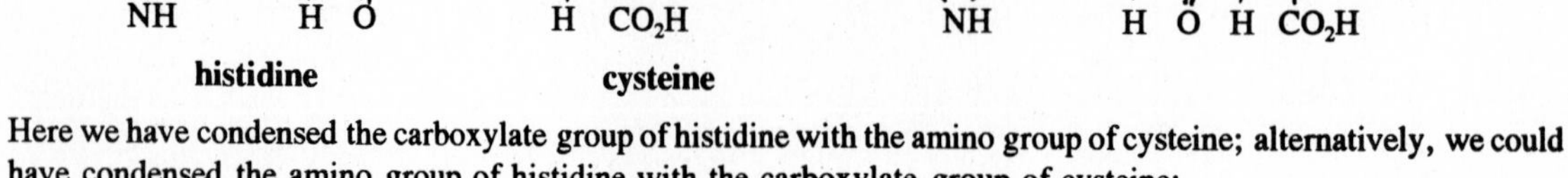

```
(a)   N—CH      NH2            H                       N—CH      NH2      H
     //    \\   |              |                      //    \\   |        |
   HC        C-CH2-C—C—OH + H—N—C—CH2-SH  →  H-C        C-CH2-C—C—N—C—CH2-SH + H2O
     \\    /       |  ||       |  |                   \\    /       |  || |  |
       NH          H  O        H  CO2H                  NH          H  O  H  CO2H
```

histidine **cysteine**

Here we have condensed the carboxylate group of histidine with the amino group of cysteine; alternatively, we could have condensed the amino group of histidine with the carboxylate group of cysteine:

```
 N—CH       CO2H           H                    N—CH       CO2H     H
HC//  \\C-CH2-C—N—H + HO—C—C—CH2-SH  →  H-C//  \\C-CH2-C—N—C—C-CH2-SH + H2O
  \   /       |  |        ||  |                \   /       |  |  ||  |
   NH         H  H        O  NH2                NH         H  H  O  NH2
     histidine              cysteine
```

(b)

```
        NH2                                        CO2H        CH—N
        |                                          |         //    \\
CH3—C—C—OH  +  H—N—CH2—C—OH  +  H—N—C—CH2—C           CH
    |  ||          |       ||          |  |          \      /
    H  O           H       O           H  H            NH
    alanine         glycine                  histidine

                         ↓

        NH2                  CO2H        CH—N
        |                    |         //    \\
CH3—C—C—N—CH2—C—N—C—CH2—C           CH
    |  || |       || |  |          \      /
    H  O  H       O  H  H            NH
```

As in part (a) there is an alternative, which is to start by condensing the -NH_2 group of alanine with the -CO_2H group of glycine.

(c) For example: **His-Cys** and **Ala-Gly-His**

---

6. How many tripeptides are possible from the condensation of (a) three, (b) four, (c) six different amino acids?

Note that there are two ways of starting the polymer chain: (1) By condensation between the -NH_2 of amino acid A with the -CO_2H group of amino acid B, and (2) by condensation between the -CO_2H of A and the -NH_2 group of B, but once started the order of peptide bonds is set. In what follows we will label the polymer units as 1, 2, 3, ... etc., as appropriate, and remember that each sequence occurs **twice, giving two different polymers**; depending on whether the peptide link sequence is **—N(H)—C(O)—** or **—C(O)—N(H)— starting from the left-hand end**; i.e., 1—2—3 and 1'—2'—3', are both possible:

(a) **1—2—3, 1—3—2** and **2—1—3,** *twice*, for a total of **6** (1 x 2 x 3).

(b) Starting with **four** different amino acids, there are **four ways** of choosing the **three** amino acids, and we showed in *part (a)* that each group of **three** amino acids gives **six** different tripeptides, so **four** amino acids give (4 x 6) = **24 tripeptides.**

(c) Following the result of *part (b)*, where we saw that there are **four** ways of selecting **three** amino acids from among **four**. Let us ask the question first: "How many ways can **three** amino acids be selected from **among six**? Clearly there are **five** ways of selecting **four** from **five**, and six ways of selecting **five** from six, giving a total of (5 x 6 x 24) = **720 tripeptides.** This problem illustrates the enormous number of simple polypeptides that can be formed from small numbers of amino acids.

---

7. (a) Draw the structure for the segment -Ala-Gly-Asp-Glu- of a polypeptide, and show how two such segments can be linked by hydrogen bonds. (b) Draw the structure for the segments Val-Cys-Gly and Tyr-Cys-Asn of two polypeptide chains, and show how they can be connected by a disulfide bridge.

(a)

```
       CH3               H                  H                   H
       |                 |                  |                   |
   H—N—C—C—OH  +  H—N—C—C—OH  +  H—N— C - C—OH  +  H—N—C - C—OH
     | | ||          | | ||          |   |   ||         |  |   ||
     H H O           H H O           H  CH2  O          H CH2  O
                                         |                 |
                                         C                CH2
                                        // \               |
                                       O   OH              C
                                                          // \
       Ala              Gly               Asp            O   OH   Glu
```

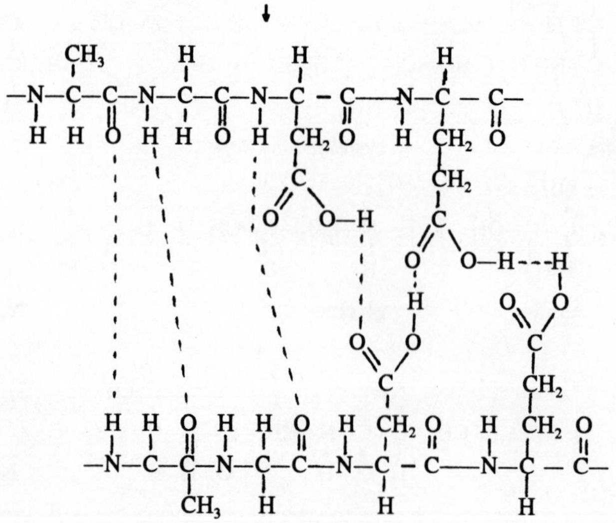

(b)

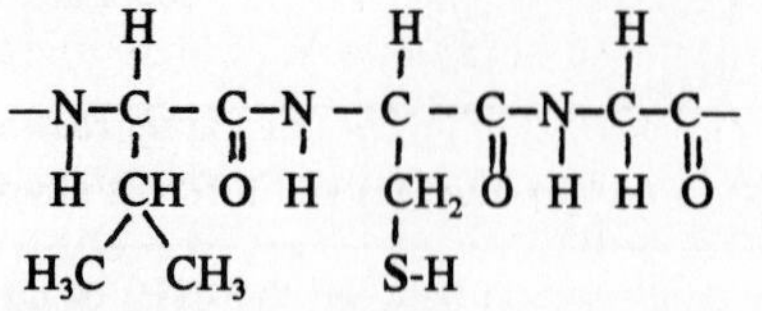

Val—Cys—Gly

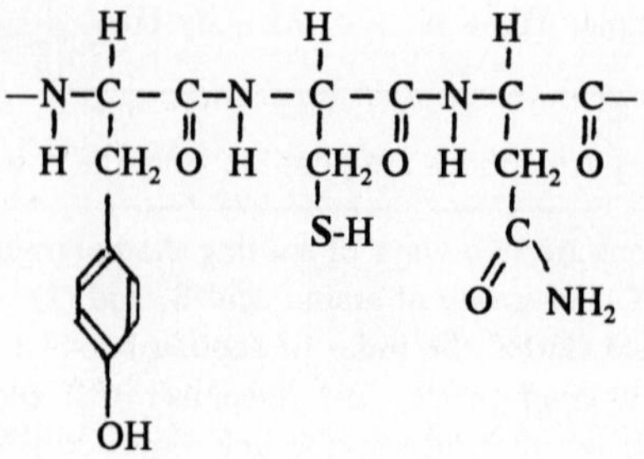

Tyr—Cys—Asn

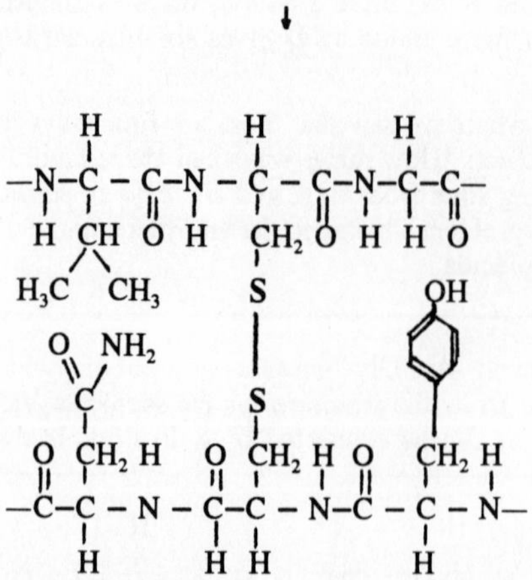

11. (a) Explain the meaning of the terms "hydrophobic" and "hydrophilic." (b) Give two examples each of hydrophobic and hydrophilic side chains found in polypeptides. (c) How are hydrophobic and hydrophilic interactions revealed in the tertiary structure of a globular protein?

(a) **Hydrophobic** literally means "water hating," and **hydrophilic means "water loving." Hydrophilic groups are** *non-polar* groups that do not form hydrogen bonds with water. Hydrophilic groups are polar groups such as —NH_2, —OH, and >C=O, that can form hydrogen bonds with water.

(b) Hydrocarbon side-chains (alkyl or aryl groups) are **hydrophobic**, whereas chains containing one or more —NH—, —NH_2, —OH, or carbonyl group, >C=O, are **hydrophilic.**

(c) The tertiary structure of a globular protein is determined mainly by the behaviour of its side-chains. Hydrophobic side chains remain on the outside of the structure, where they attract surrounding water molecules by hydrogen bonding with them, whereas the hydrophobic side chains are repelled to the inside of the structure where they attract each other by London (dispersion) forces.

Enzymes

13. (a) Explain the principal differences between how the rates of chemical reactions inside and ouside the body can be regulated; (b) What conditions in the body require that enzymes be very efficient?

(a) Inside the body the temperature is regulated at a constant value, close to 37°C, and the only solvent is water. Whereas the reaction rates of reactions outside the body are readily regulated by large factors by simply changing the temperature to slow them down or speed them up, by using any appropriate catalyst to increase the rate, or, for reactions in solution, by changing the solvent. In contrast, reactions in the healthy body not only occur at relatively low temperature but require constant pH, and can only be regulated by biologically stable catalysts (enzymes).

(b) Constant body temperature and pH mean that the only means available to the body for controlling reaction rates is *catalysis*. Hence, enzymes must be very efficient catalysts, and since processes in the body occur as the result of many successive reactions enzymes must be very specific for particular reactions.

---

15. (a) The enzyme in blood that catalyzes the decomposition of hydrogen peroxide in aqueous solution is catalase, which speeds up the reaction at 25°C by a factor of 5 x 10^6. If this is due entirely to a change in the activation energy, which is 23 kJ·mol^{-1} for the catalyzed reaction, what is the activation energy for the uncatalyzed reaction? (b) To what value would the temperature have to be raised to increase the rate of the uncatalyzed reaction to that of the catalyzed reaction at 25°C?

(a) Using the Arrhenius equation: $k_{cat} = Ae^{-\frac{(E_a)_{cat}}{RT}}$; $k_{uncat} = Ae^{-\frac{(E_a)_{uncat}}{RT}}$

$$\ln \frac{k_{cat}}{k_{uncat}} = \ln(5 \times 10^6) = \underline{15.42} = \frac{(E_a)_{uncat} - (E_a)_{cat}}{RT} = \frac{(E_a)_{uncat} - (23\ \text{kJ mol}^{-1})}{(8.31\ \text{J K}^{-1}\ \text{mol}^{-1})(\frac{1\ \text{kJ}}{10^3\ \text{J}})(298\ \text{K})}$$

$$38.2\ \text{kJ mol}^{-1} = (E_a)_{uncat} - (23\ \text{kJ mol}^{-1})\ ;\quad (E_a)_{uncat} = \underline{61\ \text{kJ mol}^{-1}}$$

(b) $$\ln \frac{k_T}{k_{298}} = \frac{E_a}{R}\left(\frac{1}{298\ \text{K}} - \frac{1}{T}\right) = \left(\frac{(61\ \text{kJ mol}^{-1})}{(8.31\ \text{J K}^{-1}\ \text{mol}^{-1})(\frac{1\ \text{kJ}}{10^3\ \text{J}})}\right)\left(\frac{1}{298\ \text{K}} - \frac{1}{T}\right) = \ln(5 \times 10^6) = \underline{15.42}$$

$$\left(\frac{1}{298\ \text{K}} - \frac{1}{T}\right) = 2.101 \times 10^{-3}\ \text{K}^{-1}\ ;\quad \frac{1}{T} = (3.356 - 2.101) \times 10^{-3}\ \text{K}^{-1} = \underline{1.255 \times 10^{-3}\ \text{K}^{-1}}$$

$$T = \underline{797\ \text{K}}\ ;\ (524°\text{C})$$

The temperature would have to be raised to **5.2 x 10^2 °C.**(520°C)

Carbohydrates

17. (a) Sugars are carbohydrates. Why was the name "carbohydrate" originally given to such compounds, and how appropriate is it? (b) Give one example each of a monosaccharide, a disaccharide, and a polysaccharide. (c) Both glucose and fructose have the molecular formula $C_6H_{12}O_6$. How do they differ structurally?

(a) **Carbohydrates** all have empirical formulas that can be written in the form $C_x(H_2O)_y$, that is as "hydrates of carbon," hence the name *carbohydrate*. Nevertheless, this formal representation has no connection with the true structures, which contain no H_2O units, but -OH groups attached to 5 and 6 membered rings.

(b) **Fructose** and **glucose** are examples of *monosaccharides*, **sucrose** is a *disaccharide*, and **starch** and **cellulose** are *polysaccharides*.

(c) The structures of **fructose** and **glucose** were drawn in Problem 16. They are both cyclic ethers with —OH functional groups; they differ in that **fructose** has a **five-membered ring** containing **4 C atoms and an O atom**, whereas **glucose** has a **six-membered ring** containing 5 C atoms and an O atom.

---

19. Amylose (starch) is a polysaccharide formed from α glucose and has an average molar mass of about 3.0 x 10^5 g mol^{-1}. (a) Approximately how many glucose units does an amylose molecule contain? (b) What is the structure of starch?

(a) Starch is a condensation polymer formed from α glucose, $C_6H_{12}O_6$, hence the repeat unit is $C_6H_{12}O_6$ - H_2O, or —$C_6H_{10}O_5$— (formula mass 162.1 u):

$$\text{Average number of polymer chain units} = (3.00 \times 10^5\ \text{g mol}^{-1})\left(\frac{1\ \text{unit}}{1.621 \times 10^2\ \text{g mol}^{-1}}\right) = \underline{1.85 \times 10^3\ \text{units}}$$

(b) **Starch** is a mixture of straight-chain and branched polymer (polysaccharide) formed from *α glucose.*

---

20. (a) Using Table 13.1, calculate ΔH° for the photosynthesis reaction $6CO_2(g) + 6H_2O(l) \rightarrow C_6H_{12}O_6(s) + 6O_2(g)$ (a) per mole, and (b) per gram of glucose. (c) If leaves of plants absorb 2.0 x 10^{-2} kJ·cm^{-2} of appropriate light energy per day, what leaf area is required to synthesize 1 g of glucose per day?

(a) $\Delta H° = [\Delta H_f°(C_6H_{12}O_6,s)] - [6\Delta H_f°(CO_2,g) + 6\Delta H_f°(H_2O,\ell)] = [(-1268)] - [6(-393.5) + 6(-285.8)]$ kJ
$= \mathbf{2808\ kJ\ mol^{-1}}$

(b) Molar mass of glucose = 180.2 g mol^{-1}

$$\Delta H°\ \text{gram}^{-1} = \frac{2808\ \text{kJ mol}^{-1}}{180.2\ \text{g mol}^{-1}} = \underline{15.58\ \text{kJ g}^{-1}}$$

(c) Assuming 100% efficiency, synthesis of 1 g of glucose requires 15.58 kJ of energy. Thus:

$$\text{Leaf area} = \frac{15.58\ \text{kJ day}^{-1}}{2.0 \times 10^{-2}\ \text{kJ cm}^{-2}\ \text{day}^{-1}} = \underline{7.8 \times 10^2\ \text{m}^2}$$

---

21. (a) Draw the structures of mono-, di-, and triphosphoric acids and the monoester that each forms with ethanol. (b) Why are ATP and ADP described as esters of di- and triphosphoric acid, respectively.

(a) Polyphosphoric acids result from condensation reactions of **monophosphoric acid**, H_3PO_4:

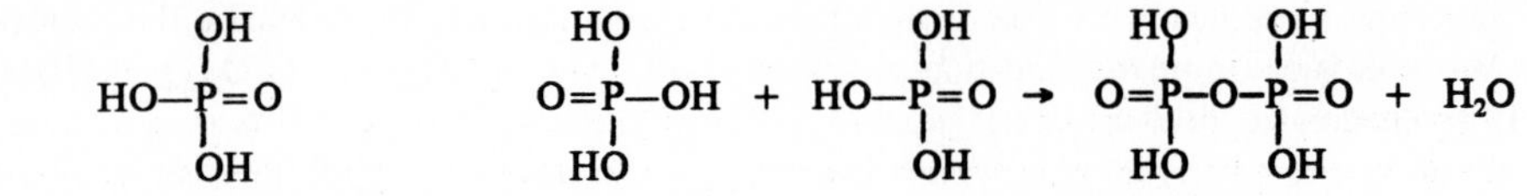

```
   OH                 HO              OH          HO   OH
   |                  |               |           |    |
HO—P=O             O=P—OH  +  HO—P=O  →  O=P-O-P=O  +  H2O
   |                  |               |           |    |
   OH                 HO              OH          HO   OH
```

monophosphoric acid **diphosphoric acid**

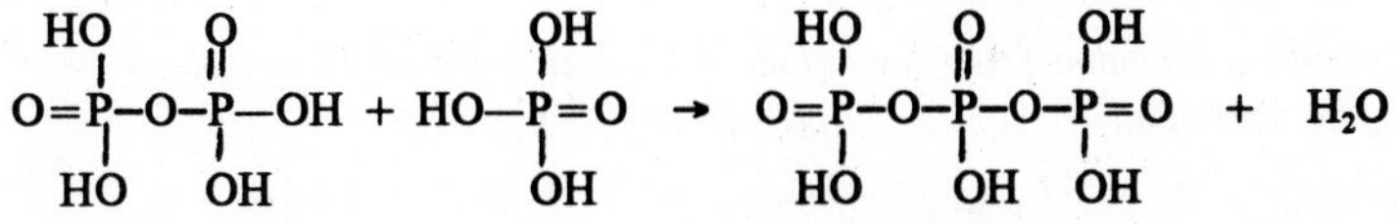

```
 HO   O           OH          HO   O    OH
 |    ||          |           |    ||   |
O=P-O-P—OH + HO—P=O  →  O=P-O-P-O-P=O  +  H2O
 |    |           |           |    |    |
 HO   OH          OH          HO   OH   OH
```

triphosphoric acid

A P—OH group of a "phosphoric acid" takes part in a condensation reaction with **ethanol,** C_2H_5OH, to give an **ethoxy-** group, P—OH + HO—CH_2CH_3 → P—OC_2H_5 + H_2O so the structures of the **monoesters** are:

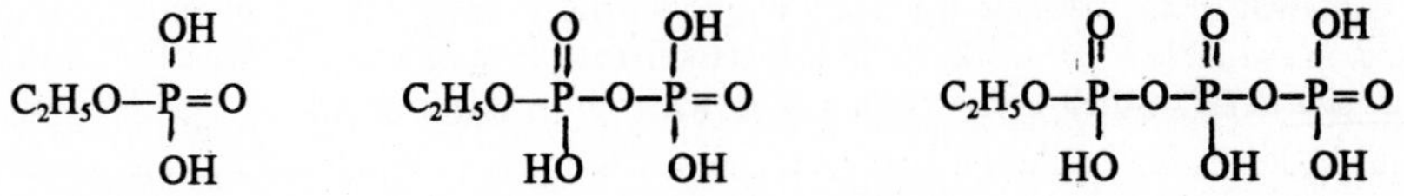

```
      OH              O    OH               O    O    OH
      |               ||   |                ||   ||   |
C2H5O—P=O      C2H5O—P-O-P=O        C2H5O—P-O-P-O-P=O
      |               |    |                |    |    |
      OH              HO   OH               HO   OH   OH
```

(b) **ADP** is the monoester formed between diphosphoric acid and the sugar *ribose* to which the heterocyclic **amine** *adenine* is attached, and **ATP** is the monoester formed between triphosphoric acid and the sugar *ribose* to which the heterocyclic amine *adenine* is attached. Such molecules are called **nucleotides.**

---

23. (a) To what class of compounds does ATP belong? (b) How does ATP differ from DNA?

(a) **ATP** is a **nucleotide**; it is an ester formed from condensing triphosphoric acid with a sugar (ribose) to which the heterocyclic base **adenine** is also attached, so it contains the structural units **adenine — sugar — triphosphate.**

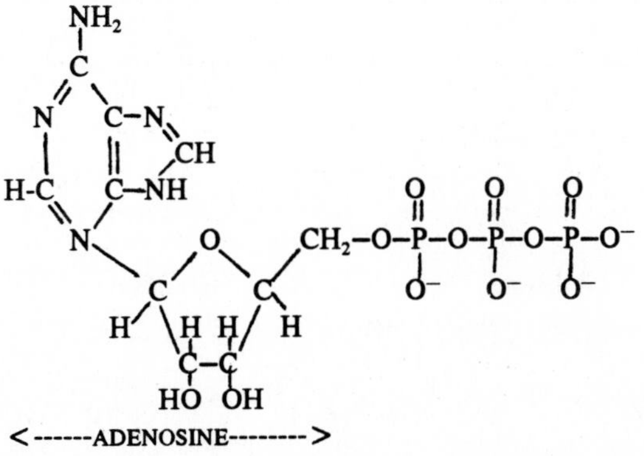

```
      NH2
      |
      C
    //  \
   N     C-N
   |     ||   \\
   |     ||    CH
   |     ||   /
  H-C    C-NH              O    O    O
    \\  /                  ||   ||   ||
     N         O       CH2-O-P-O-P-O-P-O⁻
      \      /   \    /       |    |    |
         C         C          O⁻   O⁻   O⁻
       H/ \H   H/ \H
           C-C
          HO OH
  <------ADENOSINE-------->
```

(b) **ATP** differs from **DNA** in that, while ATP is a monomeric nucleotide consisting of a **triphosphate group,** the sugar ribose, and the heterocyclic base adenine condensed together, DNA (**deoxyribonucleic acid**) is a **polynucleotide**; in other words, a polymer made up of nucleotide units. In DNA, the nucleotide units are made up from the condensation of a molecule of monophosphoric acid, a molecule of the sugar **deoxyribose,** and a molecule of one of the heterocyclic nitrogen bases: **adenine,** A, **guanine,** G, **cytosine,** C, or **thymine,** T. Thus, there are *four* different nucleotide units in DNA. The backbone of a DNA molecule consists of deoxyribose sugar units linked together by phosphate groups, with one of the bases A, C, G, or T attached to each sugar. It is the enormous number of different possible sequences of these bases that give so many different DNA molecules.

---

25. (a) How is the maximum work that can be performed by a reaction measured? (b) Calculate the maximum amount of work that can be obtained from the oxidation of glucose, $C_6H_{12}O_6(s) + 6O_2(g) \rightarrow 6CO_2(g) + 6H_2O(\ell)$. (c) How is this reaction achieved in metabolism?

(a) The maximum amount of work that can be performed (obtained from) a chemical reaction is given by the **Gibbs Free Energy change**, ΔG, for the reaction.

(b) Free energy changes are calculated from the standard free energies of formation of the products and reactants of the reaction, using the formula: $\Delta G^\circ = \Sigma\, n_p(\Delta G_f^\circ)_p - \Sigma\, n_p(\Delta G_f^\circ)_p$

Thus, for the reaction: $C_6H_{12}O_6(s) + 6O_2(g) \rightarrow 6CO_2(g) + 6H_2O(\ell)$

$$\Delta G^\circ = [6(\Delta G_f^\circ,CO_2,g) + 6(\Delta G_f^\circ,H_2O,\ell)] - [(\Delta G_f^\circ,C_6H_{12}O_6,s) + 6(\Delta G_f^\circ,O_2,)]$$

$$= [6(-394.4) + 6(-237.2)] - [(-919.2) + (0)]\ kJ = \mathbf{-2870.4\ kJ}$$

The **maximum work** that can be obtained is **+2870 kJ mol^{-1}**.

(c) In metabolism, this reaction cannot occur directly; the oxidation of glucose cannot occur in the body as it does when glucose is burned in air, because if it did a large amount of energy would be produced as heat amd almost no energy would be available to do work. Rather, metabolism occurs in a long series of steps. Much of the energy is used to synthesize **ATP**, which transfers energy to where it is required.

DNA and RNA

27. DNA and m-RNA have three common nitrogenous bases and both contain one additional base. (a) Name these bases, (b) indicate how they are usually represented, and (c) draw their structures.

(a) The **five** heterocyclic nitrogen bases in question are **Adenine, Guanine, Cytosine, Thymine,** and **Uracil.** The difference between **DNA** and **m-RNA** is that cytosine in DNA is replaced by uracil in m-RNA.

(b) They are usually represented by the letters: **A** - adenine; **G** - guanine; **C** - cytosine; **T** - thymine, and **U** - **uracil.**

(c)

Adenine **Guanine**

Cytosine

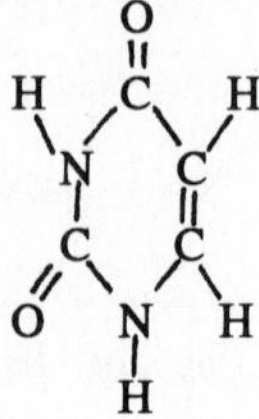

Thymine

Uracil

---

29. (a) Why does DNA form a double-helix? (b) Are the base pairs in DNA found inside the helix or outside the helix, or do they constitute part of the backbone of the structure?

(a) **DNA** contains two polynucleotide strands in which the amount of the nitrogenous base adenine, A, is the same as the base thymine, T, and the amount of the base cytosine, C, is the same as the amount of the base guanine, G.

The double helix forms because A units and T units, and C units and G units, on opposite strands form A---T and C---G hydrogen bonds that give exactly the same spacing between the two polynucleotide strands when they are wound together.

(b) The A---T and C---G base pairs are found on the *inside* of the double helix, because it is the hydrogen bonds formed between each pair that holds the double helix together.

---

30. Explain why two strands of DNA are linked by interactions between two specific pairs of bases rather than between all six possible combinations of these four bases.

Although any pair of the bases adenine, A, thymine, T, cytosine, C, and guanine, G, can form hydrogen bonds, a double helix can form only if these interactions between base units on opposite strands give the same constant spacing between the two polynucleotide chains so that the double helix of a DNA molecule can form. **A and T**, joined by two hydrogen bonds, and **C and G**, joined by three hydrogen bonds, are found to be the only combinations of the six possible base pairs with the right size and shape that satisfy the above condition.

---

33. If a section of a single strand of DNA contains the base sequence ACTCGC, what is the corresponding base sequence in (a) the complementary DNA strand and (b) the mRNA strand formed by transcription?

To solve this problem we simply have to remember that the only possible base pairings in the DNA double helix are adenosine–thymine, A—T, and cytosine–guanine, C—G. But a mRNA strand is formed from the single **complementary** DNA strand by transcription, and then uracil replaces thymine. Thus, we simply have to pair the appropriate bases:

(a) For **replication:**

—A—C—T—C—G—C—

↓

—T—G—A—G—C—G—

(b) For **transcription:**

—T—G—A—G—C—G—

↓

—A—C—U—C—G—C—

Lipids

34. (a) Draw the structure of a typical triacylglycerol. (b) What is the function of triacylglycerols in the body?

(a) **Triacylglycerols** (also called triglycerides) are **triesters** of **glycerol** (propane-1,2,3,-triol) with long chain *fatty acids* such as the *saturated* acids **palmitic acid,** $CH_3(CH_2)_{14}CO_2H$, and **stearic acid,** $CH_3(CH_2)_{16}CO_2H$, and *unsaturated* acids such as **oleic acid,** $CH_3(CH_2)_7CH{=}CH(CH_2)_7CO_2H$. They are the major constituents of fats and oils found in living organisms.

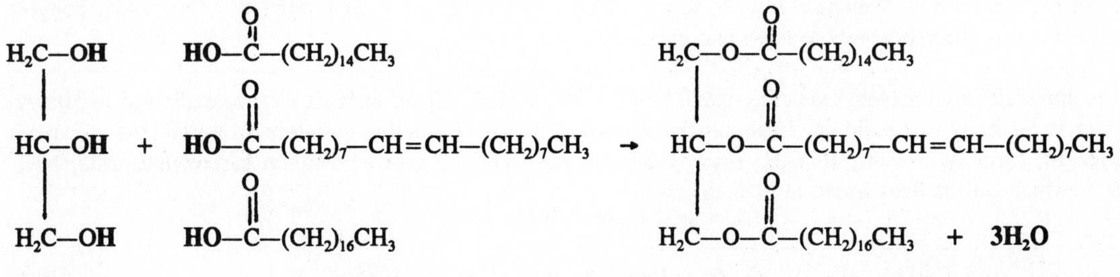

$$\begin{array}{l} H_2C\text{—}OH \\ HC\text{—}OH \\ H_2C\text{—}OH \end{array} + \begin{array}{l} HO\text{—}C(=O)\text{—}(CH_2)_{14}CH_3 \\ HO\text{—}C(=O)\text{—}(CH_2)_7\text{—}CH{=}CH\text{—}(CH_2)_7CH_3 \\ HO\text{—}C(=O)\text{—}(CH_2)_{16}CH_3 \end{array} \rightarrow \begin{array}{l} H_2C\text{—}O\text{—}C(=O)\text{—}(CH_2)_{14}CH_3 \\ HC\text{—}O\text{—}C(=O)\text{—}(CH_2)_7\text{—}CH{=}CH\text{—}(CH_2)_7CH_3 \\ H_2C\text{—}O\text{—}C(=O)\text{—}(CH_2)_{16}CH_3 \end{array} + 3H_2O$$

glycerol **fatty acids** **triacylglycerol**

(b) **Triacylglycerols** (triglycerides) are highly concentrated stores of metabolic energy, the complete oxidation of which yields about 40 kJ g^{-1} of energy, compared to about 17 kJ g^{-1} of energy for carbohydrates and proteins. In mammals they are synthesized from glucose and stored in adipose (fat) cells. The triacylglycerols are hydrolyzed first to glycerol and their constituent fatty acids, and eventually to carbon dioxide and water, thereby providing energy for the synthesis of ATP.

Nutrition

36. (a) What five classes of substances constitute the essential components of a healthy diet? (b) What class of substances is the primary source of energy for humans, and in what units is this energy usually expressed? (c) In what form is energy stored in the body?

(a) The essential components of a healthy diet are conveniently classified as **carbohydrates, fats, proteins** (α amino acids), **vitamins**, and **"minerals."**

(b) **Carbohydrates** including starch are the primary source of energy for humans. Starch is hydrolyzed in stages, by enzymes in saliva, and then in the stomach and the small intestine, to glucose. Eventually the glucose is oxidized in further steps to CO_2 and water. For historical reasons, the energy obtained from food is measured in **Calories:**

$$1 \text{ Cal} = 1 \text{ kcal} = 4.187 \text{ kJ}$$

(c) In the body, unused carbohydrates are stored as fats, which are better suppliers of energy than carbohydrates.

General Problems

39. Photosynthesis in plants requires light of maximum wavelength 700 nm, the wavelength of red light. (a) To what energy does this correspond? (b) Is this energy sufficient to break directly the O-H bond in water, BE(OH) = 463 kJ·mol^{-1}, to produce oxygen?

(a) We can calculate the energy corresponding to 700 nm photons from $E = h\nu$, and for convenience we will convert this to the energy per mole of photons:

$$E_{photon} = h\nu = \frac{hc}{\lambda}\ ; \quad E_{mol^{-1}} = \frac{N_A hc}{\lambda} = \frac{(6.022 \times 10^{23}\ \text{mol}^{-1})(6.63 \times 10^{34}\ \text{J s})(3.00 \times 10^{8}\ \text{m s}^{-1})}{(700\ \text{nm})(\frac{1\ \text{m}}{10^{9}\ \text{nm}})} = \underline{171\ \text{kJ mol}^{-1}}$$

(b) Red light (E = 171 kJ mol^{-1}) is not sufficiently energetic to break the O-H bond in water to produce oxygen directly.

---

41. (a) Ammonia is poisonous, so most mammals eliminate nitrogen, a result of the oxidation of proteins, as urea. With what substance must ammonia react to form urea? (b) Fish usually eliminate nitrogen as ammonia. Suggest a reason for this difference between fish and mammals.

(a) The ammonia from protein oxidation would be very poisonous to higher animals if it were allowed to dissolve in water in the body and build up. **Urea** has the formula $(NH_2)_2C{=}O$, and is therefore an **amide**. As we saw in Chapter 14, carboxylic acids, RCO_2H, react with ammonia, NH_3, to give ammonium carboxylate salts, NH_4^+ RCO_2^-, which can be dehydrated to give an amide:

$$R{-}\underset{\underset{O}{\|}}{C}{-}O^- \ NH_4^+ \rightarrow R{-}\underset{\underset{O}{\|}}{C}{-}NH_2 + H_2O$$

Thus, we can assume that a similar reaction probably occurs in the body between ammonia and carbonic acid:

$$CO_2 + H_2O \rightleftarrows (HO)_2C{=}O; \quad (HO)_2C{=}O + 2NH_3 \rightarrow 2NH_4^+\ CO_3^{2-} \rightarrow (NH_2)_2C{=}O + 2H_2O$$

That is, ammonia from the oxidation of proteins reacts with carbon dioxide in the presence of water to give urea.

(b) Fish live in an aqueous environment, so it is convenient for ammonia from the oxidation of their proteins to pass into the water that surrounds them, and under normal conditions it is immediately diluted and swept away, so it never achieves a lethal concentration.

---

42. Air is bubbled through a solution of iron(II) sulfate and through a solution of hemoglobin. What observations would you expect to make in the two cases? Account for the different behavior.

A solution of **iron(II) sulfate** in water, $FeSO_4(aq)$, contains the $Fe^{2+}(aq)$ ion that is readily oxidized to $Fe^{3+}(aq)$ by the $O_2(g)$ in air:

$$4[Fe^{2+}(aq) \rightarrow Fe^{3+}(aq) + e^-] \qquad \text{oxidation}$$

$$O_2(g) + 4e^- + 2H_2O(\ell) \rightarrow 4OH^-(aq) \qquad \text{reduction}$$

$$4Fe^{2+}(aq) + O_2(g) + 2H_2O(\ell) \rightarrow 4Fe^{3+}(aq) + 4OH^-(aq) \qquad \text{overall}$$

and, since $Fe(OH)_3(s)$ is insoluble, the $Fe^{3+}(aq)$ formed will be precipitated as brown $Fe(OH)_3(s)$.

Although iron is also present in **hemoglobin** as Fe(II), it is coordinated to *five* N atoms, leaving the sixth position of the octahedral coordination sphere empty. When $O_2(g)$ is bubbled through a hemoglobin solution the O_2 molecule bonds to the iron atom at this site, giving the bright red complex **oxyhemoglobin**, rather than oxidizing Fe(II) to Fe(III).

CHAPTER 18

Radioactivity

1. Give the number of protons, neutrons, and nucleons in each of the following nuclei: (a) $^{11}_{5}B$ (b) $^{13}_{6}C$ (c) $^{94}_{40}Zr$ (d) $^{137}_{56}Ba$ (e) $^{57}_{26}Fe$

In the symbol $^{x}_{y}E$, y is the number of protons and x is the number of nucleons (number of protons + number of neutrons). Thus:

Symbol	$^{11}_{5}B$	$^{13}_{6}C$	$^{94}_{40}Zr$	$^{137}_{56}Ba$	$^{57}_{26}Fe$
nucleons	11	13	94	137	57
protons	5	6	40	56	26
neutrons	6	7	54	81	31

3. What is the product nucleus when a ^{80}Br nucleus decays by (a) β emission; (b) positron emission; (c) electron capture?

A β-particle is an electron, $^{0}_{-1}e$; a positron has the same mass as an electron but the opposite charge of +1, $^{0}_{1}e$, and capture of an electron, $^{0}_{-1}e$, converts a nuclear proton, $^{1}_{1}p$, into a neutron, $^{1}_{0}n$.

(a) β emission: $^{80}_{35}Br \rightarrow {}^{80}_{36}Kr + {}^{0}_{-1}e$; the product nuclide is $^{80}_{36}Kr$

(b) positron emission: $^{80}_{35}Br \rightarrow {}^{80}_{34}Se + {}^{0}_{1}e$; the product nuclide is $^{80}_{34}Se$

(c) electron capture: $^{80}_{35}Br + {}^{0}_{-1}e \rightarrow {}^{80}_{34}Se$; the product nuclide is $^{80}_{34}Se$

5. The ^{207}Po isotope of polonium can decay in three ways: (a) by electron capture, (b) by positron emission, and (c) by α-particle. Write a balanced equation for each reaction.

(a) electron capture: $^{207}_{84}Po + {}^{0}_{-1}e \rightarrow {}^{207}_{83}Bi$; the product nuclide is $^{207}_{83}Bi$

(b) positron emission: $^{207}_{84}Po \rightarrow {}^{207}_{83}Bi + {}^{0}_{1}e$; the product nuclide is $^{207}_{83}Bi$

(c) α particle emission: $^{207}_{84}Po \rightarrow {}^{203}_{82}Pb + {}^{4}_{2}He$; the product nuclide is $^{203}_{82}Pb$

7. Complete each of the following equations: (a) $^{32}_{15}P \rightarrow {}^{32}_{16}S$ + ? (b) $^{15}_{8}O \rightarrow {}^{15}_{7}N$ + ? (c) $^{52}_{26}Fe \rightarrow {}^{52}_{25}Mn$ + ? (d) $^{218}_{87}Fr \rightarrow ? + {}^{4}_{2}He$ (e) $^{50}_{26}Fe \rightarrow ? + {}^{0}_{-1}e$ (f) $^{122}_{53}I \rightarrow {}^{122}_{54}Xe$ + ?

(a) $^{32}_{15}P \rightarrow {}^{32}_{16}S + {}^{0}_{-1}e$ (b) $^{15}_{8}O \rightarrow {}^{15}_{7}N + {}^{0}_{1}e$ (c) $^{52}_{26}Fe \rightarrow {}^{52}_{25}Mn + {}^{0}_{1}e$

(d) $^{218}_{87}Fr \rightarrow {}^{214}_{85}At + {}^{4}_{2}He$ (e) $^{50}_{26}Fe \rightarrow {}^{50}_{27}Co + {}^{0}_{-1}e$ (f) $^{122}_{53}I \rightarrow {}^{122}_{54}Xe + {}^{0}_{-1}e$

9. Thorium-231 decays to lead-207 by emitting the following particles in successive steps. Write the symbol for the isotope formed in each step: (a) β (b) α (c) α (d) β (e) α (f) α (g) α (h) β (i) β (j) α.

(a) $^{231}_{90}Th \rightarrow {}^{231}_{91}Pa + {}^{0}_{-1}e$ (b) $^{231}_{91}Pa \rightarrow {}^{227}_{89}Ac + {}^{4}_{2}He$ (c) $^{227}_{89}Ac \rightarrow {}^{223}_{87}Fr + {}^{4}_{2}He$

(d) $^{223}_{87}Fr \rightarrow {}^{223}_{88}Ra + {}^{0}_{-1}e$ (e) $^{223}_{88}Ra \rightarrow {}^{219}_{86}Rn + {}^{4}_{2}He$ (f) $^{219}_{86}Rn \rightarrow {}^{215}_{84}Po + {}^{4}_{2}He$

(g) $^{215}_{84}Po \rightarrow {}^{211}_{82}Pb + {}^{4}_{2}He$ (h) $^{211}_{82}Pb \rightarrow {}^{211}_{83}Bi + {}^{0}_{-1}e$ (i) $^{211}_{83}Bi \rightarrow {}^{211}_{84}Po + {}^{0}_{-1}e$

(j) $^{211}_{84}Po \rightarrow {}^{207}_{82}Pb + {}^{4}_{2}He$

11. **For each of the following unstable nuclides, suggest what type of single-step radioactive decay each will undergo, and name the product nuclide: (a) ^{10}Be (b) ^{12}N (c) ^{14}C (d) ^{87}Br**

As we saw in Problem 11, only β emission increases the neutron/proton ratio for nuclides above the band of stability; conversely, for nuclides below the band of stability, the neutron/proton is increased by positron emission or by electron capture (which both give the same daughter nuclide):

Nuclide	Neutron/Proton Ratio	Reaction	Product Nuclide	Neutron/Proton Ratio
(a) $^{10}_{4}Be$	6 : 4 = 1.50	$^{10}_{4}Be \rightarrow {}^{10}_{5}B + {}^{0}_{-1}e$	$^{10}_{5}B$	5 : 5 = 1.00
(b) $^{12}_{7}N$	5 : 7 = 0.71	$^{12}_{7}N \rightarrow {}^{12}_{6}C + {}^{0}_{1}e$ or $({}^{12}_{7}N + {}^{0}_{-1}e \rightarrow {}^{12}_{6}C)$	$^{12}_{6}C$	6 : 6 = 1.00
(c) $^{14}_{6}C$	8 : 6 = 1.33	$^{14}_{6}C \rightarrow {}^{14}_{7}N + {}^{0}_{-1}e$	$^{14}_{7}N$	7 : 7 = 1.00
(d) $^{87}_{35}Br$	52:35 = 1.49	$^{87}_{35}Br \rightarrow {}^{87}_{36}Kr + {}^{0}_{-1}e$	$^{87}_{36}Kr$	51:36 = 1.42

Summary:

(a) β emission; $^{10}_{4}Be \rightarrow {}^{10}_{5}B + {}^{0}_{-1}e$; boron-10 **(b) positron emission; $^{12}_{7}N \rightarrow {}^{12}_{6}C + {}^{0}_{1}e$; carbon-12**

(c) β emission; $^{14}_{6}C \rightarrow {}^{14}_{7}N + {}^{0}_{-1}e$; nitrogen-14 **(d) β emission; $^{87}_{35}Br \rightarrow {}^{87}_{36}Kr + {}^{0}_{-1}e$; krypton-87**

---

13. **Express the activity of the radioactive sources with each of the following disintegration rates in curies, Ci, and in millicuries, mCi: (a) 3.70×10^{10} disintegrations s^{-1}; (b) 3.50×10^{6} disintegrations s^{-1}; (c) 7.00×10^{12} disintegrations s^{-1}; (d) 4.80×10^{16} disintegrations s^{-1}.**

To solve this problem we use the identities: 1 Ci (curie) = 3.7×10^{10} disintegrations s^{-1} and 1 mCi = 10^{-3} Ci

$$\text{(a)}\quad (3.70 \times 10^{10}\ \text{disintegrations s}^{-1})\left(\frac{1\ \text{Ci}}{3.70 \times 10^{10}\ \text{disintegrations s}^{-1}}\right) = \underline{1.00\ \text{Ci}}$$

$$= (1.00\ \text{Ci})\left(\frac{10^3\ \text{mCi}}{1\ \text{Ci}}\right) = \underline{1 \times 10^3\ \text{mCi}}$$

$$\text{(b)}\quad (3.50 \times 10^{6}\ \text{disintegrations s}^{-1})\left(\frac{1\ \text{Ci}}{3.70 \times 10^{10}\ \text{disintegrations s}^{-1}}\right) = \underline{9.46 \times 10^{-5}\ \text{Ci}}$$

$$= (9.46 \times 10^{-5}\ \text{Ci})\left(\frac{10^3\ \text{mCi}}{1\ \text{Ci}}\right) = \underline{9.46 \times 10^{-2}\ \text{mCi}}$$

$$\text{(c)}\quad (7.00 \times 10^{12}\ \text{disintegrations s}^{-1})\left(\frac{1\ \text{Ci}}{3.70 \times 10^{10}\ \text{disintegrations s}^{-1}}\right) = \underline{1.89 \times 10^{2}\ \text{Ci}}$$

$$= (1.89 \times 10^{2}\ \text{Ci})\left(\frac{10^3\ \text{Ci}}{1\ \text{Ci}}\right) = \underline{1.89 \times 10^{5}\ \text{mCi}}$$

$$\text{(d)}\quad (4.80 \times 10^{16}\ \text{disintegrations s}^{-1})\left(\frac{1\ \text{Ci}}{3.70 \times 10^{10}\ \text{disintegrations s}^{-1}}\right) = \underline{1.30 \times 10^{6}\ \text{Ci}}$$

$$= (1.30 \times 10^{6}\ \text{Ci})\left(\frac{10^3\ \text{mCi}}{1\ \text{Ci}}\right) = \underline{1.30 \times 10^{9}\ \text{mCi}}$$

---

15. Calculate the dose of radiation (in rad) and the dose equivalent (in rem) when a 65.0-kg man absorbs 1 x 10^{-2} J of energy as the result of exposure to (a) β radiation, and (b) α radiation. Assume that the relative biological effectiveness (RBE) is 1 for β and for γ radiation and 20 for α radiation.

Here, we use: **1 rad ≡ 10^{-2} J per kg of body tissue**

<u>and</u> **Dose equivalent (in rem) = Q x dose (in rad) - where Q is the relative biological effectiveness, RBE.**

(a) <u>β rays</u>: $\text{rad dose} = \left(\frac{1 \times 10^{-2}\ \text{J}}{65.0\ \text{kg}}\right)\left(\frac{1\ \text{rad}}{10^{-2}\ \text{J}}\right) = \underline{1.54 \times 10^{-2}\ \text{rad}}$

$\text{dose equivalent} = Q(1.54 \times 10^{-2}\ \text{rad}) = \left(\frac{1\ \text{rem}}{1\ \text{rad}}\right)(1.54 \times 10^{-2}\ \text{rad}) = \underline{1.54 \times 10^{-2}\ \text{rem}}$

(b) <u>α rays</u>: $\text{rad dose} = \left(\frac{1 \times 10^{-2}\ \text{J}}{65.0\ \text{kg}}\right)\left(\frac{1\ \text{rad}}{10^{-2}\ \text{J}}\right) = \underline{1.54 \times 10^{-2}\ \text{rad}}$

$\text{dose equivalent} = Q(1.54 \times 10^{-2}\ \text{rad}) = \left(\frac{20\ \text{rem}}{1\ \text{rad}}\right)(1.54 \times 10^{-2}\ \text{rad}) = \underline{0.308\ \text{rem}}$

<u>Radioactive Decay Rates</u>

17. A container holds 1.00 Ci of a radioactive isotope that decays to stable products. How many curies of radiation are left after five half-lives?

Radioactive disintegration is a *first order* process, so the half-life is independent of the initial concentration (amount of radioactivity), and Rate of Disintegration ∝ Number of curies:

After 5 half-lives 1.00 Ci of radioactive isotope will have decayed to (1.00 Ci)($\frac{1}{2}$)^5 = 3.13 x 10^{-2} Ci.

---

19. A sample of a radioactive nuclide is recorded to have 668 disintegrations per minute. After 60 min, the number of disintegrations diminished to 25 per minute. What is the half-life of the radioactive substance?

For a *first order* process, $\ln [A]_t = \ln [A]_o - kt$, where k is the rate constant, $[A]_o$ is the initial concentration, and $[A]_t$ is the concentration after time t, which becomes $\ln N_t = \ln N_o - kt$, for disintegrating nuclei

The first order equation gives: $\ln \frac{N_o}{N_t} = kt$, whence for $N_t = \frac{N_o}{2}$, $kt_{\frac{1}{2}} = \ln 2 = 0.693$

The given data gives: $\ln \frac{N_o}{N_t} = \ln \frac{668}{25} = \ln 26.7 = 3.285 = k(60\ \text{min})$; $\underline{k = 5.48 \times 10^{-2}\ \text{min}^{-1}}$

Thus, $kt_{\frac{1}{2}} = t_{\frac{1}{2}}(5.48 \times 10^{-2}\ \text{min}^{-1}) = 0.693$; $t_{\frac{1}{2}} = \underline{13\ \text{min}}$

---

21. A cobalt bomb in a cancer hospital contains a mass of 500 g of ^{60}Co (half-life 5.26 yr). What mass of ^{60}Co will remain after (a) 1.00 yr, (b) 5.00 yr, (c) 10.00 yr?

$$kt = \ln \frac{N_o}{N_t} = \ln \frac{N_o}{N_o/2} = \ln 2 = 0.693 = kt_{\frac{1}{2}} \; ; \quad k = \frac{0.693}{5.26 \text{ yr}} = \underline{0.1317 \text{ yr}^{-1}}$$

(a) $\ln \frac{500 \text{ g}}{x \text{ g}} = (0.1317 \text{ yr}^{-1})(1.00 \text{ yr}) = 0.1317 \; ; \quad \frac{500 \text{ g}}{x \text{ g}} = 1.141 \; ; \quad \underline{x = 438 \text{ g}}$

(b) $\ln \frac{500 \text{ g}}{x \text{ g}} = (0.1317 \text{ yr}^{-1})(5.00 \text{ yr}) = 0.6585 \; ; \quad \frac{500 \text{ g}}{x \text{ g}} = 1.932 \; ; \quad \underline{x = 259 \text{ g}}$

(c) $\ln \frac{500 \text{ g}}{x \text{ g}} = (0.1317 \text{ yr}^{-1})(10.00 \text{ yr}) = 1.317 \; ; \quad \frac{500 \text{ g}}{x \text{ g}} = 3.732 \; ; \quad \underline{x = 134 \text{ g}}$

---

23. Two of the radioactive isotopes found in the fallout from nuclear explosions or accidents are ^{131}I (half-life 8.0 days), and ^{90}Sr (half-life 19.1 yr)*. How long will it take each isotope to decay to (a) 10% and (b) 1% of its initial concentration? (c) Which isotope has the most serious long-term effects?

This problem is similar to Problems 20 to 22. We first have to calculate the respective rate constants from the half-lives:

$$kt_{1/2} = \ln 2 = 0.693$$

For Iodine-131: $k = \frac{0.693}{8.0 \text{ days}} = \underline{8.66 \times 10^{-2} \text{ day}^{-1}}$

For Strontium-90: $k = \frac{0.693}{19.1 \text{ yr}} = \underline{3.628 \times 10^{-2} \text{ yr}^{-1}}$

(a) For $\frac{N_t}{N_o} = 0.10$: $\ln \frac{N_t}{N_o} = \ln 0.10 = -2.302 = -kt$

For I-131: $t = \frac{2.302}{8.66 \times 10^{-2} \text{ day}^{-1}} = \underline{27 \text{ day}}$; For Sr-90: $t = \frac{2.302}{3.628 \times 10^{-2} \text{ yr}^{-1}} = \underline{63.5 \text{ yr}}$

(b) For $\frac{N_t}{N_o} = 0.01$: $\ln \frac{N_o}{N_t} = \text{lnn } 0.01 = -4.605 = -kt$

For I-131: $t = \frac{4.605}{8.66 \times 10^{-2} \text{ day}^{-1}} = \underline{53 \text{ day}}$; For Sr-90: $t = \frac{4.605}{3.628 \times 10^{-2} \text{ yr}^{-1}} = \underline{127 \text{ yr}}$

*Note: This value is in error. Table 18.3 gives $t_{1/2} = 28.8$ yr, corresponding to $k = 2.406 \times 10^{-2}$ yr, and times of (a) 95.7 yr, and (b) 191 yr, but this does not change the answer to part (c).

(c) ^{90}Sr has the most serious long-term effect of the two isotopes because of its much greater half-life.

---

25. A sample known to contain 3.40 mg of radioactive ^{32}P originally was found to contain only 2.09 mg of ^{32}P after 10.0 days. What is the half-life of ^{32}P?

$$\ln \frac{N_o}{N_t} = \ln \frac{3.40}{2.09} = \ln 1.627 = 0.4866 = k(10.0 \text{ days}) \; ; \quad \underline{k = 4.87 \times 10^{-2} \text{ day}^{-1}}$$

Thus, $k.t_{1/2} = t_{1/2}(4.87 \times 10^{-2} \text{ day}^{-1}) = \ln 2 = 0.693 \; ; \quad t_{1/2} = \underline{14.2 \text{ days}}$

---

27. A sample containing ^{42}K, a radioactive isotope with a half-life of 12.4 h, has an initial activity of 1.10 x 10^9 disintegrations per minute: (a) Calculate the activity after 30.0 h. (b) How long will it be before the initial activity drops to 1.0 x 10^5 disintegrations per minute?

(a) $$kt_{\frac{1}{2}} = \ln 2 = 0.693\ ; \quad k = \frac{0.693}{12.4\ h} = \underline{5.589 \times 10^{-2}\ h^{-1}}$$

$$\ln\frac{N_o}{N_t} = kt\ ; \quad \ln\frac{1.10 \times 10^9}{N_t} = (5.589 \times 10^{-2}\ h^{-1})(30.0\ h) = \underline{1.677}$$

$$\frac{1.10 \times 10^9}{N_t\ \text{disintegrations min}^{-1}} = 5.348; \quad N_t = \underline{2.06 \times 10^9\ \text{disintegrations min}^{-1}}$$

(b) $$\ln\frac{N_o}{N_t} = kt\ ; \quad \ln\frac{1.10 \times 10^9}{1.00 \times 10^5} = \ln(1.10 \times 10^4) = 9.306 = (5.589 \times 10^{-2}\ h^{-1})t$$

$$t = \underline{1.67 \times 10^2\ h}\ ; \quad (6.94\ \text{days})$$

---

29. A sample of charcoal from the Lascaux cave in France has a count rate of 2.4 disintegrations per gram per minute. Assuming that the fire that produced the charcoal was lit by the artists of the renowned cave paintings, when did these artists live?

This problem is similar to Problem 28.

$$kt_{1/2} = \ln 2 = 0.693; \quad k(5730\ yr) = 0.693; \quad k = \underline{1.209 \times 10^{-4}\ yr^{-1}}$$

For $$\frac{N_o}{N_t} = \frac{15.3}{2.4} = 6.38\ ; \quad \ln\frac{N_o}{N_t} = \ln 6.38 = 1.85 = kt = (1.209 \times 10^{-4}\ yr^{-1})t\ ; \quad t = \underline{1.5 \times 10^4\ yr}$$

The cave dwellers lived around **13000 BC**.

Artificial Radioisotopes

31. Complete each of the following equations: (a) $^{238}_{92}U + ^{1}_{0}n \rightarrow ^{141}_{56}Ba + ^{92}_{36}Kr + ?$ (b) $^{238}_{92}U + ^{1}_{0}n \rightarrow ^{103}_{42}Mo + ^{131}_{50}Sn + ?$

(a) The nuclear charges balance, but the mass numbers differ by **6 mass units**, so we have to add **six neutrons** (6^0_1n) to the right-hand side of the equation to balance the masses, which gives:

$$^{238}_{92}U + ^{1}_{0}n \rightarrow ^{141}_{56}Ba + ^{92}_{36}Kr + \mathbf{6^1_0n}$$

(b) Again, the nuclear charges balance, but here the mass numbers differ by **5 mass units**, so we have to add **five neutrons** (5^0_1n) to the right-hand side of the equation, to balance the masses, which gives:

$$^{238}_{92}U + ^{1}_{0}n \rightarrow ^{103}_{42}Mo + ^{131}_{50}Sn + \mathbf{5^1_0n}$$

---

33. If the nuclei $^{209}_{83}Bi$ and $^{58}_{26}Fe$ were fused, what isotope would be produced?

$$^{209}_{83}Bi + ^{58}_{26}Fe \rightarrow ^{267}_{109}Unn$$ **Unnilennium—267**

35. The transuranium radioactive isotope of curium ^{240}Cm is synthesized by bombarding the isotope ^{232}Th of thorium with the nuclei of carbon atoms. Write a balanced equation for this reaction.

$$^{232}_{90}Th + ^{12}_{6}C \rightarrow ^{240}_{96}Cm + 4\,^{1}_{0}n$$

37. The following equation depicts the reaction that occurs when sparingly soluble lead(II) iodide dissolves in water: $PbI_2(s) \rightleftarrows Pb^{2+}(aq) + 2I^-(aq)$. How would you use a radioactive iodine isotope to show that when the reaction has achieved equilibrium, it is still proceeding at the same rate in both the forward and reverse directions?

The object of any experiment would be to show that at a given temperature the rates of

the forward reaction: $PbI_2(s) \rightarrow Pb^{2+}(aq) + 2I^-(aq)$

and the reverse reaction: $Pb^{2+}(aq) + 2I^-(aq) \rightarrow PbI_2(s)$

are the same. Initially, radioactive lead iodide (lead iodide enriched in the iodine-131 isotope), $Pb^*I_2(s)$, could be synthesized. We could then start with a *saturated* solution of ordinary lead iodide, $PbI_2(aq)$, containing excess $PbI_2(s)$ - that is $PbI_2(s)$ already in equilibrium with $Pb^{2+}(aq)$ and $I^-(aq)$. A sample of $PbI^*_2(s)$ of known radioactivity could then be added to this solution. It will not disturb the position of the heterogeneous equilibrium but the first thing it would demonstrate is that the system is in *dynamic equilibrium*, because it could be shown immediately that that the solution had become radioactive due to $^*I^-(aq)$ ions going into solution, as the result of the forward reaction, even though the total concentration of $I^-(aq)$ would be found to be unchanged. After a given time, the level of radioactivity in the solution could be determined and thus the rate of the forward reaction found. Correspondingly, the rate of the reverse reaction could be found by starting with a saturated solution of radioactive $Pb^*I_2(aq)$, adding unlabeled $PbI_2(s)$ to it, and determining the rate at which the radioactivity of the solid changes with time.

Nuclear Energy

40. The mass of a chlorine-35 atom is 34.9689 u. Calculate its (a) binding energy and (b) binding energy per nucleon.

In solving this type of problem we need the masses of the constituent particles:

mass of proton (p) = 1.007 28 u; mass of neutron (n) = 1.008 66 u; mass electron (e) = 0.000 548 58 u

We also need to use the identities: **1 u = 1.6606 x 10^{-27} kg** and **1 J = 1 kg m^2 s^{-2}**

(a) A ^{35}Cl (Z = 17) nucleus contains 17 p, 18 n, and 17 e, and the sum of their masses is:

$$(17\ p)(1.007\ 28\ u\ p^{-1}) + (18\ n)(1.008\ 66\ u\ n^{-1}) + (17\ e^-)(0.000\ 548\ 58\ u\ e^{-1}) = \mathbf{35.289\ 00\ u}$$

Thus, the mass defect is: $\Delta m = (35.289\ 0 - 34.968\ 9) = \mathbf{0.3201\ u}$

Converting this to binding energy, E, using $E = \Delta mc^2$, gives:

$$E = (0.3201\ u)\left(\frac{1.6606 \times 10^{-27}\ kg}{1\ u}\right)(2.998 \times 10^8\ m\ s^{-1})^2\left(\frac{1\ J}{1\ kg\ m^2\ s^{-2}}\right) = \underline{4.778 \times 10^{-11}\ J}$$

(b) Since there are 35 nucleons: $$E\ nucleon^{-1} = \frac{4.778 \times 10^{-11}\ J}{35\ nucleons} = \underline{1.365 \times 10^{-12}\ J\ nucleon^{-1}}$$

42. Every second, the sun radiates 3.9 x 10^{23} J of energy into space. By how much does the mass of the sun decrease each year?

We can apply Einstein's equation, E = Δmc^2, directly:

$$\Delta m = \frac{E}{c^2} = (\frac{3.9 \times 10^{23}\ J\ s^{-2}}{(2.998 \times 10^8\ m\ s^{-1})^2})(1\ yr)(\frac{365\ day}{1\ yr})(\frac{24\ h}{1\ day})(\frac{3600\ s}{1\ h})(\frac{1\ kg\ m^2\ s^{-2}}{1\ J}) = \underline{1.368 \times 10^{14}\ kg\ yr^{-1}}$$

---

44. (a) Calculate the binding energy per nucleon for ^{14}N and for ^{15}N, which have atomic masses are 14.003 07 u and 15.000 11 u, respectively. (b) What do you conclude about the relative stability of these two isotopes? (b) Which isotope has the greater natural abundance?

This problem is similar to Problem 40.

(a) $^{14}_{7}N$ contains 7 p, 7 n, and 7 e, and the **mass of the constituent particles** is given by:

(7 p)(1.007 28 u p^{-1}) + (7 n)(1.008 66 u n^{-1}) + (7 e)(0.000 548 58 u e^{-1}) = **14.115 4 u**

Thus, the mass defect, **Δm**, is (14.115 4 - 14.003 07) = **0.1123 u**, and

$$E_{nucleon} = (\frac{0.1123\ u}{14\ nucleons})(\frac{1.6606 \times 10^{-27}\ kg}{1\ u})(2.998 \times 10^8\ m\ s^{-1})^2(\frac{1\ J}{1\ kg\ m^2\ s^{-2}}) = \underline{1.197 \times 10^{-12}\ J\ nucleon}$$

$^{15}_{7}N$ has one neutron more than ^{14}N, so its constituent particle mass is: (14.115 4 + 1.008 66) = **15.124 1 u**

Thus, the mass defect, **Δm**, is (15.124 1 - 15.000 1) = **0.1240 u**, and

$$E_{nucleon} = (\frac{0.1240\ u}{15\ nucleons})(\frac{1.6606 \times 10^{-27}\ kg}{1\ u})(2.998 \times 10^8\ m\ s^{-1})^2(\frac{1\ J}{1\ kg\ m^2\ s^{-2}}) = \underline{1.234 \times 10^{-12}\ J\ nucleon}$$

(b) The binding energy per nucleon is greater for ^{15}N than for ^{14}N; **^{15}N is the more stable.**

(c) **^{14}N is the more abundant of these isotopes** (because the formation of an isotope with an *even* mass number is more probable).

---

46. Determine which of the ^{12}C, ^{13}C, and ^{14}C isotopes of carbon is the most stable.

The most stable isotope is the one with the **greatest binding energy per nucleon**, which we can calculate from the isotopic masses (^{12}C = 12.000 00 u; ^{13}C = 13.003 35 u; ^{14}C = 14.003 07 u) by the method of Problem 43:

	p n e	Mass	Isotope mass	Δm
$^{12}_{6}C$	6 6 6	12.098 9 u	12.000 00 u	**0.098 9 u**
$^{13}_{6}C$	6 7 6	13.107 6 u	13.003 35 u	**0.104 3 u**
$^{14}_{6}C$	6 8 6	14.116 3 u	14.003 24 u	**0.113 1 u**

C-12: $$E_{nucleon} = (\frac{0.0989\ u}{12\ nucleons})(\frac{1.6606 \times 10^{-27}\ kg}{1\ u})(2.998 \times 10^8\ m\ s^{-1})^2(\frac{1\ J}{1\ kg\ m^2\ s^{-2}}) = \underline{1.23 \times 10^{-12}\ J\ nucleon^{-1}}$$

<u>C-13</u>: $E_{nucleon} = (\frac{0.1043\ u}{13\ nucleons})(\frac{1.6606 \times 10^{-27}\ kg}{1\ u})(2.998 \times 10^{8}\ m\ s^{-1})^2(\frac{1\ J}{1\ kg\ m^2\ s^{-2}}) = \underline{1.197 \times 10^{-12}\ J\ nucleon^{-1}}$

<u>C-14</u>: $E_{nucleon} = (\frac{0.1131\ u}{14\ nucleons})(\frac{1.6606 \times 10^{-27}\ kg}{1\ u})(2.998 \times 10^{8}\ m\ s^{-1})^2(\frac{1\ J}{1\ kg\ m^2\ s^{-2}}) = \underline{1.206 \times 10^{-12}\ J\ nucleon^{-1}}$

The order of stability is $^{12}C > {}^{14}C > {}^{13}C$; **Carbon-12 is the most stable** of these carbon isotopes.

<u>General Problems</u>

48. Uranium-235 undergoes radioactive decay by α emission to give A; A decays by β emission to give B; B decays by α emission to give C; and C decays by β emission to give D. Write balanced equations for these four steps in the decay scheme for ^{235}U.

$^{235}_{92}U \rightarrow {}^{231}_{90}Th + {}^{4}_{2}He$ $\qquad$ $^{231}_{90}Th \rightarrow {}^{231}_{91}Pa + {}^{0}_{-1}e$ $\qquad$ $^{231}_{91}Pa \rightarrow {}^{227}_{89}Ac + {}^{4}_{2}He$

$^{227}_{89}Ac \rightarrow {}^{227}_{90}Th + {}^{0}_{-1}e$

---

51. For a controlled fusion process, a possible nuclear reaction is the conversion of deuterium ^2H to the helium isotope ^3He, $^2_1H + {}^2_1H \rightarrow {}^3_2He + {}^0_1n$. Compare the energy released per mole of deuterium consumed in the nuclear reaction with that obtained by burning deuterium in oxygen to give heavy water, ^2H_2O. Assume that the enthalpy of combustion of deuterium is the same as that of natural $H_2(g)$.

The energy from the **fusion reaction** comes from the change in mass, and it is convenient to work in terms of moles of reactants and products:

$$\Delta m = (1\ mol\ He\text{-}3)(\frac{3.016\ 03\ g}{1\ mol\ He\text{-}3}) + (1\ mol\ n)(\frac{1.008\ 66\ g}{1\ mol\ n}) - (2\ mol\ H\text{-}2)(\frac{2.014\ 10\ g}{1 mol\ H\text{-}2})$$

$$= [(3.016\ 03) + (1.008\ 66) - (4.028\ 20)]\ g = \underline{-3.51 \times 10^{-3}\ g}$$

$$E = \Delta mc^2 = (\frac{-3.51 \times 10^{-3}\ g}{2\ mol\ H\text{-}2})(\frac{1\ kg}{10^3\ g})(2.998 \times 10^8\ m\ s^{-1})^2(\frac{1\ J}{1\ kg\ m^2\ s^{-2}})(\frac{1\ kJ}{10^3\ J}) = \underline{-1.58 \times 10^8\ kJ\ mol^{-}}$$

We assume that the energy from the **chemical reaction**, $\Delta H°$ for the reaction $^2_1H_2(g) + \frac{1}{2}O_2(g) \rightarrow (^2_1H)_2O(g)$, is the same as that for ordinary (light) water, $^1_1H_2O$:

$$\Delta H° = [\Delta H_f°(H_2O,g)] - [\Delta H_f°(H_2,g) + \frac{1}{2}\Delta H_f°(O_2,g)] = -241.8\ kJ\ mol^{-1}$$

(or, more simply, just the standard enthalpy of formation of $H_2O(g)$).

Both reactions are *exothermic*, giving:

$$\frac{E_{nuclear\ fusion}}{E_{combustion}} = \frac{1.58 \times 10^8\ kJ\ mol^{-1}}{2.42 \times 10^2\ kJ\ mol^{-1}} = \underline{6.5\ x 10^5}$$

---

53. The following series of reactions is the *Bethe chain* responsible for the energy production of some stars: $^{12}C + {}^{1}H \rightarrow {}^{13}N$; $^{13}N \rightarrow {}^{0}_{1}e + X$; $X + {}^{1}H \rightarrow Y$; $Y + {}^{1}H \rightarrow Z$; $Z \rightarrow {}^{0}_{1}e + {}^{15}N$; $^{15}N + {}^{1}H \rightarrow {}^{12}C + {}^{4}He$. (a) Identify the nuclides X, Y, and Z in these equations. (b) Give the equation for the overall reaction.

(a) In Step 2: $^{13}_{7}N \rightarrow {}^{0}_{1}e + {}^{13}_{6}C$; in step 3: $^{13}_{6}C + {}^{1}_{1}H \rightarrow {}^{14}_{7}N$; in step 4: $^{14}_{7}N + {}^{1}_{1}H \rightarrow {}^{15}_{8}O$

(b) Overall: $4{}^{1}_{1}H \rightarrow {}^{4}_{2}He + 2{}^{0}_{1}e$

55. Calculate (i) the total binding energy and (ii) the binding energy per nucleon for each of the following nuclides: (a) ^{20}Ne, mass 19.992 44 u; (b) ^{64}Zn, mass 63.929 14 u; (c) ^{61}Ni, mass 60.930 06 u; (d) ^{226}Ra, mass 226.025 4 u.

	p n e	Mass	Isotope mass	Δm	BE* (J)	E (nucleon^{-1}) (J)
(a) $^{20}_{10}Ne$	10 10 10	20.164 89 u	19.992 42 u	0.172 47 u	**2.574 x 10^{-11}**	**1.287 x 10^{-12}**
(b) $^{64}_{30}Zn$	30 34 30	64.529 30 u	63.928 14 u	0.601 16 u	**8.973 x 10^{-11}**	**1.402 x 10^{-12}**
(c) $^{61}_{28}Ni$	28 33 28	61.505 00 u	60.930 06 u	0.574 94 u	**8.581 x 10^{-11}**	**1.409 x 10^{-12}**
(d) $^{226}_{88}Ra$	88 138 88	227.883 4 u	226.025 4 u	1.858 0 u	**2.773 x 10^{-10}**	**1.227 x 10^{-12}**

* Binding energy

58. Technetium is in the same group of transition elements as manganese. (a) Predict the highest oxidation state of technetium. (b) Write the formula and name of the oxoacid and the fluoride with technetium in this oxidation state.

(a) Among the first series of transition metals, in Period 4, **manganese**, Mn, with the valence shell electron configuration **$3d^5\ 4s^2$** utilizes all of these electrons in forming its highest oxidation state, which is therefore **+VII**.

(b) Thus, the Mn(VII) oxoacid is $HMnO_4$, **permanganic acid**, and the Mn(VII) fluoride is MnF_7, **manganese heptafluoride**. By analogy, **technetium**, Tc, in Period 5, has the valence shell electron configuration **$4d^5\ 5s^2$** and will utilize all of these electrons in forming its highest oxidation state, which is therefore also **+VII**. The corresponding oxoacid is $HTcO_4$, **pertechnetic acid**, and the fluoride would be TcF_7, **technetium heptafluoride**.

CHAPTER 19

The Origin of Atoms and Molecules

1. (a) Why does nuclear fusion occur only at very high temperatures? (b) How much energy is evolved in the multi-step process in which four ^1H atoms combine to give a ^4He nucleus?

(a) In nuclear fusion (*nucleosynthesis*) lighter nuclei are fused together to form heavier nuclei. For two nuclei to undergo a fusion reaction, they must be brought together with sufficient energy to overcome the strong electrostatic coulombic repulsion between their positively charged nuclei, which is very large when they are very close to one another. The kinetic energy is given by $\frac{1}{2}mv^2$, so this occurs only when they have very high speeds. The kinetic energy, $\frac{1}{2}mv^2$ is proportional to the absolute temperature, so sufficient energy results only at high temperatures. As the nuclei become heavier, that is with increasing nuclear charge, the internuclear repulsions increase in magnitude, which means that even greater energy is required for fusion. In other words, the nuclei have to collide with ever increasing velocities as their masses increase, which requires ever increasing temperatures.

(b) The overall nuclear reaction in question is:

$$4\,^1_1H \rightarrow \,^4_2He + 2\,^0_1e$$

in which four ^1_1H atoms (mass 4 x 1.007 83 u) are converted to one ^4_2He atom (mass 4.002 60 u) and two positrons (mass 2 x 0.000 548 6 u). Thus, the *overall* change in mass between products and reactants is:

$$\mathbf{\Delta m} = [4.002\ 60 + 2(0.000\ 548\ 6)]\ u - 4(1.007\ 83\ u) = \mathbf{-0.0276\ u}$$

which corresponds to:

$$E = \Delta mc^2 = (-2.763 \times 10^{-2}\ u)\left(\frac{1.660\ 6 \times 10^{-27}\ kg}{1\ u}\right)(2.998 \times 10^8\ m\ s^{-1})^2\left(\frac{1\ J}{1\ kg\ m^2\ s^{-2}}\right)\left(\frac{6.022 \times 10^{23}\ He\ atoms}{1\ mol\ He\ atoms}\right)$$

$$= \underline{-2.483 \times 10^{12}\ J\ mol^{-1}\ \text{of helium formed}}$$

The energy evolved is close to **2.5 trillion joules** per mole of helium-4 formed, which maintains the high temperature required for continuing nucleosynthesis.

---

3. (a) Why does a "hydrogen" star collapse when most of its hydrogen has been converted to helium? (b) Why does this collapse initiate the synthesis of heavier nuclei? (c) Why are no nuclei heavier than ^{56}Fe formed by this process? (d) How does a second-generation star differ from a first-generation star?

The solution to this problem was partially given in the answer to Problem 2.

(a) A **hydrogen star** collapses towards the end of its life because the rate of the nucleosynthesis of hydrogen nuclei to form helium nuclei eventually slows down as the hydrogen is depleted. Since this nucleosynthesis is exothermic, the temperature also falls, decreasing the kinetic energies of the nuclei and electrons, thus allowing gravitational forces to contract the star.

(b) Contraction of the star, with a consequent increase in temperature, increases the kinetic energies of the helium nuclei to the point where they have sufficient energy to collide and initiate nucleosynthesis reactions between themselves to form nuclei heavier than ^4_2He.

(c) Nucleosyntheses of nuclei heavier than ^{56}Fe are *endothermic* reactions and only occur when colliding nuclei have very large kinetic energies, which are not provided by the temperatures available in hydrogen or neutron stars.

(d) A first-generation star (hydrogen or helium star) contains nuclides up to ^{56}Fe, the nuclide of greatest stability, because they are incapable of forming nuclides heavier than ^{56}Fe. In contrast, a second-generation star forms from interstellar dust formed from explosion of a neutron star and contains the whole range of possible nuclides freom the lightest that are formed in the exothermic processes that take place in first-generation stars and the endothermic processes that take place when a neutron star explodes, forms a supernova, and spreads all ther nuclides over a large volume of space.

---

5. Is more energy released in making ^4He from ^1H, or in making ^{16}O from ^{12}C plus ^4He? Calculate the energy released in the latter nuclear reaction.

In any nucleosynthesis, the amount of energy released is proportional to the mass loss in the reaction, ($E = \Delta mc^2$), and the two reactions in question are:

$$4\,{}^1_1H \rightarrow {}^4_2He + 2\,{}^1_1e \quad ..... (1) \quad \text{and} \quad {}^{12}_6C + {}^4_2He \rightarrow {}^{16}_8O \quad ..... (2)$$

For (1): $\Delta m = [4.002\,60\text{ u} + 2(0.000\,548\,6)\text{ u}] - 4[1.007\,83\text{ u}] = \mathbf{-2.76 \times 10^{-2}\text{ u}}$

For (2): $\Delta m = [15.994\,91\text{ u}] - [12.000\,00 + 4.002\,60]\text{ u} = \mathbf{-7.69 \times 10^{-3}\text{ u}}$

Reaction (1) has the larger mass loss and releases more energy than reaction (2). The energy released in (1) is given by:

$$E = \Delta mc^2 = (2.76 \times 10^{-2}\text{ u})\left(\frac{1.660\,6 \times 10^{-27}\text{ kg}}{1\text{ u}}\right)(2.998 \times 10^8\text{ m s}^{-1})^2\left(\frac{1\text{ J}}{1\text{ kg m}^2\text{ s}^{-2}}\right)\left(\frac{6.022 \times 10^{23}}{1\text{ mol}}\right)$$

$$= \underline{2.49 \times 10^{11}\text{ J mol}^{-1}}$$

Interstellar Molecules

7. (a) In the laboratory, a vacuum pump that reduces the pressure inside an apparatus to 1.00 x 10^{-4} mm Hg is regarded as fairly efficient. To how many molecules per liter does this pressure correspond at 25°C? (b) What is the range of gas pressures in interstellar clouds containing from 10^5 to 10^9 H_2 molecules per liter at -270 K?

(a) We first use the ideal gas law to calculate the number of moles of gas per liter that give this pressure at 25°C, and then multiply by Avogadro's number to give the number of molecules:

$$\text{Molecules} = \frac{N_A PV}{RT} = \frac{\left(\frac{6.022 \times 10^{23}\text{ molecules}}{1\text{ mol}}\right)(10^{-4}\text{ mm Hg})\left(\frac{1\text{ atm}}{760\text{ mm Hg}}\right)(1\text{ L})}{(0.0821\text{ atm L mol}^{-1}\text{ K}^{-1})(298\text{ K})} = \underline{3.24 \times 10^{15}\text{ molecules}}$$

(b) (i) $$P = \frac{nRT}{V} = \frac{(10^5\text{ molecules})\left(\frac{1\text{ mol}}{6.022 \times 10^{23}\text{ molecules}}\right)(0.0821\text{ atm L mol}^{-1}\text{ K}^{-1})(3.15\text{ K})}{1\text{ L}}$$

$$= (4.295 \times 10^{-20}\text{ atm})\left(\frac{760\text{ mm Hg}}{1\text{ atm}}\right) = \underline{3.26 \times 10^{-17}\text{ mm Hg}}$$

(ii) $$P = \left(\frac{10^9\text{ molecules}}{10^5\text{ molecules}}\right)(3.26 \times 10^{-17}\text{ mm Hg}) = \underline{3.26 \times 10^{-13}\text{ mm Hg}}$$

The range of pressures is: **3 x 10^{-13} to 3 x 10^{-17} mm Hg.**

The Formation and Composition of the Solar Sytem

10. Using data from Table 19.3, calculate the partial pressures of (a) CO_2 and N_2 in the atmospheres of Venus and Mars, and (b) N_2, O_2, and Ar on Earth. (c) How many molecules per liter do each of these partial pressures represent?

(a) Table 19.3 gives the atmospheric pressures of the planets at their surfaces as: **90 atm for Venus, and 0.006 atm for Mars.** The atmospheric compositions are percentages by *volume*, so the partial pressures of their gases, assuming the only gases present are those given:

$$\underline{\text{Venus}}: \quad p_{CO_2} = \frac{96.5}{100}(90\text{ atm}) = \underline{87\text{ atm}}\ ; \qquad p_{N_2} = \frac{3.5}{100}(90\text{ atm}) = \underline{3.2\text{ atm}}$$

$$\underline{\text{Mars}}: \quad p_{CO_2} = \frac{95.3}{100}(0.006\text{ atm}) = \underline{0.006\text{ atm}}\ ; \quad p_{N_2} = \frac{2.7}{100}(0.006\text{ atm}) = \underline{2 \times 10^{-4}\text{ atm}}$$

(b) $$\underline{\text{Earth}}: \quad p_{N_2} = \frac{78.1}{100}(1.00\text{ atm}) = \underline{0.781\text{ atm}}\ ; \quad p_{O_2} = \frac{20.9}{100}(1.00\text{ atm}) = \underline{0.209\text{ atm}}\ ;$$

$$p_{Ar} = \frac{0.9}{100}(1.00\text{ atm}) = \underline{9 \times 10^{-3}\text{ atm}}$$

(c) The value of the **gas constant R** is normally taken as **R = 0.0821 atm L mol^{-1} K^{-1}** on earth, but we have to remember that the unit of 1 atm *on earth* is relative to the *force that the earth's gravity exerts on its atmosphere.* In other words, the value of R on another planet depends on the force exerted by the planet's gravity relative to the earth's gravitational force. Since gravitational force is proportional to *planetary mass*, this difference can be taken into account by multiplying R by the mass of a planet relative to that of the earth. Note that whereas R is proportional to a planet's mass, the units of *moles, liters, and kelvins, (and* Avogadro's number*)* are *invariable* (the same on every planet), although we have to take into account the different surface temperatures of the planets. We need the relationship that the number of molecules is proportional to the partial pressure ($nN_A \propto p$). Thus:

$$R' = R\left(\frac{\text{mass of planet}}{\text{mass of earth}}\right); \qquad PV = nR'T\ ; \quad n = \frac{PV}{R'T}\ ; \qquad \text{Number of molecules L}^{-1} = \frac{N_A P}{R'T}$$

(i) $$\underline{\text{Venus}}: \quad CO_2\text{ molecules} = \frac{(6.022 \times 10^{23}\text{ molecules mol}^{-1})(87\text{ atm})}{(0.0821\text{ atm L mol}^{-1}\text{ K}^{-1})(0.81)(732\text{ K})} = \underline{1.08 \times 10^{24}\text{ molecules L}^{-1}}$$

$$N_2\text{ molecules} = (1.08 \times 10^{24}\text{ molecules L}^{-1})\left(\frac{3.2\text{ atm}}{87\text{ atm}}\right) = \underline{4.0 \times 10^{23}\text{ molecules L}^{-1}}$$

(ii) $$\underline{\text{Mars}}: \quad CO_2\text{ molecules} = \frac{(6.022 \times 10^{23}\text{ molecules mol}^{-1})(0.006\text{ atm})}{(0.0821\text{ atm L mol}^{-1}\text{ K}^{-1})(0.11)(223\text{ K})} = \underline{1.8 \times 10^{21}\text{ molecules L}^{-1}}$$

$$N_2\text{ molecules} = (1.8 \times 10^{21}\text{ molecules})\left(\frac{2 \times 10^{-4}\text{ atm}}{0.006\text{ atm}}\right) = \underline{6.0 \times 10^{19}\text{ molecules L}^{-1}}$$

(iii) $$\underline{\text{Earth}}: \quad N_2\text{ molecules} = \frac{(6.022 \times 10^{23}\text{ molecules mol}^{-1})(0.781\text{ atm})}{(0.0821\text{ atm L mol}^{-1}\text{ K}^{-1})(1.00)(288\text{ K})} = \underline{1.989 \times 10^{22}\text{ molecules L}^{-1}}$$

$$O_2\text{ molecules} = (1.989 \times 10^{22}\text{ molecules})\left(\frac{0.209\text{ atm}}{0.781\text{ atm}}\right) = \underline{5.32 \times 10^{21}\text{ molecules L}^{-1}}$$

$$\text{Ar molecules} = (1.989 \times 10^{23}\text{ molecules})\left(\frac{9 \times 10^{-3}\text{ atm}}{0.781\text{ atm}}\right) = \underline{2 \times 10^{19}\text{ molecules L}^{-1}}$$

PLANET	Molecules L^{-1}			
	CO_2	N_2	O_2	Ar
Venus	1.1×10^{24}	4.0×10^{23}	-	-
Mars	2×10^{21}	6×10^{19}	-	-
Earth	-	1.99×10^{22}	5.32×10^{21}	2×10^{19}

The Structure and Composition of the Earth

13. The average density of the earth 5.5 g·cm^{-3}, whereas the surface rocks have a density of 2.8 g·cm^{-3}. Explain this difference in terms of the structure of the earth.

The earth has a thin crust, an upper and lower mantle, and a central inner and outer core. The crust and upper mantle constitute the solid *lithosphere*, and the lower mantle is called the *asthenosphere*, and is composed of rocks near their melting point which are plastic and easily deformed. Since the main constituent of the lithosphere and the asthenosphere is silica, SiO_2, or related silicates and aluminosilicates, its density is approximately that of SiO_2(s) (2.6 g cm^{-3}). The solid inner core and the outer liquid core are both composed of iron (density ≈ 7.9 g cm^{-3}, and nickel (density ≈ 8.9 g cm^{-3}). Thus, the density of the surface rocks of 2.8 g cm^{-3} is close to that of silica, whereas, since the core constitutes the major component of the earth's structure, the overall density is much greater than 2.8 g cm^{-3}.

Silicon and Its Compounds

15. (a) Write the ground state electronic structure of silicon, and deduce the expected valence of silicon. Explain why silicon exhibits a different valence in most of its compounds.

(a) Silicon is in Period 3 and Group IV (below carbon) and therefore has the electronic structure: $1s^22s^22p^63s^23p^2$

(b) In the ground state, the n = 3 valence shell of Si has the electron arrangement: 3s [↑↓] 3p [↑][↑][ ] with two unpaired electrons. Thus, the expected valence = 2, but in most of its compounds, such as SiO_2, $Si(OH)_4$, $SiCl_4$, and SiH_4, silicon as a valence of four. In this respect silicon resembles carbon (also in Group IV) in that most of its compounds are formed from an excited valence state with the electron configuration:

3s [↑] 3p [↑][↑][↑]

with four unpaired electrons. (In a molecule such as SiH_4, silane, the one 3s and three 3p orbitals are mixed together to form four equivalent sp^3 hydridized orbitals to account for the tetrahedral shape.)

---

17. (a) Give an example of a compound of silicon that has a very different structure and properties from the corresponding compound of carbon, and give a reason (or reasons) for this difference. (b) Give an example of a compound (or compounds) of silicon that has a similar structure to that of the corresponding carbon compound, but has at least one very different chemical property, and give reasons for this difference.

(a) Perhaps the greatest contrast in structure and properties is that between the oxides SiO_2 and CO_2. While carbon dioxide is a gas under normal conditions and composed of small linear AX_2 triatomic molecules, with the structure O=C=O, containing strong C=O double bonds, silicon dioxide, silica, forms a three-dimensional network solid with a structure related to that of diamond. In silica, the Si atoms are joined together by Si—O—Si bonds, and each Si atom forms four Si—O single bonds and is surrounded in the crystal by an AX_4 arrangement of four O atoms. This difference is a manifestation of the strong tendency of carbon to form multiple bonds, whereas silicon rarely forms even double bonds. This difference is probably related to the relative electronegativities and sizes of the C and Si atoms. Carbon, in Period 2, is limited to an octet of electrons in its valence shell and, because of its high electronegativity relative to that of silicon, four electron pairs in single bonds in a tetrahedral arrangement to other electronegative atoms, such as oxygen, are very crowded, so there is a strong tendency to form multiple bonds where the electron pairs are somewhat less crowded, because the arrangement of bonds is now trigonal planar or linear. In contrast, four electron pairs in the valence shell of the larger and less electronegative silicon atom suffer no such crowding. This difference is also manifested in the markedly different structures of carbonic acid and silicic acid. Silicic acid has the structure $Si(OH)_4$, whereas carbonic acid has the structure $(HO)_2C=O$, rather than the structure $C(OH)_4$. (Hypothetical $C(OH)_4$ readily loses a water molecule to form AX_3 trigonal planar $(HO)_2C=O$), whereas when $Si(OH)_4$ molecules lose water they form *polymeric* silicic acids in which the AX_4 arrangement of bonds to Si is retained, and all the bonds are single).

(b) Silicon and carbon form structurally similar compounds with univalent ligands such as H, *hydrides*, or a halogen, *halides*. These are composed of tetrahedral AX_4 covalent molecules, for example, $SiCl_4(\ell)$ and $CCl_4(\ell)$. However, $SiCl_4$ and CCl_4 have quite different chemical properties. For example, $SiCl_4$ fumes in moist air, and when dropped into water reacts immediately and vigorously to give insoluble silicic acid, $Si(OH)_4$, and polymeric silicic acids, whereas CCl_4 is stable in water and does not react under normal conditions.

$$SiCl_4(\ell) + 4H_2O(\ell) \rightarrow Si(OH)_4(s) + 4HCl(aq); \qquad CCl_4(\ell) + H_2O(\ell) \rightarrow \textbf{no reaction}$$

The difference in behavior is explained in terms of the sizes of the valence shells of C and Si. Carbon in Period 2 is limited to an octet of electrons and when surrounded by four Cl atoms in CCl_4 it is not possible to insert the unshared electron pair of a water molecule, $:OH_2$, into this valence shell; in other words CCl_4 is not subject to nucleophilic attack by water because it cannot behave as a Lewis acid. In contrast, the valence shell of the silicon atom in $SiCl_4$ can be expanded to accept up to a total of *six pairs* of electrons. It readily accepts a lone pair from the O atom of a water molecule to form an Si-O bonds; $SiCl_4$ behaves as a Lewis acid and is readily hydrolyzed by the Lewis base H_2O.

```
     :Cl:                       :Cl:  ..              :Cl:
      |                           | / Cl:               |
 :Cl—Si—Cl:  +  :O—H   →   :Cl—Si<  ..        →   :Cl—Si—OH  +  HCl        etc.,
      |          |                |  \O+—H              |
     :Cl:        H              :Cl:  |                :Cl:
                                      H
```

Alternatively, you could contrast $CF_4(g)$ and $SiF_4(g)$. The former is inert and unreactive, while SiF_4 reacts with fluoride ions to give SiF_5^- and SiF_6^{2-}.

---

19. Silicon reacts with sulfur at elevated temperature. (a) If 0.0932 g of silicon react with sulfur to give 0.3060 g of silicon sulfide, determine the empirical formula of silicon sulfide. (b) Suggest a possible structure for silicon sulfide. (c) Would you expect silicon sulfide to be a gas, a liquid, or a solid?

(a) By difference, 0.0932 g of Si combine with (0.3060 - 0.0932) g = **0.2128 g S**. Thus, for silicon sulfide we have:

	Si	S
Combining masses	0.0932 g	0.2098 g
Combining moles	$\frac{0.0932\ g}{28.09\ g\ mol^{-1}}$ = $\mathbf{3.32 \times 10^{-3}}$	$\frac{0.2128\ g}{32.07\ g\ mol^{-1}}$ = $\mathbf{6.64 \times 10^{-3}}$
Ratio of moles (atoms)	3.32/3.32 = 1.00	: 6.64/3.32 = 2.00

Thus, the **empirical formula** of silicon sulfide is $\mathbf{SiS_2}$

(b) By analogy with the difference between the structures of CO_2 and SiO_2, SiS_2 would not be expected to be a simple covalent liquid containing linear AX_2 :S=Si=S: molecules. Because of the inability of silicon to form multiple bonds, the structure of SiS_2 is expected to be analogous to that of SiO_2, with each Si atom bonded by single bonds to two S atoms and each Si atom forming Si-S bonds to four S atoms, in an AX_4 tetrahedral arrangement. Thus SiS_2 is expected to be a three-dimensional covalent network **solid**.

---

22. From the data in Table 9.3 and the following ΔH_f° values: $SiH_4(g)$ 34.3; $Si_2H_6(g)$ 80.3; Si(g) 450 kJ mol^{-1}, (a) calculate and compare the average bond energies of the C-H and Si-H bonds in $CH_4(g)$ and $SiH_4(g)$, respectively. (b) Using these values, calculate and compare the C-C and Si-Si bond energies in $C_2H_6(g)$ and $Si_2H_6(g)$.

(a) $\mathbf{CH_4}$ For the reaction $CH_4(g) \rightarrow C(g) + 4H(g)$,

$$\Delta H^\circ = 4BE(C\text{-}H) = [\Delta H_f^\circ(C,g) + 4\Delta H_f^\circ(H,g)] - [\Delta H_f^\circ(CH_4,g)] = [716.7 + 4(218)] - [-74.5]\ kJ = \mathbf{1663\ kJ}$$

BE(C-H) = 416 kJ mol⁻¹

SiH_4 For the reaction $SiH_4(g) \rightarrow Si(g) + 4H(g)$,

$\Delta H° = 4BE(Si\text{-}H) = [\Delta H_f°(Si,g) + 4\Delta H_f°(H,g)] - [\Delta H_f°(SiH_4,g)] = [450 + 4(218)] - [34]$ kJ = **1288 kJ**

BE(Si-H) = 322 kJ mol⁻¹

(b) **$C_2H_6(g)$** For the reaction $C_2H_6(g) \rightarrow 2C(g) + 6H(g)$, six C-H bonds and one C-C bonds are broken:

$\Delta H° = BE(C\text{-}C) + 6BE(C\text{-}H) = [2\Delta H_f°(C,g) + 6\Delta H_f°(H,g)] - [\Delta H_f°(C_2H_6,g)] = [2(716.7) + 6(218)] - [-84.7]$

= 2826 kJ

BE(C-C) = (2826 - 6(416) kJ mol⁻¹ = **330 kJ mol⁻¹**

$Si_2H_6(g)$ For the reaction $Si_2H_6(g) \rightarrow 2Si(g) + 6H(g)$, six Si-H bonds and one Si-Si bonds are broken:

$\Delta H° = BE(Si\text{-}Si) + 6BE(Si\text{-}H) = [2\Delta H_f°(Si,g) + 6\Delta H_f°(H,g)] - [\Delta H_f°(Si_2H_6,g)] = [2(450) + 6(218)] - [80]$ kJ

= 2128 kJ

BE(Si-Si) = (2128 - 6(322) kJ mol⁻¹ = **196 kJ mol⁻¹**

---

24. (a) How can silicic acid be prepared from $SiCl_4$? (b) How does silicic acid differ from carbonic acid? (c) Why cannot carbonic acid be prepared by an analogous reaction of CCl_4?

(a) **Silicic acid**, $Si(OH)_4$, results from the hydrolysis of $SiCl_4(\ell)$ by reacting it with water:

$$SiCl_4(\ell) + 4H_2O(\ell) \rightarrow Si(OH)_4(aq) + 4HCl(aq)$$

The silicic acid produced rapidly condenses to give a mixture of insoluble polysilicic acids.

(b) Silicic acid has the structure $Si(OH)_4$, whereas carbonic acid has the structure $(HO)_2C{=}O$, because of the greater tendency of carbon to form multiple bonds. Apparently hypothetical $C(OH)_4$ is unstable with respect to $(HO)_2C{=}O$, and even this molecule is very unstable and readily loses another water molecule to give carbon dioxide, $O{=}C{=}O$, so that in an aqueous solution of carbonic acid very little $(HO)_2C{=}O$, (H_2CO_3) is present. $Si(OH)_4(aq)$ is also unstable but in this case condensation reactions give a mixture of a variety of polysilicic acid, all containing single Si-O bonds. Both silicic acid and carbonic acid are weak acids in water, but silicic acid is weaker than carbonic acid because silicon is less electronegative than carbon.

(c) In contrast to $SiCl_4$, carbon tetrachloride, CCl_4, does not react with water, or indeed take part easily in any nucleophilic type reaction. In our discussion in Chapter 14 of nucleophilic substitution reactions of carbon, we saw that this type of reaction comes about by two possible mechanisms: (1) A single step bimolecular substitution reaction, which is not possible for CCl_4. Carbon is in Period 2 and its valence shell is filled with four pairs of electrons (an octet); moreover, the bulky Cl atoms prevent water from getting close to the carbon atom and forming a new bond that would lead to a transition state. (2) A two step mechanism in which the rate determining step is the initial formation of a carbocation, (in this case it would be CCl_3^+), which then reacts rapidly to give products. This also is not possible for CCl_4 in which the C atom attached to four Cl atoms is very electronegative, so the C-Cl bonds are insufficiently polar to ionize to form a Cl^- ion.

---

26. Predict the structure of the aluminosilicate anion (chain, ring, sheet, or network) in each of the following aluminosilicates: (a) anorthite, $Ca(AlSiO_4)_2$ (b) muscovite, $KAl_2(Si_3AlO_{10})(OH)_2$ (c) amesite, $Mg_2Al(AlSiO_5)(OH)_4$ (d) thomsonite, $NaCa_2(AlSiO_4)_5{\bullet}6H_2O$

(a) **Anorthite**: The formula $Ca(AlSiO_4)_2$ may be rewritten as $Ca^{2+}(AlSiO_4^-)_2$, and therefore contains the $(AlSiO_4^-)_n$ ion obtained by replacing ¼ of the Si atoms in $(SiO_2)_n$ by Al^-. Thus, it has an **infinite three-dimensional** network aluminosilicate lattice.

(b) **Muscovite**: The formula $KAl_2(Si_3AlO_{10})(OH)_2$ may be rewritten as $(K^+)(Al^{3+})_2(Si_3AlO_{10}^{5-})(OH^-)_2$, and therefore contains the $(Si_3AlO_{10}^{5-})_n$ aluminosilicate ion, derived from the silicate anion $[(Si_2O_5^{2-})_2]_n$ ion by replacing ¼ of the Si atoms by Al^-. Thus, it has an **infinite sheet structure**.

(c) **Amesite**: The formula $Mg_2Al(AlSiO_5)(OH)_4$ may be rewritten as $(Mg^{2+})_2(Al^{3+})(AlSiO_5^{3-})(OH^-)_4$, and therefore contains the $(AlSiO_5^{3-})_n$ aluminosilicate ion, derived from the $(Si_2O_5^{2-})_n$ silicate anion by replacing ½ of the Si atoms by Al^-. Thus, it has an **infinite sheet structure.**

(d) **Thomsonite**: The formula $NaCa_2(AlSiO_4)_5 6H_2O$ may be rewritten as $(Na^+)(Ca^{2+})_2(AlSiO_4^-)_5 6H_2O$, and therefeore contains the $(AlSiO_4^-)_n$ aluminosilicate ion, derived from silica, $(SiO_2)_{2n}$, by replacing ½ of the Si atoms by Al^-. Thus, it has an **infinite three-dimensional** aluminosilicate network lattice.

---

27. Describe the structure of the silicate or aluminosilicate anion in each of the following minerals: (a) diopside, $CaMgSi_2O_6$ (b) orthoclase, $KAlSi_3O_8$ (c) hardystonite, $Ca_2ZnSi_2O_7$ (d) dentitoite, $BaTiSi_3O_9$

(a) **Diopside**: We write the empirical formula $CaMgSi_2O_6$ as $(Ca^{2+})(Mg^{2+})(Si_2O_6^{4-})$, or $(Ca^{2+})(Mg^{2+})(SiO_3^{2-})_2$, which must contain the $(SiO_3^{2-})_n$ anion, which is an **infinite chain silicate ion.**

(b) **Orthoclase**: We have a choice of writing the empirical formula $KAlSi_3O_8$ as $(K^+)(Al^{3+})(Si_3O_8^{4-})$, or as $K^+(Si_3AlO_8^-)$. The $(Si_3O_8^{4-})_4$ is unrecognizable, but the $(Si_3AlO_8^-)_n$ is an **aluminosilicate** anion derived by replacing ¼ of the Si atoms in silica, $(SiO_2)_{2n}$ by Al^-. Thus, the aluminosilicate ion in **orthoclase** has an **infinite three dimensional covalent network lattice.**

(c) **Hardystonite**: We write the empirical formula as $(Ca^{2+})_2(Zn^{2+})(S_2O_7^{6-})$, containing the **disilicate ion,** $Si_2O_7^{6-}$. The parent acid of disilicate ion is the rather simple acid **disilicic acid** obtained from the condensation reaction of two $Si(OH)_4$ molecules. Thus, **hardystonite** is **calcium zinc disilicate.**

(d) **Dentitoite**: We write the empirical formula as $(Ba^{2+})(Ti^{4+})(Si_3O_9^{6-})$. The fact that the anion is given as $(Si_3O_9^{6-})_n$, rather than simply as $[(SiO_3^{2-})_3]_n$, implies that dentitoite contains discrete *cyclic* $Si_3O_9^{6-}$ ions.

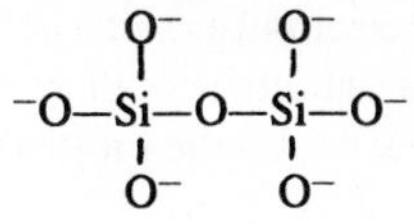

```
    O⁻    O⁻                          ⁻O    O⁻
    |     |                             \  /
⁻O—Si—O—Si—O⁻                   ⁻O     O—Si
    |     |                        \  /     \
    O⁻    O⁻                        Si       O
                                   /  \     /
                                ⁻O     O—Si
                                          / \
                                       ⁻O    O⁻
```

$Si_2O_7^{6-}$ ion 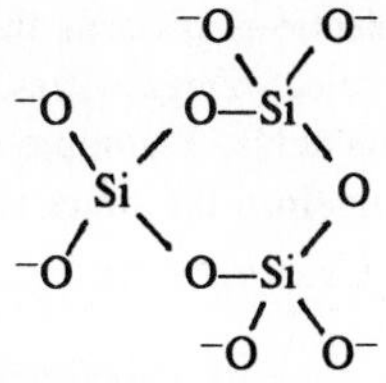 $\qquad\qquad$ $Si_3O_9^{6-}$ ion

---

Minerals and Other Geologic Resources

29. Some iron ores are sulfides, such as pyrite, $FeS_2(s)$, and some are oxides, such as hematite, $Fe_2O_3(s)$. Describe how and where these ores probably originated.

Iron is plentiful in the earth's core; indeed it is probably the most abundant element in the earth as a whole. Iron is undoubtedly forced up by the earth's internal pressure into fissures in the earth's crust, where at high temperature and pressure it met water percolating down from the oceans. Under these *hydrothermal* conditions, the direct reaction of the iron with water gives $Fe(OH)_2(s)$:

$$Fe(\ell) + 2H_2O(\ell) \rightarrow 2Fe(OH)_2(s) + H_2(g)$$

When Fe(II) containing rocks are eventually exposed to weathering and oxygen from the air, the Fe(II) is oxidized to Fe(III) and insoluble $Fe_2O_3(s)$, **hematite**, sometimes hydrated, or Fe_3O_4, $(FeOFe_2O_3)$, **magnetite**, is formed. Iron is also found associated with magnesium in igneous rocks in the silicates olivine, pyroxene, amphibole, and mica. Weathering of these also gives iron oxides. Iron sulfides, such as **pyrite**, $FeS_2(s)$, are normally mined for sulfur rather than iron. Sulfide minerals are also formed by *hydrothermal processes* involving magma, igneous, metamorphic, or sedimentary rocks. Pyrite is one of the most abundant sulfides. Since it contains Fe(II) it must have been formed under reducing conditions because it is easily oxidized to hydrated iron (III) oxide. It must therefore have precipitated from hydrothermal solution. Thus, it is unlikely to be formed at the surface where rocks are exposed to oxygen. When water meets magma exposed at diverging tectonic plate boundaries under the sea, the mineral dissolves at high temperature and pressure and is precipitated when the hot metal sulfide solution comes in contact with cold sea water. Another possible process is thought to be formation at the same time as sedimentation processes by microbial sulfate reduction.

General Problems

33. Name each of the oxoacids with the following molecular formulas. Place them in order of increasing strength, and explain this order. (a) $Si(OH)_4$ (b) $PO(OH)_3$ (c) $SO_2(OH)_2$ (d) ClO_3OH

(a) **silicic acid**; (b) **phosphoric acid**; (c) **sulfuric acid**; (d) **perchloric acid.**

To place them in order of increasing strength, we categorize them in terms of the general formula $XO_m(OH)_n$. **Acids with m = 0, or 1, are weak acids in water, and those with m > 1 are strong acids in water, with the absolute strength increasing for weak or strong acids as the value of m increases.** Thus:

(a) $Si(OH)_4$	(b) $PO(OH)_3$	(c) $SO_2(OH)_2$	(d) ClO_3OH
m = 0	m = 1	m = 2	m = 3

And in order of increasing acid strength: $Si(OH)_4$ < $PO(OH)_3$ < $SO_2(OH)_2$ < ClO_3OH

Each of these acids has four oxygen atoms bonded to the central atom and since O is highly electronegative, these O atoms polarize the bonding electrons towards them, thus increasing the electronegativity of the central atom. However, since the order of the electronegativities of these third period central atoms is Si < P < S < Cl, the same order will be retained in the acids. The more electronegative the central atom, the greater the polarity of an adjacent $^{\delta-}O—H^{\delta+}$ bond, and therefore the more acidic it becomes.

---

35. (a) Explain why silicon tetrafluoride reacts with fluoride ion to form the complex ions SiF_5^- and SiF_6^{2-}, but carbon tetrafluoride does not react with fluoride ion. (b) Predict the structures of SiF_5^- and SiF_6^{2-}.

(a) **Carbon** in Period 2 obeys the octet rule, and the C atom in CF_4 forms very strong C—F bonds and already has a completed valence shell octet. Thus, it is not expected to be able to accept any more electron pairs, such as those from fluoride ions. In contrast, **silicon** is in Period 3 and can expand its valence shell octet in SiF_4 to a maximum of six pairs of electrons. Thus, it can react with fluoride ions to form SiF_5^- (five pairs) and SiF_6^{2-} (six pairs).

(b) SiF_5^- is an AX_5 type molecule with **trigonal bipyramidal** geometry, while SiF_6^{2-} is an AX_6 type molecule with **octahedral** geometry.

---

36. Write balanced equations to show how a mixture of formaldehyde, ammonia, HCN, and water can form an amino acid, and name this acid.

$$H_2C{=}O + NH_3 \rightarrow H_2C(OH)(NH_2) \rightarrow H_2C{=}NH + H_2O \xrightarrow{HCN} H_2C(NH_2)–C{\equiv}N \xrightarrow{2H_2O} H_2C(NH_2)–C(OH){=}O$$

glycine

CHAPTER 20

Addition Polymers

1. Define: (a) addition polymer, (b) condensation polymer, and give an example of each type of polymer

---

(a) **An addition polymer** is a macromolecule consisting of large numbers of repeating units (monomers), which contains all of the atoms of the monomers from which they are made. Examples include: **polyethylene,** $(CH_2)_n$, formed from ethene (ethylene), C_2H_4, and **polytetrafluoroethene,** $(CF_2)_n$, (Teflon) formed from tetrafluoroethene, C_2F_4.

(b) **A condensation polymer** is a macromolecule formed from monomers by a reaction in which a small molecule, such as H_2O, is eliminated between monomer units. Examples include: **polyamides,** such as *nylon,* formed by condensing dicarboxylic acid and diamine monomers, and **polyesters** formed by condensing diol and dicarboxlic acid monomers (e.g., *Dacron* from 1,2-ethanediol and 1,4-benzenedicarboxylic acid).

---

3. What is meant by the terms: (a) "linear-chain polymer", (b) "branched-chain polymer", and (c) "cross-linked polymer chain"?

(a) **"Linear (continuous) chain polymer"** describes a polymer containing macromolecules formed by head to tail addition of ethene, substituted ethene, or polyene monomers, in a single continuous chain.

(b) **"Branched chain polymer"** describes a polymer containing macromolecules that are formed not only by continuous chain growth but also by intermittent addition of monomers at various points in the chain already formed, which produces branched chains that also continue to grow linearly and can themselves take part in further branching.

(c) **"Cross-linked polymer"** refers to the linking together of polymer chains, as occurs for example in the process of *vulcanization,* where natural rubber (poly-2-methyl-1,3-butadiene) or a synthetic rubber such as poly-2-chloro-1,3-butadiene is heated with sulfur to form disulfide, —S—S—, cross-links between neighboring polymer molecules. Such cross-linking is also important in polypeptides formed from sulfur containing α amino acids such as *cysteine,* $HSCH_2—C(H)(NH_2)CO_2H$.

---

5. Draw a diagram to show the mechanism of chain branching in the free-radical polymerization of styrene.

Styrene is phenylethene, $C_6H_5CH{=}CH_2$, and its free radical polymerization first gives a continuous chain polymer molecule containing units such as that shown below, **I**, which can further react with a RO free radical:

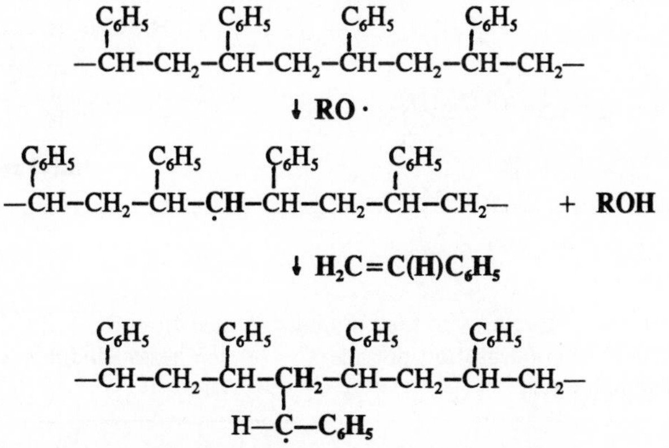

$$-\underset{}{\overset{C_6H_5}{\overset{|}{C}}}H-CH_2-\overset{C_6H_5}{\overset{|}{C}}H-CH_2-\overset{C_6H_5}{\overset{|}{C}}H-CH_2-\overset{C_6H_5}{\overset{|}{C}}H-CH_2- \qquad \textbf{I.}$$

$$\downarrow \textbf{RO}\cdot$$

$$-\overset{C_6H_5}{\overset{|}{C}}H-CH_2-\overset{C_6H_5}{\overset{|}{C}}H-\underset{\bullet}{C}H-\overset{C_6H_5}{\overset{|}{C}}H-CH_2-\overset{C_6H_5}{\overset{|}{C}}H-CH_2- \quad + \textbf{ ROH} \qquad \textbf{II.}$$

$$\downarrow \mathbf{H_2C{=}C(H)C_6H_5}$$

$$-\overset{C_6H_5}{\overset{|}{C}}H-CH_2-\overset{C_6H_5}{\overset{|}{C}}H-\underset{\underset{H-\underset{\bullet}{C}-C_6H_5}{|}}{C}H_2-\overset{C_6H_5}{\overset{|}{C}}H-CH_2-\overset{C_6H_5}{\overset{|}{C}}H-CH_2- \qquad \textbf{III.}$$

Reaction of **I** with RO gives a free radical **II** with the odd electron on a C atom in the middle of the chain, which can grow to form a branched chain by reacting with another styrene molecule to give the free radical **III**, which continues to grow by adding further styrene molecules.

---

7. Draw the repeating unit in the structures of the polymers formed from each of the following monomers: (a) propene, (b) acrylonitrile, $H_2C{=}C(H)C{\equiv}N$, (c) 1,3-butadiene

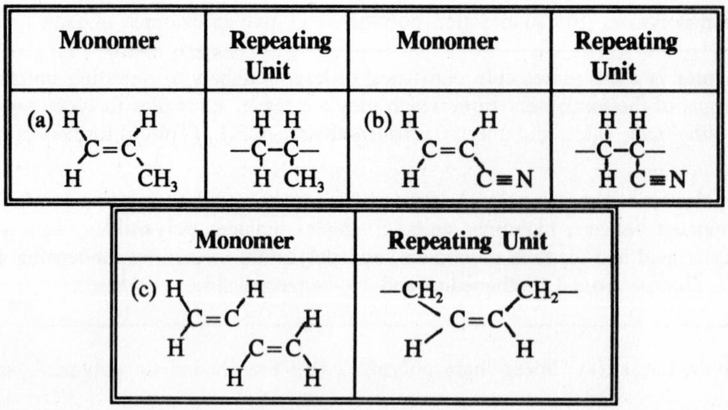

Monomer	Repeating Unit	Monomer	Repeating Unit
(a) $H_2C{=}C(H)CH_3$	$-CH_2-CH(CH_3)-$	(b) $H_2C{=}C(H)C{\equiv}N$	$-CH_2-CH(C{\equiv}N)-$

Monomer	Repeating Unit
(c) $H_2C{=}CH-CH{=}CH_2$	$-CH_2-CH{=}CH-CH_2-$

9. The synthetic fiber "Orlon" has the polymeric chain structure $-[-CH_2-CH(C{\equiv}N)-CH_2-CH(C{\equiv}N)-CH_2-CH(C{\equiv}N)-]-_n$. From what monomer is this synthesized?

"Orlon" is clearly an addition polymer, and the monomer from which it is formed must be:

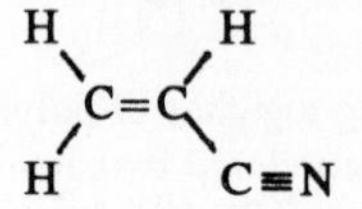

$$H_2C{=}C(H)C{\equiv}N$$

cyanoethene
(acrylonitrile)

11. Isobutylene (2-methylpropene) and isoprene (2-methylbutadiene) copolymerize to give butyl rubber. Draw the structure of the repeating unit; asssume that the two monomers alternate.

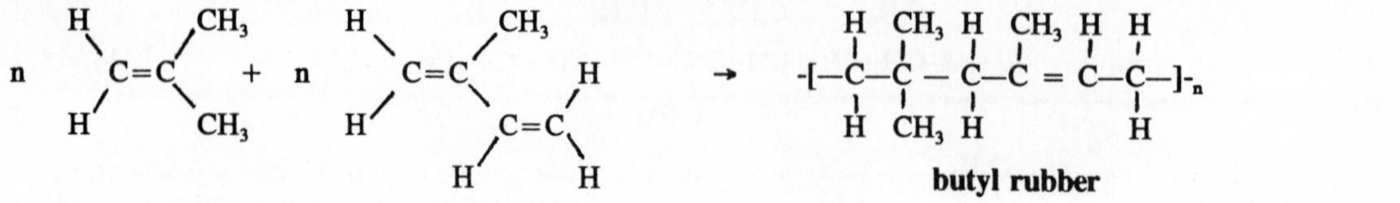

$$n\ H_2C{=}C(CH_3)_2 + n\ H_2C{=}C(CH_3)-CH{=}CH_2 \rightarrow -[-CH_2-C(CH_3)_2-CH_2-C(CH_3){=}CH-CH_2-]-_n$$

butyl rubber

Condensation Polymers

13. Draw the repeating unit in the structures of the polymers formed from the following monomers, and explain why the polymers are described as condensation polymers? (a) 1,6-hexanediamine and 1,6-hexanedioic acid; (b) 1,2-ethanediol and terephthalic acid

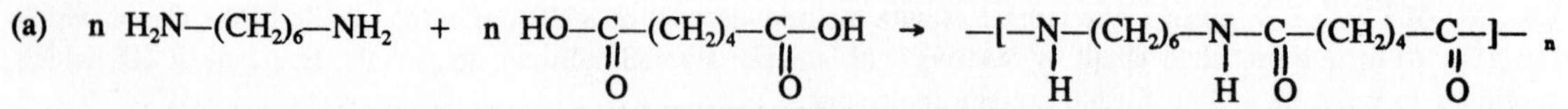

(a) $$n\ H_2N-(CH_2)_6-NH_2 + n\ HO-C(=O)-(CH_2)_4-C(=O)-OH \rightarrow -[-NH-(CH_2)_6-NH-C(=O)-(CH_2)_4-C(=O)-]-_n + 2n\ H_2O$$

1,6-hexanediamine **1,6-hexanedioic acid**

(b) n HO—CH_2—CH_2—OH + n HO—C(=O)—C_6H_4—C(=O)—OH → —[—O—CH_2—CH_2—O—C(=O)—C_6H_4—C(=O)—]— $_n$

1,2-ethanediol **terephthalic acid** **+ 2n H_2O**

These are **condensation polymers** because they are formed by elimination of water between polymer units.

---

14. Draw the repeating unit of condensation polymers made from each of the following pairs of monomers and name the alcohol that is eliminated: (a) 1,2-ethanediol and dimethylpropanedioate, (b) 1,3-propanediol and diethyl-1,4-butanedioate

---

(a) n HO—CH_2—CH_2—OH + n H_3CO—CH_2—OCH_3 → —[—O—CH_2—CH_2—C(=O)—CH_2—C(=O)—]— $_n$ + 2n CH_3OH

1,2-ethanediol **dimethylpropanedioate** **methanol**

(b) n HO—$(CH_2)_3$—OH + n H_5C_2O—$(CH_2)_2$—OC_2H_5 → —[—O—$(CH_2)_3$—C(=O)—$(CH_2)_2$—C(=O)—]— $_n$ + 2n C_2H_5OH

1,2-propanediol **diethylbutanedioate** **ethanol**

---

16. What intermolecular forces are present between polymer chains in (a) an addition polymer such as polyethylene; (b) a condensation polymer, such as Dacron; (c) a polyamide?

(a) The chains of a nonpolar **addition polymer** such as polyethylene, —[—CH_2—CH_2—CH_2—CH_2—]— $_n$, attract each other only by weak *London forces.*

(b) A **condensation polymer** such as Dacron (ethanediol - terephthalic acid polymer) is a polyester. It cannot form hydrogen bonds but the benzene rings add rigidity to the structure, and the carbonyl groups, >C=O, are polar. Its polymer chains attract each other by *dipole-dipole* and *London forces.*

—[—O—CH_2—CH_2—O—C(=O)—C_6H_4—C(=O)—]— $_n$

(c) The *amide* linkages, —C(=O)—N(H)—, in polymer chains of a **polyamide** attract each other by *hydrogen bonds.*

---

18. From what monomers are each of the following formed? (a) Teflon, (b) PVC, (c) nylon-6, (d) Dacron

---

Polymer	Structure	Monomers
(a) Teflon	—[—CF_2—CF_2—CF_2—CF_2—]— $_n$	F_2C=CF_2 tetrafluoroethene
(b) PVC	—[—CH_2—CHCl—CH_2—CHCl—]— $_n$	H_2C=CHCl vinyl chloride

(c) Nylon 6	$-[-N-(CH_2)_5-C(=O)-]-_n$ with H on N	caprolactam: ring $O=C$, CH_2, CH_2, CH_2, H_2C-CH_2, $H-N$
(d) Dacron	$-[-O-CH_2-CH_2-O-C(=O)-C_6H_4-C(=O)-]-_n$	$HO-CH_2-CH_2-OH$ 1,2-ethanediol; $HO-C(=O)-C_6H_4-C(=O)-OH$ terephthalic acid

Materials Science

20. (a) How do aluminosilicates differ from silicates? (b) How is sodalite, $Na_8(Al_6Si_6O_{24})Cl_2(s)$, related to silica? (c) Why are anhydrous zeolites excellent drying agents?

(a) Silicates have structures based upon SiO_4 tetrahedra sharing one, two, or three corners. **Aluminosilicates** have similar structures to the silicates, or to silica, except that Al^- ions are substituted for some proportion of the Si atoms and there are additional cations to neutralize the additional negative charge on the aluminosilicate anions.

(b) **Sodalite,** $Na_{16}(Al_{12}Si_{12}O_{48})Cl_4(s)$, contains the $Al_{12}Si_{12}O_{48}{}^{12-}$ anion, which is related to $(SiO_2)_{24}$, with one-half the Si atoms replaced by Al^-.

(c) In anhydrous **zeolites**, the water molecules of the original zeolite have been removed, and the channels are receptive to absorbing a variety of small molecules, including water. Thus, they are excellent drying agents.

22. Why would you expect the catalytic properties of a zeolite to be greater than those of either silica, SiO_2, or aluminum oxide, Al_2O_3?

Both silica, $SiO_2(s)$, and aluminium oxide, alumina, $Al_2O_3(s)$, have compact structures and their catalytic behaviour is best described as that of heterogeneous *surface* catalysts. In contrast, zeolites have open structures with channels and cavities which give them an enormous surface area and resulting greater catalytic activity. However, they are also more *selective* because the sizes and shapes of the tunnels and cavities limit the types of molecules that can fit into them. They also have the advantage that their AlO^- groups can be protonated to give highly acidic Al—OH groups, so that they can also be used as *acid catalysts*. An example of this kind of usage is in the synthesis of ethylbenzene from ethene and benzene, in which protonation of $H_2C=CH_2$ gives the carbocation $CH_3CH_2{}^+$, with which benzene can then undergo electrophilic substitution: $C_2H_5{}^+ + C_6H_6 \rightarrow C_6H_5-C_2H_5 + H^+$.

Semiconductors

23. Explain why the electrical conductivity of a metal decreases with increasing temperature, whereas that of a semiconductor increases.

The electrical conductivity of a **metal** is due to the movement of the electrons of its *electron gas* under the influence of a potential difference. The conductivity decreases with increasing temperature as a consequence of the increasingly greater amplitudes of the thermal vibration of the metal ions in the metal lattice, which impedes more and more their free movement and increases the resistance of the metal. In contrast, the electrical conduction of a

semimetal is a function of the number of electrons in its *conduction band*. This number of electrons increases with increasing temperature because more and more electrons are promoted from the valence band to the conduction band. Thus, the electrical conductivity of a semimetal increases with increasing temperature.

---

25. (a) How is silicon prepared from silica, $SiO_2(s)$? (b) How is ultrapure silicon prepared?

(a) **Silicon** is prepared from silica, $SiO_2(s)$, by reducing it with coke at a temperatre of about 3000°C in an electric furnace:

$$SiO_2(s) + 2C(s) \rightarrow Si(\ell) + 2CO(g)$$

The molten silicon produced is sufficiently pure for many purposes.

(b) To make *ultrapure* silicon, the industrial product is first converted to volatile **silicon tetrachloride**, $SiCl_4(\ell)$, (bp 57.6°C) by reaction with chlorine:

$$Si(s) + 2Cl_2(g) \rightarrow SiCl_4(\ell)$$

After purification by distillation, the $SiCl_4$ is reduced to silicon with $H_2(g)$, or magnesium:

$$SiCl_4(g) + 2H_2(g) \rightarrow Si(s) + 4HCl(g); \qquad SiCl_4(g) + 2Mg(s) \rightarrow Si(s) + 2MgCl_2(s)$$

Finally, the silicon prepared in this way is purified by *zone refining*, in which a rod of silicon is slowly moved through a heater so that a molten zone traverses the length of the rod. This concentrates any impurities in the molten zone, which eventually reaches the rod and is cut off after it solidifies.

---

27. Which of the following are n-type conductors and which are p-type conductors? (a) Si doped with P, (b) Si doped with B, (c) Si doped with As, (d) Ge doped with As, (e) Ge doped with Ga.

An **n-type** semiconductor is one containing impurities the atoms of which have valence electrons in excess of the number required to bond them to the atoms of the semiconductor, thus introducing an excess of (negative) electrons that enter the conduction band and move freely around the lattice. Typically, for a Group IV semiconductor such as Si or Ge, introduction of a small quantity of a Group V element makes the Si or Ge an *n-type semiconductor*. A **p-type** semiconductor is one containing impurities the atoms of which have insufficient valence electrons to bond them to the atoms of the semiconductor, and they become part of the lattice by stealing electrons from the atoms of the semiconductor, which creates *positive holes* in the semiconductor which move between the semiconductor atoms. Typically, for a Group IV semiconductor such as Si or Ge, introduction of a small quantity of a Group III element makes the Si or Ge a *p-type semiconductor*. Thus, for these examples we have:

Semiconductor	**(a) Si**	(b) Si	(c) Si	(d) Ge	(e) Ge
Impurity	P (Group V)	B (Group III)	As (Group V)	As (Group V)	Ga (Group III)
Type	**n-type**	**p-type**	**n-type**	**n-type**	**p-type**

General Problems

29. Styrene (phenylethene) may be copolymerized with maleic acid (2-butenedioic acid). What is the structure of the product in which the two monomers provide the alternating monomer units of a copolymer?

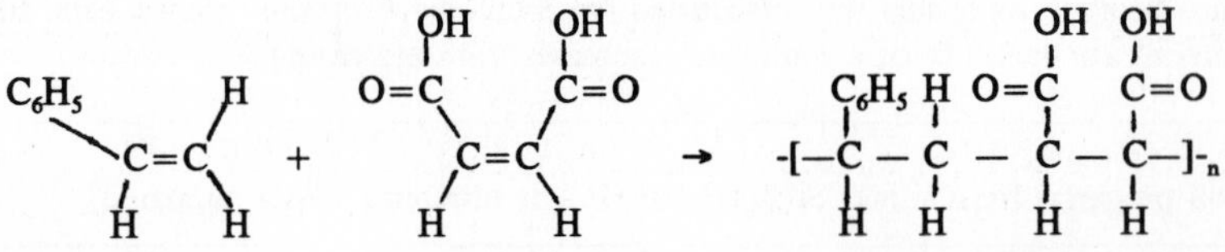

$$C_6H_5(H)C{=}CH_2 + HOOC\text{-}CH{=}CH\text{-}COOH \rightarrow -[-CH(C_6H_5)-CH_2-CH(COOH)-CH(COOH)-]_n-$$

styrene **maleic acid**

(Note that such polymer chains could be *cross-linked* by condensation to give an **addition-condensation polymer.**)

---

31. How, and from what monomers, would you synthesize the following polymers? (a) Tedlar (polyvinylfluoride) (b) Kel-F (polychlorotrifluoroethylene)

Both of these polymers are *addition polymers* formed from substituted ethene monomers: (a) vinyl fluoride, $H_2C{=}CH(F)$, 1-fluoroethene, I, and (b) 1-chloro-1,2,2-trifluoroethene, $F_2C{=}C(Cl)F$, II.

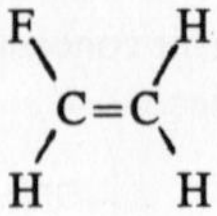

$$\underset{\textbf{I}}{(F)(H)C{=}C(H)(H)} \qquad \underset{\textbf{II}}{(Cl)(F)C{=}C(F)(F)}$$

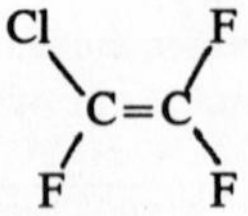

Both are formed by *free-radical* initiated polymerization reactions.

---

33. Why do polyamides such as nylon have a much greater mechanical strength than do polyalkanes such as polyethylene?

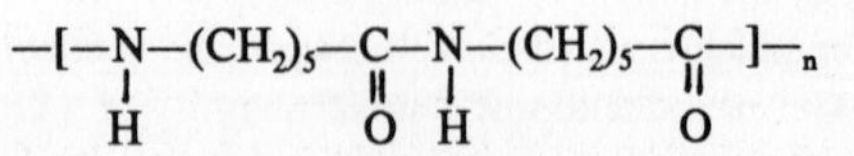

$$-[-N(H)-(CH_2)_5-C(=O)-N(H)-(CH_2)_5-C(=O)-]_n \qquad -[-CH_2-CH_2-CH_2-CH_2-]_n$$

Nylon 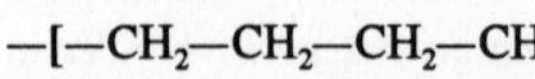 **polyethylene**

Nylon contains *peptide* groups that can form many *hydrogen bonds* between C=O and N-H groups on different polymer chains, giving them considerable mechanical strength, whereas the intermolecular forces between polyethylene chains are limited to only weak *London forces*.

---

35. The energy differences between the valence bands and conduction bands of carbon (diamond), silicon, and germanium are 502, 105, and 59 $kJ\cdot mol^{-1}$, respectively. In each case, calculate the maximum wavlength of light needed to bring about this excitation.

For an energy gap ΔE in $kJ\ mol^{-1}$, the maximum wavelength light to bring about the excitation is given by:

$$\Delta E = N_A h\nu = \frac{N_A hc}{\lambda} \; ; \; \lambda = \frac{N_A hc}{\Delta E} = \frac{(6.022 \times 10^{23}\ mol^{-1})(6.63 \times 10^{-34}\ J\ s)(3.00 \times 10^{8}\ m\ s^{-1})}{\Delta E}$$

$$= \left(\frac{0.1197\ J\ mmol^{-1}}{\Delta E}\right)\left(\frac{1\ kJ}{10^3\ J}\right)\left(\frac{10^9\ nm}{1\ m}\right) = \frac{1.197 \times 10^5\ kJ\ nm\ mol^{-1}}{\Delta E}$$

Diamond : $\lambda_{max} = \frac{1.197 \times 10^5\ kJ\ nm\ mol^{-1}}{502\ kJ\ mol^{-1}} = \underline{238\ nm}$

Silicon : $\lambda_{max} = \frac{1.197 \times 10^5\ kJ\ nm\ mol^{-1}}{105\ kJ\ mol^{-1}} = \underline{1.14 \times 10^3\ nm}$

Germanium : $\lambda_{max} = \frac{1.197 \times 10^5\ kJ\ nm\ mol^{-1}}{59\ kJ\ mol^{-1}} = \underline{2.0 \times 10^4\ nm}$

---

37. Why are some metal oxides metallic conductors?

Although most metal oxides are insulators, some transition metal oxides, such as TiO(s) and VO(s), both with the sodium chloride structure, are exceptions because the metal cations they contain, such as Ti^{2+} with the ground state electron configuration [Ar] $3d^2$, and V^{2+} with the ground state electron configuration [Ar] $3d^3$, have moderately low core charges and their *d electrons* are not held very strongly, so they are available to enter the conduction band where they form a delocalized electron cloud, as in a metal. Thus, TiO and VO behave as metal conductors. Other transition metal oxides, such as FeO(s), with cations with higher core charges than Ti^{2+} or V^{2+}, such as Fe^{2+}, with the ground state electron configuration [Ar] $3d^6$, are *not* metallic conductors because their d electrons are too strongly held to enter the conduction band. However, FeO(s) behaves as *p-type semiconductor* because it does not in fact have the exact composition Fe^{2+} O^{2-} but is a *nonstoichiometric compound* with compositions varying from $Fe_{0.89}O$ to $Fe_{0.96}O$ containing Fe^{3+} ions as well as Fe^{2+} ions, with one O^{2-} ion missing from the crystal lattice for every two Fe^{3+} ions. Each Fe^{3+} ion creates a positive hole in the crystal lattice, and when an electron is transferred from an Fe^{2+} ion to a neighboring Fe^{3+} ion the positive hole moves, giving nonstoichiometric FeO(s) the properties of a p-type semiconductor.

CHAPTER 1

1.1 Write the symbols for each of the following elements: (a) hydrogen, (b) helium, (c) lithium, (d) beryllium, (e) boron, (f) carbon, (g) nitrogen, (h) oxygen, (i) fluorine, (j) neon.

(a) H, (b) He, (c) Li, (d) Be, (e) B, (f) C, (g) N, (h) O, (i) F, (j) Ne.

1.2 Name the elements with each of the following symbols: (a) H, (b) He, (c) B, (d) Be, (e) C, (f) Ca, (g) N, (h) Na, (i) F, (j) Fe, (k) K, (ℓ) Kr.

(a) **hydrogen,** (b) **helium,** (c) **boron,** (d) **beryllium,** (e) **carbon,** (f) **calcium,** (g) **nitrogen,** (h) **sodium,** (i) **fluorine,** (j) **iron,** (k) **potassium,** (ℓ) **krypton.**

1.3 Name each of the substances with the following molecular formulas and classify each as a constituent of an element (E) or a compound (C): (a) H_2, (b) N_2, (c) O_2, (d) NO, (e) S_8, (f) Cl_2, (g) H_2O, (h) NH_3, (i) CH_4, (j) CO_2.

(a) Hydrogen (E), (b) nitrogen (E), (c) oxygen (E), (d) nitrogen monoxide (C), (e) sulfur (E), (f) chlorine (E), (g) water (C), (h) ammonia (C), (i) methane (C), (j) carbon dioxide (C).

1.4 Write the empirical formulas for the substances with each of the following molecular formulas: (a) H_2O_2, (b) CH_4, (c) C_2H_6, (d) S_8, (e) CO_2, (f) N_2, (g) $C_{12}H_{22}O_{11}$, (h) C_4H_8.

(a) HO, (b) CH_4, (c) CH_3, (d) S, (e) CO_2, (f) N, (g) $C_{12}H_{22}O_{11}$, (h) CH_2.

1.5 Classify each of the underlined terms in the following statement as a physical property (P) or a chemical property (C): "Hydrogen peroxide is (a) a colorless (b) viscous, (c) liquid, that readily (d) decomposes to water and oxygen. It has (e) a density of 1.44 g mL^{-1}, (f) a melting point of —0.89°C, and (g) a boiling point of 151°C. A solution in water is used as (h) a mild antiseptic and as (i) a bleaching agent".

(P) a, b, c, e, f, g; (C) d, h, i.

1.6 A 10.000-g sample of table salt, NaCl(s), contains 0.0030 g of impurities. What is the purity of the sample expressed as the mass percentage of NaCl?

The sample contains (10.000 - 0.0030) g = **9.997 g** NaCl(s) in the 10.000-g sample:

$$\text{Mass \% NaCl} = (9.997 \text{ g NaCl})\left(\frac{100 \text{ g sample}}{10.000 \text{ g sample}}\right)(100 \%) = \underline{99.97 \text{ mass \%}}$$

1.7 Classify each of the following as a heterogeneous mixture (H) or as a homogeneous mixture (solution) (S): (a) gasoline, (b) smog, (c) milk, (d) household bleach, (e) soil, (f) vinegar, (g) natural gas.

(a) **Gasoline (S)** is a homogeneous mixture of liquid hydrocarbons.

(b) **Smog (H)** is a heterogeneous mixture of gases, water vapor, and solid particles.

(c) **Milk (H)** is a heterogeneous mixture of water, dissolved solids and fat globules.

(e) **Household Bleach (S)** is a homogeneous solution of bleaching agents in water.

(f) **Soil (H)** is a heterogeneous mixture of rock, minerals, and decayed organic matter.

(g) **Vinegar (S)** is a solution of acetic acid in water.

(h) **Natural gas (S)** is a homogeneous mixture of gaseous hydrocarbons (mainly methane, CH_4(g)).

1.8 Write balanced equations for the following reactions: (a) the reaction of carbon monoxide, CO(g), and oxygen, O_2(g), to give carbon dioxide, CO_2(g); (b) the reaction of methane, CH_4(g), and water vapor, H_2O(g), at high temperature to give a mixture of carbon monoxide, CO(g), and hydrogen gas, H_2(g).

(a) $CO(g) + O_2(g) \rightarrow CO_2(g)$ **unbalanced;** $2CO(g) + O_2(g) \rightarrow 2CO_2(g)$ **balanced.**

(b) $CH_4(g) + H_2O(g) \rightarrow CH_4(g) + H_2(g)$ **unbalanced;** $CH_4(g)+H_2O(g) \rightarrow CO(g)+3H_2(g)$ **balanced**

1.9 The nucleus of an atom of phosphorus contains 15 protons and 16 neutrons. What are (a) the charge on the nucleus, (b) the approximate mass relative to that of the proton, and (c) the number of electrons?

(a) The charge on the nucleus is **+15**, due to the charges of all the protons in the nucleus.

(b) The mass of the nucleus is given to a good approximation by the sum of the masses of its protons and neutrons (nucleons); since the neutron has almost the same mass as the proton, the mass in this case is approximately **31** times that of the proton.

(c) The number of electrons in a neutral atom is the same as the number of nuclear protons, in this case **15.**

1.10 Boron has isotopes ^{10}B and ^{11}B with relative abundances of 19.9% and 80.1% and masses of 10.012 94 u and 11.009 31 u respectively. What is the average mass of a boron atom?

Isotope	Fraction	Mass	Contribution
^{10}B	0.199	10.012 94 u	1.993 u
^{11}B	0.811	11.009 31 u	8.929 u
	Average Atomic Mass:		**10.9 u.** (3 sig. fig.)

1.11 How many moles of water, and how many molecules, are there in 1.00 g of water?

$$\text{Molar mass of water } (H_2O) = \{2(1.008)+16.00\}\ u = 18.02\ u$$

$$\text{mol } H_2O = (1.00\ g\ H_2O)\left(\frac{1\ mol\ H_2O}{18.02\ g\ H_2O}\right) = \underline{5.55 \times 10^{-2}\ mol}$$

$$\text{molecules } H_2O = (1.00\ g\ H_2O)\left(\frac{1\ mol\ H_2O}{18.02\ g\ H_2O}\right)\left(\frac{6.022 \times 10^{23}\ molecules}{1\ mol}\right) = \underline{3.34 \times 10^{22}\ molecules}$$

1.12 What are the molecular masses and molar masses of each of the following molecules?
(a) CH_4, (b) H_2O_2, (c) S_8, (d) $C_{12}H_{22}O_{11}$

(a) Molecular mass CH_4 = {12.01 + 4(1.008)} = **16.04 u;** Molar mass = **16.04 g mol^{-1}.**

(b) Molecular mass H_2O_2 = {2(1.008) + 2(16.00)} = **34.02 u;** Molar mass = **34.02 g mol^{-1}.**

(c) Molecular mass S_8 = 8(32.07) = **256.6 u;** Molar mass = **256.6 g mol^{-1}.**

(d) Molecular mass $C_{12}H_{22}O_{11}$ = {12(12.01)+22(1.008)+11(16.00)} = **342.3 u;** Molar mass = **342.3 g mol^{-1}.**

1.13 What are the molar masses of (a) silicon dioxide, and (b) sodium chloride?

These are network solids for which only the empirical formula can be given; their molar masses are their formula masses in grams:

(a) SiO_2, molar mass = {28.09+2(16.00)} = **60.09 g mol^{-1}**

(b) NaCl, molar mass = (22.99+35.45) = **58.44 g mol^{-1}**

CHAPTER 2

2.1 Name the oxides with the following empirical formulas: (a) ZnO, (b) Na_2O, (c) SO_3, (d) CaO, (e) Cl_2O, (f) Al_2O_3, (g) P_4O_6

(a) ZnO, **zinc oxide**; (b) Na_2O, **disodium oxide** (sodium oxide); (c) SO_3, **sulfur trioxide**; (d) CaO, **calcium oxide**; (e) Cl_2O, **dichlorine monoxide**; (f) Al_2O_3, **dialuminum trioxide** (aluminum oxide); (f) P_4O_6, **tetraphosphorus hexaoxide.**

2.2 Write balanced equations for both of the following reactions and state which reactant is oxidized and which is reduced: (a) the reaction of methane with water to give carbon monoxide and hydrogen. (b) the reaction of diiron trioxide with carbon giving carbon dioxide and iron.

(a) $CH_4(g) + H_2O \rightarrow CO(g) + 3H_2(g)$; $CH_4(g)$, **oxidized**; $H_2O(g)$, **reduced.**

(b) $Fe_2O_3(s) + 3C(s) \rightarrow 2Fe(s) + 3CO(g)$; C(s), **oxidized**; $Fe_2O_3(s)$, **reduced.**

2.3 Express the following pressures in atmospheres, and temperatures in kelvins: 745 mm Hg, 783 torr, 100 kPa, 25.0°C, 100.0°C, -150°C.

$$745 \text{ mm Hg} = (745 \text{ mm Hg})\left(\frac{1 \text{ atm}}{760 \text{ mm Hg}}\right) = \underline{0.980 \text{ atm}}$$

$$783 \text{ torr} = (783 \text{ torr})\left(\frac{1 \text{ atm}}{760 \text{ torr}}\right) = \underline{1.03 \text{ atm}}\ ; \qquad 100 \text{ kPa} = (100 \text{ kPa})\left(\frac{1 \text{ atm}}{101.325 \text{ kPa}}\right) = \underline{0.987 \text{ atm}}$$

25.0°C = (25.0 + 273.1) K = **298.1 K;** 100.0°C = (100.0 + 273.1) K = **373.1 K.**

-150°C = (-150 + 273.1) K = **123 K.**

2.4 Consider an experiment in which water is electrolyzed. The resulting hydrogen and oxygen are collected separately at 25°C and 1 atm pressure, and the volume of oxygen collected is 158.9 mL. What volume of hydrogen is collected?

According to Avogadro's law, the volume of a sample of a gas at a given temperature and pressure is proportional to the number of moles of molecules, n, in the sample. The reaction when water is electrolyzed is:

$$2H_2O(\ell) \rightarrow 2H_2(g) + O_2(g)$$

Thus: $$V_{H_2(g)} = (158.9 \text{ mL } O_2)\left(\frac{2 \text{ mol } H_2}{1 \text{ mol } O_2}\right) = \underline{317.8 \text{ mL}}$$

2.5 What would be the molar volume of an ideal gas at 1 atm and 25.00°C?

$$PV = nRT; \quad V = \frac{nRT}{P} = \frac{(1 \text{ mol})(0.08206 \text{ atm L mol}^{-1} \text{ K}^{-1})(298.1 \text{ K})}{1 \text{ atm}} = \underline{24.46 \text{ L}}$$

2.6 How many moles and how many molecules, are there in a 1.00-L sample of $H_2O(g)$ at 100 °C and exactly 1 atm pressure.

$$PV = nRT; \quad n_{H_2O} = \frac{PV}{RT} = \frac{(1 \text{ atm})(1.00 \text{ L})}{(0.0821 \text{ atm L mol}^{-1} \text{ K}^{-1})(373.1 \text{ K})} = \underline{3.26 \times 10^{-2} \text{ mol}}$$

$$H_2O \text{ molecules} = (3.264 \times 10^{-2} \text{ mol})\left(\frac{6.022 \times 10^{23} \text{ molecules}}{1 \text{ mol}}\right) = \underline{1.97 \times 10^{22} \text{ molecules}}$$

2.7 A 3.525 g sample of $CO_2(g)$ occupies a volume of 2.00 L at a pressure of 745 mm Hg and a temperature of 25.0°C. How many moles of CO_2 are present, and what is the molar mass of $CO_2(g)$?

$$PV = nRT; \quad n_{CO_2} = \frac{PV}{RT} = \frac{(745 \text{ mm Hg})\left(\frac{1 \text{ atm}}{760 \text{mm Hg}}\right)(2.00 \text{ L})}{(0.0821 \text{ L atm mol}^{-1} \text{ K}^{-1})(298.1 \text{ K})} = \underline{8.01 \times 10^{-2} \text{ mol}}$$

$$\text{molar mass} = \frac{3.525 \text{ g } CO_2}{8.01 \times 10^{-2} \text{ mol } CO_2} = \underline{44.0 \text{ g mol}^{-1}}$$

2.8 The empirical formula of benzene is CH. What is its molecular formula, if 0.638 g of benzene occupies a volume of 250 mL at 100°C and a pressure of 1.00 atm?

$$n = \frac{PV}{RT} = \frac{(1.00 \text{ atm})(250 \text{ mL})\left(\frac{1 \text{ L}}{10^3 \text{ mL}}\right)}{(0.0821 \text{ atm L mol}^{-1} \text{ K}^{-1})(373.1 \text{ K})} = \underline{8.162 \times 10^{-3} \text{ mol}}$$

$$\text{molar mass benzene} = \frac{0.638 \text{ g}}{8.162 \times 10^{-3} \text{ mol}} = \underline{78.2 \text{ g mol}^{-1}}$$

The molecular formula is $(CH)_x$, and since the **empirical formula mass** is (12.01+1.008) u = **13.02 u:**

$$x = \frac{78.2\ u}{13.02\ u} = \underline{6.01}\ ;\quad \text{i.e., the } \underline{\text{molecular formula is } C_6H_6}$$

2.9 What is the density of water vapor, molar mass 18.02 g mol^{-1}, at 100°C and a pressure of 1.00 atm?

Density is g of water vapor per liter (g L^{-1}); first we evaluate mol L^{-1}, and then convert to g L^{-1}:

$$PV = nRT,\ \textbf{mol L}^{-1} = n/V = P/RT$$

$$\text{density} = \frac{\text{mass}}{\text{volume}} = \frac{PM}{RT} = \frac{(1\ \text{atm})(\frac{18.02\ g}{1\ mol})}{(0.0821\ \text{atm L mol}^{-1}\ K^{-1})(373.1\ K)} = \underline{0.588\ g\ L^{-1}}$$

2.10 The density of a hydrocarbon found in natural gas at 25.0 °C and a pressure of 1.00 atm is 2.375 g L^{-1}. What are its molar mass and molecular formula?

$$\text{density} = \frac{\text{mass}}{V} = \frac{nM}{V};\quad PV = nRT\ ;\quad \frac{n}{V} = \frac{P}{RT}\ ;\quad \underline{\text{density}} = \frac{PM}{RT};$$

$$M = \frac{(\text{density})(RT)}{P} = \frac{(\frac{2.375\ g}{1\ L})(0.0821\ \text{atm L mol}^{-1}\ K^{-1})(298.1\ K)}{1\ \text{atm}} = \underline{58.1\ g\ L^{-1}}$$

which is consistent with the expected molar mass of **butane, C_4H_{10},** (58.12 g mol^{-1}).

2.11 The volume of gas inside an automobile cylinder is 0.50 L a pressure of 1.00 atm and a temperature of 20.0°C. What is the volume when the gas is compressed to a pressure of 5.00 atm at a temperature of 200°C?

For a given mass (amount) of gas

$$\frac{PV}{T} = \text{constant}\ ;\quad \frac{P_1V_1}{T_1} = \frac{P_2V_2}{T_2}\ ;\quad V_2 = \frac{P_1V_1T_2}{P_2T_1} = \frac{(1.00\ \text{atm})(0.50\ L)(473.1\ K)}{(5.00\ \text{atm})(293.1\ K)} = \underline{0.16\ L}$$

2.12 In the electrolysis of water a dry (water-free) sample of the mixture of $H_2(g)$ and $O_2(g)$ produced is collected in a single vessel at a total pressure of 736.2 mm Hg. What is the partial pressure of each gas in the mixture?

Water decomposes according to the equation: $2H_2O(\ell) \rightarrow 2H_2(g) + O_2(g)$, so the mixture of gases produces contains $H_2(g)$ and $O_2(g)$ in the **mole ratio 2:1** (2 mol H_2 for every mol of O_2). Since the partial pressure of a gas is proportional to the number of moles of the gas:

$$p_{H_2} = (\frac{2\ mol}{3\ mol})(736.2\ \text{mm Hg}) = \underline{490.8\ \text{mm Hg}}\ ;\quad p_{O_2} = (\frac{1\ mol}{3\ mol})(736.2\ \text{mm Hg}) = \underline{245.4\ \text{mm Hg}}$$

2.13 At a given temperature, by what factor are the average speeds of each of the following gases slower than that of hydrogen? (a) He(g), (b) N_2(g), (c) O_2(g), (d) CO_2(g), (e) SO_2(g), (f) Br_2(g).

For any two gases, $\frac{1}{2}mv_1^2 = \frac{1}{2}m_2v_2^2$, or $\frac{1}{2}Mv_1^2 = \frac{1}{2}M_2v^2$, so average velocity is proportional to the square root of molar mass:

$$\frac{v_{He}}{v_{H_2}} = \frac{2.016\ \text{g mol}^{-1}}{4.003\ \text{g mol}^{-1}} = \underline{0.7097};\quad \frac{v_{N_2}}{v_{H_2}} = \sqrt{\frac{2.016\ \text{g mol}^{-1}}{28.02\ \text{g mol}^{-1}}} = \underline{0.2682}\ ;\quad \frac{v_{O_2}}{v_{H_2}} = \sqrt{\frac{2.016\ \text{g mol}^{-1}}{32.00\ \text{g mol}^{-1}}} = \underline{0.2510}$$

$$\frac{v_{CO_2}}{v_{H_2}} = \sqrt{\frac{2.016\ \text{g mol}^{-1}}{44.01\ \text{g mol}^{-1}}} = \underline{0.2140}\ ;\quad \frac{v_{SO_2}}{v_{H_2}} = \sqrt{\frac{2.016\ \text{g mol}^{-1}}{64.07\ \text{g mol}^{-1}}} = \underline{0.1774}\ ;\quad \frac{v_{Br_2}}{v_{H_2}} = \sqrt{\frac{2.016\ \text{g mol}^{-1}}{159.8\ \text{g mol}^{-1}}} = \underline{0.1123}$$

2.14 Nitrogen diffuses through a porous barrier about 50% faster than a gaseous oxide of sulfur diffuses under the same conditions. What are the approximate molar mass of the oxide of sulfur and its likely molecular formula?

From the molar mass of N_2 (28.02 g mol^{-1}) and the relative rates of diffusion (N_2 approximately 1½ times as fast as the other gas, X):

$$\frac{r_{N_2}}{r_X} = 1.50 = \sqrt{\frac{M_X}{28.02\ \text{g mol}^{-1}}}\ ;\qquad M_X = (28.02\ \text{g mol}^{-1})(1.50)^2 = \underline{63.0\ \text{g mol}^{-1}}$$

and the approximate molar mass of X (**63 g mol^{-1}**) is close to that expected for SO_2, **sulfur dioxide**, (64.07 g mol^{-1}).

CHAPTER 3

3.1 Name and classify each of the following elements as a metal or a nonmetal: (a) Li, (b) Mg, (c) C, (d) P, (e) F, (f) Na, (g) Cl, (h) N, (i) Ca.

Metals are on the left and nonmetals on the right of the periodic table, separated by a diagonal line that runs from boron in Group III and Period 2, to polonium in Group VI and Period 6; elements at the interface between metals and nonmetals are metalloids (semimetals). Thus:

(a) **Li, lithium**, in Group I and Period 2, is a **metal**;
(b) **Mg, magnesium** in Group II and Period 3, is a **metal**;
(c) **C, carbon**, in Group IV and Period 2, is a **nonmetal**;
(d) **P, phosphorus**, in Group V and Period 3, is a **nonmetal**;
(e) **F, fluorine**, in Group VII and Period 2, is a **nonmetal**;
(f) **Na, sodium**, in Group I and Period 3, is a **metal**;
(g) **Cl, chlorine**, in Group VII and Period 3, is a **nonmetal**;
(h) **N, nitrogen**, in Group V and Period 2, is a **nonmetal**;
(i) **Ca, calcium**, in Group II and Period 4, is a **metal**.

3.2 Name each of the following elements. Without consulting the periodic table, place each in its appropriate group and period: (a) C, (b) Na, (c) N, (d) Cl, (e) Mg, (f) Si, (g) Al, (h) O, (i) F, (j) Ca.

(a) **Carbon**, group IV, period 2; (b) **Sodium**, group I, period 3; (c) **Nitrogen**, group V, period 2;
(d) **Chlorine**, group VII, period 3; (e) **Magnesium**, group II, period 3; (f) **Silicon**, group IV, period 3;
(g) **Aluminum**, group III, period 3; (h) **Oxygen**, group VI, period 2; (i) **Fluorine**, group VII, period 2;
(j) **Calcium**, group II, period 4.

3.3 Write the formulas of the compounds formed between each of the following pairs of atoms: (a) Ca and H, (b) K and O, (c) Mg and S, (d) Na and N, (e) Al and S.

(a) Ca (Group II, valence 2) and H (Group I, valence 1) form **CaH_2**.
(b) K (Group I, valence 1) and O (Group VI, valence (8 - 6) = 2) form **K_2O**.
(c) Mg (Group II, valence 2) and S (Group VI, valence (8 - 6) = 2) form **MgS**.
(d) Na (Group I, valence 1) and N (Group V, valence (8 - 5) = 3) form **Na_3N**.
(e) Al (Group III, valence 3) and S (Group VI, valence (8 - 6) = 2) form **Al_2S_3**.

3.4 Predict the following properties of the halogen astatine, At; its physical state, molecular formula, and the empirical formulas of the compounds its forms with H, O, Na, and Mg.

Astatine, At, is the halogen in Group VII below iodine. In descending Group VII, fluorine and chlorine are gases, bromine is a liquid, and iodine is a solid. Thus, the **physical state** of astatine is expected to be **solid**. All the other halogens form diatomic molecules with the molecular formula X_2. Thus, the **molecular formula** of astatine is predicted to be **At_2**. Like the other halogens, the common valence of astatine is expected to be 1:
The **empirical formula** of the compounds formed with H (Group I, valence 1) is predicted to be **HAt**.
The **empirical formula** of the compound formed with O (Group VI, valence (8 - 6) = 2) is predicted to be **At_2O**.
The **empirical formula** of the compound formed with Na (Group I, valence 1) is predicted to be **NaAt**.
The **empirical formula** of the compound formed with Mg (Group II, valence 2) is predicted to be **$MgAt_2$**.

3.5 Give the shell-model electron arrangements for each of the following atoms: (a) C, (b) O, (c) F, (d) Mg, (e) K, (f) P, (g) S, (h) Cl.

The **period** in which the element is located gives the number of shells, and its **group** gives the number of electrons in the outer (valence) shell:

(a) **C**,	(period 2, group IV)	**2.4**	(b) **O**,	(period 2, group VI)	**2.6**
(c) **F**,	(period 2, group VII)	**2.7**	(d) Mg,	(period 3, group II)	**2.8.2**
(e) **K**,	(period 4, group I)	**2.8.8.1**	(f) **P**,	(period 3, group V)	**2.8.5**
(g) **S**,	(period 3, group VI)	**2.8.6**	(h) **Cl**,	(period 3, group VII)	**2.8.7**

3.6 Give the number of the group to which each of the following elements belongs. How many valence electrons does an atom of each of these elements have? (a) F, (b) N, (c) Cl, (d) B, (e) Na, (f) S, (g) Ca, (h) P, (i) C, (j) Al.

These elements are all among the first *twenty elements* in the periodic table, the position of which we need to learn.

(a) F, group VII, 7;	(b) N, group V, 5;	(c) Cl, group VII, 7;	(d) B, group III, 3;
(e) Na, group I, 1;	(f) S, group VI, 6;	(g) Ca, group II, 2;	(h) P, group V, 5;
(i) C, group IV, 4;	(j) Al, group III, 3.		

3.7 What is the core charge of each of the following atom? (a) F, (b) N, (c) Cl, (d) B, (e) Na, (f) S, (g) Ca, (h) P, (i) C, (j) Al.

For the main group elements, the core charge is positive and numerically equal to the **number of the group**.

(a) F, group VII, **+7**; (b) N, group V, **+5**; (c) Cl, group VII, **+7**; (d) B, group III, **+3**; (e) Na, group I, **+1**;
(f) S, group VI, **+6**; (g) Ca, group II, **+2**; (h) P, group V, **+5**; (i) C, group IV, **+4**; (j) Al, group III, **+3**.

3.8 Write the Lewis symbol for each of the following atoms: (a) H, (b) F, (c) C, (d) Al, (e) N, (f) Ne, (g) Ca, (h) O.

The number of valence electrons is equal to the number of the group in which the element is located; in the Lewis symbol these electrons are arranged singly up to a maximum of 4; valence electrons in excess of 4 are paired with existing electrons. Thus:

(a) H, group I, 1 valence electron, H·

(b) F, group VII, 7 valence electrons, :F̤̈·

(c) C, group IV, 4 valence electrons, ·Ċ̣·

(d) Al, group III, 3 valence electrons, ·Al·

(e) N, group V, 5 valence electrons, :Ṅ·

(f) Ne, group VIII, 8 valence electrons, :N̤̈e:

(g) Ca, group II, 2 valence electrons, ·Ca·

(h) O, group VI, 6 valence electrons, :Ȯ̤·

3.9 Draw the Lewis structures of each of the following molecules: (a) F_2, (b) CCl_4, (c) OF_2, (d) SiF_4, (e) SCl_2

In each case, the atoms forming the molecules are nonmetals, so all the bonds are covalent bonds. We draw the Lewis symbols of the atoms and combine unpaired electrons to form electron pair covalent bonds.

(a) F_2 F is in group VII and has 7 valence electrons:

:F̤̈· + ·F̤̈: → :F̤̈–F̤̈:

(b) CCl_4 C is in group IV and has 4 valence electrons, and Cl is in group VII and has 7 valence electrons

·Ċ̣· + 4 ·C̤̈l: → :C̤̈l–C(–C̤̈l:)(:C̤̈l:)(:C̤̈l:)

(c) OF_2 O is in group VI and has 6 valence electrons, and F is in group VII and has 7 valence electrons

:F̤̈· + ·Ö̤· + ·F̤̈: → :F̤̈–Ö̤–F̤̈:

(d) SiF_4 Si is in group IV and has 4 valence electrons, and F is in group VII and has 7 valence electrons

·Ṩi· + 4 ·F̤̈: → :F̤̈–Si(–F̤̈:)(:F̤̈:)(:F̤̈:)

(e) SCl_2 S is in group VI and has 6 valence electrons, and Cl is in group VII and has 7 valence electrons

:C̤̈l· + ·S̤̈· + ·C̤̈l: → :C̤̈l–S̤̈–C̤̈l:

3.10 Draw the Lewis structures of each of the following compounds. In each case the carbon atom or atoms are in the center of the molecule: (a) H_2CO, (b) CS_2, (c) C_2F_4, (d) FCCF, (e) HCN.

In each case, the atoms forming the molecules are nonmetals, so all the bonds are covalent bonds. We draw the Lewis symbols of the atoms and combine unpaired electrons to form **single** electron pair covalent bonds, **double** bonds with two shared electron pairs, or **triple** bonds with three shared electron pairs, as appropriate, and to satisfy **the *Lewis octet rule*:**

(a) H_2CO

2 H· + ·Ċ̣· + ·Ö̤: → (H)(H)C=Ö:

(b) CS_2

:S̤̈· + ·Ċ̣· + ·S̤̈: → :S̈=C=S̈:

(c) C_2F_4

$$2\,:\ddot{F}\cdot + \cdot\dot{C}\cdot + \cdot\dot{C}\cdot + 2\cdot\ddot{F}: \rightarrow (:\ddot{F}:)_2C = C(:\ddot{F}:)_2$$

(d) FCCF

$$:\ddot{F}\cdot + \cdot\dot{C}\cdot + \cdot\dot{C}\cdot + \cdot\ddot{F}: \rightarrow :\ddot{F}-C \equiv C-\ddot{F}:$$

(e) HCN

$$H\cdot + \cdot\dot{C}\cdot + \cdot\dot{N}: \rightarrow H-C \equiv N:$$

3.11 Classify the bonds in the following substances as covalent, polar covalent, or ionic: (a) Br_2, (b) ClF, (c) BF_3, (d) CaF_2, (e) O_2, (f) K_2S, (g) $SiCl_4$, (h) PF_3.

Metal—nonmetal bonds are ionic, nonmetal—nonmetal bonds are covalent (pure covalent when the atoms are **identical, and thus have the same electronegativity, and polar covalent** when the atoms forming the bond have **different electronegativities).**

(a) Br_2, covalent; (b) ClF, polar covalent; (c) BF_3, polar covalent; (d) CaF_2, ionic; (e) O_2, covalent;

(f) K_2S, ionic; (g) $SiCl_4$, polar covalent; (h) PF_3, polar covalent.

3.12 Classify each of the following molecules in terms of the AX_nE_m nomenclature, and give the geometric shape for each: (a) CF_4, (b) NF_3, (c) BCl_3, (d) F_2O, (e) HCl.

The first step is to draw the Lewis structure of a molecule, which can then be classified in terms of the AX_nE_m **nomenclature by counting up the number of bonding pairs, n, and the number of non-bonding (lone) pairs, m,** on the central atom A. The arrangement of the **m + n** electron pairs will be one of the following: **four - tetrahedral; three - trigonal planar;** and **two - linear**. Finally, the arrangement of the **n bonds** gives the molecular geometry.

Molecule	Lewis Structure	AX_nE_m Type	Geometry
(a) CF_4	C bonded to four :F̤:	AX_4	tetrahedral
(b) NF_3	:F̤–N̈–F̤: with :F̤: below N	AX_3E	triangular pyramid
(c) BCl_3	:C̤l–B–C̤l: with :C̤l: below B	AX_3	triangular planar
(d) F_2O	:F̤–Ö–F̤:	AX_2E_2	angular
(e) HCl	H–C̤̈l:	AX_3E	linear

3.13 Classify each of the following molecules in terms of the AX_nE_m nomenclature, and give the geometric shape for each: (a) OCS, (b) Cl_2CO, (c) ClCN, (d) FCCF, (e) ONCl.

Molecule	Lewis Structure	AX_nE_m Type	Geometry
(a) OCS	:O=C=S:	AX_2	linear
(b) Cl_2CO	:Cl–C–Cl: :O:	AX_3	triangular planar
(c) ClCN	:Cl–C≡N:	AX_2	linear
(d) FCCF	:F–C ≡ C–F:	AX_2*	linear
(e) ONCl	:O=N–Cl:	AX_2E	angular

*At each C atom.

CHAPTER 4

4.1 Classify each of the following reactions as a synthesis reaction (S) or a decomposition reaction (D): (a) $2KClO_3(s) \rightarrow 2KVl(s) + 3O_2(g)$; (b) $N_2(g) + 3H_2(g) \rightarrow 2NH_3(g)$; (c) $C_2H_4(g) + H_2(g) \rightarrow C_2H_6(g)$; (d) $CaCO_3(s) \rightarrow CaO(s) + CO_2(g)$

(a) $KClO_3(s)$ decomposes to KCl(s) and oxygen; (D); (b) $NH_3(g)$ is synthesized from $N_2(g)$ and $H_2(g)$; (S).
(c) $C_2H_6(g)$ is synthesized from $C_2H_4(g)$ and $H_2(g)$; (S); (d) $CaCO_3(s)$ decomposes to CaO(s) and $CO_2(g)$; (D).

4.2 Write balanced equations for the reactions in Demonstration 4.4.

(i) $2Na(s) + Cl_2(g) \rightarrow 2NaCl(s)$; (ii) $2Fe(s) + 3Cl_2(g) \rightarrow 2FeCl_3(s)$;
(iii) $2Sb(s) + 3Cl_2(g) \rightarrow 2SbCl_3(s)$; (iv)$Mg(s) + Cl_2(g) \rightarrow 2MgCl_2(s)$.

4.3 Write balanced equations for each of the following reactions: (a) strontium with bromine; (b) arsenic with chlorine; (c) rubidium with fluorine; (d) bromine with chlorine. In each case state whether you expect the product to be (i) a solid or (ii) a liquid or gas at room temperature and atmospheric pressure.

(a) Sr (a metal in group II, valence 2) reacts with bromine (a nonmetal in group VII, valence 1):

$$Sr(s) + Br_2(\ell) \rightarrow Sr^{2+}[:Br:^-]_2(s)$$

and the product $SrBr_2$ is an ionic compound and therefore expected to be a solid.
(b) As (a nonmetal in group V, valence 3) reacts with chlorine (a nonmetal in group VII, valence 1):

$$2As(s) + 3Cl_2(g) \rightarrow 2AsCl_3(g \text{ or } \ell)$$

and the product $AlCl_3$ is a covalent compound and therefore expected to be a **liquid or gas**.
(c) Rb (a metal in group I, valence 1) reacts with fluorine (a nonmetal in group VII, valence 1):

$$2Rb(s) + F_2(g) \rightarrow 2Rb^+F^-(s)$$

and the product RbF is an ionic compound, and therefore expected to be a **solid**.

(d) Bromine (a nonmetal in group VII, valence 1) reacts with chlorine (a nonmetal also in group VII, valence 1):

$$Br_2(\ell) + Cl_2(g) \rightarrow 2BrCl(g \text{ or } \ell)$$

and the product BrCl is a covalent compound and therefore expected to be a **liquid or gas.**

4.4 Among the species (molecules or ions) I_2, I^-, Cl_2, F_2, and F^-, which is the strongest reducing agent?

The strongest **reducing agent** will be the weakest **oxidizing agent**. Among the halogens the order of oxidizing strength is $F_2 > Cl_2 > Br_2 > I_2$, so their order of reducing strengths is the reverse order: $I_2 > Br_2 > Cl_2 > F_2$, but the order of reducing strengths among the halide ions is $I^- > Br^- > Cl^- > F^-$, so I_2 is a stronger reducing agent than either Cl_2 or F_2, but I^- is a stronger reducing agent than F^-, so among these species **I^- is the strongest reducing agent.**

4.5 Among the species Br_2, Br^-, Cl^-, I_2, and F^-, which is the strongest oxidizing agent?

Among the halogens the order of oxidizing strength is $F_2 > Cl_2 > Br_2 > I_2$, and all are weaker oxidizing agents than the halide ions, so among these species **F_2 is the strongest oxidizing agent.**

4.6 In each of the following reactions, identify the oxidizing reagent, the reducing reagent, the substance that is oxidized, and the substance that is reduced: (a) $2Cu(s)+O_2(g) \rightarrow 2CuO(s)$; (b) $2Cs(s)+I_2(s) \rightarrow 2CsI(s)$ (c) $Zn(s)+S(s) \rightarrow ZnS(s)$; (d) $Zn(s)+Cu^{2+}(aq) \rightarrow Zn^{2+}(aq)+Cu(s)$.

In (a), (b), and (c), the reactions are between a metal and a nonmetal to give an ionic compound consisting of metal cations and nonmetal anions. The metals lose electrons, behave as reducing agents, and are oxidized; the nonmetals gain electrons, behave as oxidizing agents, and are reduced. Thus, in terms of half-reactions:

(a) $Cu \rightarrow Cu^{2+} + 2e^-$, Cu, the **reducing agent, is oxidized;** $O_2 + 4e^- \rightarrow 2O^{2-}$, O_2, the **oxidizing agent,** is **reduced**

(b) $Cs \rightarrow Cs^+ + e^-$, Cs, the **reducing agent,** is **oxidized,** $I_2 + 2e^- \rightarrow 2I^-$, I_2, the **oxidizing agent,** is **reduced**

(c) $Zn \rightarrow Zn^{2+} + 2e^-$, **Zn,** the **reducing agent,** is **oxidized,** $S + 2e^- \rightarrow S^{2-}$, S, the **oxidizing agent,** is **reduced**

(d) $Zn \rightarrow Zn^{2+} + 2e^-$, **Zn,** the **reducing agent,** is **oxidized,** $Cu^{2+} + 2e^- \rightarrow Cu$, Cu^{2+}, **oxidizing agent,** is **reduced**

4.7 Write an equation for an acid-base reaction by which a solution of each of the following salts could be prepared: (a) LiI (b) $CaBr_2$ (c) $Mg(NO_3)_2$.

A solution of each could be prepared by reacting the appropriate metal hydroxide (or oxide) with a solution of the appropriate acid, according to the amounts given by the respective balanced equation:

(a) $LiOH(s) + HI(aq) \rightarrow LiI(aq) + H_2O(\ell)$

(b) $Ca(OH)_2(s) + 2HBr(aq) \rightarrow CaBr_2(aq) + 2H_2O(\ell)$

(c) $Mg(OH)_2(s) + 2HNO_3(aq) \rightarrow Mg(NO_3)_2(aq) + 2H_2O(\ell)$

4.8 Use the solubility rules to predict which of the following substances are soluble and which are insoluble: (a) $CaCO_3$, (b) PbI_2, (c) LiBr, (d) CuO, (e) $MgSO_4$.

(a) **Calcium carbonate** is not among the soluble carbonates; it is **insoluble.**

(b) **Lead iodide** is among the insoluble halides. It is **insoluble.**

(c) **Lithium bromide** is not among the insoluble halides. It is **soluble.**

(d) **Copper oxide** is not among the soluble oxides. It is **insoluble.**

(e) **Magnesium sulfate** is not among the insoluble sulfates. It is **soluble.**

4.9 Use the solubility rules to predict whether a precipitate will form when the following pairs of aqueous solutions are mixed: (a) $MgCl_2$ and NaF, (b) $MgSO_4$ and NaCl, (c) $FeCl_2$ and Na_2S.

(a) The possible salts to precipitate are MgF_2(s) or NaCl(s); the former is **insoluble and forms a precipitate.**

(b) The possible salts to precipitate are $MgCl_2$(s) or Na_2SO_4(s); both are soluble; **no precipitate forms.**

(c) The possible precipitates are FeS(s) or NaCl(s); FeS(s) is **insoluble and forms a precipitate.**

4.10 Identify each of the following reactions as an oxidation-reduction reaction (OR), an acid-base reaction (AB), or a precipitation reaction (P). In each case give reasons for your choice:

(a) $Mg(s)+H_2(g) \rightarrow MgH_2(s)$; (b) $Cu(s)+S(s) \rightarrow CuS(s)$; (c) $NH_3(g)+HBr(g) \rightarrow NH_4Br(s)$;
(d) $CuSO_4(aq)+K_2S(aq) \rightarrow CuS(s) + K_2SO_4(aq)$; (e) $Mg(s)+2HBr(aq) \rightarrow MgBr_2(aq) + H_2(g)$;
(f) $NaH(s)+H_2O(\ell) \rightarrow NaOH(aq) + H_2(g)$; (g) $Ba(OH)_2(aq)+H_2SO_4(aq) \rightarrow BaSO_4(s) + 2H_2O(\ell)$

(a) $Mg(s) + H_2(g) \rightarrow Mg^{2+}(H^-)_2(s)$; **OR** - electrons are transferred from Mg, the oxidizing agent, to H_2, the reducing agent.

(b) $Cu(s) + S(s) \rightarrow Cu^{2+}S^{2-}(s)$; **OR**, - electrons are transferred from Cu, the oxidizing agent, to S, the reducing agent.

(c) $NH_3(g) + HBr(g) \rightarrow NH_4^+Br^-(s)$; **AB**, - protons are transferred from the acid HBr to the base NH_3.

(d) $CuSO_4(aq) + K_2S(aq) \rightarrow CuS(s) + K_2SO_4(aq)$; **P**, - mixing aqueous solutions of $CuSO_4$ and K_2S gives a precipitate of insoluble CuS(s) and soluble K_2SO_4.

(e) $Mg(s) + 2HBr(aq) \rightarrow Mg^{2+}(Br^-)_2(aq) + H_2(g)$; **OR**, - electrons are transferred from Mg, the oxidizing agent, to H^+ ions from HBr, the reducing agent.

(f) $Na^+H^-(s) + H_2O(\ell) \rightarrow Na^+OH^-(aq) + H_2(g)$; **AB**, - protons are transferred from the acid water to H^- ions, the base, - but also **OR**, because H^+ from water is reduced to H_2 and H^- is oxidized to H_2.

(g) $Ba(OH)_2(aq) + H_2SO_4(aq) \rightarrow BaSO_4(s) + 2H_2O(\ell)$; **AB**, - protons are transferred from the acid H_2SO_4 to the base OH^-, - but also **P** because the insoluble salt $BaSO_4$(s) is precipitated from the solution.

CHAPTER 5

5.1 What mass of $CaCl_2$(s) is produced when 0.2500 g of Ca metal is burned completely in Cl_2(g)?

Calcium is in Group 2 (valence 2) and chlorine is in Group VII (valence 1) - the balanced equation for the reaction is: $Ca(s) + Cl_2(g) \rightarrow CaCl_2(s)$

$$\text{mass of } CaCl_2 \text{ produced} = (0.2500 \text{ g Ca})\left(\frac{1 \text{ mol Ca}}{40.08 \text{ g Ca}}\right)\left(\frac{111.0 \text{ g } CaCl_2}{1 \text{ mol } CaCl_2}\right) = \underline{0.6922 \text{ g}}$$

5.2 When iron oxide is heated with aluminum powder, a very vigorous reaction occurs iron oxide is reduced to molten iron (Demonstration 10.6) according to the equation: $Fe_2O_3(s) + 2Al(s) \rightarrow Al_2O_3(s) + 2Fe(\ell)$. A mixture of 30.0 g of aluminum and 100.0 g of Fe_2O_3 is heated: (a) Which is the limiting reactant? (b) How much iron will form? (c) How much of the nonlimiting reactant remains when the reaction is complete?

(a) Molar mass of Fe_2O_3 = 2(55.85)+3(16.00) = 159.7 g mol^{-1}.

Initially: moles of Fe_2O_3 = $(100.0 \text{ g } Fe_2O_3)(\frac{1 \text{ mol } Fe_2O_3}{159.7 \text{ g } Fe_2O_3})$ = <u>0.6262 mol</u>

molesl of Al = $(30.0 \text{ g Al})(\frac{1 \text{ mol Al}}{26.98 \text{ mol Al}})$ = <u>1.11 mol</u>

1 mol Fe_2O_3 reacts with 2 mol Al; **aluminum** is the **limiting reactant** because 1.11 mol Al reacts with ½(1.11) = 0.555 mol Fe_2O_3; i.e., Fe_2O_3 is in excess:

(b) mass Fe produced = $(1.11 \text{ mol Al})(\frac{2 \text{ mol Fe}}{2 \text{ mol Al}})(\frac{55.85 \text{ g Fe}}{1 \text{ mol Fe}})$ = <u>62.0 g Fe</u>

(c) Fe_2O_3 is in excess, and the amount remaining after complete reaction is: (0.626 - 0.555) = **0.071 mol Fe_2O_3**

Excess mass of $Fe_2O_3(s)$ = $(0.071 \text{ mol } Fe_2O_3)(\frac{159.7 \text{ g } Fe_2O_3}{1 \text{ mol } Fe_2O_3})$ = <u>11 g</u>

5.3 In an experiment where chlorine was passed into a solution containing 1.000 g of NaI(aq), 0.250 g of sodium chloride crystals was obtained after the following steps occurred: evaporating the solution, recrystallizing the sodium chloride from water, and drying it in an oven at 120°C. What are (a) the theoretical yield and (b) the percent yield of NaCl(s)?

The reaction in question is $2NaI(aq) + Cl_2(g) \rightarrow 2NaCl(s) + I_2(s)$: Molar masses: NaI, 149.9; NaCl, 58.44 g mol^{-1}

(a) Theoretical yield = $(1.000 \text{ g NaI})(\frac{1 \text{ mol NaI}}{149.9 \text{ g NaI}})(\frac{1 \text{ mol NaCl}}{1 \text{ mol NaI}})(\frac{58.44 \text{ g NaCl}}{1 \text{ mol NaCl}})$ = <u>0.389 86 g NaCl(s)</u>

(b) Percent yield NaCl(s) = $(\frac{0.250 \text{ g NaCl}}{0.3899 \text{ g NaCl}}) \text{x } 100$ = <u>64.1%</u>

5.4 Phosphoric acid has the formula H_3PO_4. What are its molar mass and its theoretical composition in mass percentage?

Molar mass of H_3PO_4 = 3(1.008) + 30.97 + 4(16.00) = **97.99 g mol^{-1}**

Mass % H = $(\frac{3(1.008 \text{ g mol}^{-1})}{97.99 \text{ g mol}^{-1}}) \text{ x } 100\%$ = <u>3.086%</u>

Mass % P = $(\frac{30.97 \text{ g mol}^{-1}}{97.99 \text{ g mol}^{-1}}) \text{ x } 100\%$ = <u>31.61%</u>

Mass % O = $(\frac{4(16.00 \text{ g mol}^{-1})}{97.99 \text{ g mol}^{-1}}) \text{ x } 100\%$ = <u>65.31%</u>

5.5 Phenol, an important product of the chemical industry is used in the production of certain plastics. It contains only carbon, hydrogen, and oxygen. When a sample of phenol of mass 0.8874 g was burned in excess oxygen 2.491 g of CO_2 and 0.510 g of H_2O were obtained. What is the empirical formula of phenol?

Sample contains: $(2.491\ g\ CO_2)(\frac{1\ mol\ CO_2}{44.01\ g\ CO_2})(\frac{1\ mol\ C}{1\ mol\ CO_2})(\frac{12.01\ g\ C}{1\ mol\ C}) = \underline{0.6784\ g\ C}$

$(0.510\ g\ H_2O)(\frac{1\ mol\ H_2O}{18.02\ g\ H_2O})(\frac{2\ mol\ H}{1\ mol\ H_2O})(\frac{1.008\ g\ H}{1\ mol\ H}) = \underline{0.05706\ g\ H}$

and the mass of **oxygen** is found by difference: Mass of O = (0.8874-0.6784-0.0571)g = **0.1519 g**

Now we can use these masses to calculate moles of C, H, and O in the sample, and hence the empirical formula:

	<u>C</u>	H	<u>O</u>
	0.6784 g	0.0571 g	0.1519 g
<u>Moles</u>	$\frac{0.6784\ g\ C}{12.01\ g\ mol^{-1}}$	$\frac{0.0571\ g\ H}{1.008\ g\ mol^{-1}}$	$\frac{0.1519\ g\ O}{16.00\ g\ mol^{-1}}$
=	$\underline{5.65 \times 10^{-2}}$	$\underline{5.66 \times 10^{-2}}$	$\underline{9.49 \times 10^{-3}}$
<u>Atom ratio</u>	$\frac{5.65 \times 10^{-2}}{9.49 \times 10^{-3}}$	$\frac{5.66 \times 10^{-2}}{9.49 \times 10^{-3}}$	$\frac{9.49 \times 10^{-3}}{9.49 \times 10^{-3}}$
=	<u>5.95</u>	<u>5.96</u>	<u>1.00</u>

Hence the **empirical formula** of **phenol** is **C_6H_6O.**

5.6 The molecular mass of glucose is 180 u and its empirical formula is CH_2O. Deduce its molecular formula.

The **empirical formula mass** of CH_2O is [12.01+2(1.008)+16.00]u = **30.03 u**, and the **molecular mass** is **180 u.** Hence:

For the molecular formula $(CH_2O)_n$: $n = \frac{\text{molecular mass}}{\text{empirical formula mass}} = \frac{180\ u}{30.03\ u} = \underline{5.99}$

Thus, the **molecular formula** of glucose is $(CH_2O)_6$, or **$C_6H_{12}O_6$.**

5.7 The gas propane, C_3H_8, burns in oxygen to give carbon dioxide and water. Write the balanced equation for the reaction. How many liters of carbon dioxide are formed at 25°C and 1.00 atm when 5.00 L of propane at 25°C and 1 atm is burned in oxygen?

The balanced equation for the reaction is: $C_3H_8(g) + 5O_2(g) \rightarrow 3CO_2(g) + 4H_2O(g)$

Under constant conditions of temperature and pressure, from Avogadro's law, 1 volume of $C_3H_8(g)$ gives 3 volumes of $CO_2(g)$. Hence, volume of $CO_2(g)$ produced from 5.00 L of propane at 25°C and 1 atm pressure = **15.00 L.**

5.8 What volume of hydrogen at 26°C and 740 mm Hg would be obtained from the reaction of 2.00 g of magnesium with excess hydrochloric acid, HCl(aq)?

The reaction of magnesium with HCl(aq) is: $Mg(s) + 2HCl(aq) \rightarrow MgCl_2(aq) + H_2(g)$

$$\text{moles of } H_2 \text{ produced} = (2.00 \text{ g Mg})\left(\frac{1 \text{ mol Mg}}{24.30 \text{ g Mg}}\right)\left(\frac{1 \text{ mol } H_2}{1 \text{ mol Mg}}\right) = \underline{0.0823 \text{ mol } H_2}$$

$$\text{Using } PV = nRT \; ; \quad V = \frac{nRT}{P}$$

$$\text{Volume } H_2 \text{ produced} = \frac{(0.0823 \text{ mol})(0.0821 \text{ atm L mol}^{-1} \text{ K}^{-1})(303 \text{ K})}{(740 \text{ mm Hg})\left(\frac{1 \text{ atm}}{760 \text{ mm Hg}}\right)} = \underline{2.10 \text{ L}}$$

5.9 Suppose that 5.000 g of $BaCl_2(s)$ are dissolved in water to give 100.0 mL of solution. What are the molar concentrations of $BaCl_2(s)$, $Ba^{2+}(aq)$, and $Cl^-(aq)$?

$$\text{Molar mass } BaCl_2 = [137.3 + 2(35.45)] = \mathbf{208.2 \text{ g mol}^{-1}}$$

$$\text{moles of } BaCl_2 \text{ per } \ell = (5.000 \text{ g})\left(\frac{1 \text{ mol } BaCl_2}{208.2 \text{ g } BaCl_2}\right)\left(\frac{1}{100 \text{ mL}}\right)\left(\frac{1000 \text{ mL}}{1 \text{ L}}\right) = \underline{0.240 \text{ mol L}^{-1}} \quad (0.240 \text{ M})$$

and, since $BaCl_2(s) \rightarrow Ba^{2+}(aq) + 2Cl^-(aq)$, we have:

0.240 M $BaCl_2(aq)$, <u>or</u> 0.240 M $Ba^{2+}(aq)$ and 0.480 M $Cl^-(aq)$

5.10 How many grams of solute are needed to prepare 500 mL of each of the following aqueous solutions? (a) 0.100-M silver nitrate, $AgNO_3(aq)$; (b) 1.00-M sodium bromide, NaBr(aq); (c) 0.200-M barium chloride, $BaCl_2(aq)$.

(a) Molar mass of $AgNO_3(s)$ = [107.9 + 14.01 + 3(16.00)] = **169.9 g mol^{-1}**

$$\text{grams of } AgNO_3(s) = (500 \text{ mL})\left(\frac{1 \text{ L}}{10^3 \text{ mL}}\right)\left(\frac{0.100 \text{ mol } AgNO_3}{1 \text{ L}}\right)\left(\frac{169.9 \text{ g } AgNO_3}{1 \text{ mol } AgNO_3}\right) = \underline{8.50 \text{ g}}$$

(b) Molar mass of NaBr(s) = [22.99 + 79.90] = **102.9 g mol^{-1}**

$$\text{grams of NaBr(s)} = (500 \text{ mL})\left(\frac{1 \text{ L}}{10^3 \text{ mL}}\right)\left(\frac{1.00 \text{ mol NaBr}}{1 \text{ L}}\right)\left(\frac{102.9 \text{ g NaBr}}{1 \text{ mol NaBr}}\right) = \underline{51.5 \text{ g}}$$

(c) Molar mass of $BaCl_2(s)$ = [137.3 + 2(35.45)] = **208.2 g mol^{-1}**

$$\text{grams of } BaCl_2(s) = (500 \text{ mL})\left(\frac{1 \text{ L}}{10^3 \text{ mL}}\right)\left(\frac{0.200 \text{ mol } BaCl_2}{1 \text{ L}}\right)\left(\frac{208.2 \text{ g } BaCl_2}{1 \text{ mol } BaCl_2}\right) = \underline{20.8 \text{ g}}$$

5.11 A 25.00-mL sample of KOH(aq) reacted completely with 38.60 mL of a 0.0500-M solution of hydrobromic acid, HBr(aq). What was the concentration of the KOH(aq) solution?

The acid-base reaction that occurs is:

$$KOH(aq) + HBr(aq) \rightarrow KBr(aq) + H_2O(\ell)$$

Thus:

$$\text{mol KOH} = (25.00\text{ mL})(x\text{ mol L}^{-1}) = \text{mol HBr} = (38.60\text{ mL})(0.0500\text{ mol L}^{-1})$$

$$x = \mathbf{0.0772\ M}$$

5.12 A 35.00-mL sample of $Ba(OH)_2$(aq) reacted completely with 46.25 mL of a 0.0750-M solution of perchloric acid, $HClO_4$(aq). What was the concentration of the $Ba(OH)_2$(aq)?

The reaction that occurs is: $Ba(OH)_2(aq) + 2HClO_4(aq) \rightarrow Ba(ClO_4)_2(aq) + 2H_2O(\ell)$

Thus:

$$\text{mol } HClO_4 = (46.25\text{ mL})(0.0750\text{ mol L}^{-1}) = \underline{3.4688\text{ mmol}}$$

$$\text{mol } Ba(OH)_2 = (25.00\text{ mL})(x\text{ mol L}^{-1}) = \underline{25.00x\text{ mmol}}$$

$$3.4688\text{ mmol } HClO_4 = (25.00x\text{ mmol } Ba(OH)_2)\left(\frac{2\text{ mol } HClO_4}{1\text{ mol } Ba(OH)_2}\right); \quad x = \underline{0.0694\text{ M}}$$

CHAPTER 6

6.1 The colors that make up the visible spectrum range from 400 nm (violet) to 750 nm (red). What is the corresponding range of frequencies?

$$\lambda\nu = c; \quad \nu = \frac{c}{\lambda}$$

$$\nu_{violet} = \frac{c}{\lambda_{violet}} = \frac{3.00 \times 10^8\text{ m s}^{-1}}{(400\text{ nm})\left(\frac{1\text{ m}}{10^9\text{ nm}}\right)} = \underline{7.50 \times 10^{14}\text{ s}^{-1}\text{ (Hz)}}$$

$$\nu_{red} = \frac{c}{\lambda_{red}} = \frac{3.00 \times 10^8\text{ m s}^{-1}}{(750\text{ nm})\left(\frac{1\text{ m}}{10^9\text{ nm}}\right)}) = \underline{4.00 \times 10^{14}\text{ s}^{-1}\text{ (Hz)}}$$

6.2 What is the maximum wavelength of light that will dissociate O_2 molecules into O atoms, given that the dissociation energy of O_2 is 498 kJ·mol^{-1}?

$$E = 498\text{ kJ mol}^{-1} = N_A h\nu = \frac{N_A hc}{\lambda}; \quad \lambda = \frac{N_A hc}{E}$$

$$\lambda = \frac{(6.022 \times 10^{23}\text{ mol}^{-1})(6.63 \times 10^{-34}\text{ J s})(3.00 \times 10^8\text{ m s}^{-1})}{(498\text{ kJ mol}^{-1})\left(\frac{10^3\text{ J}}{1\text{ kJ}}\right)} = \underline{2.41 \text{x} 10^{-7}\text{ m}} \quad (242\text{ nm})$$

6.3 (a) What is the energy change associated with the transition of a hydrogen atom in the n = 5 excited state to the ground state? (b) What are the frequency and the wavelength of the corresponding line in the spectrum? (c) In what region of the spectrum is this line found?

(a) $E_{photon} = (2.18 \times 10^{-18}\ \text{J})(\frac{1}{n_f^2} - \frac{1}{n_i^2}) = (2.18 \times 10^{-18}\ \text{J})(\frac{1}{1^2} - \frac{1}{5^2}) = \underline{2.09_3 \times 10^{-18}\ \text{J}}$

$N_A E_{photon} = (\frac{6.022 \times 10^{23}\ \text{photons}}{1\ \text{mol}})(\frac{2.093 \times 10^{-18}\ \text{J}}{1\ \text{photon}})(\frac{1\ \text{MJ}}{10^6\ \text{J}}) = \underline{1.26\ \text{MJ mol}^{-1}}$

(b) $E_{photon} = h\nu\ ;\quad \nu = \frac{E_{photon}}{h} = \frac{2.093 \times 10^{-18}\ \text{J}}{6.63 \times 10^{-34}\ \text{J s}} = 3.15_7 \times 10^{15}\ \text{s}^{-1}$ $\quad \underline{(3.16 \times 10^{15}\ \text{Hz})}$

$\lambda = \frac{c}{\nu} = \frac{3.00 \times 10^8\ \text{m s}^{-1}}{3.157 \times 10^{15}\ \text{s}^{-1}} = \underline{9.50 \times 10^{-8}\ \text{m}}$ $\quad$ (95.0 nm)

(c) Ultraviolet

6.4 Using only the periodic table, and without reference to Figure 6.18, give the ground state electron configurations of the atoms of the following elements: (a) Li (b) C (c) Ne (d) Al (e) P (f) Cl (g) Ca

---

Element	Group	Period	Atomic No.	Configuration
(a) Li	I	2	3	$1s^2\ 2s^1$
(b) C	IV	2	6	$1s^2\ 2s^2\ 2p^2$
(c) Ne	VIII	2	10	$1s^2\ 2s^2\ 2p^6$
(d) Al	III	3	13	$1s^2\ 2s^2\ 2p^6\ 3s^2\ 3p^1$
(e) P	V	3	15	$1s^2\ 2s^2\ 2p^6\ 3s^2\ 3p^3$
(f) Cl	VII	3	17	$1s^2\ 2s^2\ 2p^6\ 3s^2\ 3p^5$
(g) Ca	II	4	20	$1s^2\ 2s^2\ 2p^6\ 3s^2\ 3p^6\ 4s^2$

6.5 Draw box diagrams for the electron configurations of the atoms in Exercise 6.4.

		1s	2s	2p	3s	3p	4s
(a)	Li	[↑↓]	[↑]				
(b)	C	[↑↓]	[↑↓]	[↑][↑][ ]			
(c)	Ne	[↑↓]	[↑↓]	[↑↓][↑↓][↑↓]			
(d)	Al	[↑↓]	[↑↓]	[↑↓][↑↓][↑↓]	[↑↓]	[↑][ ][ ]	
(e)	P	[↑↓]	[↑↓]	[↑↓][↑↓][↑↓]	[↑↓]	[↑][↑][↑]	
(f)	Cl	[↑↓]	[↑↓]	[↑↓][↑↓][↑↓]	[↑↓]	[↑↓][↑↓][↑]	
(g)	Ca	[↑↓]	[↑↓]	[↑↓][↑↓][↑↓]	[↑↓]	[↑↓][↑↓][↑↓]	[↑↓]

6.6 Consider the methyl cyanide molecule, CH_3CN. (a) Draw the Lewis structure of CH_3CN. (b) Predict the approximate bond angles at each carbon atom. (c) Describe the bonding in terms of hybrid orbitals and the σ—π model.

(a)
```
    H
    |
  H-C-C≡N:
    |
    H
```

(b) C has AX_4 geometry, so the approximate bond angles expected are:

$<HCH = <HCC = 109.5°$

C has AX_2 geometry, so $<CCN = 180°$

(c) In terms of hybrid orbitals and the σ—π model, the C-H bonds in the CH_3 group are described as being formed from sp^3 hybrid orbitals on the carbon atom and hydrogen 1s orbitals. The C≡N triple bond is described as consisting of a σ-bond formed from an sp orbital on its carbon atom and an sp orbital on the N atom, and two π-bonds formed from 2p orbitals on the C and N atoms. The C-C single bond is described as formed from a sp^3 orbital on the C atom of the CH_3 group and an sp orbital on the carbon atom of the CN group. The lone pair on the N atom is described as occupying an sp orbital on the N atom.

CHAPTER 7

7.1 Write balanced equations for the reaction of each of the bases CsOH, NH_3, $Ba(OH)_2$, and MgO with excess sulfuric acid in aqueous solution.

$$2CsOH(s) + H_2SO_4(aq) \rightarrow 2Cs^+(aq) + SO_4^{2-}(aq) + 2H_2O(\ell)$$

$$2NH_3(g) + H_2SO_4(aq) \rightarrow 2NH_4^+(aq) + SO_4^{2-}(aq)$$

$$Ba(OH)_2(aq) + H_2SO_4(aq) \rightarrow BaSO_4(s)^* + 2H_2O(\ell)$$

$$MgO(s) + H_2SO_4(aq) \rightarrow Mg^{2+}(aq) + SO_4^{2-}(aq) + H_2O(\ell)$$

*$BaSO_4$ is insoluble.

7.2 What is the oxidation number of each of the atoms in the following compounds? (a) Na_2O, (b) Al_2O_3, (c) BaH_2, (d) CaS, (e) SF_6.

(a) Na_2O is ionic, $[Na^+]_2O^{2-}$, so the oxidation numbers of the elements are just the charges on the respective ions:

Na, +1; O, -2

(b) Al_2O_3 is ionic, $[Al^{3+}][O^{2-}]_3$, so the oxidation numbers of the elements are just the charges on the respective ions:

Al, +3; O, -2

(c) BaH_2 is ionic, Ba^{2+} $[H^-]_2$, so the oxidation numbers of the elements are just the charges on the respective ions:

Ba, +2; H,-1

(d) CaS is ionic $Ca^{2+}S^{2-}$, so the oxidation numbers of the elements are just the charges on the respective ions:

Ca, +2; S, -2

(e) SF_6 is covalent and the oxidation number of F, the more electronegative element, is -1, so that of sulfur, x, is given by $x + 6(-1) = 0$, i.e., $x = +6$.

S, +6; F, -1

7.3 Assign oxidation numbers to each of the atoms in the following: (a) SO_3^{2-} (b) S_2^{2-} (c) ClO_4^- (d) $Al_2(SO_4)_3$.

(a) SO_3^{2-} The oxidation number of O is -2, so $x + 3(-2) = -2$, i.e., $x = ON(S) = +4$. O, -2; S, +4.

(b) S_2^{2-} The sum of the oxidation numbers of the two S atoms is equal to the charge on the ion, -2, so ON(S) 2[ON(S)] = -2, **S, -1.**

(c) ClO_4^- The oxidation number of O is -2, so x + 4(-2) = -1, i.e., x = ON(Cl) = +7. **O -2; Cl +2**

(d) $Al_2(SO_4)_3$ contains Al^{3+} and SO_4^{2-} ions. The ON(Al) is +3, and the ON(O) in SO_4^{2-} is -2, so x + 4(-2) = -2, i.e., x = ON(S) = +6. **Al +3; S +6; O -2**

7.4 By adding the appropriate half equations derive the equations for the oxidation of $H_2S(aq)$ with $O_2(g)$, and with $Br_2(aq)$.

(a) S(-2) in H_2S is oxidized to S(0) in S(s), and O(0) in O_2 is reduced to O(-2) in H_2O:

$2(H_2S \rightarrow S + 2e^- + 2H^+)$	oxidation
$O_2 + 4e^- \rightarrow 2O^{2-}$	reduction
$2H_2S + O_2 \rightarrow 2S + 2H_2O$	**overall**

(b) S(-2) in H_2S is oxidized to S(0) in S(s), and Br(0) in Br_2 is reduced to Br(-1) in HBr:

$H_2S \rightarrow S + 2e^- + 2H^+$	oxidation
$Br_2 + 2e^- \rightarrow 2Br^-$	reduction
$H_2S + Br_2 \rightarrow S + 2HBr$	**overall**

7.5 Write the formula of each of the following acids, and classify each as strong or weak in aqueous solution: (a) phosphoric acid (b) chloric acid (c) hypochlorous acid (d) silicic acid (e) boric acid, $B(OH)_3$.

We first write the formulas of the acids in the form $XO_m(OH)_n$; acids with m = 0, or 1, are weak, those with m = 2, or greater are strong. Hence:

(a)	phosphoric acid	$PO(OH)_3$	m = 1	**weak**
(b)	chloric acid	ClO_2OH	m = 2	**strong**
(c)	hypochlorous acid	ClOH	m = 0	**weak**
(d)	silicic acid	$Si(OH)_4$	m = 0	**weak**
(e)	boric acid	$B(OH)_3$	m = 0	**weak**

7.6 Write a complete balanced equation for the oxidation of $SO_3^{2-}(aq)$ to $SO_4^{2-}(aq)$ by hypochlorous acid.

S(+4) in SO_3^{2-} is oxidized to S(+6) in SO_4^{2-}, and Cl(+1) in HOCl is reduced to Cl(-1) in Cl^-:

$SO_3^{2-} + H_2O \rightarrow SO_4^{2-} + 2e^- + 2H^+$	oxidation
$HOCl + 2e^- + H^+ \rightarrow Cl^- + H_2O$	reduction
$SO_3^{2-} + HOCl \rightarrow SO_4^{2-} + HCl$	

7.7 Assign formal charges to the atoms in (a) the ammonium ion, NH_4^+; (b) the hydroxide ion, OH^-.

Formal charge = (core charge) - (number of unshared electrons) - ½(number of shared electrons)

(a) N (Group V) has a core charge of +5, and forms four electron pair bonds to H atoms:

Formal charge = +5 - ½(8) = **+1**

H (Group I) has a core charge of +1, and forms one electron pair bond to the N atom:

Formal charge = +1 - ½(2) = **0**

```
   H
   |
H–N⁺–H
   |
   H
```

(b) O (Group VI) has a core charge of +6, has three unshared pairs, and forms one electron pair bond to the H atom:

Formal charge = +6 - (6) - ½(2) = **-1**

H (Group I) has a core charge of +1, and forms one electron pair bond to the N atom:

Formal charge = +1 - ½(2) = **0**

```
  ••
⁻:O–H
  ••
```

7.8 Draw Lewis structures for (a) the sulfite ion, SO_3^{2-}; (b) the phosphate ion, PO_4^{3-}; (c) the phosphorus pentachloride molecule. Assign formal charges where appropriate.

(a) (1) We first connect the atoms by single bonds, with the single unique S atom as the central atom connected to the three O atoms. (2) We count the number of valence electrons [S(Group VI) + 3 O(Group VI) + 2e⁻] = **6 + 3(6) + 2 = 26 electrons.** (3) We subtract the **6 electrons** needed to form three S–O bonds, which leaves **20 electrons** (10 pairs) to be used in completing *octets* around the atoms. Completing the octets around the O atoms first uses up **nine pairs**, and the tenth pair completes the octet on the S atom. Then we assign formal charges to each atom, to give:

```
  ••  ••  ••
⁻:O–S⁺–O:⁻
  ••  |  ••
    :O:⁻
     ••
```

(5) Since the central atom is in Period 3, an unshared pair is then delocalized from one of the O atoms to remove the +1 formal charge on the S atom, and forms one sulfur-oxygen double bond, giving finally:

```
 ••  ••  ••
:O=S–O:⁻
     |  ••
   :O:⁻
    ••
```

(b) PO_4^{3-} has P(5) + 4 O(6) + 3e⁻ = **32 electrons**, and following rules (1) to (3) gives:

```
     ••
    :O:⁻
  ••  |  ••
⁻:O–P⁺–O:⁻
  ••  |  ••
    :O:⁻
     ••
```

but since phosphorus is in Period 3, its formal +1 charge is then removed by delocalizing a lone pair from one of the oxygen atoms, to form a double bond:

```
     ••
    :O:⁻
  ••  |  ••
⁻:O–P=O:
  ••  |
    :O:⁻
     ••
```

(c) **PCl_5 has P(5) + 5 Cl(7) = 40 electrons,** and following rules (1) to (3) gives

$$\begin{array}{c} :\ddot{Cl}: \\ | \\ :\ddot{Cl}-P(\ddot{Cl}:)_2 \\ | \\ :\ddot{Cl}: \end{array}$$

CHAPTER 8

8.1 Predict the products and write the balanced equation for each of the following reactions. Which are redox reactions, and which are acid-base reactions?

(a) $MgCO_3(s) \xrightarrow{Heat}$ (b) $PbO(s) + CO(g) \xrightarrow{Heat}$

(c) $MgO(s) + C(s) \xrightarrow{Heat}$ (d) $MgC_2(s) + H_2O(l) \longrightarrow$

(a) On heating, the carbonate of magnesium in Group II above calcium, $MgCO_3(s)$, like $CaCO_3(s)$, is expected to decompose to the oxide, with the evolution of $CO_2(g)$:

$$CaCO_3(s) \rightarrow CaO(s) + CO_2(g)$$

The reaction involves no change in the *oxidation states* of any of the reactant atoms, nor does it involve the transfer of protons; it is *neither* an oxidation-reduction reaction *nor* an acid-base reaction.

(b) CO(g) is readily oxidized to $CO_2(g)$, so on heating will reduce PbO(s) to lead metal, Pb(s):

$$PbO(s) + CO(g) \rightarrow Pb(\ell) + CO_2(g)$$

Pb(ON +2) in PbO is reduced to Pb(ON 0) in Pb, and C (ON +2) in CO is oxidized to C(ON +4) in CO_2; this is an **oxidation-reduction** reaction.

(c) Carbon is a good reducing agent for many oxides, but as we saw for CaO(s), oxides of Group I and Group II are difficult to reduce; rather they react with carbon to give carbides:

$$MgO(s) + 3C(s) \rightarrow MgC_2(s) + CO(g)$$

The oxidation states of Mg and O remain unchanged, but C (ON 0) in C(s) is oxidized to C(ON +2) in CO and reduced to C (ON -1) in C_2^{2-}. This is an **oxidation-reduction** reaction.

(d) Magnesium carbide, $MgC_2(s)$ contains the carbide ion, $^-:C\equiv C:^-$, a strong base that is protonated by water to give $H\text{-}C\equiv C\text{-}H$, acetylene:

$$C_2^{2-} + 2H^+ \rightarrow C_2H_2$$

or overall:

$$MgC_2(s) + 2H_2O(\ell) \rightarrow Mg(OH)_2(s) + C_2H_2(g)$$

There is no change in the oxidation states of any of the atoms; because this is a proton transfer reaction, it is an **acid-base** reaction.

8.2 Which of the following bases are **strong** bases, which are **weak** bases, and which have **no basic properties** in water? (a) CN^-, (b) CO_3^{2-}, (c) Cl^-, (d) HCO_3^-, (e) C_2^{2-}, (f) OH^-, (g) F^-.

The acidity and basicity of **conjugate acid-base pairs**, e.g., an acid HA and its conjugate base A^-, or a base B and its conjugate acid BH^+, are related; the stronger the acid the weaker its conjugate base, and vice-versa. Thus, a strong acid HA is strong in water because its conjugate base A^- is too weak a base to be protonated by water, (so the ionization of HA goes essentially to completion). In contrast, a weak acid HA is weak because its conjugate base A^- behaves as a weak base in water, so that both the forward reaction where HA ionizes as an acid, and the reverse reaction of A^- as a base, are significant. We have learned the handful of common acids that are strong, so any acid not on this list is a weak acid and has a conjugate base that is a weak base.

(a) HCN(aq) is a weak acid, so its conjugate base, **CN^-(aq),** is a **weak base** in water.

(b) Hydrogen carbonate ion, HCO_3^-(aq) is a weak acid, so its conjugate base, **CO_3^{2-}(aq),** is a **weak base** in water.

(c) HCl(aq) is a strong acid, so its conjugate base, **Cl^-(aq),** has **no basic properties** in water.

(d) Acetylene is a very weak acid, so the carbide ion, **C_2^{2-}(aq),** behaves as a moderately **strong base. Its reaction with water goes to completion** since the HC≡CH formed bubbles off because it is a not very soluble gas.

(e) Water, H_2O, is a very weak acid, so its conjugate base, **OH^-(aq),** is a **strong base**; it is the strongest base that can exist in water.

(g) HF(aq) is a weak acid, so its conjugate base, **F^-(aq),** is a **weak base** in water.

8.3 Draw the Lewis structures and give the names of three molecules or ions that are isoelectronic with the cyanide ion.

The cyanide ion, CN^- has the Lewis structure $^-$:C≡N: and any species that is **isoelectronic** with CN^- will have the same number of valence electrons, i.e., **ten** ; thus, any diatomic molecule or ion made up two atoms or monatomic ions with a total of ten valence electrons will do, for example:

(1) :N + N: → :N≡N: — **nitrogen**

(2) C + O: → $^-$:C≡O:$^+$ — **carbon monoxide**

(3) :C$^-$ + $^-$C: → $^-$:C≡C:$^-$ — **carbide ion**

(4) :N + O:$^+$ → :N≡O:$^+$ — **nitrosonium ion**

8.4 What is the IUPAC name of the following alkane?

```
   CH2–CH3                  CH3
   |                        |
CH3–CH–CH2–CH2–CH2–CH–CH3
```

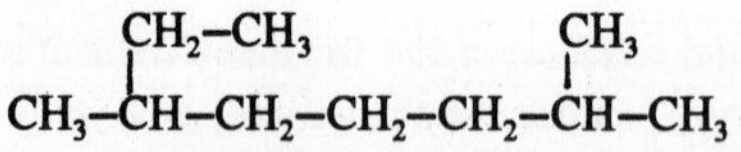

The longest continuous chain of C atoms contains **eight** atoms, so this is the molecular formula of a substituted **octane:**

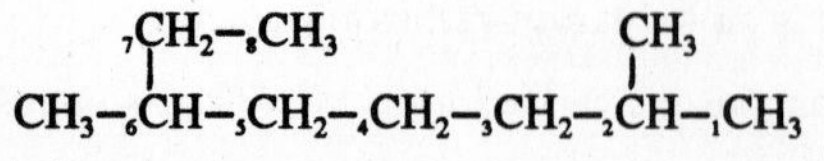

```
    7CH2–8CH3                    CH3
    |                            |
CH3–6CH–5CH2–4CH2–3CH2–2CH–1CH3
```

with $-CH_3$, methyl, substituents at positions 2 and 6, so that the IUPAC name is **2,6-dimethyloctane.**

8.5 Draw the structure of the following alkanes: (a) 3-methylhexane, (b) 3,4-dimethyloctane (c) 2,2,4-trimethylpentane.

```
(a) CH3—CH2—CH—CH2—CH2—CH3
            |
            CH3

(b) CH3—CH2—CH—CH—CH2—CH2—CH2—CH3
            |   |
            CH3 CH3

          CH3
          |
(c) CH3-C-CH2-CH-CH3
          |     |
          CH3   CH3
```

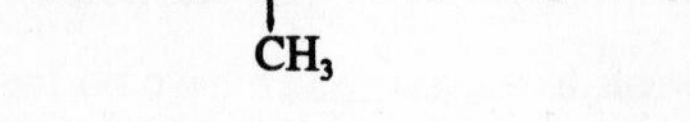

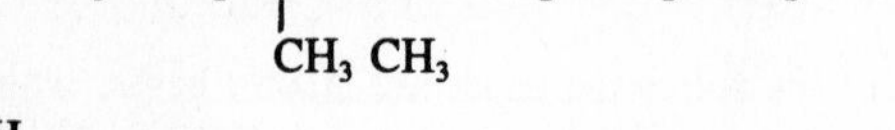

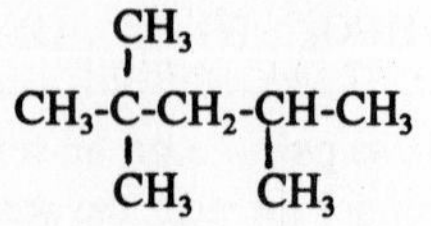

8.6 Draw the structures of (a) *cis*-3-hexene, (b) *trans*-3-hexene (c) 2-butyne.

(a) $CH_3{-}CH_2(H)C{=}C(H)CH_2{-}CH_3$ — *cis*-3-hexene

(b) $CH_3{-}CH_2(H)C{=}C(H)CH_2{-}CH_3$ — *trans*-3-hexene

(c) $H_3C{-}C{\equiv}C{-}CH_3$ — 2-butyne

8.7 Give the IUPAC names for the compounds that result from the addition of Cl_2 to (a) propene, and (b) propyne.

(a) $H_3C{-}CH{=}CH_2 + Cl{-}Cl \rightarrow H_3C{-}CH(Cl){-}CH_2Cl$ 1,2-dichloropropane
propene

(b) $H_3C{-}C{\equiv}C{-}H + Cl{-}Cl \rightarrow$ $(H_3C)(Cl)C{=}C(H)(Cl)$ and $(H_3C)(Cl)C{=}C(Cl)(H)$
propyne

cis-1,2-dichloropropene *trans*-1,2-dichloropropene

Chlorine will then add to *cis*-1,2-dichloropropene and *trans*-1,2-dichloropropene, to give in *each* case the same product, 1,1,2,2-tetrachloropropane:

$(H_3C)(Cl)C{=}C(H)(Cl)$ and $(H_3C)(Cl)C{=}C(Cl)(H)$ $\xrightarrow{Cl_2}$ $H_3C{-}CCl_2{-}CCl_2{-}H$

cis-1,2-dichloropropene *trans*-1,2-dichloropropene 1,1,2,2-tetrachloropropane

8.8 The sulfate ion has a tetrahedral structure in which all four SO bonds have the same length (149 pm). Draw six resonance structures for the sulfate ion, and determine the bond order and the charge on each oxygen atom.

First we write the basic Lewis structure, which gives a structure with two single S-O bonds and two double S=O bonds. The six resonance structures are the different possible arrangements of these pairs of S-O and S=O bonds, as follow:

[six resonance structures of the sulfate ion, each with two S=O and two S–O⁻ bonds, connected by ↔]

Taking any particular bond, say the one at the top:

$$\text{sulfur-oxygen bond order} = \frac{\text{total bond order}}{\text{number of bonds}} = \frac{2+2+2+1+1+1}{6} = 1\frac{1}{2}$$

$$\text{charge per O atom} = \frac{\text{total charge}}{\text{number bonds}} = \frac{0+0+0-1-1-1}{6} = -\frac{1}{2}$$

CHAPTER 9

9.1 A 25-mL sample of a 0.10-M aqueous H_2SO_4 solution and 50 mL of an aqueous 0.10-M KOH solution were mixed in a coffee-cup calorimeter. The temperature of the solution rose from 21.20°C to 22.10°C as a result of the reaction that occurred upon mixing the solutions. Calculate the enthalpy change for the reaction: 2KOH(aq) + H_2SO_4(aq) → K_2SO_4(aq) + $2H_2O(\ell)$. Compare the value you obtain with that obtained from the neutralization of HCl(aq) in Example 9.1, and explain any difference.

The total volume of solution is 75 mL and, assuming a density of 1.00 g·mL^{-1}, the mass of solution is 75 g. The temperature rise was (22.10-21.20) = 0.90°C = 0.90 K, and the specific heat capacity of water is 4.18 J·K^{-1}·g^{-1}, so

$$\text{Heat released} = (75\ g)(4.18\ J\cdot K^{-1}\cdot g^{-1})(0.90\ K) = 282\ J$$

In the reaction (25 mL)(0.10 mol·L^{-1}) = 2.5 mmol H_2SO_4 reacted with (50 mL)(0.10 mol·L^{-1}) = 5.0 mmol KOH, so, from the balanced equation, the reaction was complete, and for the reaction of 1 mol H_2SO_4(aq):

$$\text{heat released} = (282\ J)\left(\frac{1\ mol\ H_2SO_4}{(2.5\ mmol\ H_2SO_4)\left(\frac{1\ mol}{10^3\ mmol}\right)}\right)\left(\frac{1\ kJ}{10^3\ J}\right) = \underline{1.1\text{x}10^2\ kJ}\ \text{per mol}\ H_2SO_4$$

The enthalpy of reaction of -1.1x10^2 kJ·mol^{-1} is to only two significant figures but is close to **twice** the enthalpy of neutralization of -54 kJ·mol^{-1} for the strong acid HCl(aq) obtained in Example 9.1, which is close to **108 kJ mol^{-1}** anticipated for the neutralization of a **strong diprotic acid**, such as H_2SO_4, because for a strong monoprotic acid, such as HCl(aq), the reaction is

$$H_3O^+(aq) + OH^-(aq) \rightarrow 2H_2O(\ell)$$

while for a strong diprotic acid, such as H_2SO_4(aq), it is

$$2H_3O^+(aq) + 2OH^-(aq) \rightarrow 4H_2O(\ell)$$

9.2 When 1.3 g of butane, C_4H_{10}, was burned in oxygen in a flame calorimeter containing 1800 g of water, the temperature of the water rose from 20.2°C to 28.2°C. What is the enthalpy change for the combustion of 1 mol of butane?

The temperature rise was (28.2-20.2) = 8.0°C = 8.0 K, and the specific heat capacity of water is 4.18 J·K^{-1}·g^{-1}, so:

$$\text{Heat change} = -(1800\ g)(4.18\ J\cdot K^{-1}\cdot g^{-1})(8.0\ K) = 6.02\text{x}10^4\ J$$

The molar mass of C_4H_{10} is **58.12 g mol^{-1}**, so for 1 mol of butane:

$$\text{heat released} = \left(\frac{(6.02\ \text{x}\ 10^4\ J)\left(\frac{1\ kJ}{10^3\ J}\right)}{(1.3\ g\ C_4H_{10})}\right)\left(\frac{58.12\ g\ C_4H_{10}}{1\ mol\ C_4H_{10}}\right) = \underline{2.7\ \text{x}\ 10^3\ kJ\ mol^{-1}}$$

Thus, the **reaction enthalpy** is **ΔH = -2.7 x 10^3 kJ·mol^{-1}**

9.3 Given the standard enthalpy changes for the two reactions: 2P(s) + $3Cl_2$(g) → $2PCl_3(\ell)$ ΔH° = -574 kJ; $PCl_3(\ell)$ + Cl_2(g) → PCl_5(s) ΔH° = -87.9 kJ; find the standard enthalpy change for the reaction 2P(s) + $5Cl_2$(g) → $2PCl_5(\ell)$

Doubling the second equation and adding it to the first equation gives:

$2P(s) + 3Cl_2(g) \rightarrow 2PCl_3(\ell)$	$\Delta H°$	$= -574$ kJ
$2(PCl_3(\ell) + Cl_2(g) \rightarrow PCl_5(s))$	$\Delta H° = 2(-87.9)$	$= -176$ kJ
$2P(s) + 5Cl_2(g) \rightarrow 2PCl_5(\ell)$	$\Delta H° = -(574\text{-}176)$	$=$ **-750 kJ**

9.4 Calculate the standard enthalpy change, $\Delta H°$, for the reaction, $2SO_2(g) + O_2(g) + 2H_2O(g) \rightarrow 2H_2SO_4(\ell)$, from the standard enthalpies of formation in Table 9.1. How much heat would be liberated if 5.20 g of SO_2 was converted to H_2SO_4?

$$\Delta H° = \Sigma\, n_p(\Delta H_f°)_p - \Sigma\, n_r(\Delta H_f°)_r$$

$$\Delta H° = [2\Delta H_f°(H_2SO_4,\ell)] - [2\Delta H_f°(SO_2,g) + \Delta H_f°(O_2,g) + 2\Delta H_f°(H_2O,g)]$$

$$= [2(-814.0)] - [2(-296.8)+0+2(-241.8)] = \mathbf{-550.8\ kJ}$$

Molar mass SO_2 = [32.07+2(16.00)] = **64.07 g mol^{-1}**, so amount of heat liberated from the conversion of 5.20 g SO_2 to $H_2SO_4(\ell)$ is given by:

$$\text{heat liberated} = (5.20\ g\ SO_2)\left(\frac{1\ mol\ SO_2}{64.07\ g\ SO_2}\right)\left(\frac{2\ mol\ H_2SO_4}{2\ mol\ SO_2}\right)\left(\frac{550.8\ kJ}{2\ mol\ H_2SO_4}\right) = \underline{44.7\ kJ}$$

9.5 From the enthalpy of formation of ethene, C_2H_4, given in Table 9.1, and the bond energy of the C-H bond in CH_4 (416 kJ · mol^{-1}), calculate the C=C bond energy.

From Table 9.1 $\Delta H_f°(C_2H_4,g) = +52.3$ kJ·mol^{-1}, and for the dissociation of $C_2H_4(g)$ into gaseous atoms:

$$H_2C{=}CH_2 \rightarrow 2C(g,atomic) + 4H(g,atomic)$$

$$\Delta H° = \Sigma\, n_p(\Delta H_f°)_p - \Sigma\, n_r(\Delta H_f°)_r = [2\Delta H_f°(C,g) + 4\Delta H_f°(H,g)] - [\Delta H_f°(C_2H_4,g)]$$

$$= [2(716.7) + 4(218.0)] - [+52.3] = 2253\ kJ$$

and we can now equate this value with the bond energies of all the bonds broken:

$\Delta H° = 2253$ kJ = 4BE(C-H) + BE(C=C) = 4(416 kJ) + BE(C=C); **BE(C=C) = 589 kJ·mol^{-1}**, which is to be compared with BE(C=C) = 619 kJ·mol^{-1} given in Table 9.2.

9.6 From the bond energies in Table 9.2, estimate the standard reaction enthalpy for the combustion of 1 mol of ethane. Compare this value with the value calculated from standard enthalpies of formation (Table 9.1).

$$C_2H_6(g) + 3\tfrac{1}{2}O_2(g) \rightarrow 2CO_2(g) + 3H_2O(g)$$

$$H_3C{-}CH_3 + 3\tfrac{1}{2}\ O{=}O \rightarrow 2\ O{=}C{=}O + 3\ H{-}O{-}H$$

<u>From bond energies</u>: $\Delta H°$ = Σ BE(bonds broken) - Σ BE(bonds formed) = [6BE(C-H) + 3½BE(O=O)] - [4BE(C=O) + 6BE(O-H)] = [6(413) + 3½(494)] - [4(707) + 6(463)] = **-1399 kJ·mol^{-1}**

<u>From standard enthalpies of formation</u>: $\Delta H° = \Sigma\, n_p(\Delta H_f°)_p - \Sigma\, n_r(\Delta H_f°)_r = [2\Delta H_f°(CO_2,g) + 3\Delta H_f°(H_2O,g)] - [\Delta H_f°(C_2H_6,g) + 3\tfrac{1}{2}\Delta H_f°(O_2,g)]$ = [2(-393.5) + 3(-241.8)] - [-84.7 + 3½(0)] = **-1428 kJ·mol^{-1}**

9.7 Use the data in Table 9.3 to find the entropy change for the reduction of iron(III) oxide to iron, using carbon monoxide as the reducing agent: $Fe_2O_3(s) + 3CO(g) \rightarrow 2Fe(s) + 3CO_2(g)$.

$$\Delta S° = \Sigma\, n_p(S°)_p - \Sigma\, n_r(S°)_r$$

$= [2S°(Fe,s)+3S°(CO_2,g)] - [S°(Fe_2O_3,s)+3S°(CO,g)] = [2(27.3)+3(213.7)] - [87.4+3(197.6)] =$ **15.5 kJ·mol⁻¹**

9.8 Predict the sign of the entropy change for each of the following reactions: (a) $NH_3(g) + HCl(g) \rightarrow NH_4(s)$; (b) $BaO(s) + CO_2(g) \rightarrow BaCO_3(s)$; (c) $CH_4(g) + 2O_2(g) \rightarrow CO_2(g) + 2H_2O(\ell)$

(a) 2 mol gas → 1 mol solid, so we expect a decrease in entropy in this reaction; **ΔS is negative.**

(b) 1 mol solid + 1 mol gas → 1 mol solid, so we expect a decrease in entropy in this reaction; **ΔS is negative.**

(c) 3 mol gas → 1 mol gas + 2 mol liquid, so we expect a decrease in the entropy in this reaction; **ΔS is negative.**

9.9 Calculate the standard reaction Gibbs free energy for the following reactions, and deduce whether they are spontaneous or not standard conditions: (a) $CaCO_3(s) \rightarrow CaO(s) + CO_2(g)$; (b) $CH_3OH(\ell) + 1\frac{1}{2}O_2(g) \rightarrow CO_2(g) + 2H_2O(\ell)$

$$\Delta G° = \Sigma\, n_p(\Delta G_f°)_p - \Sigma\, n_r(\Delta G_f°)_r$$

(a) $\Delta G° = [\Delta G_f°(CaO,s) + \Delta G_f°(CO_2,g)] - [\Delta G_f°(CaCO_3,s)] = [-603.5 + (-394.4)] - [-1128.8] =$ **+130.9 kJ**

ΔG° is **positive**, so the reaction is **not spontaneous** under standard conditions.

(b) $\Delta G° = [\Delta G_f°(CO_2,g) + 2\Delta G_f°(H_2O,\ell)] - [\Delta G_f°(CH_3OH,\ell) + 1\frac{1}{2}\Delta G_f°(O_2,g)] = [-394.4 + 2(-237.2)] - [-166.4 + 1\frac{1}{2}(0)] =$ **-702.4 kJ;**

ΔG° is **negative,** so the reaction is **spontaneous** under standard conditions.

CHAPTER 10

10.1 Starting with the element magnesium. explain how you would prepare MgO(s), convert it to $MgCl_2(aq)$, and then to $Mg(OH)_2(s)$. Classify the reactions you use as redox, acid-base, or precipitation.

---

(1) **Magnesium oxide**, $Mg^{2+}O^{2-}(s)$, results from burning magnesium, Mg, in oxygen or air. In this reaction, electrons are transferred from Mg, $Mg \rightarrow Mg^{2+} + 2e^-$, to O_2, $O_2 + 4e^- \rightarrow 2O^{2-}$, so this is a **redox** reaction:

$$2Mg(s) + O_2(g) \rightarrow 2MgO(s)$$

(2) **Aqueous magnesium chloride**, $Mg^{2+}(Cl^-)_2(aq)$, results from reacting MgO(s) with HCl(aq). In this reaction, $H_3O^+(aq)$ ions from the acid transfer protons to O^{2-} ions from MgO(s), $O^{2-} + 2H_3O^+ \rightarrow 3H_2O$, so this is an **acid-base** reaction:

$$MgO(s) + 2HCl(aq) \rightarrow MgCl_2(aq) + H_2O(\ell)$$

(3) **Magnesium hydroxide**, $Mg^{2+}(OH^-)_2(s)$, is insoluble and results from mixing excess of an aqueous solution of a soluble alkali metal or alkaline metal hydroxide, such as NaOH(aq), to the $MgCl_2(aq)$ solution, when the reaction $Mg^{2+}(aq) + 2OH^-(aq) \rightarrow Mg(OH)_2(s)$ takes place, which is a **precipitation** reaction, so overall:

$$MgCl_2(aq) + 2OH^-(aq) \rightarrow Mg(OH)_2(s) + 2Cl^-(aq)$$

10.2 Write balanced equations for the preparation of the following salts from aqueous solutions of the appropriate acid and base: (a) Li_2SO_4 (b) Na_2HPO_4 (c) $BaCl_2$

(a) **Li_2SO_4** is composed of Li^+ and SO_4^{2-} ions; an appropriate base is LiOH and the acid is H_2SO_4:

$$2LiOH(aq) + H_2SO_4(aq) \rightarrow Li_2SO_4(aq) + 2H_2O(\ell)$$

(b) **Na_2HPO_4** is composed of Na^+ and HPO_4^{2-} ions; an appropriate base is NaOH and the acid is H_3PO_4:

$$2NaOH(aq) + H_3PO_4(aq) \rightarrow Na_2HPO_4(aq) + 2H_2O(\ell)$$

(c) **$BaCl_2$** is composed of Ba^{2+} and Cl^- ions; an appropriate base is $Ba(OH)_2$ and the acid is HCl:

$$Ba(OH)_2(aq) + 2HCl(aq) \rightarrow BaCl_2(aq)(aq) + 2H_2O(\ell)$$

10.3 Write balanced equations for the formation of the following salts from an acidic oxide and a basic oxide: (a) $MgCO_3$, (b) $CaSO_4$, (c)Na_2SO_3.

The **anion** of a salt results from reacting O^{2-} ion (a **Lewis base**) with a nonmetal oxide (a **Lewis acid**) with the nonmetal in the same oxidation state as the anion. The oxide involved contains the same cation as the salt:

(a) **$MgCO_3$** contains Mg^{2+} and CO_3^{2-} ions, so the anion results from the reaction $CO_2 + O^{2-} \rightarrow CO_3^{2-}$, and the overall reaction is: $CO_2(g) + MgO(s) \rightarrow MgCO_3(s)$

(b) **$CaSO_4$** contains Ca^{2+} and SO_4^{2-} ions, so the anion results from the reaction $SO_3 + O^{2-} \rightarrow SO_4^{2-}$, and the overall reaction is: $SO_3(g) + CaO(s) \rightarrow CaSO_4(s)$

(c) **Na_2SO_3** contains Na^+ and SO_3^{2-} ions, so the anion results from the reaction $SO_2 + O^{2-} \rightarrow SO_3^{2-}$, and the overall reaction is: $SO_2(g) + Na_2O(s) \rightarrow Na_2SO_3(s)$

10.4 Complete and balance the following equations. In each case, state what type of reaction is occurring: (a) $Al(s) \rightarrow AlBr_3(aq)$; (b) $Al(OH)_3(s) \rightarrow Al_2O_3(s)$; (c) $Al(OH)_3(s) \rightarrow Al_2(SO_4)_3(aq)$.

(a) **$AlBr_3(aq)$** results from reacting aluminum metal with HBr(aq), in a reaction where Al is oxidized to Al^{3+} by H_3O^+ ions from the aqueous acid: $2Al + 6H_3O^+ \rightarrow 2Al^{3+} + [6H_3O] \rightarrow 2Al^{3+} + 6H_2O + 3H_2$. Thus, the reaction is a **redox** reaction, and the overall equation for the reaction is most simply written as:

$$2Al(s) + 6HBr(aq) \rightarrow 2AlBr_3(aq) + 3H_2(g)$$

(b) **$Al_2O_3(s)$** results from *strongly heating* $Al(OH)_3(s)$, when water is driven off as $H_2O(g)$, so this is a **decomposition** reaction:

$$2Al(OH)_3(s) \rightarrow Al_2O_3(s) + 3H_2O(g)$$

(c) **$Al_2(SO_4)_3(aq)$** results from dissolving $Al(OH)_3(s)$ in $H_2SO_4(aq)$. $Al^{3+}(OH^-)_3(s)$ is insoluble in water but dissolves in aqueous acid as a result of the **acid-base** reaction, $OH^- + H_3O^+ \rightarrow 2H_2O$, and the overall reaction is:

$$2Al(OH)_3(s) + 3H_2SO_4(aq) \rightarrow Al_2(SO_4)_3(aq) + 6H_2O(\ell)$$

10.5 Write the balanced equations for the reaction of (a) iron with dilute $H_2SO_4(aq)$, and (b) iron with hot concentrated $H_2SO_4(aq)$ to give $Fe_2(SO_4)_3(aq)$ and $SO_2(g)$.

(a) Fe(s) is oxidized to $Fe^{2+}(aq)$ by $H_3O^+(aq)$ from the acid, which is reduced to $H_2(g)$:

$Fe \rightarrow Fe^{2+} + 2e^-$	oxidation
$2H^+ + 2e^- \rightarrow H_2$	reduction
$Fe + 2H^+ \rightarrow Fe^{2+} + H_2$	overall

and adding SO_4^{2-} to each side of this equation gives:

$$Fe(s) + H_2SO_4(aq) \rightarrow Fe^{2+}(aq) + SO_4^{2-}(aq) + H_2(g)$$

(b) Fe(s) (ON 0) is oxidized to Fe^{3+}(aq) (ON +3), and S(+6) in H_2SO_4(conc) is reduced to S(+4) in SO_2(g):

$$2[Fe \rightarrow Fe^{3+} + 3e^-] \qquad \text{oxidation}$$

$$3[H_2SO_4 + 2e^- + 2H^+ \rightarrow SO_2 + 2H_2O] \qquad \text{reduction}$$

$$2Fe + 3H_2SO_4 + 6H^+ \rightarrow 2Fe^{3+} + 3SO_2 + 6H_2O \qquad \text{overall}$$

and, adding $3SO_4^{2-}$ to each side of this equation gives:

$$2Fe(s) + 6H_2SO_4(conc) \rightarrow Fe_2(SO_4)_3(aq) + 3SO_2(g) + 6H_2O(l)$$

10.6 Write the balanced equations for the reactions needed to carry out the following transformations:

$$Cu(s) \rightarrow CuO(s) \rightarrow CuSO_4(aq) \rightarrow CuS(s) \rightarrow CuCl_2(aq)$$

in each case, state what type of reaction is occurring.

---

(1) On strongly heating in O_2(g) or air, Cu(s) is oxidized to Cu^{2+} ions, $Cu \rightarrow Cu^{2+} + 2e^-$, and O_2(g) is reduced to O_2_ ions, $O_2 + 4e^- \rightarrow 2O^{2-}$, so this is a **redox** reaction, and the overall balanced equation is:

$$2Cu(s) + O_2(g) \rightarrow 2CuO(s)$$

(2) Insoluble $Cu^{2+}O^{2-}$(s) dissolves in H_2SO_4(aq) on warming due to the **acid-base** reaction, $2H_3O^+ + O^{2-} \rightarrow 3H_2O$, so the balanced equation for this reaction is:

$$CuO(s) + H_2SO_4(aq) \rightarrow CuSO_4(aq) + H_2O(\ell)$$

(3) Insoluble CuS(s) is precipitated when H_2S(g) is bubbled through a solution of $CuSO_4$(aq), as a result of the **precipitation** reaction, $Cu^{2+}(aq) + S^{2-}(aq) \rightarrow CuS(s)$, and the overall balanced equation is:

$$CuSO_4(aq) + H_2S(g) \rightarrow CuS(s) + H_2SO_4(aq)$$

(**or**, alternatively, CuS(s) could be precipitated by addition of a solution of a soluble sulfide, such as Na_2S(aq), to the $CuSO_4$(aq) solution).

(4) CuS(s) dissolves in HCl(aq) to give $CuCl_2$(s) in a reaction where H_2S(g) is evolved. This is an **acid-base** reaction where the reaction, $S^{2-} + 2H_3O^+ \rightarrow H_2S(g) + 2H_2O$, is driven to the right by the loss of H_2S(g) from the solution, and the overall balanced equation is:

$$CuS(s) + 2HCl(aq) \rightarrow CuCl_2(aq) + H_2S(g)$$

10.7 Write an equation for the half-reaction in which the $Cr_2O_7^{2-}$ ion is reduced to Cr^{3+}. Then write the balanced equation for the oxidation of Fe^{2+} to Fe^{3+} by $Cr_2O_7^{2-}$.

---

(1) Cr (ON +6) in $Cr_2O_7^{2-}$(aq) is reduced to Cr (ON +3) in Cr^{3+}, so 3e$^-$ are required for each Cr atom in $Cr_2O_7^{2-}$:

$$Cr_2O_7^{2-}(aq) + 6e^- + 14H^+(aq) \rightarrow 2Cr^{3+}(aq) + 7H_2O(\ell)$$

(2) Fe^{2+} (ON +2) is oxidized to Fe^{3+} (ON + 3):

$$6[Fe^{2+}(aq) \rightarrow Fe^{3+}(aq) + e^-] \qquad \text{oxidation}$$

$$Cr_2O_7^{2-}(aq) + 6e^- + 14H^+(aq) \rightarrow 2Cr^{3+}(aq) + 7H_2O(\ell) \qquad \text{reduction}$$

$$6Fe^{2+}(aq) + Cr_2O_7^{2-}(aq) + 14H^+(aq) \rightarrow 6Fe^{3+}(aq) + 2Cr^{3+}(aq) + 7H_2O(\ell) \qquad \text{overall}$$

10.8 Predict the products of each of the following reactions. Complete and balance the equations and classify the reaction as a redox, Lewis acid-base, or precipitation reaction:

(a) $FeCl_2(s) + Cl_2(g) \rightarrow$; (b) $FeCl_2(aq) + H_2S(g) \rightarrow$; (c) $CuSO_4(aq) + NaOH(aq) \rightarrow$;
(d) $Fe(s) + HClO_4(aq) \rightarrow$; (e) $Fe^{3+}(aq) + CN^-(aq) \rightarrow$.

(a) $Cl_2(g)$ is an oxidizing agent and oxidizes Fe^{2+} in $FeCl_2$ to Fe^{3+}, and is reduced to Cl^-:

$$2[Fe^{2+} \rightarrow Fe^{3+} + e^-] \quad \text{oxidation}$$
$$Cl_2(g) + 2e^- \rightarrow 2Cl^- \quad \text{reduction}$$
$$2Fe^{2+} + Cl_2 \rightarrow 2Fe^{3+} + 2Cl^- \quad \text{overall}$$

So the balanced equation is:

$$2FeCl_2(s) + Cl_2(g) \rightarrow 2FeCl_3(s) \quad \textbf{redox} \text{ reaction}$$

(b) $H_2S(g)$ behaves as a weak acid in water: $H_2S(aq) + 2H_2O(\ell) \rightarrow 2H_3O^+(aq) + S^{2-}(aq)$, and $FeCl_2(aq)$ contains $Fe^{2+}(aq)$ and $Cl^-(aq)$ ions. $H_3O^+(aq)$ is an insufficiently strong oxidizing agent to oxidize $Fe^{2+}(aq)$ to $Fe^{3+}(aq)$, but $Fe^{2+}(aq)$ and $S^{2-}(aq)$ ions combine to give a precipitate of insoluble FeS(s), so overall:

$$FeCl_2(aq) + H_2S(aq) \rightarrow FeS(s) + 2HCl(aq) \quad \textbf{precipitation} \text{ reaction}$$

(c) $CuSO_4(aq)$ contains $Cu^{2+}(aq)$ and $SO_4^{2-}(aq)$ ions, and NaOH(aq) contains $Na^+(aq)$ and $OH^-(aq)$ ions. $Cu^{2+}(aq)$ ions combine with $OH^-(aq)$ ions to give a precipitate of insoluble $Cu(OH)_2(s)$:

$$CuSO_4(aq) + 2NaOH(aq) \rightarrow Cu(OH)_2(s) + Na_2SO_4(aq) \quad \textbf{precipitation} \text{ reaction}$$

(d) $HClO_4(aq)$ is a solution of a strong acid containing $H_3O^+(aq)$ ions and $ClO_4^-(aq)$ ions. Fe(s) is oxidized by $H_3O^+(aq)$ to give $Fe^{2+}(aq)$ and $H_2(g)$:

$$Fe(s) \rightarrow Fe^{2+}(aq) + 2e^- \quad \text{oxidation}$$
$$2H^+(aq) + 2e^- \rightarrow H_2(g) \quad \text{reduction}$$
$$Fe(s) + 2H^+(aq) \rightarrow Fe^{2+}(aq) + H_2(g) \quad \text{overall}$$

and adding two ClO_4^- ions to each side gives the balanced equation:

$$Fe(s) + 2HClO_4(aq) \rightarrow Fe(ClO_4)_2(aq) + H_2(g) \quad \textbf{redox} \text{ reaction}$$

(e) $Fe^{3+}(aq)$ (a Lewis acid) coordinates with **six** $CN^-(aq)$ ions (a Lewis base) to give the complex ion $Fe(CN)_6^{3-}(aq)$:

$$Fe^{3+}(aq) + 6CN^-(aq) \rightarrow Fe(CN)_6^{3-}(aq) \quad \textbf{Lewis acid-base} \text{ reaction}$$

CHAPTER 11

11.1 Titanium has the iron (body-centered cubic) structure, and its density is 4.50 g·cm^{-3}. Calculate the edge-length of the unit cell and the atomic radius of titanium.

For the body centered cubic structure, the unit cell contains 2 atoms, and the mass of 2 Ti atoms is:

$$(2 \text{ atoms})\left(\frac{47.90 \text{ g}}{1 \text{ mol Ti}}\right)\left(\frac{1 \text{ mol Ti}}{6.022 \times 10^{23} \text{ atoms}}\right) = \underline{1.591 \times 10^{-22} \text{ g}}$$

from which using the density we can calculate the volume of the cubic unit cell, and hence the length of the cell edge, and the length of the **cell diagonal**, which is equal to four times the atomic radius:

$$\text{Volume of unit cell} = \frac{1.591 \times 10^{-22} \text{ g}}{\left(\frac{4.50 \text{ g}}{1 \text{ cm}^3}\right)\left(\frac{10^2 \text{ cm}}{1 \text{ m}}\right)^3\left(\frac{1 \text{ m}}{10^{12} \text{ pm}}\right)^3} = \underline{3.535 \times 10^7 \text{ pm}^3}$$

$$\text{Length of edge} = (35.35 \times 10^6 \text{ pm}^3)^{\frac{1}{3}} = 3.282 \times 10^2 \text{ pm} = a$$
$$\text{Cell diagonal edge} = \sqrt{3}a = (1.732)(3.282 \times 10^2 \text{ pm}) = \underline{5.684 \times 10^4 \text{ pm}} = 4r_{Ti}$$
$$r_{Ti} = \underline{1.42 \times 10^2 \text{ pm}} \quad (142.1 \text{ pm})$$

Edge-length = <u>328 pm</u>; atomic radius = <u>142 pm</u> (3 sig fig.)

11.2 Iridium has the cubic close-packed structure. The edge-length of the unit cell is 383.3 pm. The density of iridium is 22.61 $g \cdot cm^{-3}$. Calculate a value for Avogadro's number.

The cubic close-packed structure has **four** atoms per unit cell: thus:

$$\text{Mass of unit cell} = \text{Volume x Density} = (383.3\ \text{pm})^3\left(\frac{1\ \text{m}}{10^{12}\ \text{pm}}\right)^3\left(\frac{10^2\ \text{cm}}{1\ \text{m}}\right)^3\left(\frac{22.67\ \text{g}}{1\ \text{cm}^3}\right) = \underline{1.2766 \times 10^{-21}\ \text{g}}$$

$$\text{Mass of 4 mol Ir atoms} = N_A(1.2766 \times 10^{-21}\ \text{g}) = (4\ \text{mol})\left(\frac{192.2\ \text{g}}{1\ \text{mol}}\right);\quad N_A = \frac{7.688\text{x}10^2\ \text{g}}{1.2766 \times 10^{-21}\ \text{g}} = \underline{6.022 \times 10^{23}}$$

11.3 How many sodium ions and how many chloride ions are there in the unit cell of the sodium chloride structure?

The unit cell is defined by a face-centered cubic arrangement of Cl^- ions (or Na^+ ions), with 8 Cl^- ions (or Na^+ ions) at the corners of a cube, with an Na^+ ion (or Cl^- ion) between each pair of neighboring Cl^- ions (or Na^+ ions). Thus, in the unit cell:

<u>Cl^- ions</u>: 8 corner Cl^- ions each make a contribution of one-eighth, for <u>one</u> Cl^- per unit cell, and 6 Cl^- ions in the centers of the faces each make a contribution of one-half, for <u>three</u> Cl^- per unit cell, giving a total of <u>four</u> Cl^- per unit cell.

<u>Na^+ ions</u>: 12 Na^+ ions at the centers of each edge each make a contribution of one-quarter, for <u>three</u> Na^+ ions per unit cell, and 1 Na^+ ion is at the center of the cube, for <u>one</u> Na^+ ion per unit cell, giving a total of <u>four</u> Na^+ ions per unit cell.

<u>Answer</u>: **4 Na^+ ions and 4 Cl^- ions.**

11.4 How many cesium ions and how many choride ions are there in the unit cell of the cesium chloride structure?

The Cl^- ions form a simple cubic lattice, with a Cs^+ ion at the center of the cube. Thus, each Cl^- ion makes a contribution of one-eighth of a Cl^- ion, for a total of <u>one</u> Cl^- per unit cell, and <u>one</u> Cs^+ ion.

<u>Answer</u>: **1 Cs^+ ion and 1 Cl^- ion.**

11.5 At what pressure will (a) tetrachloromethane, and (b) water boil at a temperature of 60°C?

At 60°C each liquid will boil when its pressure is the same as the vapor pressure at that temperature, which we can estimate from Figure 11.20 as: (a) **400 mm Hg**, and (b) **200 mm Hg.**

11.6 Which of the following molecules have a dipole moment? (a) CS_2 (b) $SiCl_4$ (c) NH_3 (d) PCl_3 (e) SO_2 (f) SO_3 (g) SF_6

In each case, we first write the Lewis structure, and then categorize the molecule in terms of the AX_nE_m nomenclature of the VSEPR model. If it is AX_n, where n is 2 to 6, then the molecule has no dipole moment and is **nonpolar**; otherwise, it has a dipole moment and is polar:

(a) CS_2 :S̈=C=S̈: AX_2, dipole moment, NO

(b) $SiCl_4$:C̈l–Si(–C̈l:)(–C̈l:)–C̈l: AX_4, dipole moment, NO

(c) NH_3 H–N̈–H (with H below N) AX_3E, dipole moment, **YES**

(d) PCl_3 :C̈l–P̈–C̈l: (with :C̈l: below P) AX_3E, dipole moment, **YES**

(e) SO_2 :Ö=S̈=Ö: AX_2E, dipole moment, **YES**

(f) SO_3 :Ö=S=Ö: (with =Ö: below S) AX_3, dipole moment, **NO**

(g) SF_6 (S surrounded by six :F̈: atoms) AX_6, dipole moment, **NO**

Of these molecules, only **NH_3, PCl_3** and **SO_2** have **dipole moments.**

11.7 By considering the relative magnitudes of the intermolecular forces expected for the following substances, place them in order of increasing boiling point: C_2H_5Cl, C_2H_5Br, CH_4, C_2H_6.

All the molecules are **covalent**, so the most important forces will be **London forces**, the strength of which will depend on the numbers of atoms in the molecules and their polarizabilities:
CH_4 is expected to have the lowest boiling point, because it contains the least number of atoms (five) each of which has a relatively low polarizability; C_2H_6 with eight atoms would be next, followed by C_2H_5Cl, and then C_2H_5Br, because replacing a H atom of C_2H_6 by the larger Cl atom increases the polarizability of the molecule, and Br has an even higher polarizability than Cl, so C_2H_5Br is more polarizable than C_2H_5Cl and has a higher boiling point.

Answer: **CH_4 < C_2H_6 < C_2H_5Cl < C_2H_5Br**

11.8 Which of the following are expected to form strong hydrogen bonds in the liquid or solid state?
(a) HF, (b) CH_3SH, (c) CH3OH, (d) $(CH3)_3N$, (e) $(CH_3)_2NH$, (f) CH_3CO_2H

Compounds with X–H bonds, where X is N, O, or F, satisfy the criteria for formation of strong hydrogen bonds:
(a) H-F: YES; (b) H_3C-S-H NO; (c) H_3C-O-H YES; (d) $(CH_3)_3N$: NO; (e) $(CH_3)_2N$-H YES;
(f) H_3C-C(=O)-O-H YES

CHAPTER 12

12.1 What are the H_3O^+ concentration and the percent dissociation of HF in a 0.50-M solution?

	$HF(aq)$ +	$H_2O(\ell)$ ⇄	$H_3O^+(aq)$ +	$F^-(aq)$	
initially	0.50	-	0	0	mol L^{-1}
at equilibrium	0.50-x	-	x	x	mol L^{-1}

$$K_a = \frac{[H_3O^+][F^-]}{[HF]} = \frac{x^2}{0.50-x} = 3.5 \times 10^{-4} \text{ mol L}^{-1}$$

For x << 0.50; $\frac{x^2}{0.50} = 3.5 \times 10^{-4}$ mol L^{-1}; $x^2 = 1.75 \times 10^{-4}$ mol^2 L^{-2}; $[H_3O^+] = x = \underline{1.32 \times 10^{-2} \text{ mol L}^{-1}}$

$\underline{\%\ dissociation} = \frac{1.32 \times 10^{-2} \text{ mol L}^{-1}}{0.50 \text{ mol L}^{-1}} \times 100\% = \underline{2.6\%}$ (which also justifies the assumption x < 050).

12.2 What is the OH^- concentration in a 0.10-M solution of methylamine, CH_3NH_2? What is the percent dissociation of the methylamine?

$$CH_3NH_2(aq) + H_2O(\ell) \rightleftarrows CH_3NH_3^+(aq) + OH^-(aq)$$

	$CH_3NH_2(aq)$	$H_2O(\ell)$	$CH_3NH_3^+(aq)$	$OH^-(aq)$	
initially	0.10	-	0	0	mol L^{-1}
at equilibrium	0.10-x	-	x	x	mol L^{-1}

$$K_b = \frac{[C_3NH_3^+][OH^-]}{[CH_3NH_2]} = \frac{x^2}{0.10-x} = 3.9 \times 10^{-4} \text{ mol L}^{-1}$$

Assume x << 0.10; $\frac{x^2}{0.10} = 3.9 \times 10^{-4}$ mol L^{-1}; $x^2 = 3.9 \times 10^{-5}$ mol^2 L^{-2}; $[OH^-] = x =$ <u>6.2×10^{-3} mol L^{-1}</u>

<u>% dissociation</u> $= \frac{6.2 \times 10^{-3} \text{ mol L}^{-1}}{0.10 \text{ mol L}^{-1}} \times 100\% =$ <u>6.2%</u>

12.3 What is the pH of a 0.10-M solution of hypochlorous acid, HOCl?

$$HOCl(aq) + H_2O(\ell) \rightleftarrows H_3O^+(aq) + OCl^-(aq)$$

	$HOCl(aq)$	$H_2O(\ell)$	$H_3O^+(aq)$	$OCl^-(aq)$	
initially	0.10	-	0	0	mol L^{-1}
at equilibrium	0.10-x	-	x	x	mol L^{-1}

$$K_a = \frac{[H_3O^+][OCl^-]}{[HOCl]} = \frac{x^2}{0.10-x} = 3.1 \times 10^{-8} \text{ mol L}^{-1}$$

Assume x << 0.10; $\frac{x^2}{0.10} = 3.1 \times 10^{-8}$ mol L^{-1}; $x^2 = 3.1 \times 10^{-9}$ mol^2 L^{-2}

$[H_3O^+] = x =$ <u>5.6×10^{-5} mol L^{-1}</u>; pH = $-\log [H_3O^+] = -\log (5.6 \times 10^{-5}) =$ <u>4.25</u>

12.4 What is the pH of a 0.05-M solution of methylamine, CH_3NH_2?

$$CH_3NH_2(aq) + H_2O(\ell) \rightleftarrows CH_3NH_3^+(aq) + OH^-(aq)$$

	$CH_3NH_2(aq)$	$H_2O(\ell)$	$CH_3NH_3^+(aq)$	$OH^-(aq)$	
initially	0.05	-	0	0	mol L^{-1}
at equilibrium	0.05-x	-	x	x	mol L^{-1}

$$K_b = \frac{[C_3NH_3^+][OH^-]}{[CH_3NH_2]} = \frac{x^2}{0.05-x} = 3.9 \times 10^{-4} \text{ mol L}^{-1}$$

Assume x << 0.05; $\frac{x^2}{0.05} = 3.9 \times 10^{-4}$ mol L^{-1}; $x^2 = 1.95 \times 10^{-5}$ mol^2 L^{-2}

$[OH^-] = x =$ <u>4.4×10^{-3} mol L^{-1}</u>; pOH = $-\log (4.4 \times 10^{-3}) = 2.35$; pH = 14.00 - 2.35 = <u>11.65</u>

12.5 Given that K_b for the carbonate ion, CO_3^{2-}, has a value of 2.1 x 10^{-4} mol L^{-1}, calculate the value of K_a for its conjugate acid, HCO_3^-?

$$K_a(HCO_3^-)K_b(CO_3^{2-}) = K_w = 10^{-14}\ mol^2\ L^{-2}\ ;\quad K_a(HCO_3^-) = \frac{10^{-14}\ mol^2\ L^{-2}}{2.1 \times 10^{-4}\ mol\ L^{-1}} = \underline{4.8 \times 10^{-11}\ mol\ L^{-1}}$$

12.6 By comparing the K_a and K_b values for HPO_4^{2-} from Tables 12.1 and 12.2, decide whether HPO_4^{2-} behaves as a base or as an acid in aqueous solution.

From tables, K_a = 2.1x10^{-13} mol L^{-1}; K_b = 1.6x10^{-7} mol L^{-1}

$K_b > K_a$, therefore HPO_4^{2-}(aq) behaves as a **base** in aqueous solution.

12.7 Predict whether the pH of an aqueous solution of each of the following salts is less than, greater than, or equal to 7: (a) $FeCl_3$ (b) NH_4NO_3 (c) NaH_2PO_4 (d) K_2HPO_4

In each case, we consider the acid and base properties of the ions that each salt dissociates into in water:

(a) $FeCl_3(aq) \rightarrow Fe^{3+}(aq) + 3Cl^-(aq)$

Cl^-(aq) is the conjugate base of the strong acid HCl(aq) and therefore has no basic properties in water, while Fe^{3+}(aq) is present as the **weak acid** $Fe(H_2O)_6^{3+}$ in aqueous solution. A solution of $FeCl_3$(aq), therefore, is **acidic**, with **pH** < 7.

(b) $NH_4NO_3(aq) \rightarrow NH_4^+(aq) + NO_3^-(aq)$

NO_3^-(aq) is the conjugate base of the strong acid HNO_3(aq) and therefore has no basic properties in water, while NH_4^+(aq) is the conjugate acid of the **weak base** NH_3(aq) and therefore behaves as a **weak acid**. A solution of NH_4NO_3 is therefore acidic, with **pH** < 7.

(c) $NaH_2PO_4(aq) \rightarrow Na^+(aq) + H_2PO_4^-(aq)$

Sodium is a group I metal and Na^+(aq) has no acidic or basic properties, while $H_2PO_4^-$ has pK_a 7.21 and pK_b 11.88 and therefore behaves as a **weak acid** in water ($pK_a < pK_b$); NaH_2PO_4(aq) is a **weak acid**, with **pH** < 7.

(d) $K_2HPO_4(aq) \rightarrow 2K^+(aq) + HPO_4^{2-}(aq)$

Potassium is a group I metal and K^+(aq) has no acidic or basic properties, while HPO_4^{2-} has pK_a 12.68 and pK_b 6.79 and therefore behaves as a **weak base** in water ($pK_b < pK_a$); K_2HPO_4(aq) is a **weak base**, with **pH** > 7.

12.8 What is the pH of a 0.05-M solution of NH_4Cl?

	$NH_4(aq)$ +	$H_2O(\ell)$ ⇄	$H_3O^+(aq)$ +	$NH_3(aq)$
initially	0.05	-	0	0 mol L^{-1}
at equilibrium	0.05-x	-	x	x mol L^{-1}

$$K_a(NH_4^+) = \frac{[H_3O^+][NH_3]}{[NH_4^+]} = \frac{x^2}{0.05 - x} = \frac{K_w}{K_b(NH_3)} = \frac{10^{-14}\ mol^2\ L^{-2}}{1.8 \times 10^{-5}\ mol\ L^{-1}} = 5.6 \times 10^{-10}\ mol\ L^{-1}$$

Assume x << 0.05 ; $\frac{x^2}{0.05} = 5.6 \times 10^{-10}\ mol\ L^{-1}$; $x^2 = 2.8 \times 10^{-11}\ mol^2\ L^{-2}$

$[H_3O^+] = x = \underline{5.29 \times 10^{-6}\ mol\ L^{-1}}$; pH = $-\log [H_3O^+]$ = $\underline{5.28}$

12.9 Calculate the pH of each of the following buffer solutions: (a) A solution made by dissolving 0.500 g of sodium acetate in 50.00 mL of 0.120-M acetic acid. (b) A solution obtained by mixing 15.00 mL of 0.150-M HCl(aq) with 25.00 mL of 0.180-M NH_3(aq). You should first calculate the moles of NH_4Cl(aq) and the moles of NH_3(aq) in the solution.

For each, we use: $K_a = \frac{[H_3O^+][A^-]}{[HA]} = [H_3O^+]\frac{n_{base}}{n_{acid}}$; $pK_a = pH - \log\frac{n_{base}}{n_{acid}}$ i.e., $pH = pK_a + \log\frac{n_{base}}{n_{acid}}$

(a) $$\text{mol } CH_3CO_2H = n_{acid} = (50.00 \text{ mL})(\frac{0.120 \text{ mol}}{1 \text{ L}}) = \underline{6.00 \text{ mmol}}$$

$$\text{mol } CH_3CO_2Na = n_{base} = (0.500 \text{ g})(\frac{1 \text{ mol } CH_3CO_2Na}{82.03 \text{ g } CH_3CO_2Na}) = 6.095 \times 10^{-3} \text{ mol} = \underline{6.095 \text{ mmol}}$$

$$pH = pK_a + \log\frac{n_{base}}{n_{acid}} = 4.74 + \log\frac{6.10 \text{ mmol}}{6.00 \text{ mmol}} = 4.74 + 0.01 = \underline{4.75}$$

(b) In this case we start with (15.00 mL)(0.150 mol L^{-1}) = **2.25 mmol HCl(aq)**, and the buffer solution results from adding (25.00 mL)(0.180 mol L^{-1}) = **4.50 mmol NH_3(aq)**, which forms forms the salt NH_4Cl in situ:

	NH_3(aq) +	HCl(aq) →	NH_4^+(aq) +	Cl^-(aq)	
initially	4.50	2.25	0	0	mmol
at equilibrium	2.25	0	2.25	2.25	mmol

Thus, moles NH_4^+(aq) = moles of NH_3(aq) = 2.25 mmol, <u>or</u> $n_{acid} = n_{base}$

$$pH = pK_a(NH_4^+) = pK_w - pK_b(NH_3) = 14.00 - 4.74 = \underline{9.26}$$

12.10 Lactic acid is a monoprotic acid that accumulates in the muscles during strenuous exercise and can lead to cramps. The pH of an 0.10-M solution is found to be 2.43. What is the K_a of lactic acid?

$[H_3O^+]$ in solution = $10^{-pH} = 10^{-2.43} = 3.72 \times 10^{-3}$ mol·L^{-1}

	HA(aq) +	$H_2O(\ell)$ ⇄	H_3O^+(aq) +	A^-(aq)	
initially	0.10	-	0	0	mol L^{-1}
at equilibrium	0.10 - 3.72 x 10^{-3}	-	3.72 x 10^{-3}	3.72 x 10^{-3}	mol L^{-1}

$$K_a = \frac{[H_3O^+][A^-]}{[HA]} = \frac{(3.72 \times 10^{-3})(3.72 \times 10^{-3})}{0.10} = \underline{1.4 \times 10^{-4} \text{ mol L}^{-1}}$$

12.11 Dimethylamine, $(CH_3)_2NH$, is a weak base used in the manufacture of detergents. The pH of a 1.00-M aqueous solution of dimethylamine is 12.36. What is the K_b for dimethylamine?

---

pOH = 14.00 - pH = 1.64; $[OH^-] = 10^{-1.64} = 0.0229$ mol L^{-1}

$$(CH_3)_2NH(aq) + H_2O(\ell) \rightleftarrows (CH_3)_2NH_2^+(aq) + OH^-(aq)$$

initially	1.00	-	0	0	mol L^{-1}
at equilibrium	1.00 - 0.023	-	0.023	0.023	mol L^{-1}

$$K_b = \frac{[(CH_3)_2NH_2^+][OH^-]}{[(CH_3)_2NH]} = \frac{(0.023)(0.023)}{0.977} = \underline{5.4 \times 10^{-4}\ \text{mol L}^{-1}}$$

12.12 Write equilibrium constant expressions in terms of concentrations and in terms of partial pressures for both of the following reactions:
(a) The reaction of ethane with steam at high pressure: $C_2H_6(g) + 2H_2O(g) \rightleftarrows 2CO(g) + 5H_2(g)$
(b) The decomposition of NO_2 at high temperature: $2NO_2(g) \rightleftarrows 2NO(g) + O_2(g)$

(a) $K_c = \left(\frac{[CO]^2[H_2]^5}{[C_2H_6][H_2O]^2}\right)_{eq}$; $K_p = \left(\frac{p_{CO}^2\, p_{H_2}^5}{p_{C_2H_6}\, p_{H_2O}^2}\right)_{eq}$ (b) $K_c = \left(\frac{[NO]^2[O_2]}{[NO_2]^2}\right)_{eq}$; $K_p = \left(\frac{p_{NO}^2\, p_{O_2}}{p\ NO_2^2}\right)_{eq}$

12.13 Predict how the position of the equilibrium shifts when the total pressure is increased by decreasing the volume of the reacting system in each of the following reactions: (a) $PCl_5(g) \rightleftarrows PCl_3(g) + Cl_2(g)$; (b) $N_2 + O_2(g) \rightleftarrows 2NO(g)$; (c) $2SO_2(g) + O_2(g) \rightleftarrows 2SO_3(g)$

Pressure increase favors a shift of equilibrium towards products if there are **less** gaseous molecules on the left of the balanced equation than on the right, and towards reactants if there are **more** gaseous molecules on the right than on the left; i.e., the position of equilibrium shifts in favor of whichever of the forward and reverse reactions tends to counter the increase in pressure, so:

(a) The formation of more $PCl_5(g)$, **the reverse reaction is favored.**

(b) Change in pressure favors neither the forward nor the reverse reaction because neither can affect the pressure of the system.

(c) The formation of more $SO_3(g)$, **the forward reaction**, is favored.

12.14 For each of the following reactions, predict which conditions of temperature and pressure would favor a high yield of product: (a) $N_2 + O_2(g) \rightleftarrows 2NO(g)$, $\Delta H° = 173$ kJ; (b) $CO(g) + 3H_2(g) \rightarrow H_2O(g) + CH_4(g)$, $\Delta H° = -206$ kJ

Increase in pressure favors whichever of the forward reaction or reverse reaction gives a decrease in the number of gaseous molecules, because this works to restore the original pressure, whereas increase in temperature favors whichever of the forward reaction and the reverse reaction is endothermic (absorbs heat) because this works to restore the original temperature:

(a) For formation of NO(g), the forward reaction, is favored by **high temperature** (because it is endothermic), but **pressure** change is unimportant, except in as far as it might increase the reaction rate, because there is no change of pressure on reaction in either direction.

(b) Formation of products is favored by **low temperature**, because the forward reaction is exothermic, **and high pressure**, because the forward reaction leads to a decrease in the number of molecules, and hence the pressure.

12.15 Write the expressions for the equilibrium constants for each of the following reactions, in terms of concentrations and in terms of partial pressures:
(a) $C(s) + CO_2(g) \rightleftarrows 2CO(g)$; (b) $FeO(s) + CO(g) \rightleftarrows Fe(s) + CO_2(g)$; (c) $PCl_5(s) \rightleftarrows PCl_3(\ell) + Cl_2(g)$

Neither pure solids nor pure liquids appear in the equilibrium constant expressions for heterogeneous equilibria:

(a) $K_c = \left(\frac{[CO]^2}{[CO_2]}\right)_{eq}$; $K_p = \left(\frac{p_{CO}^2}{p_{CO_2}}\right)_{eq}$

(b) $K_c = \left(\frac{[CO_2]}{[CO]}\right)_{eq}$; $K_p = \left(\frac{p_{CO_2}}{p_{CO}}\right)_{eq}$

(c) $K_c = ([Cl_2])_{eq}$; $K_p = (p_{Cl_2})_{eq}$

12.16 The equilibrium constant for the reaction $NH_4Cl(s) \rightleftarrows NH_3(g) + HCl(g)$ is 1.04×10^{-2} atm^2 at 275°C. What will be the partial pressures of NH_3 and HCl in equilibrium with $NH_4Cl(s)$ in a closed vessel at 275°C?

	$NH_4Cl(s) \rightleftarrows$	$NH_3(g)$ +	$HCl(g)$	
at equilibrium	-	x	x	atm

$K_p = p_{NH_3} \cdot p_{HCl} = x^2 = 1.04 \times 10^{-2}$ atm^2 ; $p_{NH_3} = p_{HCl} = \underline{0.102 \text{ atm}}$

CHAPTER 13

13.1 The standard cell potential for the cell **Zn(s)|ZnSO₄(aq)||AgNO₃(aq)|Ag(s)** is **1.56 V**. What is the standard reduction potential for the reaction $Ag^+(aq) + e^- \rightarrow Ag(s)$?

E°_{red} for $Zn^{2+}(aq) + 2e^- \rightarrow Zn(s)$, is -0.76 V, and $E^\circ_{cell} = E^\circ_{ox} + E^\circ_{red}$
Since the Zn electrode is on the **left** of the cell diagram, it is Zn that is oxidized, $Zn(s) \rightarrow Zn^{2+}(aq) + 2e^-$, thus:

$$1.56\text{ V} = E^\circ_{Zn(s) \rightarrow Zn^{2+} 2e^-} + E^\circ_{Ag^+ + e^- \rightarrow Ag(s)} = +0.76\text{ V} + E^\circ_{Ag^+ + e^- \rightarrow Ag(s)} \;;\quad E^\circ_{Ag^+ + e^- \rightarrow Ag(s)} = \underline{+0.80\text{ V}}$$

13.2 Place Zn^{2+}, Co^{3+}, H_2O, and NO_3^- in order of their strengths as oxidizing agents in aqueous solution.

The order of oxidizing strengths (under standard conditions) is given by the order of the reduction potentials, the substance with the most positive reduction potential being the strongest oxidizing agent. From Table 13.1:

$Zn^{2+} + 2e^- \rightarrow Zn(s)$	$E^\circ_{red} = -0.76$ V
$Co^{3+} + e^- \rightarrow Co^{2+}$	$E^\circ_{red} = +1.81$ V
$2H_2O + 2e^- \rightarrow H_2(g) + 2OH^-$	$E^\circ_{red} = -0.83$ V
$NO_3^- + 2H^+ + e^- \rightarrow NO_2(g) + H_2O$	$E^\circ_{red} = +0.80$ V

i.e., $\mathbf{Co^{3+} > NO_3^- > Zn^{2+} > H_2O}$

13.3 Place H_2S, Al(s), H_2O_2, and Fe^{2+} in order of their strengths as reducing agents in aqueous solution.

The order of reducing strengths (under standard conditions) is given by the order of the oxidation potentials, the substance with the most positive oxidation potential being the strongest reducing agent. From Table 13.1 (taking the reverse of the reduction potentials listed):

$$H_2S \rightarrow S(s) + 2H^+ + 2e^- \qquad E^\circ_{ox} = -0.14\ V$$

$$Al(s) \rightarrow Al^{3+} + 3e^- \qquad E^\circ_{ox} = +1.66\ V$$

$$H_2O_2 \rightarrow O_2 + 2H^+ + 2e^- \qquad E^\circ_{ox} = -0.68\ V$$

$$Fe^{2+} \rightarrow Fe^{3+} + e^- \qquad E^\circ_{ox} = -0.77\ V$$

i.e., **Al > H_2S > H_2O_2 > Fe^{2+}**

13.4 What will be the spontaneous reaction, and what is the value of E°_{cell} when the following half-cells are combined: (i) $Cr_2O_7^{2-} + 14H^+ + 6e^- \rightarrow 2Cr^{3+} + 7H_2O$; (ii) $Cu^{2+} + 2e^- \rightarrow Cu(s)$

For the spontaneous reaction, (i) and (ii) must be combined to give a positive value for E°_{cell}, which requires:

$$Cr_2O_7^{2-} + 14H^+ + 6e^- \rightarrow 2Cr^{3+} + 7H_2O \qquad E^\circ_{red} = +1.33\ V$$

$$3(Cu(s) \rightarrow Cu^{2+} + 2e^-) \qquad E^\circ_{ox} = -0.34\ V$$

$$Cr_2O_7^{2-} + 14H^+ + 3Cu \rightarrow 2Cr^{3+} + 3Cu^{2+} + 7H_2O \qquad E^\circ_{cell} = \mathbf{+0.99\ V}$$

13.5 What is the equilibrium constant for the following reaction at 25°C?

$$5Fe^{2+} + MnO_4^- + 8H^+ \rightarrow 5Fe^{3+} + Mn^{2+} + 4H_2O$$

First we write the two half cell potentials and calculate the standard cell potential:

$$5(Fe^{2+} \rightarrow Fe^{3+} + e^-) \qquad E^\circ_{ox} = -0.77\ V$$

$$MnO_4^- + 8H^+ + 5e^- \rightarrow Mn^{2+} + 4H_2O \qquad E^\circ_{red} = +1.49\ V$$

$$5Fe^{2+} + MnO_4^- + 8H^+ \rightarrow 5Fe^{3+} + Mn^{2+} + 4H_2O \qquad E^\circ_{cell} = \mathbf{+0.72\ V}$$

For this reaction, **n = 5**, so we have:

$$\log K = \frac{nE^\circ_{cell}}{0.0592} = \frac{5(0.72)}{0.0592} = \underline{61}; \quad K = \underline{1\text{x}10^{61}}$$

13.6 What mass of magnesium and what volume of chlorine at STP will be produced by the electrolysis of molten magnesium chloride, using a current of 3.00 A for 24.0 h?

The half reactions in the electrolysis reaction of $Mg^{2+}(Cl^-)_2(\ell)$ are:

$$Mg^{2+} + 2e^- \rightarrow Mg(s), \text{ and } 2Cl^- \rightarrow Cl_2(g) + 2e^-$$

$$\text{moles of electrons} = n = \frac{Q}{F} = \frac{(3.00\ A)(24.0\ h)(\frac{60\ s}{1\ h})(\frac{1\ C}{1\ A\ s})}{(1\ F)(\frac{96\ 500\ C}{1\ F})} = \underline{4.477\ mol}$$

$$\text{mass of Mg} = (4.477\ F)\left(\frac{1\ \text{mol Mg}}{2\ F}\right)\left(\frac{24.30\ \text{g Mg}}{1\ \text{mol Mg}}\right) = \underline{54.0\ \text{g Mg}}$$

$$\text{Volume of } Cl_2 \text{ at STP} = \frac{n_{Cl_2}RT}{P} = \frac{(4.447\ F)\left(\frac{1\ \text{mol } Cl_2}{2\ F}\right)(0.0821\ \text{atm L mol}^{-1}\ K^{-1})(273.1\ K)}{1\ \text{atm}} = \underline{49.9\ L}$$

CHAPTER 14

14.1 Draw the structures of the following alcohols, and identify each as primary, secondary, or tertiary.
(a) 3-methyl-3-pentanol, (b) cyclohexanol, (c) 2-methyl-4-heptanol, (d) 3-chloro-1-propanol

---

Alcohols with the C of the C-OH group attached to 1, 2, or 3 other C atoms, respectively, are primary, secondary, and tertiary:

(a) CH_3CH_2-C(OH)(CH_3)-CH_2CH_3 (b) 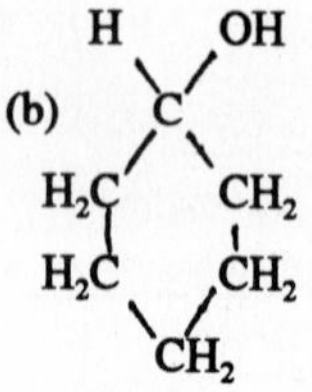 cyclohexanol: C(H)(OH) in a ring with H_2C, CH_2, H_2C, CH_2, CH_2 (c) $CH_3CH(CH_3)CH_2CH(OH)CH_2CH_2CH_3$ (d) Cl-$CH_2CH_2CH_2$-OH

TERTIARY SECONDARY SECONDARY PRIMARY

14.2 Write the formula for the organic product of each of the following reactions:
(a) $(CH_3)_2C(H)I(\ell) + NaOH(aq) \rightarrow$ (b) $CH_3OH(\ell) + K(s) \rightarrow$ (c) $(CH_3)_2C(H)OH + HI(g) \rightarrow$
(d) $C_6H_{11}\text{-}OH \xrightarrow{\text{heat, } H_2SO_4}$

(a) This is a substitution reaction: $(CH_3)_2C(H)I(\ell) + NaOH(aq) \rightarrow (CH_3)_2C(H)OH(\ell) + NaI(aq)$
2-propanol

(b) Potassium reduces an alcohol to alkoxide ion and $H_2(g)$: $2CH_3OH(\ell) + 2K(s) \rightarrow 2[CH_3O^-K^+] + H_2(g)$
potassium methoxide

(c) This is a substitution reaction: $(CH_3)_2C(H)OH + HI(g) \rightarrow (CH_3)_2C(H)I + H_2O(g)$
2-iodopropane

(d) Elimination of H_2O gives an alkene: cyclohexanol (C(H)(OH) ring with H_2C, CH_2, H_2C, CH_2, CH_2) $+ H_2SO_4 \rightarrow$ cyclohexene (C(H)=C–H ring with H_2C, H_2C, CH_2, CH_2) $+ H_3O^+ + HSO_4^-$
cyclohexene

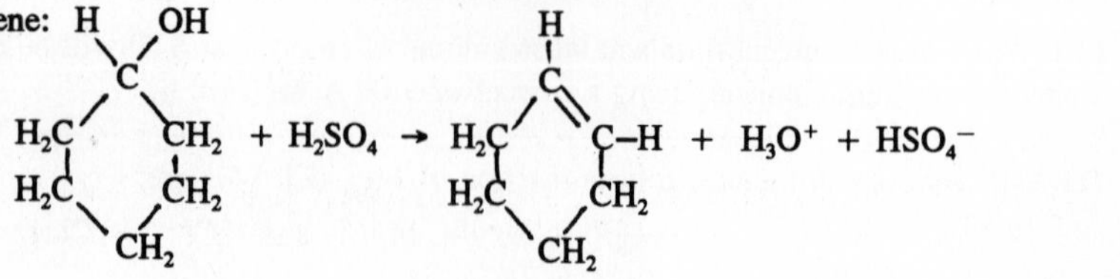

14.3 Identify each of the following compounds as an alcohol, a thiol, a phenol, or an ether, and name each.

(a) $CH_3CH_2CH(OH)CH_3$ (b) 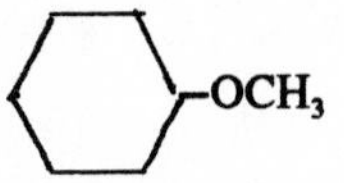 [cyclohexane ring]$-OCH_3$ (c) 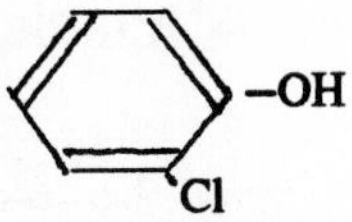 [benzene ring with Cl]$-OH$ (d) $CH_3CH_2CH_2CH_2SH$

(a) **alcohol**; 2-butanol (b) **ether**; methyl cyclohexyl ether, or cyclohexylmethoxide (c) **phenol**; 2-chlorophenol (d) **thiol**; 1-butanethiol

---

14.4 Draw the structure of 2,4-dimethyl-3-pentanone and 2,2-dimethyl butanal.

2,4-dimethyl-3-pentanone, I, contains a continuous chain of <u>five</u> C atoms, with the carbon atom of the >C=O group at position 3, and a methyl substituent at each of positions 2 and 4. 2,2-dimethyl butanal, II, contains a continuous chain of 4 C atoms with the carbon of the aldehyde group at position 1, and two methyl substituents at position 2:

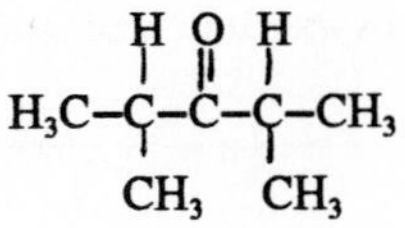

```
     H  O  H                      H  CH3
     |  ‖  |                      |  |
H3C-C--C--C-CH3             H3C-C--C----C=O
     |     |                      |  |    |
    CH3   CH3                     H  CH3  H

        I                            II
```

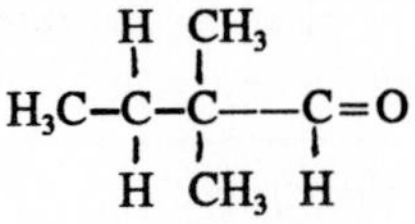

14.5 Draw the structures of the products of oxidation of each of the following: (a) cyclohexanol, (b) 2-butanol, (c) 1-butanol

---

(c) is a primary alcohol, which is oxidized to an aldehyde and then to a carboxylic acid; (a) and (b) are **secondary** alcohols, which are oxidized only to ketones, and (d) is an aldehyde, which is oxidized to a carboxylic acid.

```
        CH2-CH2    H     [O]          CH2-CH2
(a)  H2C        C          →      H2C        C=O + H2O
        CH2-CH2    OH                 CH2-CH2

        cyclohexanol                  cyclohexanone

         H  H            [O]           H
         |  |                          |
(b)  H3C-C--C-CH3         →       H3C-C--C-CH3
         |  |                          |  ‖
         H  OH                         H  O

        2-butanol                     2-butanone

                       [O]                       [O]
(c)   CH3CH2CH2CH2OH    →   CH3CH2CH2-C=O    →   CH3CH2CH2-C=O
                                      |                    |
                                      H                    OH

        1-butanol                butanal             butanoic acid
```

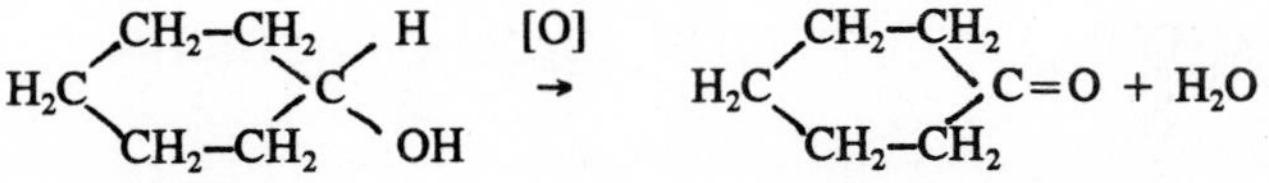

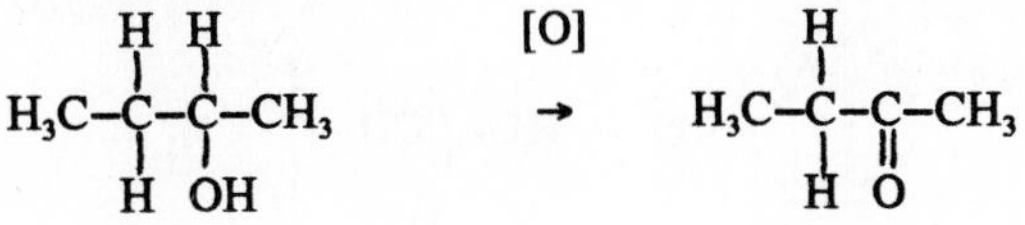

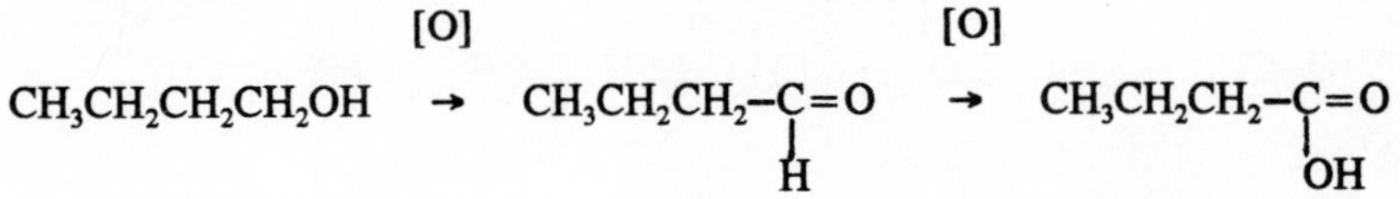

14.6 From what alcohol can pentanoic acid be prepared by oxidation? Write the structures of both compounds.

A carboxylic acid is the product of oxidation, via the corresponding aldehyde, of a primary alcohol. Pentanoic acid has 5 C atoms, so the corresponding alcohol is 1-pentanol:

$$CH_3CH_2CH_2CH_2CH_2OH \rightarrow CH_3CH_2CH_2CH_2\text{–}\underset{\displaystyle H}{\underset{|}{C}}\text{=O} \rightarrow CH_3CH_2CH_2CH_2\text{–}\underset{\displaystyle OH}{\underset{|}{C}}\text{=O}$$

1-pentanol **pentanal** **pentanoic acid**

14.7 Give the names and structures of the products of the following reactions: (a) $CH_3CH_2CH_2CO_2H$ + KOH → (b) 2-methylpentanoic acid + $Ba(OH)_2$ →

Carboxylic acids are weak acids that will react with a metal hydroxide to give the corresponding carboxylate salt:

(a) $CH_3CH_2CH_2CO_2H + KOH \rightarrow CH_3CH_2CH_2CO_2^- K^+ + H_2O$

potassium butanoate

(b) $2(CH_3)_2CHCH_2CH_2CO_2H + Ba(OH)_2 \rightarrow [(CH_3)_2CHCH_2CH_2CO_2^-]_2 Ba^{2+} + 2H_2O$

barium 2-methylpentanoate

14.8 Oil of wintergreen can be made by the reaction of o-hydroxybenzoic acid with methanol. What is its structure and its systematic name?

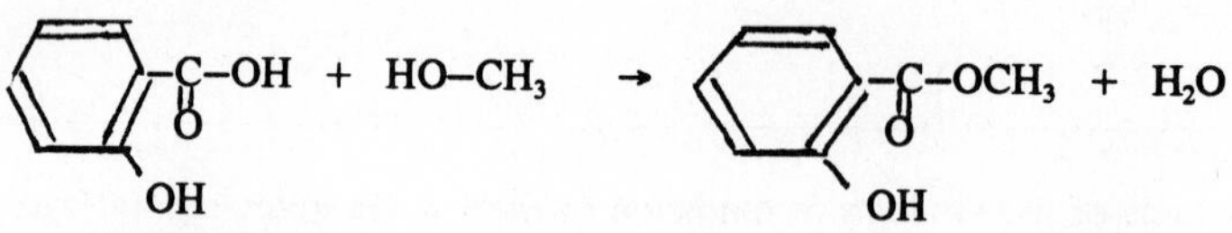

$$C_6H_4(OH)\text{–}\underset{\displaystyle O}{\underset{\|}{C}}\text{–OH} + HO\text{–}CH_3 \rightarrow C_6H_4(OH)\text{–}\underset{\displaystyle O}{\underset{\|}{C}}\text{–OCH}_3 + H_2O$$

o-hydroxybenzoic acid **methyl o-hydroxybenzoate**

14.9 What carboxylic acid and what alcohol are needed to make each of the following esters?

(a) 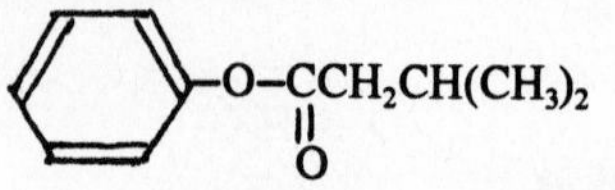 $C_6H_5\text{–O–}\underset{\displaystyle O}{\underset{\|}{C}}CH_2CH(CH_3)_2$ (b) $CH_3CH_2CH_2\text{–}\underset{\displaystyle O}{\underset{\|}{C}}\text{–O–CH(CH}_3)_2$

Hydrolysis of an ester gives the carboxylic acid and alcohol from which it is formed by the *reverse* reaction:

(a) 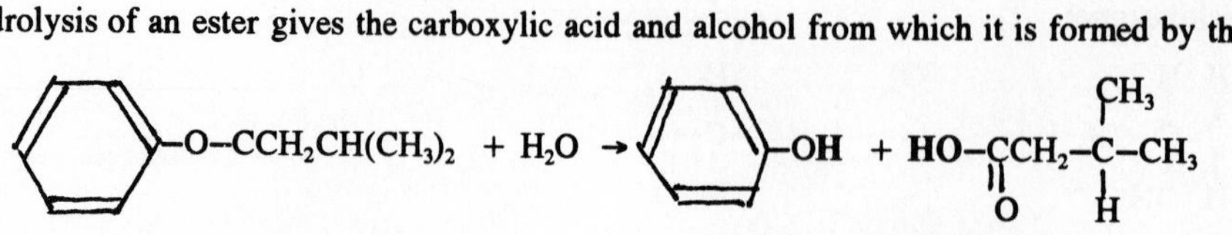 $C_6H_5\text{–O–}CCH_2CH(CH_3)_2 + H_2O \rightarrow C_6H_5\text{–OH} + HO\text{–}\underset{\displaystyle O}{\underset{\|}{C}}CH_2\text{–}\overset{\displaystyle CH_3}{\underset{\displaystyle H}{C}}\text{–}CH_3$

phenol **2-methylbutanoic acid**

(b) $CH_3CH_2CH_2\text{–}\underset{\displaystyle O}{\underset{\|}{C}}\text{–O–CH(CH}_3)_2 + H_2O \rightarrow CH_3CH_2CH_2\text{–}\underset{\displaystyle O}{\underset{\|}{C}}\text{–OH} + HO\text{–}\underset{\displaystyle H}{\underset{|}{C}}(CH_3)_2$

butanoic acid **2-propanol**

14.10 What products would you obtain from the hydrolysis of each of the following esters? (a) ethylformate, (b) propyl-p-bromobenzoate?

(a) $CH_3CH_2{-}O{-}\underset{\underset{O}{\|}}{C}{-}H + H_2O \rightarrow CH_3CH_2{-}OH + HO{-}\underset{\underset{O}{\|}}{C}{-}H$

ethyl formate — ethanol — methanoic (formic) acid

(b) 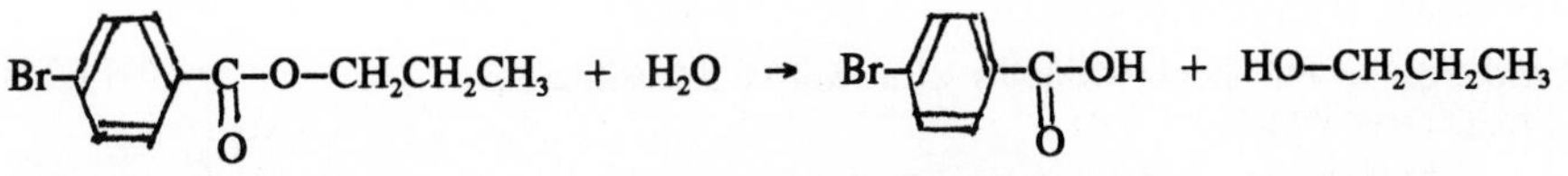 $Br{-}C_6H_4{-}\underset{\underset{O}{\|}}{C}{-}O{-}CH_2CH_2CH_3 + H_2O \rightarrow Br{-}C_6H_4{-}\underset{\underset{O}{\|}}{C}{-}OH + HO{-}CH_2CH_2CH_3$

propyl-p-bromobenzoate — p-bromobenzoic acid — 1-propanol

14.11 Write an equation for the acid-base equilibrium of dimethylamine with water.

$$H_3C{-}\overset{\overset{CH_3}{|}}{\underset{\cdot\cdot}{N}}{-}H + H_2O \rightleftarrows H_3C{-}\overset{\overset{CH_3}{|}}{\underset{\underset{H}{|}}{N^+}}{-}H + OH^-$$

14.12 Complete the following equations: (a) $C_6H_5NH_2(l) + HCl(aq) \rightarrow$ (b) $CH_3CH_2NH_2(l) + CH_3CO_2H(aq) \rightarrow$ (c) $CH_3NH_3^+ Cl^-(s) + NaOH(aq) \rightarrow$

(a) $C_6H_5NH_2(l) + HCl(aq) \rightarrow C_6H_5NH_3^+(aq) + Cl^-(aq)$

(b) $CH_3CH_2NH_2(l) + CH_3CO_2H(aq) \rightarrow CH_3CH_2NH_3^+(aq) + CH_3CO_2^-(aq)$

(c) $CH_3NH_3^+ Cl^-(s) + NaOH(aq) \rightarrow CH_3NH_2(aq) + NaCl(aq) + H_2O(\ell)$

14.13 (a) Which of the aminoacids $CH_3CH_2CH(NH_2)CO_2H$ and $(CH_3)_2C(NH_2)CO_2H$ are chiral? (b) Is lactic acid chiral?

A compound is **chiral** if it contains a C atom (C) bonded to four different groups:

(a) $CH_3{-}CH_2{-}\overset{\overset{H}{|}}{\underset{\underset{NH_2}{|}}{C}}{-}CO_2H$; **YES.** $\quad H_3C{-}\overset{\overset{CH_3}{|}}{\underset{\underset{NH_2}{|}}{C}}{-}CO_2H$; **NO**

(b) Lactic acid is 2-hydroxypropanoic acid $\quad H_3C{-}\overset{\overset{OH}{|}}{\underset{\underset{H}{|}}{C}}{-}CO_2H$

with 4 different groups attached to C, and is thus **CHIRAL.**

14.14 Draw diagrams of the low-resolution proton NMR spectra of the following molecules:
(a) C_2H_6 (b) $CH_3CH_2CH_3$ (c) $(CH_3)_2CHCHCl_2$

(a) $H-\overset{H}{\underset{H}{C}}-\overset{H}{\underset{H}{C}}-H$ has all chemically identical H atoms and would give a single NMR peak.

(b) In $H-\overset{H}{\underset{H}{C}}-\overset{H}{\underset{H}{C}}-\overset{H}{\underset{H}{C}}-H$ the end CH_3 groups are identical and the CH_2 group is chemically different; the NMR spectrum would show two peaks with relative intensities 3:1.

(c) $H_3C-\overset{H}{\underset{H_3C}{C}}-\overset{Cl}{\underset{Cl}{C}}-H$ has two chemically equivalent -CH_3 groups and two CH groups that are chemically nonequivalent the NMR spectrum would have three peaks with relative intensities 6:1:1.

14.15 How could NMR and IR spectroscopy be used to show whether a compound with the molecular formula C_4H_6 is 1,3-butadiene or 2-butyne?

The structures of these two isomers are:

$$H_2C=\underset{H}{C}-\underset{H}{C}=CH_2 \quad \text{and} \quad H_3C-C\equiv C-CH_3$$

A. 1,3-butadiene B. 2-butyne

In the NMR spectra, **compound A** should have **two** main **peaks** of relative intensities **2:1**, while **compound B** should have **one peak**. In the infrared spectra, as well as C-H stretching frequencies, **compound A** should have a stretching frequency close to **1650 cm^{-1}** diagnostic of a **C=C bond**, and **compound B** should have a stretching frequency close to **2200 cm^{-1}** diagnostic of a **C≡C bond**.

CHAPTER 15

15.1 For the reaction $N_2(g) + 3H_2(g) \rightarrow 2NH_3(g)$, if the rate of formation of $NH_3(g)$ in a given time interval was $0.0010\ mol \cdot L^{-1} \cdot min^{-1}$, what was the rate of disappearance of $N_2(g)$ and $H_2(g)$, respectively in the same time interval?

The rate may be expressed in terms of the consumption of any reactant, or the formation of any product, given by:

$$\frac{\Delta[NH_3]}{\Delta t} = -2\frac{\Delta[N_2]}{\Delta t} = -\frac{2}{3}\frac{\Delta[H_2]}{\Delta t}$$

$$\text{Thus,} \quad -\frac{\Delta[N_2]}{\Delta t} = \frac{1}{2}(0.0010\ mol\ L^{-1}\ min^{-1}) = \underline{0.0005\ mol\ L^{-1}\ min^{-1}}$$

$$\frac{\Delta[H_2]}{\Delta t} = \frac{3}{2}(0.0010\ mol\ L^{-1}\ min^{-1}) = \underline{0.0015\ mol\ L^{-1}\ min^{-1}}$$

(Otherwise, more simply, we can simply argue that 1 N_2 molecule is removed for every 2 NH_3 molecules formed, and 3 H_2 molecules are removed for every 2 NH_3 molecules formed.)

15.2 What are the rate laws and the units of the rate constants for the following reactions?
(a) The isomerization of cyclopropane to propene at 500°C, $C_3H_6(g) \rightarrow CH_3CH{=}CH_2(g)$, which is first order in cyclopropane; (b) The reaction of hydrogen with nitrogen monoxide at 800°C, $2H_2(g) + 2NO(g) \rightarrow 2H_2O(g) + N_2(g)$, which is first order in H_2 and second order in NO

(a) Rate = $k[C_3H_6]$; (b) Rate = $k[H_2][NO]^2$.

15.3 A knowledge of the reaction between oxygen, $O_2(g)$, in air and nitrogen monoxide, NO(g), from automobile exhausts, $2NO(g) + O_2(g) \rightarrow 2NO_2(g)$, is important in air pollution studies. Several experiments gave the following data:

Experiment	[NO] (mol L^{-1})	[O_2] (mol L^{-1})	Initial Rate (mol L^{-1} s^{-1})
1	0.001	0.001	7.0×10^{-6}
2	0.001	0.002	1.4×10^{-5}
3	0.001	0.003	2.1×10^{-5}
4	0.002	0.003	8.4×10^{-5}

What are (a) the rate law for the reaction, and (b) the value of the rate constant?

(a) The rate equation has the form Rate = $k[NO]^a[O_2]^b$.
From experiments 1 and 2, the rate doubles when [O_2] doubles, thus, the rate is proportional to [O_2], i.e., **b** = 1.
From experiments 3 and 4, the rate is quadrupled when [NO] doubles, thus the rate is proportional to $[NO]^2$, i.e., **a** = 2. And the overall rate equation is:

$$\text{rate} = k[NO]^2[O_2].$$

(b) From experiment 1, Rate = 7.0×10^{-6} mol L^{-1} s^{-1} = $k(0.001 \text{ mol } L^{-1})^2(0.001 \text{ mol } L^{-1})$

$$k = \frac{7.0 \times 10^{-6} \text{ mol } L^{-1} s^{-1}}{1 \times 10^{-9} \text{ mol}^3 L^{-3}} = \underline{7 \times 10^3 \text{ mol}^{-2} L^2 s^{-1}}$$

Alternatively, using data from any of the other experiments, we obtain the same value for the rate constant k.

15.4 If the first step of the three-step mechanism for the oxidation of $I^-(aq)$ by $Fe^{3+}(aq)$, equation (2), is the rate-determining step, what is the rate law for the reaction?

The first step is $Fe^{3+}(aq) + I^-(aq) \rightarrow FeI^{2+}(aq)$, which determines the overall rate, and

$$\text{rate} = k[Fe^{3+}][I^-]$$

which is first order in $Fe^{3+}(aq)$, first order in $I^-(aq)$, and second order overall.

15.5 The rate constant for the first-order isomerization of cyclopropane to propene at 500°C is 6.7×10^{-4} s^{-1}. Find the time for the concentration of cyclopropane to drop to (a) half and (b) one-quarter its initial value.

For a first order reaction: $kt_{\frac{1}{2}} = \ln 2 = 0.693$; $t_{\frac{1}{2}} = \dfrac{0.693}{6.7 \times 10^{-4} s^{-1}} = 1.034 \times 10^3 s^{-1}$

(a) $t_{\frac{1}{2}} = (1.034 \times 10^3 s^{-1})(\dfrac{1 \text{ min}}{60 \text{ s}}) = \underline{17.2 \text{ min}}$

(b) Two half-lives $= 2(t_{\frac{1}{2}}) = \underline{34.5 \text{ min}}$

15.6 For the hydrolysis of methyl bromide by OH^- ion, the activation energy is 83.7 kJ·mol^{-1}, and the rate constant is 3.44 x 10^{-2} mol^{-1} L s^{-1} at 55°C. What is the value of the rate constant at 25°C?

$$\ln\frac{k_2}{k_1} = \frac{E_a}{R}\left(\frac{1}{T_1} - \frac{1}{T_2}\right) = \ln\frac{k_2}{3.44 \times 10^{-2}\ \text{mol}^{-1}\ \text{L s}^{-1}} = \frac{(83.7\ \text{kJ mol}^{-1})\left(\frac{10^3\ \text{J}}{1\ \text{kJ}}\right)}{8.31\ \text{J K}^{-1}\ \text{mol}^{-1}}\left(\frac{1}{328\ \text{K}} - \frac{1}{298\ \text{K}}\right) = -3.091$$

$$\frac{k_2}{3.44 \times 10^{-2}\ \text{mol}^{-1}\ \text{L s}^{-1}} = e^{-3.091} = 4.546 \times 10^{-2}\ ; \quad k_2 = \underline{1.56 \times 10^{-3}\ \text{mol}^{-1}\ \text{L s}^{-1}}$$

15.7 For the hydrolysis of ethyl chloride by OH^- ion, the activation energy is 96 kJ·mol^{-1}. By what factor will the rate increase from 25°C to 35°C?

$$\ln\frac{k_2}{k_1} = \frac{E_a}{R}\left(\frac{1}{T_1} - \frac{1}{T_2}\right) = \frac{(96\ \text{kJ mol}^{-1})\left(\frac{10^3\ \text{J}}{1\ \text{kJ}}\right)}{8.31\ \text{J K}^{-1}\ \text{mol}^{-1}}\left(\frac{1}{308\ \text{K}} - \frac{1}{298\ \text{K}}\right) = 1.25\ ; \quad \frac{k_2}{k_1} = e^{1.25} = \underline{3.5}$$

Thus, the rate increases by a factor of 3.5 from 25°C to 35°C.

15.8 Find the activation energy for the reaction between ethyl bromide and hydroxide ions, using the rates at 25°C and 35°C given in Table 15.4. Compare this value with that obtained from Figure 15.9.

$$\ln\frac{k_2}{k_1} = \frac{E_a}{R}\left(\frac{1}{T_1} - \frac{1}{T_2}\right) = \ln\frac{2.8\text{x}10^{-4}\ \text{mol}^{-1}\ \text{L s}^{-1}}{8.8\text{x}10^{-5}\ \text{mol}^{-1}\ \text{L s}^{-1}} = \ln 3.18 = \underline{1.16}$$

$$= \frac{(E_a\ \text{kJ mol}^{-1})\left(\frac{10^3\ \text{J}}{1\ \text{kJ}}\right)}{8.31\ \text{J K}^{-1}\ \text{mol}^{-1}}\left(\frac{1}{298\ \text{K}} - \frac{1}{308\ \text{K}}\right) = 0.0131\ E_a;\quad E_a = \underline{89\ \text{kJ}}$$

E_a = 89 kJ mol^{-1} (2 significant figures) compares favorably with that from Figure 15.9 of 89.2 kJ mol^{-1}.

15.9 Dinitrogen monoxide, $N_2O(g)$, decomposes at 600°C, $2N_2O(g) \rightarrow 2N_2(g) + O_2(g)$, according to the following mechanism:

$N_2O \rightarrow N_2 + O$	Slow	..... Step 1.
$O + N_2O \rightarrow N_2 + O_2$	Fast	..... Step 2.

The reaction is catalyzed by a trace of $Cl_2(g)$, and the catalyzed reaction follows the mechanism

$Cl_2 \rightleftarrows 2Cl$	Fast equilibrium	..... Step 1.
$N_2O + Cl \rightarrow N_2 + ClO$	Slow	..... Step 2
$2ClO \rightarrow Cl_2 + O_2$	Fast	..... Step 3

(a) Derive the overall equation for the catalyzed reaction, and thus confirm that Cl_2 behaves as a catalyst for this reaction. (b) Derive the rate laws for the uncatalyzed and the catalyzed reactions.

(a) For the catalyzed reaction, if we add the equations for steps 1 and 3 and twice the equation for step 2, and thus eliminate any intermediates, we obtain:

$$Cl_2 + 2N_2O \rightarrow 2N_2O + Cl_2$$

Cl_2 appears on both sides of the equation, and is thus both a reactant and a product that it not consumed in the reaction. In other words it does indeed behave as a catalyst in this reaction.

(b) In each case the rate equation is derived from the reactants in the slowest step in the reaction:

Uncatalyzed reaction The slowest step is $N_2O \rightarrow N_2 + O$, and thus the rate equation is: **rate = k[N$_2$O].**

Catalyzed reaction The slowest step is $N_2O + Cl \rightarrow N_2 + ClO$, and thus the rate equation is: rate = $k_1[N_2O][Cl]^{1/2}$, but Cl is an intermediate and must be replaced, which we achieve via the equilibrium constant for the first step:

$$Cl_2 \rightleftharpoons 2Cl\ ; \quad K_{eq} = \frac{[Cl]^2}{[Cl_2]}\ ; \quad \text{i.e., } [Cl] = (K_{eq}[Cl_2])^{\frac{1}{2}} = K_{eq}^{\frac{1}{2}}[Cl_2]^{\frac{1}{2}}$$

$$\text{Thus:} \quad \text{rate} = k_1[N_2O][Cl] = k_1K_{eq}^{\frac{1}{2}}[N_2O][Cl_2]^{\frac{1}{2}} = \underline{k_2[N_2O][Cl_2]^{\frac{1}{2}}}$$

where the constant $k_1K_{eq}^{1/2}$ is replaced by the rate constant k_2.

15.10 Given that the activation energy is 76 kJ·mol^{-1} for the uncatalyzed and 57 kJ mol^{-1} for the catalyzed decomposition of $H_2O_2(aq)$ by $I^-(aq)$, by what factor does the Arrhenius equation predict that the rate of the catalyzed reaction will be greater than that of the uncatalyzed reaction at 25°C?

Let k_1 be the rate for the uncatalyzed reaction, and k_2 the rate for the catalyzed reaction at 25°C. Then from the Arrhenius equation:

$$k = Ae^{-\frac{E_a}{RT}}; \quad \frac{k_2}{k_1} = \frac{Ae^{\frac{-(E_a)_2}{RT}}}{Ae^{\frac{-(E_a)_1}{RT}}}$$

$$\ln\frac{k_2}{k_1} = \frac{-[(E_a)_2 - (E_a)_1]}{RT} = \frac{-[(57 - 76 \text{ kJ mol}^{-1}](\frac{10^3 \text{ J}}{1 \text{ kJ}})}{(8.31 \text{ J K}^{-1} \text{ mol}^{-1})(298 \text{ K})} = 7.67\ ; \quad \frac{k_2}{k_1} = e^{7.67} = \underline{2.1\text{x}10^3}$$

CHAPTER 16

16.1 At low temperature, NO and NO_2 combine to form the deep-blue liquid oxide N_2O_3. Draw a Lewis structure for N_2O_3.

N_2O_3 contains 2(5) + 3(6) = 28 valence electrons. We have already seen that O-O bonds are rather weak, so it seems likely that the structure does not contain O—O bonds but is ONNO$_2$ (I), in which the unpaired electron of the N atom of NO forms a bond with the unpaired electron of the N atom of NO_2:

I. We write first: O—N—N—O, and completing the octets of electrons on the O atoms gives

```
   ..   ..        ..
-:O—N+—N2+—O:-
   ..      |     ..
          :O:-
           ..
```

Then we reduce the formal charges by having each N atom form a double bond with an O atom, giving:

$$:\ddot{O}=\ddot{N}-N^+=\ddot{O}:$$
$$|$$
$$:\ddot{O}:^-$$

II. Another possibility is ONONO (II), the Lewis structure of which is: :O=N—O—N=O: However, experimental determination of the structure shows that the structure is in fact I above.

16.2 Write equations for the half-reaction in which nitrate ion is reduced to (a) $NO_2(g)$, (b) NO(g), and (c) NH_4^+ in acidic aqueous solution.

(a) $NO_3^-(aq) + e^- + 2H^+(aq) \rightarrow NO_2(g) + H_2O(\ell)$

(b) $NO_3^-(aq) + 3e^- + 4H^+(aq) \rightarrow NO(g) + 2H_2O(\ell)$

(c) $NO_3^-(aq) + 8e^- + 10H^+(aq) \rightarrow NH_4^+(aq) + 3H_2O(\ell)$

16.3 Draw Lewis structures for (a) hydrogen peroxide, (b) the hydroxyl radical, (c) the hydroperoxyl radical, and (d) the ethyl radical. Assign oxidation numbers to the oxygen and carbon atoms in these molecules.

(a) H_2O_2 contains 2(1) + 2(6) = 14 valence electrons; joining the atoms with single bonds and completing the octets around the O atoms gives H—O—O—H

(b) Two **OH radicals** are formed by breaking the O—O bond in hydrogen peroxide, H—O—O—H → 2 H—O·

(c) The **hydroperoxyl radical** is derived from hydrogen peroxide by removing a *hydrogen atom*

H—O—O—H → H—O—O· + ·H

(d) The **ethyl radical** is formed from ethane by removing a hydrogen atom from an ethane molecule

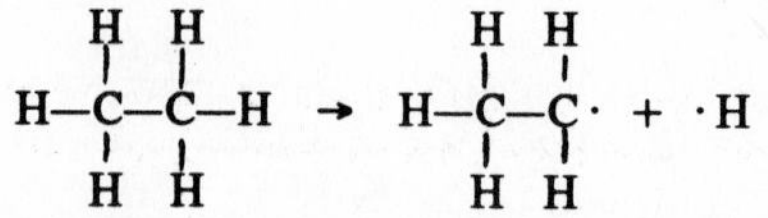

```
  H H           H H
  | |           | |
H—C—C—H  →   H—C—C·  +  ·H
  | |           | |
  H H           H H
```

Oxidation Numbers: H_2O_2 H, +1; O, -1.

H : O· — Assigning all the electrons to the more electronegative O atom gives :O·, ON -1.

H : O : O· — Assigning all the electrons to the O atoms, gives :O:O·, ON = ½(—1) = —½

C_2H_5· — Assigning all the electrons to the more electronegative C atoms gives

:C:C· ON = ½[—15 + 2(+4)] = —3½

16.4 What is the bond order for the bonds in the ozone molecule? Show that the observed bond length of 128 pm is consistent with this bond order.

From the two *resonance* structures: $:\ddot{O}=\ddot{O}^+-\ddot{O}:^- \leftrightarrow {}^-:\ddot{O}-\ddot{O}^+=\ddot{O}:$

oxygen-oxygen bond order = ½(2 + 1) = 1½

and this bond of order 1½ should be intermediate in length between that of the oxygen-oxygen double bond in the O_2 molecule (bond order 2, length 123 pm) and the oxygen-oxygen single bond in hydrogen peroxide, H_2O_2, or the peroxide ion, O_2^{2-}, (bond order 1, length 149 pm).

16.5 Write half-equations for the reactions in which (a) ozone is reduced to oxygen and (b) hydrogen peroxide is reduced to water in acidic aqueous solution.

(a) $O_3(g) + 2e^- + 2H^+(aq) \rightarrow O_2(g) + H_2O(\ell)$

(b) $H_2O_2(aq) + 2e^- + 2H^+(aq) \rightarrow 2H_2O(\ell)$

16.6 (a) Derive the relationship

$$\lambda_{max} = \frac{1.20 \times 10^5 \text{ nm}}{BE}$$

between the bond energy, BE (the energy in kilojoules needed to break one mole of bonds), and the maximum wavelength of radiation, λ_{max}, (in nanometers), that will break the bond. (b) Use this relationship to show that the bonds in CO_2 cannot be broken in the troposphere but may be broken in the stratosphere.

(a) For one photon, $E_{photon} = h\nu = hc/\lambda$, where h is Planck's constant and c is the speed of light. Thus, for 1 mole of photons:

$$E = \frac{N_A hc}{\lambda} = \frac{(6.022 \times 10^{23} \text{ mol}^{-1})(6.626 \times 10^{-34} \text{ J s})(\frac{1 \text{ kJ}}{10^3 \text{ J}})(2.998 \times 10^8 \text{ m s}^{-1})}{(\lambda \text{ nm})(\frac{1 \text{ m}}{10^9 \text{ nm}})} = \frac{1.196 \times 10^5 \text{ kJ mol}^{-1}}{\lambda_{max}}$$

$$\text{i.e.,} \quad \lambda_{max} = \frac{1.20 \times 10^5 \text{ nm}}{BE}$$

(b) Using the equation from part (a), we have

$$\lambda_{maximum} = \frac{1.20 \times 10^5 \text{ nm kJ}}{707 \text{ kJ}} = 170 \text{ nm}$$

Radiation of wavelength 170 nm or shorter will break C=O bonds, so bonds in CO_2 will be broken in (i) the stratosphere, where solar radiation of this wavelength penetrates, but not in (ii) the troposphere, where no radiation of this wavelength penetrates (Figure 16.2).

16.7 (a) What is the maximum wavelength radiation that will dissociate a Cl atom from a CF_2Cl_2 molecule if the energy needed to dissociate the Cl atom is 318 kJ mol^{-1}? (b) Why is CF_2Cl_2 not dissociated in the troposphere?

(a) Using the equation from Example 16.6, we have

$$\lambda_{maximum} = \frac{1.20 \times 10^5 \text{ nm kJ}}{318 \text{ kJ}} = 377 \text{ nm}$$

(b) Although 377-nm light should break the C-Cl bond, CF_2Cl_2 must absorb the much greater energy of 250-nm photons, to first raise it to a suitable excited state, before the dissociation to give a Cl atom can occur, and radiation of this wavelength does not penetrate into the troposphere.

16.8 Molecules absorb infrared radiation only if their dipole moment changes during at least one of their vibrations. Explain why CO, CO_2, and H_2O absorb infrared radiation but O_2 and N_2 do not.

CO Because of the electronegativity differences of carbon and oxygen, carbon monoxide has a dipole moment. When it vibrates, the CO bond lengthens and shortens; the dipole moment changes and fluctuates, so CO absorbs infrared radiation.

CO_2 $:\ddot{O}{=}C{=}\ddot{O}:$, an AX_2 type linear molecule, has polar bonds but no dipole moment, because the centers of negative and positive charge coincide. However, in (i) the vibrational mode in which one C=O bond is stretched while the other C=O bond is compressed, or (ii) the two C=O bonds move up and down perpendicular to the axis of the molecule, the centers of positive and negative charge no longer coincide, which gives the molecule undergoing this vibration a changing and fluctuating dipole moment, so CO_2 absorbs infrared radiation.

H_2O is an AX_2E type angular molecule and has a dipole moment. This dipole moment changes in the vibrational modes where (i) both O-H bonds stretch, moving the H atoms simultaneously in and out of their equilibrium positions, (ii) one O-H bond stretches while the other is compressed, and (iii) when the O-H bonds bend so that the H atoms move in and out first apart and then closer to each other. Thus, H_2O absorbs infrared radiation when it undergoes these vibrations.

In contrast, both O_2 and N_2 are nonpolar and have zero dipole moments. Their only vibrational modes involve the shortening and lengthening of their bonds about their average positions, during which the dipole moments remain zero. Thus, neither O_2 nor N_2 absorb infrared radiation.

CHAPTER 17

17.1 Write the structure of the polypeptide formed by condensation of four glycine molecules.

Glycine is $H_2N\text{-}CH_2\text{-}CO_2H$, so the condensation reaction is

$$H_2N\text{-}CH_2\text{-}\overset{\overset{\displaystyle O}{\|}}{C}\text{-}OH + H\text{-}\overset{\overset{\displaystyle H}{|}}{N}\text{-}CH_2\text{-}\overset{\overset{\displaystyle O}{\|}}{C}\text{-}OH + H\text{-}\overset{\overset{\displaystyle H}{|}}{N}\text{-}CH_2\text{-}\overset{\overset{\displaystyle O}{\|}}{C}\text{-}OH + H\text{-}\overset{\overset{\displaystyle H}{|}}{N}\text{-}CH_2\text{-}CO_2H$$

$$\downarrow$$

$$H_2N\text{-}CH_2\text{-}\overset{\overset{\displaystyle O}{\|}}{C}\text{-}\overset{\overset{\displaystyle H}{|}}{N}\text{-}CH_2\text{-}\overset{\overset{\displaystyle O}{\|}}{C}\text{-}\overset{\overset{\displaystyle H}{|}}{N}\text{-}CH_2\text{-}\overset{\overset{\displaystyle O}{\|}}{C}\text{-}\overset{\overset{\displaystyle H}{|}}{N}\text{-}CH_2\text{-}CO_2H + 3H_2O$$

17.2 Draw the primary structure of a segment of a polypeptide containing the sequence -Gly-His-Val-Glu-.

Gly represents a glycine, His a histidine, Val a valine, and Glu a glutamic acid residue. Thus, we have:

$$HO\text{-}\underset{\underset{\displaystyle O}{\|}}{C}\text{-}\overset{\overset{\displaystyle H}{|}}{\underset{\underset{\displaystyle H}{|}}{C}}\text{-}NH_2 + HO\text{-}\underset{\underset{\displaystyle O}{\|}}{C}\text{-}\overset{\overset{\displaystyle H}{|}}{\underset{\underset{\displaystyle CH_2\text{-}(\text{imidazolyl: N-H, N})}{|}}{C}}\text{-}NH_2 + HO\text{-}\underset{\underset{\displaystyle O}{\|}}{C}\text{-}\overset{\overset{\displaystyle H}{|}}{\underset{\underset{\displaystyle CH\text{-}C(CH_3)_2}{|}}{C}}\text{-}NH_2 + HO\text{-}\underset{\underset{\displaystyle O}{\|}}{C}\text{-}\overset{\overset{\displaystyle H}{|}}{\underset{\underset{\displaystyle CH_2\text{-}CH_2CO_2H}{|}}{C}}\text{-}NH_2$$

$$\downarrow$$

$$\cdots\underset{\underset{\displaystyle O}{\|}}{C}\text{-}\overset{\overset{\displaystyle H}{|}}{\underset{\underset{\displaystyle H}{|}}{C}}\text{-}\overset{\overset{\displaystyle H}{|}}{N}\text{-}\underset{\underset{\displaystyle O}{\|}}{C}\text{-}\overset{\overset{\displaystyle H}{|}}{\underset{\underset{\displaystyle CH_2\text{-}(\text{imidazolyl: N-H, N})}{|}}{C}}\text{-}\overset{\overset{\displaystyle H}{|}}{N}\text{-}\underset{\underset{\displaystyle O}{\|}}{C}\text{-}\overset{\overset{\displaystyle H}{|}}{\underset{\underset{\displaystyle CH\text{-}C(CH_3)_2}{|}}{C}}\text{-}\overset{\overset{\displaystyle H}{|}}{N}\text{-}\underset{\underset{\displaystyle O}{\|}}{C}\text{-}\overset{\overset{\displaystyle H}{|}}{\underset{\underset{\displaystyle CH_2\text{-}CH_2CO_2H}{|}}{C}}\text{-}\overset{\overset{\displaystyle H}{|}}{N}\cdots + 4H_2O$$

17.3 Draw a diagram to show how the two polypeptide segments -Gly-Ala- and -Gly-Ser can be joined by hydrogen bonds.

First we draw the structures of the two segments

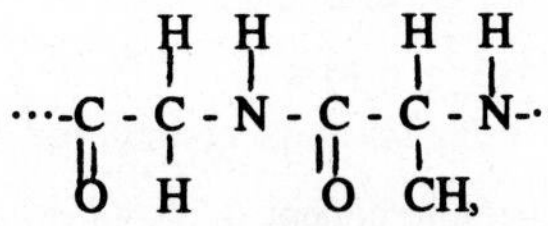

```
       H  H       H  H                  H  H          H
       |  |       |  |                  |  |          |
  ····C - C - N - C - C - N····    ····C - C - N - C - C - N····
      ||  |       ||  |                ||  |       ||  |
      O   H       O   CH3              O   H       O   CH2OH
```

-Gly-Ala- 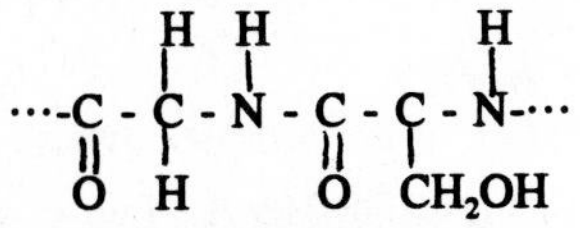 **-Gly-Ser-**

and note that the hydrogen bonds formed between the two segments are C=O:····H-N hydrogen bonds:

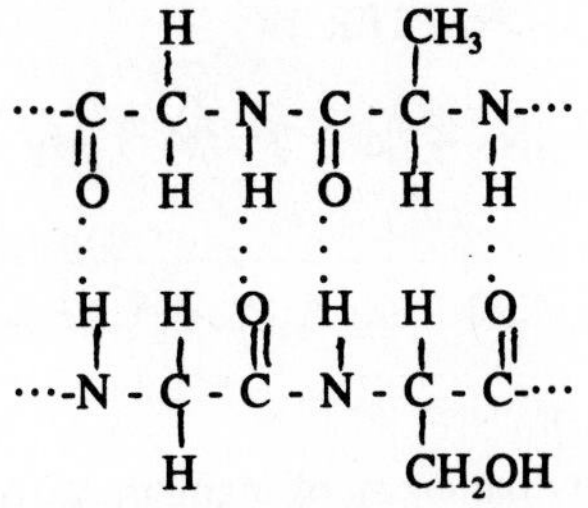

```
            H           CH3
            |            |
      ····C - C - N - C - C - N····
          ||  |   |   ||  |   |
          O   H   H   O   H   H
          :       :   :       :
          H   H   O   H   H   O
          |   |   ||  |   |   ||
      ····N - C - C - N - C - C····
              |           |
              H           CH2OH
```

17.4 An uncatalyzed reaction has an activation energy of 100 kJ mol^{-1}. If an enzyme increases the rate of this reaction by a factor of 1 million at 37°C, to what value does the enzyme reduce the activation energy?

$$k = Ae^{-\frac{E_A}{RT}} \quad ; \quad \frac{k_2}{k_1} = \frac{Ae^{-\frac{(E_a)_2}{RT}}}{Ae^{-\frac{(E_a)_1}{RT}}} = e^{\frac{(E_a)_2 - (E_a)_1}{RT}}$$

$$\ln \frac{k_2}{k_1} = \frac{(E_a)_2 - (E_a)_2}{RT} \quad ; \quad \ln 10^6 = 13.82 = \frac{[100 - (E_a)_2]\text{kJ}}{(8.314 \text{ J K}^{-1})(\frac{1 \text{ kJ}}{10^3 \text{ J}})(310 \text{ K})}$$

$$35.6 \text{ kJ} = 100 - (E_a)_2 \quad ; \quad (E_a)_2 = \underline{64.4 \text{ kJ}}$$

Thus, the activation energy of the catalyzed reaction is 64.4 kJ mol^{-1}.

17.5 (a) For what polypeptide segment does the base sequence GCGUUUGGA code? (b) Draw the structure of this segment.

(a) GCG = Ala, UUU = Phe, and GGA = Gly, so the segment is -Ala-Phe-Gly-.

(b)

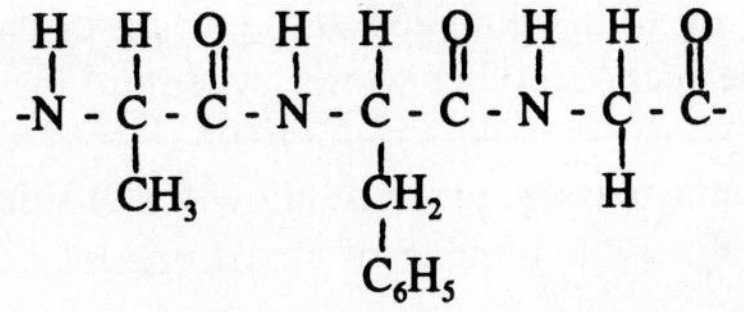

```
   H   H   O   H   H   O   H   H   O
   |   |   ||  |   |   ||  |   |   ||
  -N - C - C - N - C - C - N - C - C-
       |           |           |
       CH3         CH2         H
                   |
                   C6H5
```

17.6 How many possible genetic codes are there for (a) the —Cys—Phe— and (b) the —Trp—Lys—Glu—Met— segments of polypeptides?

(a) Cys = UGU or UGC, and Phe = UUU or UUC, so there are <u>four</u> possible combinations (genetic codes):

UGU UUU, UGC UUU, UGC UUU, and UGC UUC

(b) Trp = UGG, Lys = AAA or AAG, Glu = GAA or GAG, and Met = AUG, so possible combinations coding for Trp-Lys are UGGAAA or UGGAAG, and possible combinations coding for Glu-Met are GAAAUG or GAGAUG, giving for Trp-Lys-Glu-Met the <u>four</u> possible genetic codes:

UGG AAA GAA AUG, UGG AAA GAG AUG, UGG AAG GAA AUG, and UGG AAG GAG AUG

CHAPTER 18

18.1 Complete the following equations: (a) ${}^{22}_{11}Na \rightarrow {}^{0}_{1}e + ?$ (b) ${}^{27}_{12}Mg \rightarrow {}^{0}_{-1}e + ?$ (c) ${}^{36}_{17}Cl \rightarrow {}^{36}_{18}Ar + ?$ (d) ${}^{23}_{94}Pu \rightarrow {}^{235}_{92}U + ?$

(a) ${}^{22}_{11}Na \rightarrow {}^{0}_{1}e + {}^{22}_{10}Ne$ (b) ${}^{27}_{12}Mg \rightarrow {}^{0}_{-1}e + {}^{27}_{13}Al$ (c) ${}^{36}_{17}Cl \rightarrow {}^{36}_{18}Ar + {}^{0}_{-1}e$ (d) ${}^{23}_{94}Pu \rightarrow {}^{235}_{92}U + {}^{4}_{2}He$

18.2 The radioactive noble gas ${}^{222}_{86}Rn$ formed by the decay of uranium seeps out of the ground. It may collect in tightly closed basements and pose a health hazard, as it is believed to cause lung cancer. How much ${}^{222}_{86}Rn$ remains after 12.0 days in a sample initially containing 30 μg of ${}^{222}_{86}Rn$. The half-life of radon-222 is 3.8 days.

Example 19.2 gave $k_1 = 0.18\ \text{days}^{-1}$; $\therefore \quad \ln \frac{m_0}{m_t} = k_1 t = (0.18\ \text{days}^{-1})(12.0\ \text{days}) = 2.16$

i.e., $\frac{m_0}{m_t} = \frac{30\ \mu g}{N_t} = e^{2.16} = \underline{8.67}$; $\quad m_t = \frac{30\ \mu g}{8.67} = \underline{3.5\ \mu g\ Rn}$

18.3 A sample that originally contained 0.30 mg of ${}^{60}Co$ was found to contain only 0.25 mg of ${}^{60}Co$ 1.40 yr later. What is the half-life of ${}^{60}Co$?

For a first order process $k_1 = \frac{0.693}{t_{\frac{1}{2}}}$; thus, $\ln \frac{m_0}{m_t} = k_1 = \frac{0.693}{t_{\frac{1}{2}}}$

For t = 1.40 yr, $\frac{m_0}{m_t} = \frac{0.300\ mg}{0.250\ mg} = \underline{1.200}$, so $\ln \frac{m_0}{m_t} = \underline{0.182}$; hence: $t_{\frac{1}{2}} = \frac{(0.693)(1.40\ \text{years})}{0.182} = \underline{5.33\ \text{years}}$

18.4 A sample of rock contains 13.2 μg of uranium-238 and 3.42 μg of lead-206. If the half-life of uranium-238 is 4.51×10^9 yr, what is the age if the rock? In other words, when did the rock solidify?

Since ${}^{238}U$ and ${}^{206}Pb$ have different atomic masses, we must allow for this difference in calculating the ratio of ${}^{238}U$ at the time of formation of the rock, N_0, to the number of atoms present now, N_t:

$$\frac{m_0}{m_t} = \frac{[13.2 + 3.42(\frac{238}{206})]\mu g}{13.2\ \mu g} = \frac{17.2\ \mu g}{13.2\ \mu g} = 1.303; \quad \ln\frac{m_0}{m_t} = \underline{0.2647}$$

$$\text{But:} \quad \ln\frac{m_o}{m_t} = k_1 t \ ; \quad \text{i.e.,} \quad k_1 = \frac{\ln\frac{m_o}{m_t}}{t} = \frac{0.693}{t_{\frac{1}{2}}}$$

$$\therefore\ t = \frac{\ln\frac{m_0}{m_t}}{k_1} = \frac{(\ln\frac{m_0}{m_t})t_{\frac{1}{2}}}{0.693} = \frac{(0.2647)(4.51\text{x}10^9\ \text{yr})}{0.693} = \underline{1.72\text{x}10^9\ \text{yr}} \quad (1.72 \text{ billion yr})$$

18.5 A wooden bowl found in an archaeological excavation has a ^{14}C disintegration rate of 11.2 per gram per minute. What is the bowl's age?

The disintegration rates are $R_0 = 15.3\ g^{-1}\ min^{-1}$, and $R_t = 11.2\ g^{-1}\ min^{-1}$, and therefore:

$$\ln\frac{R_0}{R_t} = \ln\frac{15.3\ g^{-1}\ min^{-1}}{11.2\ g^{-1}\ min^{-1}} = \ln 1.366 = \underline{0.3119}$$

$$\text{Hence:} \quad t = (\frac{t_{\frac{1}{2}}}{0.693})(\ln\frac{R_0}{R_t}) = (\frac{5730\ \text{yr}}{0.693})(0.3119) = \underline{2.58\text{x}10^3\ \text{yr}} \quad (\approx 2580\ \text{yr})$$

18.6 The mass of an ^{16}O nucleus is 15.990 53 u. What are its binding energy and binding energy per nucleon?

An oxygen-16 nucleus contains 8 protons and 8 neutrons and therefore has a mass defect of

$$[\ 8(1.007\ 28 + 1.008\ 66)-15.999\ 53] = 0.127\ 99\ u = 2.049\text{x}10^{-28}\ kg.$$

$$\Delta mc^2 = (2.049\ \text{x}10^{-28}\ kg)(2.998\ \text{x}\ 10^8\ m\ s^{-1})^2 = 1.8412\ \text{x}\ 10^{-11}\ kg\ m^2\ s^{-2}$$

$$= (1.8412\ \text{x}\ 10^{-11}\ kg\ m^2\ s^{-2})(\frac{1\ J}{1\ kg\ m^2\ s^{-2}}) = \underline{1.841\ \text{x}\ 10^{-11}\ J}$$

$$\text{Binding energy nucleon}^{-1} = \frac{1.8412\ \text{x}\ 10^{-11}\ J}{16\ \text{nucleon}} = \underline{1.115\ \text{x}\ 10^{-12}\ J\ \text{nucleon}^{-1}}$$

18.7 When nitric acid oxidizes UO_2, it is reduced to NO(g). Write a balanced equation for this reaction.

As we have seen, reaction of pitchblende with nitric acid gives $UO_2(NO_3)_2(s)$, containing the UO_2^{2+} ion. Thus, the half-reactions are:

$$3[UO_2 \rightarrow UO_2^{2+} + 2e^-]$$
$$2[HNO_3 + 3e^- + 3H^+ \rightarrow NO + 2H_2O]$$
$$3UO_2 + 2HNO_3 + 6H^+ \rightarrow 3UO_2^{2+} + 2NO + 4H_2O$$

and adding $6NO_3^-$ to each side gives:

$$3UO_2(s) + 8HNO_3(aq) \rightarrow 3UO_2(NO_3)_2(aq) + 2NO(g) + 4H_2O(l)$$

18.8 Find the amount of energy released when 1 mol of deuterium undergoes fusion according to the reactions just listed.

The overall reaction is: $6\,^2_1H \rightarrow 2\,^4_2He + 2\,^1_1H + 2\,^1_0n$

and from Table 1.3, the appropriate masses are ^2_1H 2.014 10 g mol^{-1}, ^1_1H 1.007 83 g mol^{-1}, ^4_2He 4.002 60 g mol^{-1}, and ^1_0n 1.008 66 g mol^{-1}.

Thus the change in mass per mole of deuterium, $D_2(g)$, is <u>one-third</u> of

$$6(2.014\ 10)\text{ g } \underline{\text{minus}}\ 2(4.002\ 60)\text{ g} + 2(1.007\ 83)\text{ g} + 2(1.006\ 66)\text{ g}$$

$$= \tfrac{1}{3}(12.084\ 60 - 12.038\ 18)\text{ g} = \underline{1.5474 \times 10^{-2}\text{ g}}.$$

$$\Delta mc^2 = (1.5474 \times 10^{-5}\text{ kg})(2.998 \times 10^8\text{ m s}^{-1})^2 = 1.3908 \times 10^{12}\text{ kg m}^2\text{ s}^{-2}$$

$$= (1.3908 \times 10^{12}\text{ kg m}^2\text{ s}^{-2})\left(\frac{1\text{ J}}{1\text{ kg m}^2\text{ s}^{-2}}\right)\left(\frac{1\text{ kJ}}{10^3\text{ J}}\right) = \underline{1.391 \times 10^8\text{ kJ}}$$

CHAPTER 19

19.1 (a) Identify the free radicals among the two- and three-atom molecules in Table 19.2. (b) Why are so many free radicals found in space, whereas they are rare on earth?

(a) Free radicals are molecules or ions containing an **odd** number of electrons, so among the two- and three-atom molecules and ions in Table 19.2, we have:

Species	Number of electrons	Species	Number of electrons	Species	Number of electrons
H_2	1+1 = 2	CS	4+6 = 10	HCO^+	1+4+6-1 = 10
CH^+	4+1-1 = 4	SiO	4+6 = 10	N_2H^+	2(5)+1-1 = 10
CH	4+1 = **5**	SO	6+6 = 12	H_2S	2(1)+6 = 8
OH	6+1 = **7**	**NS**	5+6 = **11**	HCS^+	1+4+6-1 = 10
C_2	2(4) = 8	SiS	4+6 = 10	OCS	6+4+6 = 16
CN	4+5 = **9**	H_2O	2(1)+6 = 8	SO_2	6+2(6) = 18
CO	4+6 = 10	$\mathbf{C_2H}$	2(4)+1 = **9**	NaOH	1+6+1 = 8
$\mathbf{CO^+}$	4+6-1 = **9**	HCN	1+4+5 = 10		
NO	5+6 = **11**	HNC	1+4+5 = 10		

Summary: The free radicals are: **Diatomic** CH, OH, CN, CO^+, NO, NS; **triatomic** C_2H

(b) Most free radicals are formed by (i) breaking a covalent bond *homolytically*, or (ii) ionization of molecules where all the electrons are paired. Such processes readily occur at temperatures sufficiently high for these processes to occur, as is the case when stars fragment and give rise to interstellar dust clouds. However, since the partial pressures of the radicals in interstellar dust clouds are very low, they will react with other free radicals very slowly. In contrast, free radicals are formed on earth largely by photolysis and the solar radiation that reaches the earth is in general insufficiently energetic to break all but the weakest bonds. Also, once formed the rate of reaction of free radicals would be quite fast at the higher concentrations of matter on earth.

19.2 How do you explain that the density of the sun is 1.4 times greater than that of water, when the composition of the sun is largely hydrogen and helium?

If the sun was composed of gaseous hydrogen and helium the density would be relatively small, but because of its enormous mass the pressure in the interior of the sun is very high, so it is probably composed of solid hydrogen and helium.

19.3 Explain why basalt has a higher density than granite.

Basalt has a density of 2.8 g cm^{-3}, while **granite** has a density of 2.7 g cm^{-3}, which is due to the fact that basalt is richer in Fe (molar mass 55.85), Ti (molar mass 47.90), Mg (molar mass 24.30), and Ca (molar mass 40.08), and poorer in Na (molar mass 22.99), K (molar mass 39.10) — all molar masses in g mol^{-1} — and silicate than is granite. This higher proportion of "heavier" elements makes it more dense. The mixture of iron and magnesium silicates, for example, that mainly constitutes the mantle has a density of 3.3 g cm^{-3}.

19.4 (a) If the earth (density 5.52 g cm^{-3}) were composed entirely of iron (density 7.5 g cm^{-3}) and silicate rock (density 3.2 g cm^{-3}), what would be the relative amounts of each? (b) If the earth's core were all iron and the mantle all silicate, what would be the thickness of the mantle relative to the radius of the iron core?

(a) Let us assume that the relative fraction by mass of **Fe** is x, and **(1-x) silicate**, then:

$$\frac{5.52\ g}{1\ cm^3} = \frac{x(7.5\ g)}{1\ cm^3} + \frac{(1-x)(3.2\ g)}{1\ cm^3}\ ; \quad 5.52 = 7.5x + 3.2 - 3.2x\ ; \quad 4.3x = 2.32\ ; \quad \underline{x = 0.54}$$

The relative amounts by mass would be **54% iron and 46% silicate.**

(b) Suppose the radius of the earth is **r cm**, and the relative thickness of the mantle is **x**, and the relative radius of the iron core is **1 - x**. Then:

$$\text{Mass of earth} = \left(\frac{4}{3}\pi r^3\ cm^3\right)\left(\frac{5.52\ g}{1\ cm^3}\right) = \left(\frac{4}{3}\pi r^3(1-x)^3\ cm^3\right)\left(\frac{7.5\ g}{1\ cm^3}\right) + \left[\left(\frac{4}{3}\pi r^3 - \frac{4}{3}\pi r^3(1-x)^3\right)\ cm^3\right]\left(\frac{3.2\ g}{1\ cm^3}\right)$$

$$\text{Whence:}\quad 5.52 - 3.2 = (7.5 - 3.2)(1-x)^3\ ; \quad (1-x)^3 = \frac{2.32}{4.1} = 0.566\ ; \quad 1-x = 0.83\ ; \quad \underline{x = 0.17}$$

The relative thickness of the mantle is **17% of the earth's radius**, or 20% of the radius of the iron core.

19.5 What is the empirical formula of the mica mineral phlogopite in which one-quarter of the silicon atoms in talc are replaced by aluminum and the additional negative charge is balanced by K^+ ions?

Talc has the empirical formula $Mg_3Si_4O_{10}(OH)_2$ and contains the $Si_4O_{10}{}^{4}$ ion. Replacement of one-quarter of the Si atoms by Al^- ions gives $AlSi_3O^{10-}$. Thus the empirical formula of phlogopite is $KMg_3AlSi_3O_{10}(OH)_2$.

19.6 (a) Explain why the concentrations of the noble gases are about 10 times lower in the atmosphere than in the universe as a whole. (b) Explain how and why the isotopic composition of argon on the earth differs from that in the universe as a whole.

(a) The noble gases in the original earth's **atmosphere** have been replaced as a result of degassing.

(b) The argon found in the sun is almost entirely argon-36 and argon-38, and these isotopes were presumably present in the earth's original atmosphere. Almost all of this original atmosphere was blown away by the solar wind, but the argon that now constitutes about 1% of the atmosphere is mainly argon-40 formed over time from the radioactive decay of potassium-40 in rocks: ${}^{40}_{19}K \rightarrow {}^{40}_{18}Ar + {}^{0}_{+1}e$.

19.7 Briefly summarize the role of solar energy in the development of our present atmosphere from the primeval one.

The earth's primeval atmosphere contained little oxygen. This very reactive element would have entirely combined with other elements to give oxides in the surface rocks. The unique feature of the present atmosphere is its high steady state oxygen content that is maintained by living organisms. In respiration, animals use oxygen and generate carbon dioxide, which is used by plants in photosynthesis, which utilizes solar energy to form carbohydrates and regenerates oxygen as a by-product.

19.8 (a) Write complete balanced equations for each step in the conversion of acetaldehyde to alanine. (b) Describe each reaction step in (a) as an elimination reaction, and addition reaction, or some other type of reaction.

(a) 1. $CH_3{-}CH{=}O + NH_3 \rightarrow CH_3{-}CH(NH_2){-}OH$

2. $CH_3{-}CH(NH_2){-}OH \rightarrow CH_3{-}CH{=}NH + H_2O$

3. $CH_3{-}CH{=}NH + HC{\equiv}N \rightarrow CH_3{-}CH(NH_2){-}C{\equiv}N$

4. $CH_3{-}CH(NH_2){-}C{\equiv}N + 2H_2O \rightarrow CH_3{-}CH(H_2N){-}C(OH)_2{-}NH_2$

$\downarrow$

$CH_3{-}CH(H_2N){-}C({=}O){-}OH + NH_3$

(b) 1. addition; 2. elimination; 3. addition; 4. addition, then elimination.

19.9 The rate of lightning flashes in the (entire) atmosphere is estimated at about 100 s^{-1}, and each flash produces about 10^{27} molecules of NO. What is the approximate amount of nitrogen, in metric tons, that is fixed annually by lightning?

We calculate the number of lightning flashes per year and thus the number of NO molecules, which we convert to moles, and then to the mass of NO:

$$\text{Mass of NO} = (1\ \text{yr})\left(\frac{365\ \text{day}}{1\ \text{yr}}\right)\left(\frac{24\ \text{h}}{1\ \text{day}}\right)\left(\frac{3600\ \text{s}}{1\ \text{h}}\right)\left(\frac{100\ \text{flash}}{1\ \text{s}}\right)\left(\frac{10^{27}\ \text{NO molecules}}{1\ \text{flash}}\right)\left(\frac{1\ \text{mol NO}}{6.022 \times 10^{23}\ \text{molecules}}\right)$$

$$\times \left(\frac{30.01\ \text{g NO}}{1\ \text{mol NO}}\right)\left(\frac{1\ \text{tonne}}{10^6\ \text{g}}\right) = \underline{1.57 \times 10^8\ \text{metric ton}}$$

CHAPTER 20

20.1 What is the average molecular mass of polystyrene if the average chain consists of 2000 monomers?

Styrene is C_6H_5—CH=CH_2, (C_8H_8), molecular mass 104.1 u, and since **polystyrene** is an addition polymer, the repeating unit will also have the formula C_8H_8, so for an average chain $(C_8H_8)_{200}$,

$$\text{average molecular mass} = (2000 \text{ units})(104.1 \text{ u unit}^{-1}) = \mathbf{2.082 \times 10^5 \text{ u}}$$

20.2 Poly-4-methyl-1-pentene is a hard transparent polymer used in the manufacture of laboratory ware such as flasks and beakers. Draw the structure of the monomer and the polymer.

4-methyl-1-pentene has the structure

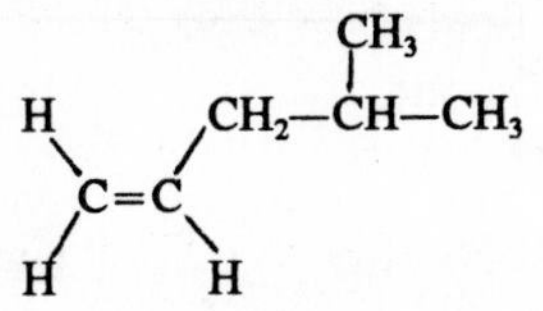

```
                  CH3
                  |
   H      CH2—CH—CH3
    \    /
     C=C
    /    \
   H      H
```

Thus, the structure of this addition polymer is:

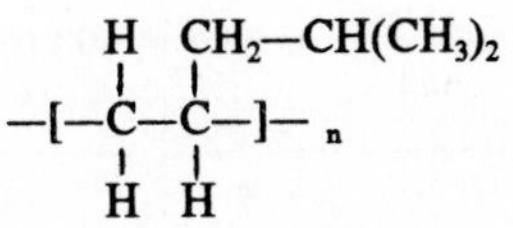

```
      H  CH2—CH(CH3)2
      |  |
  —[—C—C—]—n
      |  |
      H  H
```

20.3 Draw the structure of styrenebutadiene (SBR) rubber. Assume that the two monomers alternate in the polymer.

Styrene is C_6H_5—CH=CH_2, phenylethene, and **butadiene** is H_2C=CH-CH=CH_2, so **SBR rubber** has the structure:

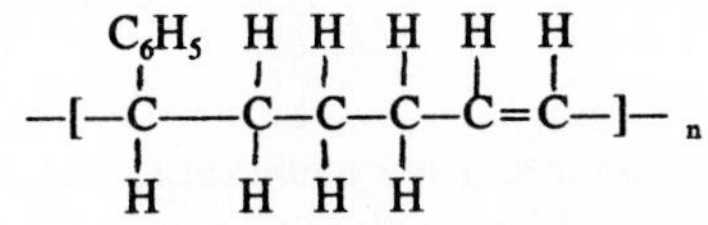

```
     C6H5   H  H  H  H  H
     |      |  |  |  |  |
  —[—C——C—C—C—C=C—]—n
     |      |  |  |
     H      H  H  H
```

20.4 If 1.30 x 10^6 kg of nylon 66 is made annually in the United States, how many kilograms of 1,6-diaminohexane are used annually in the manufacture of nylon 66?

Nylon 66 contains the repeating unit —NH—$(CH_2)_6$—NH—CO—$(CH_2)_6$—CO—, (formula mass 254.4 u), where —NH—$(CH_2)_6$—NH—, formula mass 114.2 u, is the part derived from 1,6-diaminohexane, H_2N—$(CH_2)_6$—NH_2, (molecular mass 116.2 u):

$$\text{Mass of 1,6-diaminohexane} = (1.30 \times 10^6 \text{ kg polymer})\left(\frac{\text{unit mass}}{254.4 \text{ u}}\right)\left(\frac{114.2 \text{ u}}{\text{unit mass}}\right)\left(\frac{116.2 \text{ u}}{114.2 \text{ u}}\right) = \underline{5.94 \times 10^5 \text{ kg}}$$

20.5 The nylon shown in Demonstration 20.1 is made by the reaction of 1,6-diaminohexane with decanedioyl chloride, Cl—C(O)—$(CH_2)_8$—C(O)—Cl. (a) Write the equation for the reaction, and draw the structure of the polymer. (b) Explain why it is called nylon 610.

```
(a) n H—N—(CH2)6—N—H  + n Cl—C—(CH2)8—C—Cl → —[—N—(CH2)6—N—C—(CH2)8—C—]—n + 2nHCl
        |         |            ||         ||          |         |  ||         ||
        H         H            O          O           H         H  O          O
                                                                 nylon 610
```

(b) The name **Nylon 610** provides the information that the diamine involved in the polymer (diaminohexane) has a chain of **six carbon atoms**, and the dicarboxylic acid derivative, decanedioyl chloride, has a chain of **ten carbon atoms.**

20.6 What monomers could be used to make Kevlar?

Kevlar has the structure:

$$-[-\underset{\mathrm{H}}{\underset{|}{\mathrm{N}}}-C_6H_4-\underset{\mathrm{H}}{\underset{|}{\mathrm{N}}}-\underset{\mathrm{O}}{\underset{\|}{\mathrm{C}}}-C_6H_4-\underset{\mathrm{O}}{\underset{\|}{\mathrm{C}}}-]-_n$$

and is therefore a condensation polymer; a *polyamide* that could be formed by condensation of **1,4-benzenediamine** and **1,4-benzenedicarboxylic acid** (terephthalic acid)

$$H_2N-C_6H_4-NH_2 \qquad HO-\underset{\mathrm{O}}{\underset{\|}{\mathrm{C}}}-C_6H_4-\underset{\mathrm{O}}{\underset{\|}{\mathrm{C}}}-OH$$

1,4-benzenediamine **1,4-benzenedicarboxylic acid**

20.7 Classify each of the reactions in Figure 20.7 as an addition reaction, an oxidation reaction, or some other type of reaction.

(a)

$$\text{Benzene} \xrightarrow{H_2} \text{Cyclohexane} \xrightarrow[\text{Co catalyst}]{\text{air}} \text{Cyclohexanone} \xrightarrow{HNO_3} HO-\underset{}{\overset{\mathrm{O}}{\overset{\|}{\mathrm{C}}}}-(CH_2)_4-\overset{\mathrm{O}}{\overset{\|}{\mathrm{C}}}-OH$$

Benzene **Cyclohexane** **Cyclohexanone** **Adipic acid** (1,6-Hexanedioicacid)

Benzene is hydrogenated (an *addition* reaction) to give **cyclohexane**, which is *oxidized* with air, using a catalyst, to give **cyclohexanone** (a ketone), which is *oxidized* with nitric acid to give **adipic acid.**

(b)

$$CH_2{=}CH{-}CH{=}CH_2 \xrightarrow{2HCN} N{\equiv}C(CH_2)_4C{\equiv}N \xrightarrow{4H_2} H_2N{-}CH_2(CH_2)_4CH_2{-}NH_2$$

Butadiene **Adiponitrile** (1-4,Dicyanobutane) **Hexamethylenediamine** (1,6-Diaminohexane)

Butadiene reacts with hydrogen cyanide in an *addition* reaction to give **1,4-dicyanobutane, which is hydrogenated** in an *addition* reaction to give **1,6-diaminohexane.**

20.8 Explain why a zeolite that has been used for softening water can be reused after a concentrated NaCl solution has been passed through it.

A zeolite removes and retains the Ca^{2+} and Mg^{2+} ions in hard water when it is passed through a water softener containing zeolite because it binds these ions more strongly than it does Na^+ ions. However, when a concentrated solution of NaCl(aq) is then passed through it, Ca^{2+} and Mg^{2+} ions are replaced by Na^+ ions and the zeolite is thus regenerated for further use, because the Ca^{2+} and Mg^{2+} ions are gradually swept away in the solution that passes through the zeolite.